Skeletonization
Theory, Methods, and Applications

Computer Vision and Pattern Recognition Series

Series Editors

Also in the Series:

Lin and Zhang, Low-Rank Models in Visual Analysis: Theories, Algorithms, and Applications, 2017, 9780128127315

Murino et al., Group and Crowd Behavior for Computer Vision, 2017, 9780128092767

Zheng et al., Statistical Shape and Deformation Analysis, 2017, 9780128104934

De Marsico et al., Human Recognition in Unconstrained Environments: Using Computer Vision, Pattern Recognition and Machine Learning Methods for Biometrics, 2017, 9780081007051

Saha et al., Skeletonization: Theory, Methods, and Applications, 2017, 9780081012918

Skeletonization
Theory, Methods, and Applications

Edited by
Punam K. Saha
Gunilla Borgefors
Gabriella Sanniti di Baja

Academic Press is an imprint of Elsevier
125 London Wall, London EC2Y 5AS, United Kingdom
525 B Street, Suite 1800, San Diego, CA 92101-4495, United States
50 Hampshire Street, 5th Floor, Cambridge, MA 02139, United States
The Boulevard, Langford Lane, Kidlington, Oxford OX5 1GB, United Kingdom

Notices

Knowledge and best practice in this field are constantly changing. As new research and experience broaden our understanding, changes in research methods, professional practices, or medical treatment may become necessary.

Practitioners and researchers must always rely on their own experience and knowledge in evaluating and using any information, methods, compounds, or experiments described herein. In using such information or methods they should be mindful of their own safety and the safety of others, including parties for whom they have a professional responsibility.

To the fullest extent of the law, neither the Publisher nor the authors, contributors, or editors, assume any liability for any injury and/or damage to persons or property as a matter of products liability, negligence or otherwise, or from any use or operation of any methods, products, instructions, or ideas contained in the material herein.

British Library Cataloguing-in-Publication Data
A catalogue record for this book is available from the British Library

Library of Congress Cataloging-in-Publication Data
A catalog record for this book is available from the Library of Congress

ISBN: 978-0-08-101291-8

For information on all Academic Press publications
visit our website at https://www.elsevier.com/books-and-journals

Publisher: Mara Conner
Acquisition Editor: Tim Pitts
Editorial Project Manager: Charlotte Kent
Production Project Manager: Anusha Sambamoorthy
Designer: Mark Rogers

Typeset by VTeX

Contents

Contributors xiii
About the Editors xvii
Preface xix

PART 1 THEORY AND METHODS

CHAPTER 1 Skeletonization and its applications – a review . 3
Punam K. Saha, Gunilla Borgefors, and Gabriella Sanniti di Baja

1.1 Introduction 3
1.1.1 Basic Concepts 3
1.1.2 Background 5
1.2 Different Approaches of Skeletonization 7
1.2.1 Geometric Approaches 10
1.2.2 Curve Propagation Approaches 11
1.2.3 Digital Approaches 13
1.3 Topology Preservation 19
1.4 Pruning 22
1.5 Multiscale Skeletonization 23
1.6 Parallelization 24
1.6.1 Subiterative Parallelization Schemes 25
1.6.2 Parallelization Using Minimal Nonsimple Sets 25
1.6.3 Parallelization Using P-Simple Points 26
1.7 Applications 26
1.8 Performance Evaluation 29
1.9 Conclusions 30
References 31

CHAPTER 2 Multiscale 2D medial axes and 3D surface skeletons by the image foresting transform 43
Alexandre Falcão, Cong Feng, Jacek Kustra, and Alexandru Telea

2.1 Introduction 43
2.2 Related Work 44
2.2.1 Definitions 45
2.2.2 Skeleton Regularization 45
2.3 Proposed Method 48
2.3.1 Multiscale Regularization—Strengths and Weaknesses 49

2.3.2 Image Foresting Transform 50
2.3.3 Multiscale Skeletonization—Putting It All Together. 56
2.4 Comparative Analysis 60
2.4.1 2D Medial Axes 60
2.4.2 3D Medial Surfaces 60
2.5 Conclusion 67
References 67

CHAPTER 3 **Fuzzy skeleton and skeleton by influence zones – a review** **71**
Isabelle Bloch
3.1 Introduction 71
3.2 Distance-Based Approaches 73
3.3 Morphological Approaches to Compute the Centers of Maximal Balls 75
3.4 Morphological Thinning 76
3.5 Fuzzy Skeleton of Influence Zones 78
3.5.1 Definition Based on Fuzzy Dilations 79
3.5.2 Definitions Based on Distances 80
3.5.3 Illustrative Example (Reproduced from [8]) 82
3.6 Conclusion 84
References 85

CHAPTER 4 **Unified part-patch segmentation of mesh shapes using surface skeletons** **89**
Joost Koehoorn, Cong Feng, Jacek Kustra, Andrei Jalba, and Alexandru Telea
4.1 Introduction 90
4.2 Related Work 91
4.2.1 Skeletonization 91
4.2.2 Shape Segmentation 93
4.2.3 Summary of Challenges 95
4.3 Method 96
4.3.1 Preliminaries 96
4.3.2 Regularized Surface Skeleton Computation 97
4.3.3 Cut-Space Computation 99
4.3.4 Cut-Space Partitioning 100
4.3.5 Partitioning the Full Surface Skeleton 102
4.3.6 Partition Projection to Surface 104
4.3.7 Part-Based Partition Refinement 105
4.3.8 Unified (Part and Patch) Segmentation 107
4.4 Results 112
4.5 Discussion 115

4.6 Conclusion 119
References 120

CHAPTER 5 **Improving the visual aspect of the skeleton and simplifying its structure** 123
Gabriella Sanniti di Baja and Luca Serino

5.1 Introduction 123
5.2 Preliminary Notions and Definitions 125
5.3 Tools Improving the Visual Aspect of the Skeleton 127
5.3.1 Zig-zag Straightening 127
5.3.2 Fusion of Close Branch Points 129
5.3.3 Pruning 131
5.4 Experimental Results on the Improvement of Skeleton Visual Aspect 135
5.5 Tools to Simplify Skeleton Structure 138
5.5.1 Polygonal Approximation 140
5.5.2 Building a Linearized Version of the Skeleton 144
5.6 Experimental Results on the Simplification of Skeleton Structure 145
5.7 Conclusion 147
References 149

CHAPTER 6 **Curve skeletonization using minimum-cost path** . 151
Dakai Jin, Cheng Chen, Eric A. Hoffman, and Punam K. Saha

6.1 Introduction 151
6.2 General Principles of Curve Skeletonization 153
6.2.1 Geodesic Distance Transform Based Methods 154
6.2.2 Direct Methods 155
6.2.3 Minimum-Cost Path Based Methods 156
6.3 Curve Skeletonization with Minimum-Cost Path 157
6.3.1 Overall Outline 158
6.3.2 Minimum-Cost Path 161
6.3.3 Centered Minimum-Cost Paths 162
6.3.4 Skeletal Branch Detection 164
6.3.5 Object Volume Marking 165
6.3.6 Skeletal Branch Significance 166
6.3.7 Termination Criterion 166
6.3.8 Algorithm Efficiency 167
6.4 Applications 169
6.4.1 Virtual Colonoscopy and Bronchoscopy 170
6.4.2 Anatomical Labeling of Human Airway Trees 172
6.4.3 Morphologic Assessment of Pulmonary Arteries ... 173

6.5 Conclusions 174
References 175

CHAPTER 7 **Parallel skeletonization algorithms in the cubic grid based on critical kernels** **181**
Gilles Bertrand and Michel Couprie

7.1 Introduction 181
7.2 Voxel Complexes and Simple Voxels 183
7.3 Critical Cliques 184
7.4 Decreasing Rank Strategy 186
7.5 Asymmetric Thinning 187
7.5.1 Generic Parallel Asymmetric Thinning Scheme 187
7.5.2 Isthmus-Based Asymmetric Thinning 189
7.5.3 Comparison with Other Parallel Curve Skeletonization Methods 191
7.6 Symmetric Thinning 193
7.6.1 Generic Parallel Symmetric Thinning Scheme 194
7.6.2 Isthmus-Based Symmetric Thinning 195
7.7 Characterization of Critical Cliques and k-Isthmuses 197
7.8 Isthmus Persistence and Skeleton Filtering 200
7.9 Hierarchies of Skeletons 202
7.10 Complexity 206
7.11 Conclusion 207
References 208

CHAPTER 8 **Critical kernels, minimal nonsimple sets, and hereditarily simple sets in binary images on n-dimensional polytopal complexes** **211**
T.Y. Kong

8.1 Introduction 212
8.1.1 Background and Prior Work 214
8.1.2 Contributions of This Chapter 217
8.2 Preliminaries: Complexes, Polyhedra, and Binary Images 218
8.2.1 Convex Polytopes 218
8.2.2 Complexes and Pure Complexes 219
8.2.3 Polyhedra 221
8.2.4 Binary Images on Pure Polytopal Complexes 222
8.3 Nonempty Topologically Simple (NETS) Polyhedra 222
8.3.1 Properties of NETS Polyhedra 223
8.3.2 The Axioms That Specify NETS Polyhedra 224
8.3.3 In a Strongly Normal Collection of Nonempty Convex Polytopes, the Union of Any Element with Its Neighbors Is NETS 225

8.3.4 Possible Definitions of NETS Polyhedra Using Homology Groups 226
8.4 Consequences of the NETS Axioms 228
8.4.1 A 0- or 1-Dimensional Complex Is NETS Just If It Is a Tree 229
8.4.2 Collapsible Polyhedra Are NETS, and NETS Polyhedra Never Collapse to Non-NETS Polyhedra . 231
8.4.3 A Polyhedron in the Plane or in the Boundary of a 3-Dimensional Convex Polytope Is NETS If and Only If It Is Collapsible 232
8.5 Simple, Hereditarily Simple, and Minimal Nonsimple Sets . 233
8.5.1 Simple 1s, Attachment Complexes, and Attachment Sets 233
8.5.2 Definitions and Basic Properties of Simple, Hereditarily Simple, and Minimal Nonsimple Sets .. 235
8.6 Cliques, Cores, and the Critical Kernel 236
8.6.1 $\mathbb{I}$-Induced Cliques and $\mathbb{I}$-Essential Cells 237
8.6.2 $\mathbb{I}$-Cores 238
8.6.3 $\mathbb{I}$-Regular and $\mathbb{I}$-Critical Cells; the Critical Kernel of $\mathbb{I}$ 239
8.7 Characterizations of P-Simple 1s, Hereditarily Simple Sets, and Minimal Nonsimple Sets 240
8.8 A Proof of Theorem 8.5.5 and a Second Proof of the "If" Part of Theorem 8.7.1 Based on NETS-Collapsing 245
8.8.1 NETS-Collapsing and a Proof of Theorem 8.5.5 245
8.8.2 $\mathbb{I}$-Essential Complexes and Another Proof of the "If" Part of Theorem 8.7.1 251
8.9 Concluding Remarks 253
References 254

PART 2 APPLICATIONS

CHAPTER 9 Skeletonization in natural images and its application to object recognition **259**
Wei Shen, Kai Zhao, Jiang Yuan, Yan Wang, Zhijiang Zhang, and Xiang Bai

9.1 Introduction 260
9.2 Related Works 262
9.3 Methodology 264
9.3.1 Network Architecture 264
9.3.2 Skeleton Extraction by Fusing Scale-Associated Side Output 265
9.3.3 Understanding of the Proposed Method 269
9.4 Experimental Results 269

9.4.1 Implementation Details 270
9.4.2 Performance Comparison 271
9.5 Conclusion 283
Acknowledgments 283
References 284

CHAPTER 10 Characterization of trabecular bone plate–rod micro-architecture using skeletonization and digital topologic and geometric analysis 287
Punam K. Saha

10.1 Introduction 287
10.2 Definitions and Notations 289
10.3 Skeletonization 290
10.4 Digital Topological Analysis 291
10.4.1 Surface–Surface Junction Line Extension 294
10.4.2 Detection of Junction Voxels Between Surfaces and Curves 295
10.5 Volumetric Topological Analysis 297
10.5.1 Geodesic Distance Transform 299
10.5.2 Feature Propagation and Representative 302
10.5.3 Applications 302
10.6 Conclusion 305
References 305

CHAPTER 11 Medial structure generation for registration of anatomical structures 313
Sergio Vera, Debora Gil, H.M. Kjer, Jens Fagertun, Rasmus R. Paulsen, and Miguel Á. González Ballester

11.1 Medial Maps for Reliable Extraction of Anatomical Medial Surfaces 314
11.2 Extracting Anatomical Medial Surfaces Using Medialness Maps 316
11.2.1 Gaussian Steerable Medial Maps 317
11.3 Validation Framework for Medial Anatomy Assessment ... 322
11.3.1 Synthetic Database 322
11.3.2 Medial Surface Quality Metrics 323
11.4 Validation Experiments 324
11.4.1 Medial Surface Quality 325
11.4.2 Reconstruction Power for Clinical Applications 327
11.5 Application to Cochlea Registration 327
11.5.1 Material and Methods 330
11.5.2 Skeletonization 330
11.5.3 Image Registration 331

11.5.4 Evaluation 333
11.5.5 Results 335
11.6 Discussion 337
Acknowledgments........ 340
References 340

CHAPTER 12 Skeleton-based fast, fully automated generation of vessel tree structure for clinical evaluation of blood vessel systems **345**
Kristína Lidayová, Hans Frimmel, Chunliang Wang, Ewert Bengtsson, and Örjan Smedby

12.1 Introduction 346
12.2 Medical Context 347
12.2.1 Clinical Evaluation of the Peripheral Arterial System 347
12.2.2 Need for Image Postprocessing 349
12.3 Background 350
12.3.1 Previous Methods 350
12.3.2 From Colon to Blood Vessels........ 352
12.4 Fast Skeleton-Based Generation of the Centerline Tree 354
12.4.1 Parameter Selection........ 355
12.4.2 The Artery Nodes Detection 359
12.4.3 The Artery Nodes Connection 364
12.4.4 Anatomy-Based Analysis........ 365
12.4.5 Vascular Segmentation Based on the Vascular Centerline Tree 366
12.5 Validation of the Vascular Centerline Tree........ 367
12.5.1 Computed Angiographic Images of the Lower Limbs 368
12.5.2 Evaluation of Automatically Selected Changeable Parameters 368
12.5.3 Reference and United Skeleton 369
12.5.4 Skeleton Evaluation 371
12.5.5 Computation Time 374
12.6 Adapting the Algorithm to Other Applications 375
12.7 Discussion and Conclusion 376
12.8 Future Development of the Method 379
Acknowledgments........ 379
References 379

Index 383

Contributors

Xiang Bai
School of Electronic Information and Communications, Huazhong University of Science and Technology, Wuhan, China

Ewert Bengtsson
Centre for Image Analysis, Division of Visual Information and Interaction, Uppsala University, Uppsala, Sweden

Gilles Bertrand
Université Paris-Est, LIGM, Équipe A3SI, ESIEE Paris, France

Isabelle Bloch
LTCI, Télécom ParisTech, Université Paris-Saclay, Paris, France

Gunilla Borgefors
Centre for Image Analysis, Uppsala University, Uppsala, Sweden

Cheng Chen
Department of Electrical and Computer Engineering, University of Iowa, Iowa City, IA, USA

Michel Couprie
Université Paris-Est, LIGM, Équipe A3SI, ESIEE Paris, France

Jens Fagertun
Department of Applied Mathematics and Computer Science, Technical University of Denmark, Copenhagen, Denmark

Alexandre Falcão
Department of Information Systems, Institute of Computing, University of Campinas (UNICAMP), Campinas, SP, Brazil

Cong Feng
Institute Johann Bernoulli, University of Groningen, Groningen, The Netherlands

Hans Frimmel
Division of Scientific Computing, Uppsala University, Uppsala, Sweden

Debora Gil
Computer Vision Center, Universitat Autònoma de Barcelona, Barcelona, Spain

Miguel Á. González Ballester
Department of Information and Communication Technologies, Universitat Pompeu Fabra, Barcelona, Spain;
ICREA, Barcelona, Spain

Eric A. Hoffman
Department of Radiology, University of Iowa, Iowa City, IA, USA

Andrei Jalba
Department of Mathematics and Computer Science, Technical University Eindhoven, The Netherlands

Dakai Jin
Department of Electrical and Computer Engineering, University of Iowa, Iowa City, IA, USA

H.M. Kjer
Department of Applied Mathematics and Computer Science, Technical University of Denmark, Copenhagen, Denmark

Joost Koehoorn
Institute Johann Bernoulli, University of Groningen, Groningen, The Netherlands

T.Y. Kong
Department of Computer Science, Queens College, CUNY, Flushing, NY, USA

Jacek Kustra
Philips Research, Eindhoven, The Netherlands

Kristína Lidayová
Centre for Image Analysis, Division of Visual Information and Interaction, Uppsala University, Uppsala, Sweden

Rasmus R. Paulsen
Department of Applied Mathematics and Computer Science, Technical University of Denmark, Copenhagen, Denmark

Punam K. Saha
Department of Electrical and Computer Engineering, Department of Radiology, University of Iowa, Iowa City, IA, USA

Gabriella Sanniti di Baja
Institute for High Performance Computing and Networking, CNR, Naples, Italy

Luca Serino
Institute for High Performance Computing and Networking, CNR, Naples, Italy

Wei Shen
Key Laboratory of Specialty Fiber Optics and Optical Access Networks, Shanghai University, Shanghai, China

Örjan Smedby
School of Technology and Health, KTH Royal Institute of Technology, Stockholm, Sweden

Alexandru Telea
Institute Johann Bernoulli, University of Groningen, Groningen, The Netherlands

Sergio Vera
Alma IT Systems, Barcelona, Spain;
Computer Vision Center, Universitat Autònoma de Barcelona, Barcelona, Spain

Chunliang Wang
School of Technology and Health, KTH Royal Institute of Technology, Stockholm, Sweden

Yan Wang
Rapid-Rich Object Search Lab, Nanyang Technological University, Singapore

Jiang Yuan
Key Laboratory of Specialty Fiber Optics and Optical Access Networks, Shanghai University, Shanghai, China

Zhijiang Zhang
Key Laboratory of Specialty Fiber Optics and Optical Access Networks, Shanghai University, Shanghai, China

Kai Zhao
Key Laboratory of Specialty Fiber Optics and Optical Access Networks, Shanghai University, Shanghai, China

About the Editors

Punam K. Saha received his PhD degree in 1997 from the Indian Statistical Institute, where he served as a faculty member during 1993–97. In 1997, he joined the University of Pennsylvania as a postdoctoral fellow, where he served as a Research Assistant Professor during 2001–06, and moved to the University of Iowa in 2006, where is currently serving as a tenured professor of Electrical and Computer Engineering and Radiology. His research interests include image processing and pattern recognition, quantitative medical imaging, musculoskeletal and pulmonary imaging, image restoration and segmentation, digital topology, geometry, shape and scale. He has published over 100 papers in international journals and over 300 papers/abstracts in international conferences, holds numerous patents related to medical imaging applications, has served as an Associate Editor of Pattern Recognition and Computerized Medical Imaging and Graphics journals and has served in many international conferences at various levels. Currently, he is an Associate Editor of the IEEE Transactions on Biomedical Engineering and the Pattern Recognition Letters journals. He received a Young Scientist award from the Indian Science Congress Association in 1996, has received several grant awards from the National Institute of Health, USA, and is a Fellow of International Association for Pattern Recognition (IAPR) and American Institute for Medical and Biological Engineering (AIMBE).

Gunilla Borgefors was born in 1952 and received her Master of Engineering in Applied Mathematics from Linköping University, Sweden, in 1975. After a Licentiate in Applied Mathematics from the same university, she received her PhD in Numerical Analysis from Royal Institute of Technology, Stockholm, in 1986. During a sabbatical, she received a Master of Journalism at Uppsala University, Sweden, in 2007.

She was employed at Department of Mathematics at Linköping University as a Researcher and Teacher and Researcher 1873–81. Then she moved to National Defense Research Establishment in Linköping in 1982 and ended as Director of Research and the Head of the Division of Information Systems. In 1993, she became a full professor in Image Analysis at Swedish University of Agricultural Sciences, Uppsala, dividing her time between the aforementioned university and Uppsala.

G. Borgefors is a Fellow of International Association for Pattern Recognition (IAPR) since 1998 and of Institute of Electrical and Electronics Engineers (IEEE) since 2008. Nationally she is member No. 19 of Royal Society of Sciences in Uppsala since 2000 and of Royal Swedish Academy of Engineering Sciences (IVA) since 2011. She has had many positions of trust in IAPR, such as 1st Vice President, Secretary, and several Committees' Chair and member. After being an Associate Editor in Pattern Recognition Letters since 1998, Borgefors became one of its three Editors-in-Chief in 2011. She is in the Steering committees for "Discrete Geometry for Computer Imagery" and "International Symposium on Mathematical Morphology" and has been involved in various roles in close to a hundred international conferences.

G. Borgefors research interest is in various aspects of discrete geometry and on description, manipulation, and analysis of digital shapes, preferably in 3D, and in applications of such methods to virus recognition, wood and plant cell analysis, wood fiber and blood vessel networks, areal and satellite photo analysis of woods and fields, to mention a few. She has written well over a hundred scientific articles and is cited more than 7000 times.

Gabriella Sanniti di Baja received the Laurea degree (cum laude) in physics from the "Federico II University" of Naples, Italy, in 1973, and the PhD honoris causa from the Uppsala University, Sweden, in 2002. She has been working from 1973 until 2015 at the Institute of Cybernetics E. Caianiello of the Italian National Research Council (CNR), where she has been Director of Research, and is currently an Associate Researcher at the Institute for High Performance Computing and Networking, CNR. Her research activity is in the field of image processing, pattern recognition, and computer vision. She has published more than 200 papers in international journals and conference proceedings, is the coeditor-in-chief of Pattern Recognition Letters, has been President of the International Association for Pattern Recognition (IAPR) and of the Italian Group of Researchers in Pattern Recognition (GIRPR), is IAPR Fellow and Foreign Member of the Royal Society of Sciences at Uppsala, Sweden.

Preface

Ever since the beginning of computerized image analysis, skeletonization has been a powerful tool. It started as an idea proposed by Harry Blum in 1967. It is pure chance that this book comes exactly fifty years later, but it is still a nice coincidence. Blum's purpose was to reduce a two-dimensional shape to a one-dimensional curve, and his reason for doing that was to get a simpler and more compact structure to work with. The original idea had to overcome two obstacles to become useful. The first was that either the digital shape had to be converted to a continuous one or the continuous skeleton had to be adapted for digital images. The second one was to construct efficient algorithms to compute correct skeletons. Neither of these problems is trivial, and work on them is still continuing.

Over the years, the concept of "skeleton" has been extended to higher dimensions, to grey-level images, to graphs, and to other data sets. The algorithms have become more and more sophisticated and gave more correct results. Long gone are the early days when, e.g., the skeleton of a two-dimensional shape could have a loop even though the shape had no hole. How to ensure that only significant branches (however that is defined) are part of the skeleton is still an unsolved problem, as it is almost impossible to avoid the creation of "noisy" branches. There are of course numerous methods to remove noisy branches once created, and for specific applications, these methods usually work well, but there is no general solution. Neither is there agreement on how a correct, a "true", skeleton for a specific shape should be defined, which makes evaluating and comparing skeletonization algorithms in general a difficult task. Thus there is still exciting research to be done on the theory of skeletonization.

Existing skeletonization algorithms have for many years been successfully used in numerous image processing and computer vision applications, such as shape recognition and analysis, shape decomposition, registration, interpolation, character recognition, fingerprint analysis, animation, motion tracking, and various medical imaging applications, all in two, three, or even four dimensions. In addition to general shape handling, the skeletons also create hierarchical structures, where the criterion could be the local thickness or saliency or simply the distance from the border.

For this book, we have asked our colleagues to contribute their latest results on skeletonization, both for theoretic work and for new applications, thus making it a snapshot of the research front today. To make it useful also for readers that are not yet completely familiar with skeletonization, we have included a review chapter that briefly summarizes much of the story so far.

Some of the chapters in this book are extended and updated versions of articles from a special issue of Pattern Recognition Letters, published in 2016, but there are also chapters written directly for this book. We would like to take this opportunity to warmly thank all contributors and hope that our readers will find the book useful.

Punam K. Saha
Gunilla Borgefors
Gabriella Sanniti di Baja

PART 1

Theory and Methods

CHAPTER

1 Skeletonization and its applications – a review

Punam K. Saha*, Gunilla Borgefors†, Gabriella Sanniti di Baja‡

Department of Electrical and Computer Engineering, Department of Radiology, University of Iowa, Iowa City, IA, USA Centre for Image Analysis, Uppsala University, Uppsala, Sweden† Institute for High Performance Computing and Networking, CNR, Naples, Italy‡*

Contents

1.1 Introduction ... 3
1.1.1 Basic Concepts ... 3
1.1.2 Background ... 5
1.2 Different Approaches of Skeletonization ... 7
1.2.1 Geometric Approaches ... 10
1.2.2 Curve Propagation Approaches ... 11
1.2.3 Digital Approaches ... 13
1.3 Topology Preservation ... 19
1.4 Pruning ... 22
1.5 Multiscale Skeletonization ... 23
1.6 Parallelization ... 24
1.6.1 Subiterative Parallelization Schemes ... 25
1.6.2 Parallelization Using Minimal Nonsimple Sets ... 25
1.6.3 Parallelization Using P-Simple Points ... 26
1.7 Applications ... 26
1.8 Performance Evaluation ... 29
1.9 Conclusions ... 30
References ... 31

1.1 INTRODUCTION

1.1.1 BASIC CONCEPTS

Despite the long and rich tradition of computing *skeletons* from the 1960s onward [1], in the image processing literature, we are not agreed on definitions, notation, or evaluation methods. In this section, we will define what we mean by a skeleton and the must common steps used in the most common skeletonization methods.

The underlying idea about skeletonization is to reduce the dimension of an object under consideration, so that it will be easier to further process the data. Thus, a 2-D

Skeletonization. DOI: 10.1016/B978-0-08-101291-8.00002-X

object is reduced to a set of 1-D curves, and a 3-D object is reduced to either a set of 2-D surfaces and 1-D curves or to a set of only 1-D curves. A skeleton should ideally have the following properties:

1. It should have the same topology as the object, i.e., the same number of components and holes (and tunnels);
2. It should be thin;
3. It should be centered within the object;
4. It should preserve the geometric features of the object, usually meaning that the skeleton should have components corresponding to the various parts of the object;
5. It should allow complete recovery of the original object.

A set of points fulfilling all but Property 5 is usually called a *medial axis* (but this is not universally agreed on).

A 2-D skeleton consists of three types of points: end-points with one neighboring point in the skeleton, normal points with two neighbors, and branch-points with more than two neighbors. In some cases the end-points are determined before the skeleton is computed. In 3-D case, it becomes more complex, and the skeletal points can be classified into many more types. In this chapter, digital grid elements in 2-D and 3-D images will be referred to as *pixel*s and *voxel*s, respectively; we will use *spel*s to refer to digital grid elements independently of dimensionality.

Most skeletonization methods produce a skeleton that contains many more branches than desired. This can occur for several reasons, but the most common is border noise, where every little protrusion gives rise to a skeletal branch. Unwanted parts of the skeleton are called *spurious*. Many skeletonization methods contain a postprocessing step, called *pruning*, were spurious branches are removed. In all those cases, it must be determined beforehand what defines a spurious part, and this decision is seldom simple. Even in the cases were this step is built into the original algorithm, usually some sort of parameter is used, openly or hidden.

One way of handling the different importance of different skeletal parts is to build *multiscale skeletons,* where only the most important parts remain at the lower resolution levels. For each task, the proper scale can then be chosen.

Most skeletonization methods can only compute the skeleton of a crisp object. This limits the accuracy of the skeleton compared to a continuous representation. Therefore, *fuzzy skeletonization* methods have been developed, which work for objects were each spel has a membership value μ, $0 < \mu \leq 1$.

A skeleton should allow *recovery* of the original shape. This is usually accomplished by giving each skeletal point a value corresponding to the distance to the nearest border point or the time (iteration step) at which the point was reached. This can be done by ensuring that the centers of the *Maximal Inscribed Balls* (MIBs) are part of the skeleton as the set of these balls is equal to the original object. A marked skeleton can, in general, recreate the object, as long as no (or very little) pruning was done. In 3-D case, a surface+curve skeleton can recreate the object, but a pure curve skeleton cannot. If, however, the 3-D object largely consists of long cylindrical structures, then a reasonable reconstruction is possible.

Some types of skeletons can only be computed by *sequential algorithms,* some types only by *parallel algorithms,* but for many types of skeletons, especially those computed on digital structures both possibilities occur. In the sequential case, the skeleton risks to be different depending on the order the spels are visited. In the parallel case, where computation is usually faster than in the sequential case, topology preservation is a challenge.

1.1.2 BACKGROUND

Blum [1] introduced the foundation of skeletonization in the form of medial loci of an object in R^n that forms its skeleton. Analytically, Blum's medial axis can be defined using a grassfire transform process [2], where an object is assumed to be a field of dry grass, and a fire is simultaneously lit at all boundary points of the object. The fire propagates inside the object at a uniform velocity. The skeleton is formed as the union of all quench points, where two independent fire-fronts meet [3–9]. Essentially, a skeleton formed by quench points of grassfire propagation is, in R^n, the loci of the centers of maximally inscribed balls. Although Blum's formulation of skeleton has been popularly adopted by the image processing and computer vision community, there are other forms of medial representation [9, Chapters 1 and 2]. For example, Brady and Asada [10] suggested the use of the loci of the midpoint of a cord, each making an equal angle with the tangents at the two points where the cord meets the object boundary. Leyton [11] proposed the loci of the centers of shortest geodesic paths, each connecting two points where a bitangent sphere meets the object boundary.

Blum's grassfire transform has been popularly adopted by the image processing, computer vision, and graphics community. Many computational approaches and algorithms have been reported in the literature to compute the skeleton of an object. Although most of these approaches share the common principle of Blum's grassfire propagation, often their computational pathways are vastly different. Several researchers have used continuous methods or a mix of continuous and discrete methods, whereas others have used purely digital approaches to compute the skeleton of an object. Discussion on different principles of skeletonization algorithms has been reported in [9]. Commonly, when a *continuous* approach of skeletonization is applied, the boundary of the object is approximated by a polygon or a curve, and the grassfire propagation is computationally imitated by using curve evolution or constrained mathematical morphological erosion; finally, the skeleton is formed as the quench points where the curve evolution process is terminated [4–6,8]. Several researchers have applied *geometric* tools, e.g., the Voronoi diagram, to compute symmetry structures in an object [12–15]. A large number of algorithms adopt digital techniques, where the skeletonization process directly works on a pixel or voxel representation of an object. *Digital* approaches simulate the grassfire propagation process using either iterative constrained erosion of digital objects [16–21] or adopt digital distance transform field [22] to locate maximal balls [7], which are used as primitives to construct the skeleton [23–27,7].

The loci of the centers of MIBs of an object, together with their radii, allow exact reconstruction of the object. This representation was already suggested by Blum. Although the set of MIBs and their radii also can reconstruct the exact object in digital space, their positions and radii values will be different, depending on which digital distance is used. It is also the case that the set of MIBs will not generally be minimal but will contain MIBs that are not necessary for reconstruction. The inherent discrete nature of digital objects further complicates the skeletonization task by posing major hurdles, e.g., the impossibility to achieve one-pixel skeletons in areas of even thickness, high sensitivity to small details on the object boundary, homotopy, etc. Note, however, that the high sensitivity to small details on object boundaries is also featured by nondigital or continuous approaches to skeletonization [28,29].

Over the last several decades, a diverse growth is noticeable in the literature related to skeletonization approaches and algorithms. It poses a big challenge to researchers to select the appropriate skeletonization approach for a specific application. This challenge is further amplified by the lack of a unified approach for evaluation of skeletonization algorithms. In most applications, it is desired that the computed skeleton of a digital object is robust under different conditions of digitization and imaging artifacts, consists of thin curves and surfaces to enable tracking and local topological characterization, and allows acceptable reconstruction of the original object. A major hurdle in evaluating skeletonization algorithms for digital objects is the lack of the definition of "true" skeletons [30,31]. Note that taking the continuous route does not remove these problems—the digital object in an image then has to be converted to continuous data, and the resulting skeleton may have to be digitized. Neither is trivial.

The skeleton of an object is useful in a large number of low-level and high-level image-related tasks such as object description, retrieval, manipulation, matching, registration, tracking, recognition, compression, etc. Skeletonization supports efficient characterization of local structural properties of an object, e.g., scale, orientation, topology, etc. Also, skeletonization has been popularly applied in different medical imaging applications, including thickness computation [32], topological classification [33,34], path finding [35], object shape modeling [36], etc.

A recent survey of skeletonization algorithms and their applications has been reported by Saha et al. [37] and Tagliasacchi et al. [38]. Damon [39,40] has described the smoothness and geometric properties of skeletons.

In this chapter, we present a concise survey of different skeletonization methods and algorithms, and discuss their properties, challenges, and advantages. We will pay special attention to 3-D voxel-based skeletons, as this is our area of expertise. Different important topics related to skeletonization, such as topology preservation, skeleton simplification and pruning, parallelization, and multiscale approaches, are discussed. Finally, various applications of skeletonization are briefly reviewed, and the fundamental issues related to the analysis of performance of different skeletonization algorithms are discussed.

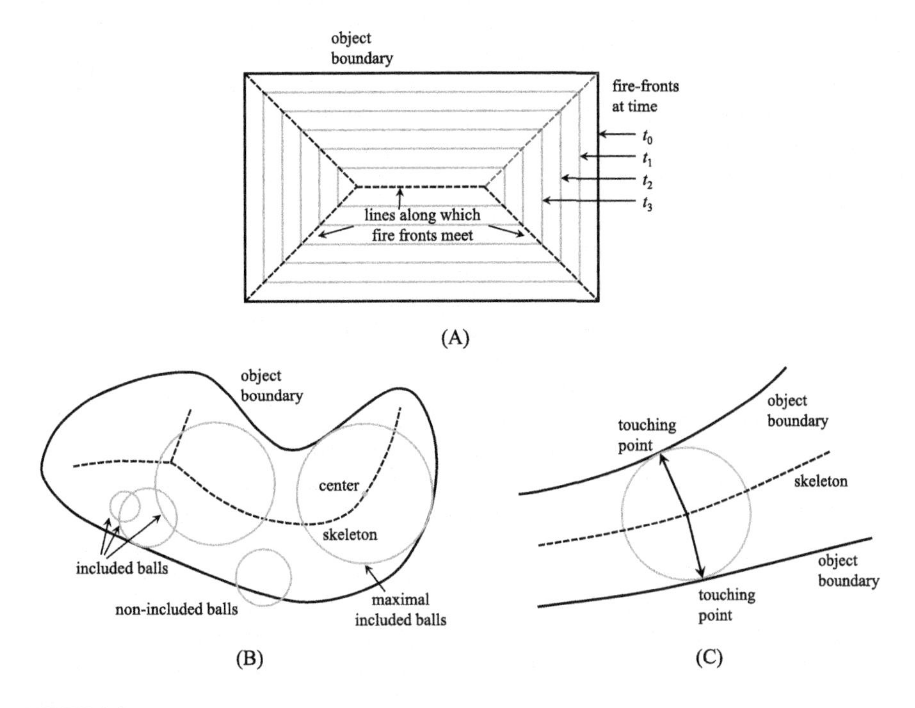

FIGURE 1.1

Different approaches of locating skeletal points in an object. (A) The grassfire propagation generates imprints of fire-fronts at different time instances, and the skeleton is formed at quench points where opposing fire-fronts meet; (B) the skeleton of an object is the loci of the centers of its maximal inscribed balls; (C) the centers of the enclosed balls touching the object boundary at two (or more) disjoint locations form the skeleton of an object. Note that, for objects in the continuous space, such a ball is always maximal; however, the same is not true in a digital grid.

1.2 DIFFERENT APPROACHES OF SKELETONIZATION

Three different basic approaches, illustrated in Fig. 1.1, may be applied to locate the skeletal points of an object: (a) the set of quench points formed at the meeting location of opposing "fire-fronts"; (b) the loci of the centers of MIBs; and (c) the centers of the enclosed balls touching the object boundary at two (or more) disjoint locations. These three approaches of locating skeletal points are equivalent for objects in R^2 or R^3 (as long as the object boundaries are Jordan curves). However, the same is not true for digital objects, and these three definitions usually produce different skeletons for the same digital object [9]. Note that computation methods for the approaches described in Figs. 1.1B and C are different, which leads to different results in digital grids. Besides the variability in locating skeletal points, a large number of skeletonization algorithms are available in the literature with wide differences in terms of their basic computational pathways. Therefore, classifying skeletonization algorithms into major categories is an important and challenging task, and the results may be different depending on the perspective of the classification mechanism. We classify skeletonization algorithms into three major categories based on their computational strategies and the underlying object representation. Three-dimensional versions are presented in parentheses.

1. *Geometric Approaches.* The object boundary is represented by discrete sets of points in continuous space, such as point-clouds or polygonal (polyhedral) representations. Algorithms are based on the Voronoi diagram or other continuous geometric approaches. Mostly, these algorithms use Voronoi edges (Voronoi planes) to locate the symmetry structures or the skeleton of an object.
2. *Curve Propagation Approaches.* The object boundary is represented by a continuous curve or a digital approximation of a continuous curve. Algorithms are based on the principle of continuous curve evolution of the object boundary, where the symmetry structures or the skeleton are formed at singularity locations, specifically, at collision points of evolving curves.
3. *Digital Approaches.* The object is represented by a set of spels in a digital space, usually by Z^n. Algorithms use the principle of digital morphological erosion or the location of singularities on a digital distance transform (DT) field to locate skeletal structures. Often, such algorithms use explicit criteria for topology preservation.

Besides the above three categories, there are a few skeletonization algorithms, which are somewhat different in their computational strategies and may not fit into any of the above three categories. For example, Pizer and his colleagues [41–43] developed algorithms to extract *zoom-invariant cores* from digital images. They defined the medial cores as generalized maxima in the scale-space produced by a medial filter that is invariant to translation, rotation, and, in particular, zoom. See [44,29,45] for early works on mathematical morphological approaches to homotopic thinning in continuous and discrete spaces. Skeletal subsets produced by such methods are dependent on structure elements used by mathematical morphology operators. Also, the resulting skeletons may not preserve topological connectedness. Lantuejoul [46] introduced a mathematical morphological approach for skeletonization using the notion of the influence zone. Maragos and Schafer [47] used mathematical morphological set operations to transform a digital binary image using parts of its skeleton containing complete information about its shape and size. Jonker [48] developed mathematical morphological operations for topology preserving skeletonization and its variants for 4-D images, which was later generalized to n dimensions [49].

It may be noted that most digital approaches to skeletonization use object representations in pixel (2-D) or voxel (3-D) grids. A major drawback with such skeletonization algorithms is that these methods do not guarantee single-pixel (or voxel) thin skeletons for all objects, especially, at busy junctions; see Fig. 1.2. In other words, if the object of Fig. 1.2 (or its 3-D version) is used as an input, then a digital skeletonization algorithm will not be able to remove any pixel (or voxel) and thus, will fail to produce a single-pixel or single-voxel thin skeleton. Note that the example

FIGURE 1.2

A 2-D example of a busy junction of digital lines forming a central blob region, which cannot be thinned any further because the removal of any pixel from this object alters its topology. It is possible to construct a similar example in 3-D.

of Fig. 1.2 may be modified to increase the size of the central blobby region. This is also true for the skeleton in a region of even thickness, as the skeletal points cannot occur between the two innermost layers. A more complex and richer approach to object representation is using simplicial or cubical complex frameworks [50–55], where volumetric and lower-dimensional grid elements are processed in a unified manner. The above problem disappears when skeletonizing objects using these frameworks [52,53,55].

Digital objects can also be represented in other grids. In 2-D case the hexagonal grid has many advantages, and skeletons have been developed for this grid [56,57]. In 3-D case the fcc and bcc grids are sometimes considered as topologically simpler than the cubic grid. Strand [58] has presented skeletons for these grids.

Several researchers have contributed to the area related to skeletonization of objects from their fuzzy representations. Pal and Rosenfeld [59] introduced a fuzzy medial axis transformation algorithm using the notion of maximally included fuzzy disks. Following their definition, a fuzzy disk centered at a given point is a fuzzy set in which membership depends only on the distance from the center. Others have used a fuzzy distance transform [60] to define centers of fuzzy maximal balls [61,62]. Also, mathematical morphological operations have been used to define centers of maximal balls [45,63]. Bloch [64] introduced the notion of skeleton by influence zones (SKIZ) or generalized Voronoi diagram for fuzzy objects. Influence zones define regions of space that are closer to a region or object, and the SKIZ of an image is the set of points that do not belong to any influence zone. She presented an algorithm to compute the fuzzy SKIZ of a fuzzy set using iterative fuzzy erosion and dilation [65]. A detail discussion and review on fuzzy skeletonization is presented in Chapter 3.

In the rest of this section, we present a brief survey of skeletonization algorithms under each of the above three categories.

1.2.1 GEOMETRIC APPROACHES

Several algorithms [12,14,15] focus on symmetry and other geometric properties of Blum's medial axis to compute the skeleton of an object. Often, such methods are applied on a mesh representation of the object or on a point-cloud generated by sampling points on the object boundary. One popular approach under this category is based on the principle of the Voronoi diagram [66,12–15,67]. The "Voronoi skeleton" is computed on a polygonal (or polyhedral) representation of an object, where the vertices are considered as sample points on the object boundary. This skeleton is obtained by computing the Voronoi diagram of its boundary vertices and then taking its intersection with the polygonal object itself. The basic principle is pictorially described in Fig. 1.3. The original object is a 2-D shape in the continuous plane (A), where the interior is light gray. In (B) a discrete representation of the object is obtained by sampling finitely many points on the object boundary and then connecting every pair of adjacent points to form a polygon. The Voronoi diagram of the sample points is then computed, light gray lines in (C). The Voronoi skeleton consist of the part of the Voronoi diagram that intersects the discrete object, dotted lines in (D). Skeletal segments "deep" inside the object, thick lines in (E), consist of all segments that do not touch the boundary.

One major challenge for Voronoi skeletonization algorithms is that each additional vertex on the polygon adds a new skeletal branch; see Fig. 1.3D. Thus, a suitable polygonal approximation of an object is crucial to generate the desired complexity of the skeleton. On the other hand, an accurate polygonal representation of an object requires a large number of vertices. This leads to a large number of spurious Voronoi skeletal branches, which are not needed to represent the overall geometry of the object. Ogniewicz [14] observed that the skeletal segments that lie deeply inside the polygonal object are less sensitive to small changes on the boundary. Such segments are essential for the description of the global topology and geometry of an object and also less sensitive to small changes on the boundary and to the sampling density; see Fig. 1.3E. Based on this observation, they derived different residual functions and used those to differentiate spurious branches from those essential to represent object topology and geometry.

Schmitt [68] showed that, as the number of generating boundary points increases, the Voronoi diagram converges in the limit to the continuous medial locus, with the exception of the edges generated by neighboring pairs of boundary points. Later, Voronoi skeletonization was generalized to 3-D polyhedral solids [66,69–71]. Amenta et al. [66] characterized inner and outer Voronoi balls for a set of boundary sample points to reconstruct its "power crust," an approximation of a polyhedral boundary, and to compute its Voronoi skeleton. Jalba et al. [72] developed a GPU-based efficient framework for extracting surface and curve skeletons from large meshes. Bucksch et al. [73] presented a graph-based approach to extract the skeletal tree from point clouds using collapsing and merging procedures in octree-graphs. This approach offers a computationally efficient solution for computing skeletons from point-clouds that is robust to varying point density, and the complexity of the skeleton may be adjusted by varying the size of the octree cell. Ma et al. [74] in-

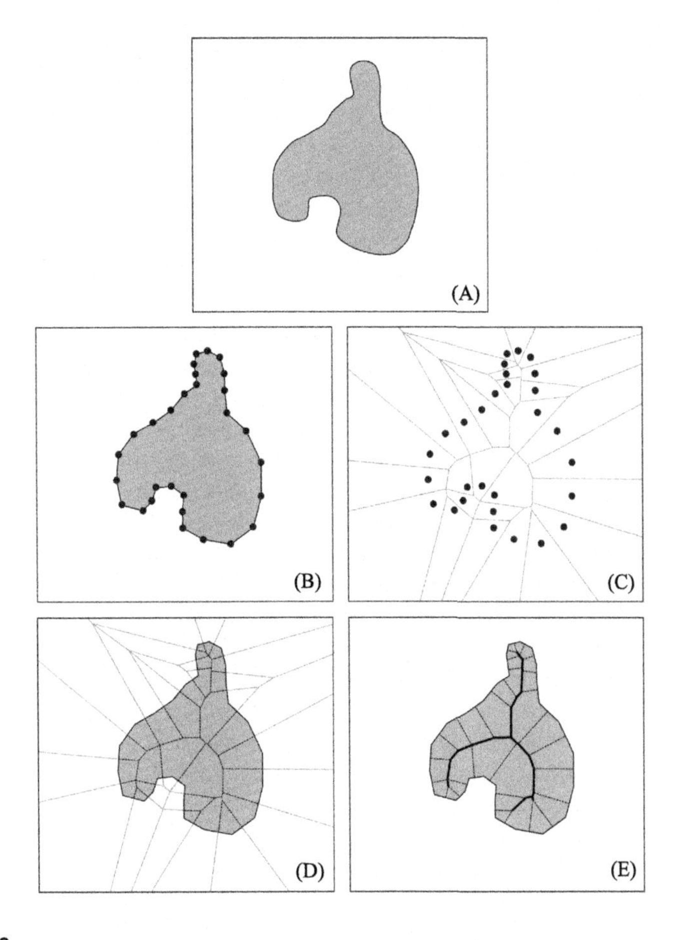

FIGURE 1.3

An illustration of Voronoi skeletonization in 2-D case. See the text.

troduced a nearest-neighbor approach to skeletonization from points sampled on the object boundary, where each point is associated with a normal vector. For each sample point, the method finds its maximal tangent ball, containing no other sample point, by iteratively reducing its radius using nearest-neighbor queries. They showed that the computed skeleton converges to the continuous skeleton as the sampling density tends to infinity.

1.2.2 CURVE PROPAGATION APPROACHES

Some researchers have simulated Blum's grassfire propagation using a curve propagation model for the object boundary [4–6,75,8]. A thorough review of continuous curve propagation-based skeletonization algorithms was presented in an earlier edited book [9]. Unlike the methods based on digital morphological peeling of object boundary layers, curve evolution algorithms are modeled using partial differential equations.

The first step is to approximate the object border by a (set of) C^n curve(s), where usually $n \geq 2$. A common choice is B-splines. The accuracy of this approximation has a large influence on the skeleton. During the following process of continuous curve evolution, certain singularities occur, which are mathematically referred to as "shocks" and used to define the symmetry axis or skeleton. Leymarie and Levine [6] modeled the curve evolution process using an active contour on a distance surface and used the singularities of curvature features to initiate skeletal branches. Kimmel et al. [5] segmented the object boundary at singularities of maximal positive curvature and computed a DT from each boundary segment. Finally, the skeleton is located at the zero level sets of distance map differences, which represent the locations where fire-fronts from different object segments meet. Siddiqi et al. [8] proposed a Hamiltonian formulation of curve evolution and computed the outward flux of the vector field of the underlying system using the Hamilton–Jacobi equation. Skeletons are located at singularities of this flux field. Additionally, they imposed the topology preservation constraints of digital grids to ensure the robustness of the computed skeletons.

In curve propagation approaches, the time $\phi(\mathbf{x})$, when a fire-front crosses the location $\mathbf{x}$, is used to locate the singularities. Essentially, the time-of-flight ϕ is the distance transform function from the object boundary. The divergence theorem is applied on the grassfire flow function $\nabla\phi$ to compute its average outward flux at a given location. For a 2-D binary digital object, which is not first approximated by delimiting continuous curves, the average outward flux is computed using the following equation:

$$F(p) = \frac{1}{8} \sum_{q \in N_8^*(p)} \mathbf{n}_q \cdot \nabla\phi(q), \tag{1.1}$$

where $N_8^*(p)$ is the 3×3 excluded neighborhood of the pixel p, and $\mathbf{n}_q$ is the outward normal at $q \in N_8^*(p)$, i.e., the unit vector along the direction from p to q. The excluded neighborhood of a pixel $p \in Z^2$, denoted by $N_8^*(p)$, is the set of all pixels in the 3×3 neighborhood of p excluding p itself. The excluded neighborhood in 3-D, $N_{26}^*(p)$, is defined similarly. Thus, in a 3-D digital image, N_{26}^* is used, and the normalizing factor 26 is used instead of 8. The operator ∇ must be approximated by a more or less accurate numerical function in digital space. Just replacing derivation by simple subtraction is seldom sufficient. Examples of outward flux computation for several 2-D binary objects are shown in Fig. 1.4.

Several algorithms [77–81] apply general fields for front propagation, which are smoother than distance transform fields, and use such fields to enhance the smoothness of the computed skeletons. Ahuja and Chuang [77] used a potential field model where the object border is assigned constant electric charges and the valleys of the resulting potential field are used to extract the skeleton.

Tari et al. [81] used an edge-strength function to extract object skeletons, which equals 1 along object boundary, and elsewhere it is computed using a linear diffusion

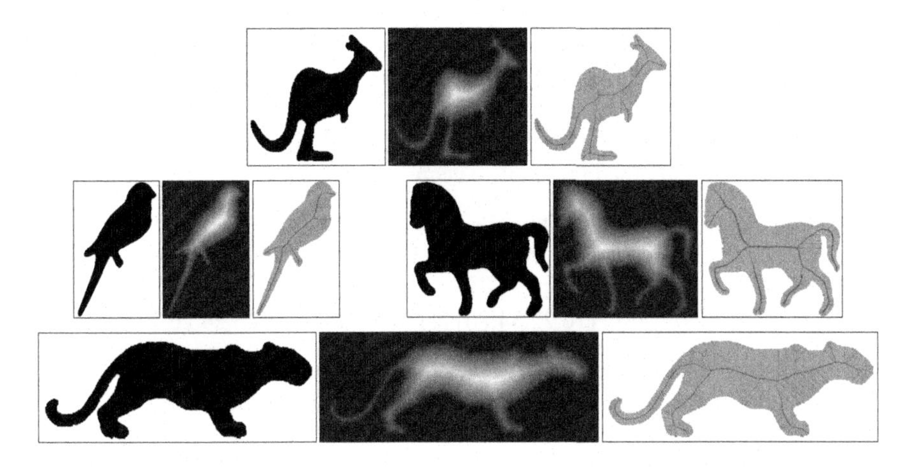

FIGURE 1.4

Illustrations of computed outward flux for 2-D binary objects. For each shape, the original binary object, the distance transform, and the outward flux maps are shown. The 3–4 weighted distance transform [76] was used for flux computation. Using the digital Euclidean distance transform could slightly improve the smoothness of the flux map. The authors thank to Dr. Jin for generating this figure.

equation. In this approach, the skeleton is defined as the projection of singular lines on the elevated surface representation of the edge-strength function and by analyzing the geometry of the level curve of the function. A major advantage of this approach is that the method can be applied to gray scale images as the computation of the edge-strength function only requires object edge points instead of a complete connected object boundary.

Aslan et al. [78] further improved this method by incorporating the notion of absolute scale related to the maximum allowable regularization that still permits morphological analysis. An important drawback with these methods [78,81] is that the resulting skeletons may not be topologically correct, as they may have breaks not corresponding to breaks in the object. However, in a positive note, such algorithms allow multiscale regularization of skeletal smoothness. Cornea et al. [79] developed a curve skeletonization algorithm using a force vector field, whereas Hassouna et al. [80] designed a curve skeletonization algorithm using a gradient vector flow.

1.2.3 DIGITAL APPROACHES

The most straightforward and popular skeletonization approach is based on simulating Blum's grassfire propagation as an iterative erosion on a digital grid under certain predefined topological and geometric rules. The literature of iterative digital erosion algorithms for skeletonization in 2-D case had matured in the early 1990s

[16]; however, the same was not true for 3-D skeletonization algorithms. Tsao and Fu [21] reported the first complete skeletonization algorithm for 3-D digital objects in a voxel grid.

From the perspective of computational principles, digital skeletonization algorithms may be also classified into three categories:

1. Fully predicate-*kernel-based* iterative algorithms [17,82,83,18,19];
2. Iterative *boundary peeling* algorithms under topological and geometric constraints [17,20,21];
3. *Distance transform* (DT) [76,22,84,85] based algorithms [23,24,26,27].

In terms of computational efficiency, skeletonization algorithms may be classified into two categories, namely, *sequential* [23–26,17,27] and *parallel* [86,82,83,87–90, 18–21] methods.

Although, fully kernel-based skeletonization was established in 2-D during the 1980s [16], the first kernel-based 3-D skeletonization algorithm was reported by Palágyi and Kuba [19], which was further improved by Németh et al. [18]. In a kernel-based approach, topological and geometric constraints are coupled together in a form of $3 \times 3 \times 3$ (or larger) window predicates defining expected colors of different neighbors including several "don't care" positions. Multiple template kernels are used to determine whether a boundary voxel can be deleted during an iteration. Although such methods are easy to implement, it is generally difficult to make any modification. Saha et al. [20] presented an iterative boundary peeling algorithm for 3-D skeletonization where topological and geometric issues are addressed separately.

Chatzis and Pitas [91] and Lantuejoul [46] applied mathematical morphological approaches to skeletonization. Several algorithms [92–95] have been tailored to directly compute curve skeletons for elongated volumetric objects.

A large number of research groups have applied DTs for skeletonization [23–27, 7]. Arcelli and Sanniti di Baja [23] introduced the notion of DT-based skeletonization in 2-D and discussed its advantages. Borgefors and her colleagues [96,26,97] applied DTs for skeletonization in 3-D. Saito and Toriwaki [98] introduced the idea of using the DT to define the sequence of voxel erosions, which has been further studied by other research groups [24,27]. A major advantage of DT-based methods is that these methods do not require repetitive image scans. Moreover, these methods do not require the management of two buffers, respectively storing the elements of the current boundary and their neighbors that will constitute the successive boundary. Thus, DT-based approaches are characterized by computation efficiency. Also, a DT-driven voxel erosion strategy, together with a suitable choice of distance metric [84], makes the skeletons more robust under image rotation. However, an apparent difficulty of this approach is that it may be hard to parallelize such algorithms.

Even though a few works on gray-scale and fuzzy skeletonization have been presented in the literature [99,81,100], a comprehensive skeletonization algorithm for fuzzy digital objects have been reported only recently [101].

A DT-driven 3-D skeletonization algorithm [24,101,34] is briefly outlined in the following, which, in principle, is applicable to both binary and fuzzy digital objects.

1. Primary skeletonization

1.1 Compute the distance transform and identify the set of quench voxels (discussed later in this section).

1.2 In decreasing order of distance value, identify and mark significant quench voxels based on their local shape significance factors.

1.3 In the increasing order of distance value, delete nonsignificant quench voxels, which are also simple points (see Section 1.3), i.e., voxels not needed for topology preservation.

2. Final skeletonization

2.1 Locate two-voxel-thick structures and convert those into one-voxel-thick structures.

2.2 Locate and remove voxels with conflicting topological and geometric properties.

3. Pruning

3.1 Identify and remove spurious branches using some global shape significance factor.

The notion of primary and final skeletonization was simultaneously introduced by Saha et al. [102,20] and Sanniti di Baja [7]. Saha et al. defined the two steps for 3-D skeletonization using an iterative boundary erosion approach, whereas Sanniti di Baja presented the idea in 2-D using a DT-based algorithm. Primary skeletonization produces a "thin set" [23] representing the overall geometry and topology of the skeleton of an object, where every voxel has at least one background neighbor except at very busy intersections of surfaces and curves, in which case some voxels may be internal voxels; see, e.g., Fig. 1.2. The thin set often includes two-voxel-thick structures. Generally, a scan-independent algorithm for locating quench voxels leads to two-voxel-thick medial structures for objects with even-voxel thickness. On the other hand, a scan-independent definition of quench voxels enhances the stability of the overall skeleton geometry as it annuls the influence of alterations in local object configurations during the erosion process. During final skeletonization, two-voxel-thick structures are eroded under topology preservation and some additional geometric constraints [24,20,7], maintaining the overall shape of the skeleton. A few examples of fuzzy skeletonization results using the above scheme are presented in Fig. 1.5.

Quench Points

Following Blum's grassfire analogy, the skeleton of an object is formed at quench points where opposing fire fronts meet. Location of *quench points* is essential to capture geometric properties of an object in its skeleton. In early works on skeletonization, locations of quench pixels or voxels had been defined algorithmically,

FIGURE 1.5

Results of fuzzy skeletonization [101] on two medical objects (leftmost in first and second rows) and several other 3-D shapes available online.

and a precise definition was missing. Arcelli and Sanniti di Baja [103] defined the *centers of maximal balls* (CMBs), equivalent to centers of MIBs, in a 3×3 weighted distance transform, using a method that is often used for locating quench pixels. It can easily be extended to weighted $3 \times 3 \times 3$ DTs in 3-D. Borgefors [104] extended the definition of CMBs to 5×5 weighted distances. For the digital Euclidean DT, precomputed look-up tables are the only way to identify CMBs [105].

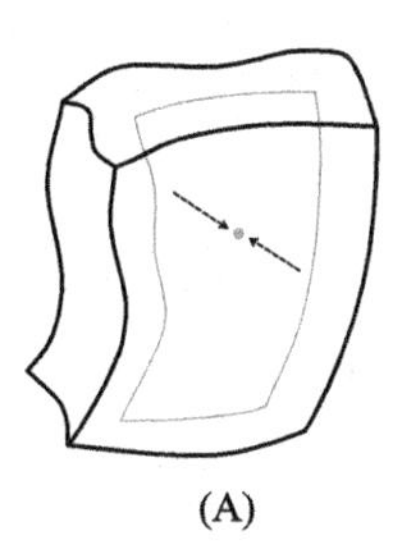

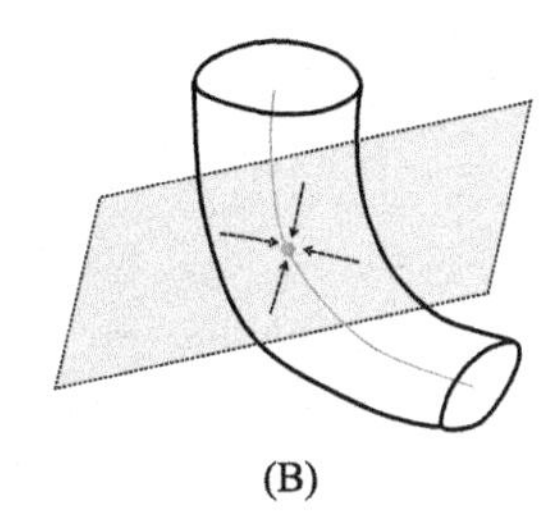

FIGURE 1.6

Surface quench points occur were two fire-fronts meet (A), whereas curve quench points occur where fire-fronts from all directions meet (B).

The identification of CMBs was generalized to fuzzy sets by Saha and Wehrli [61], Svensson [62], and Jin and Saha [101]. In a fuzzy digital object $\mathcal{O}$, a CMB, or quench voxel (in 3-D), is defined as follows.

A voxel p in a cubic grid Z^3 is a *fuzzy CMB* or a *fuzzy quench voxel* if the following inequality holds for every neighbor $q \in N_8^*(p)$:

$$FDT(q) - FDT(p) < \frac{1}{2}(\mu_{\mathcal{O}}(p) + \mu_{\mathcal{O}}(p))\,|p - q|, \tag{1.2}$$

where FDT is the fuzzy distance transform [60] value, $\mu_{\mathcal{O}}$ is the membership function for the fuzzy object $\mathcal{O}$, and $|\cdot|$ is the distance between the voxels, using the same distance definition as in the DT used.

Quench voxels (or CMBs) are singularities on the DT map forming ridges on it, where the singularity signifies that a quench voxel cannot pass its DT value, or grassfire propagation, from itself to any of its neighboring voxels. In other words, grassfire propagation terminates at a quench voxel. Definitions of fuzzy quench voxels and CMBs [103] are equivalent for binary digital objects.

A well-known challenge using quench voxels for computing the skeleton is the generation of a large number of CMBs and their sensitivity to small protrusions or dents on object-boundaries leading and spurious (noise-caused) skeletal branches. Also, the set of CMBs is dependent on the weights used in the distance transform. This challenge is further intensified for fuzzy objects where each fluctuation in the membership value adds an extra CMB. Minimizing the set of CMBs is essential to generate a compact and effective representation of primary skeleton. Borgefors et al. [106,107] suggested to remove redundant CMBs caused by the digital representation and then to use the minimal set as anchor points for a contour peeling algorithm.

In 3-D, two types of quench voxels, namely *surface* and *curve* quench voxels, occur [24,101,20]; see Fig. 1.6. A surface quench point is formed when two opposite fire-fronts meet, whereas a curve quench point is formed when fire-fronts meet from all directions on a plane. These two types of quench voxels can be handled differently when filtering the spurious ones [101]. A recently introduced notion of

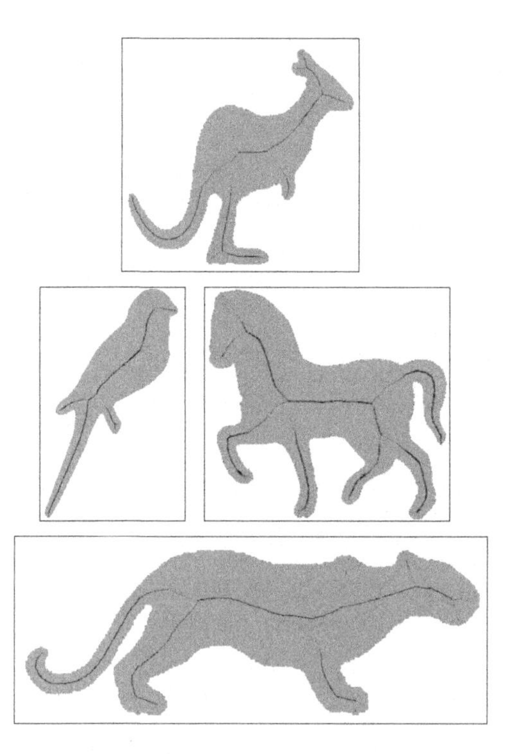

FIGURE 1.7

Examples of collision-impact skeletons for the same binary shapes as in Fig. 1.4. The weighted 3–4 distance transform was used [76]. Use of the Euclidean distance transform could slightly improve the smoothness of collision-impact structures. The authors thank to Dr. Jin for generating this figure.

collision-impact [101,108,109] of fire-fronts at individual quench points defines their shape significance, which can be used to distinguish between spurious and meaningful quench points. Although colliding fire-fronts always terminate at quench voxels, their collision-impact may vary depending upon the angle between colliding fire-fronts. The collision-impact at a given quench voxel p is defined as follows:

$$\xi_{\mathrm{D}}(p) = 1 - \max_{q \in N_{26}^{*}(p)} \frac{f_{+}(FDT(q) - FDT(p))}{\frac{1}{2}(\mu_{\mathcal{O}}(p) + \mu_{\mathcal{O}}(q))|p - q|}, \tag{1.3}$$

where the function $f_{+}(x)$ returns the value of x if $x > 0$ and 0 otherwise. Collision-impact weighted skeletons for the same binary 2-D shapes as in Fig. 1.4 are shown in Fig. 1.7. The role of collision-impact to produce an effective and compact skeleton for fuzzy object in 3-D is illustrated in Fig. 1.8.

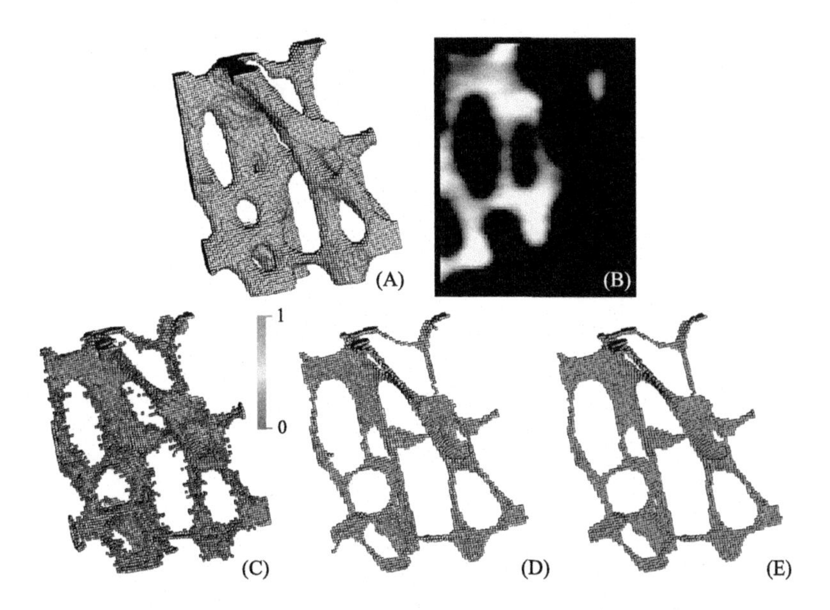

FIGURE 1.8

Collision-impact values generate compact and effective skeleton for fuzzy objects. (A) A binary visualization of a trabecular bone region in a micro-CT image of a cadaveric distal tibia specimen. (B) A sagittal image slice of the trabecular bone showing the image's fuzziness. (C) Initial skeletal voxels color-coded with their collision-impact values. The color-coding bar is shown on the right. (D) Skeleton after removing quench-voxels with low collision-impact values that do not break connectivity. The same color-coding bar of (C) is used. (E) Final skeleton after pruning.

1.3 TOPOLOGY PRESERVATION

A *simple point* is a spel that can be removed from an object without changing its topology. Therefore, 2-D or 3-D simple point constraints are often applied while deleting individual pixels or voxels in topology-preserving skeletonization.

A spel is defined as a simple point if and only if its binary conversion, i.e., conversion from object to background or vice versa, preserves the object components and cavities (and tunnels in 3-D) in the $3 \times 3(\times 3)$ neighborhood of the candidate spel.

A concise characterization of 2-D simple points was established in the early 1970s [110–113]. Characterization of simple points in 3-D is more complex than in 2-D, primarily due to the presence of tunnels in 3-D. Tourlakis and Mylopoulos [114] presented a generalized characterization of simple points that applies to any dimension. Morgenthaler [115] presented a characterization of 3-D simple points where the preservation of tunnels was imposed by adding a constraint of Euler character-

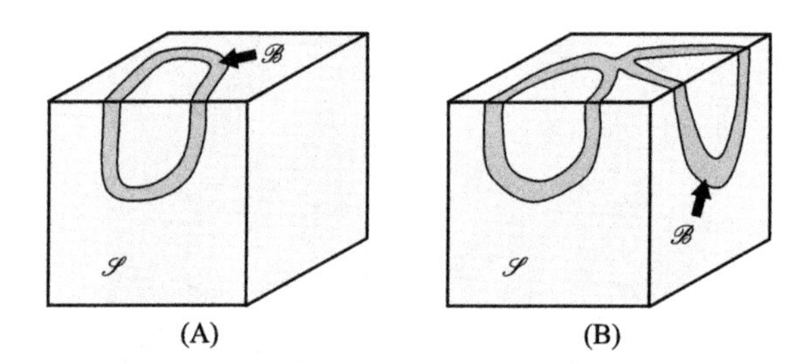

FIGURE 1.9

Examples of tunnels on the boundary surface of am R^3 cube constituting a topological sphere. (A) The object $\mathscr{B}$, darker gray, on the surface $\mathscr{S}$ of a cube, light gray, forms one tunnel; such an object always generates exactly two components on $\mathscr{S} - \mathscr{B}$. (B) Here, the object $\mathscr{B}$ forms two tunnels and generates exactly three components on $\mathscr{S} - \mathscr{B}$.

istic preservation before and after the deletion of a candidate voxel. Lobregt et al. [116] presented an efficient algorithm for 3-D simple point detection based on the Euler characteristic preservation. However, as described by Saha et al. [117], their algorithm fails to detect the violation of topology preservation when the deletion of a point simultaneously splits an object into two and creates a tunnel. Saha et al. [118,119,102,117] defined the existence and the number of tunnels in the $3 \times 3 \times 3$ neighborhood using a straightforward connectivity criterion as stated in the following theorem.

Theorem 1.1. *If a voxel p has a background 6-neighbor, the number of tunnels $\eta(p)$ in its excluded neighborhood $N_{26}^*(p)$ is one less than the number of 6-components of background 18-neighbors of p that contain a 6-neighbor or zero otherwise.*

It is easier to explain the motivation of this theorem using examples of tunnels on the surface of a topological sphere, the boundary surface $\mathscr{S} \subset R^3$ of a cube in Fig. 1.9. In both Figs. 1.9A and B, the object $\mathscr{B} \subset \mathscr{S}$ is present on the surface of the cube. Such an object forms a tunnel if it forms a loop and, thus, divides the topological sphere $\mathscr{S}$ into two or more disconnected regions, i.e., $\mathscr{S} - \mathscr{B}$ is not connected. In Theorem 1.1, this observation is translated for the digital case, where the neighborhood $N_{26}^*(p)$ forms the digital boundary surface of a $3 \times 3 \times 3$ cube, and $N_{26}^*(p) \cap \mathscr{B}$ is the set of object voxels on that surface. While defining 6-components of background points, $N_{18}^*(p)$ is used instead of $N_{26}^*(p)$ to avoid any crossing between a background voxel path and an object voxel path (Fig. 1.10). A formal proof of this theorem is available in [120].

The above definition of the number of tunnels is used to formulate a concise characterization of 3-D simple points in the following four-condition format [118, 119,102,117].

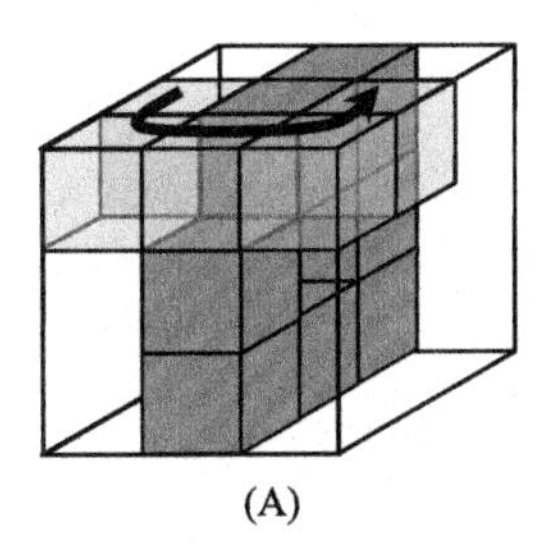

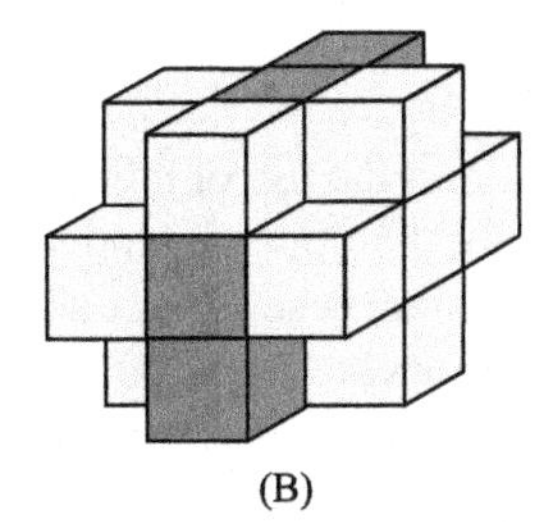

(A) (B)

FIGURE 1.10

An example of tunnels in the $3 \times 3 \times 3$ neighborhood. Object voxels are dark; background voxels are transparent and semitransparent in (A) and opaque in (B). The object voxel configurations are the same in (A) and (B). (A) shows that a background path (semitransparent voxels) crosses a nontrivial closed loop of object voxels on the boundary of $N_{26}(p)$. This phenomenon of "crossing" [119,102,117] is crucial for the characterization of tunnels and 3-D simple points. (B) demonstrates the 18-neighborhood; note that no background 6-path may cross a nontrivial closed loop of 26-path of object voxels in this neighborhood.

Theorem 1.2. *A voxel p is a (26, 6) simple point in a 3-D binary image if and only if it satisfies the following four conditions.*

Condition 1. p has a background 6-neighbor.
Condition 2. p has an object 26-neighbor.
Condition 3. The set of object 26-neighbors of p is 26-connected.
Condition 4. The set of background 6-neighbors of p is 6-connected in the set of background 18-neighbors.

In this theorem, Condition 1 ensures that no cavity is created by deletion of the point, whereas Condition 2 confirms that no isolated point is deleted. Condition 3 ensures that, after the deletion of the central point p, its neighboring points remain connected. Finally, Condition 4 guarantees that the deletion of a point does not create a tunnel in the neighborhood. This result was rediscovered by Malandain and Bertrand, which was presented in a two-condition format [121,122]. Topology preservation in a simplicial and cubical complex framework is characterized using the classical notion of homotopy [51,52,54]. The notion of "attachment set" is also used to define the simplicity of a voxel [54]. According this theory, the object voxels in $N_{26}^*(p)$ are contractible to the attachment set, i.e., intersection between the central voxel and object voxels in $N_{26}^*(p)$. In other words, the simplicity of a voxel is determined by the simplicity of its attachment set. Also, the simplicity or topology preservation in a generalized tessellation of R^n with convex tiles (equivalent to voxels) were studied [120,123,124], and it was observed that the excluded neighborhood of a tile is contractible to its attachment set if and only if the neighboring tiles satisfy a simple intersection rules among themselves.

1.4 PRUNING

A factor that can limit the use of the skeleton in applications is its sensitivity to small deformations along the object boundary. Even negligible boundary noise can cause spurious skeleton branches, so that skeleton pruning techniques are of interest. In general, skeleton branches are expected originate in correspondence with regions of the object that are perceived as individually meaningful, such as limbs and sufficiently sharp boundary convexities, whereas the actual structure of the skeleton is usually considerably more complicated. Effective pruning techniques should be based on the use of criteria for the evaluation of the significance of skeleton branches to take a decision on whether removing or keeping them. Clearly, branch removal caused by pruning modifies the skeleton in such a way that the object represented by the pruned skeleton is characterized by a smoother boundary or a different number of protrusions compared to the original object. A pruning process is adequate if the resulting skeleton structure is noticeably simplified, but the above differences are negligible for the specific application.

Whichever significance criterion is adopted, this should establish a strict relation between skeleton branches and the relevance of the object parts that they represent. In the 2-D case, some criteria involving propagation velocity, maximal thickness, radius function, axis arc length, and the boundary/axis length ratio can be found in [125]. Other pruning methods use contour partitioning via discrete curve evolution [126] and with bending potential ratio [127], where the decision on whether pruning a branch is based on the context of the contour segment corresponding to that branch. A more recent pruning criterion is in the framework of the group-wise medial axis transform that yields a fuzzy significance measure for each branch, derived from information provided by a group of shapes [128].

Pruning techniques for the 3-D skeleton are also available in the literature. In particular, for the 3-D curve skeleton, a measure of global shape significance based on the ratio between the number of CMBs present in a skeleton branch and the total number of voxels in the branch has been used [24]. A pruning method for the 3-D surface skeleton first unglues skeleton branches at junctions and then retrieves the meaningful branches lost during ungluing using a connectivity analysis on a "shape distance transform" capturing global shape significance of individual skeletal branches [34]. Local significance-weighted path length is used to define the shape significance of individual branches in a minimum-cost path-based skeletonization algorithm for tree-like objects [129].

An interesting aspect that has not received enough attention in the literature is the modality, parallel or sequential, that is used during pruning. Using the parallel modality, points classified as branch points in the initial skeleton maintain their status until all peripheral branches have been examined and possibly pruned. Using the sequential modality, the status of a branch point is checked, and possibly updated to that of normal point or end-point, as soon as branches sharing it are removed. Thus, according to the parallel modality, peripheral branches are checked for removal up to the branch points delimiting them in the initial skeleton. In turn, by following the sequential modality a peripheral branch delimited by a given branch point in the initial skeleton may be pruned up to a more internal branch point if the originally delimiting branch point has been transformed into a normal point due to the removal of other peripheral branches sharing it.

Also, the number of pruning iterations has to be taken into account. In fact, due to branch removal, branches that were internal in the initial skeleton are likely to be transformed into new peripheral branches, which are not necessarily all significant. Thus, pruning may need to be iterated, and the proper number of pruning iterations has to be established carefully to avoid that the structure of the pruned skeleton is excessively simplified.

1.5 MULTISCALE SKELETONIZATION

As has been mentioned before, skeletons are in general very sensitive to small-scale features on objects borders that generate spurious skeleton segments. An alternative to pruning as described in Section 1.4 is multiscale skeletonization. These approaches use a scale, regularization, or significance factor to control the trade-off between the smoothness and simplicity of the skeleton and the exactness of the skeleton's representation of object features. Different principles of multiscale have been adopted in different skeletonization approaches. Pizer et al. [75] laid out mathematical properties and computational abilities of different approaches and compared and contrasted those.

In Voronoi skeletonization approaches, stability and significance of Voronoi segments are used to reduce spurious branches and to improve the skeleton's robustness. Several significance measures have been proposed in the literature [15,67], which are used to establish a hierarchical representation of segments in a Voronoi skeleton. Skeletal segments or branches that are at the periphery of this hierarchy and have a low significance measure are candidates for pruning. This framework underlies the notion of multiscale skeletons based on the thresholds for hierarchical position and significance of branches.

Another perspective of multiscale skeletonization is to use different regularization levels in terms of initial smoothing of the original object boundary [126,130,131]. At higher levels of smoothing, the skeletons become less detailed and more stable to boundary changes. Essentially, this class of algorithms introduces the notion of skeleton scale-space.

Yet another view of multiscale skeletonization is to use a regularization term, or curve smoothing, during the front propagation [132,4,133,81]. Kimia et al. [4] used a regularization approach, where a curvature-dependent smoothing component is added to the uniform speed representing the morphological propagation of the fire-front. Tari et al. [81] introduced a blurring radius parameter to control the smoothness of the edge-strength function that in turn decides the smoothness of computed skeletons. This approach leads to simpler and faster implementation applicable to higher dimensions even where there are gaps among object boundary points. Aslan et al. [78] brought the notion of absolute scale to skeletonization by letting the regularization term tend to infinity and dominate the morphological component of curve evolution. Cornea et al. [79] used a different strategy, where they computed the divergence of the repulsive force field inside an object by charging the object's boundary and used a threshold on it to model a multiscale curve skeletonization.

Giesen et al. [134] introduced an interesting notion of scale axis transformation that uses a scale multiplicative dilation of maximal included balls along the medial axis of the original object to remove less important features. Miklos et al. [135] applied the discrete scale axis transform to compute multiscale skeletons from 3-D mesh representation of an object. The scale-axis transformation by Giesen et al. [134] and Miklos et al. [135] fails to guarantee that the skeleton is included in the original object. An alternative definition of scale axis transform introduced by Postolski et al. [136] guarantees the inclusion of the skeleton.

In digital approaches to skeletonization, skeletal pruning strategies have been adopted to control the trade-off between the simplicity of the skeleton and the inclusion of important object features. Borgefors et al. [137] presented a multiscale discrete skeletonization algorithm generating multiresolution decomposition and hierarchical skeletal representation of objects. Several researchers have used a global shape significance factor [24,138,101,34] and a predefined threshold for this factor to distinguish relevant skeletal branches carrying important object information from those generated by small scale protruding dents on object boundaries. The advantage of the global shape significance factor is that it ignores the section of a branch that grows only for connectedness preservation while capturing the section carrying geometric object information. Attali et al. [138] laid down the foundation of global significance measures of skeletal branches, which was further generalized and improved by others [24,138,101,34]. Németh et al. [18] have recommended an iteration-by-iteration smoothing approach to improve the quality of final skeletons.

1.6 PARALLELIZATION

Almost fifty years ago, Rutovitz [139] had proposed the first parallel skeletonization algorithm. Since then, a large number of 2-D parallel skeletonization algorithms have been reported. The first 3-D parallel skeletonization algorithm was presented by Tsao and Fu [21]. The core challenge in parallel skeletonization emerges from the fact that, although a characterization of simple point guarantees topology preservation when one simple point is deleted at a time, it fails to ensure topology preservation when multiple simple points are simultaneously deleted. Different strategies have been adopted in the literature to solve this specific problem of parallel skeletonization.

1.6.1 SUBITERATIVE PARALLELIZATION SCHEMES

In iterative skeletonization algorithms, during one iteration, one layer of object voxels is peeled off while preserving the topology and local elongatedness of the object. Here, each iteration is divided into several subiterations. Two schemes for dividing an iteration in 3-D into subiterations are available in literature: either based on the direction of open face(s) of boundary voxels [86,140,17,87,89,141,19,142,21,143] or based on subfield partitioning of the image grid [144,145,90,146,147,20]. In the first scheme, an image is divided into six directional subsets, e.g., North, South, East, West, Top, and Bottom. One iteration of the skeletonization algorithm is completed in six subiterations, one for each direction, where the voxels in the corresponding directional subset are processed in parallel. In general, directional subdivision cannot eliminate all ambiguous simple sets, and additional restricting conditions are required to ensure topology preservation. Algorithms are available under this category where different numbers of subiterations are used in one complete iteration, e.g., 12-subiteration [82,19], 8-subiteration [148], 6-subiteration [83,142], 3-subiteration [149], and also a fully parallel [150]. The premise behind the subfield-based parallelization scheme is to divide the image space such that simplicity of any voxel is independent of the object/nonobject configuration of any other voxel from the same subfield [145]. Thus, eight subfields are usually defined in the 3-D cubic grid [147, 20]. Ma et al. [90] presented a 4-subfield parallel skeletonization algorithm for the 3-D cubic grid and established its topology preservation property. Arcelli et al. [151] suggested a 2-subiteration parallel algorithm to compute the surface skeleton of the D^6 distance transform of a 3-D object without resorting to directional processes.

1.6.2 PARALLELIZATION USING MINIMAL NONSIMPLE SETS

Ronse [152,153] introduced the fundamental notion of minimal nonsimple sets in 2-D to study the conditions under which pixels may be removed in parallel while preserving the topology. In particular, it was shown that a set of pixels D and its proper subsets are all codeletable in a binary image if each singleton (single pixel) and that each pair of 8-adjacent pixels in D is codeletable. Similar results on parallel skeletonization have been reported for 2-D, 3-D, and 4-D binary digital images [154–156, 88,157]. Such strategies lead to a fully parallel skeletonization scheme, where topology preservation constraints are defined over an extended neighborhood beyond the $3 \times 3 \times 3$ neighborhood needed for characterization of singleton simple points. Ma and Sonka presented the first fully parallel skeletonization algorithm in 3-D. Lohou [158] and Wang and Basu [159] detected nontopology preservation and counterexamples to Ma and Sonka's algorithm. Lohou and Bertrand [158,160] further studied the nontopology preservation of Ma and Sonka's algorithm and presented an algorithm [161] for automatic correction of Ma and Sonka's algorithm. Recently, Kong [162] extended Ronse's work on minimal nonsimple sets to binary images on almost any polytopal complex whose union is the n-dimensional Euclidean space for $n \leq 4$. Passat et al. [163] established a characterization of minimal 3-D simple pairs in the sense that, while each voxel in the pair is independently nonsimple, deletion of both voxels leaves the topology of the object unchanged.

1.6.3 PARALLELIZATION USING P-SIMPLE POINTS

Bertrand [164,165] introduced a new interpretation of simple points, referred to as *P-simple points,* which guarantees topology preservation even when we delete those points in parallel. A parallel skeletonization scheme based on P-simple points uses two steps. In the first step, a set of voxels D is identified, which may be considered for deletion based on geometric requirements of the skeletonization algorithm. In the second step, the voxels of D that are P-simple are deleted in parallel. Depending upon the propagation strategy, a parallel skeletonization algorithm using P-simple points may be either directional [164,166,82,83,167] or symmetrical [168].

Bertrand and Couprie [169–171] designed a new parallel skeletonization scheme for the cubical complex representation of binary digital images using a new notion of *critical kernel of a complex.* The parallelization scheme utilizes the following theoretical properties of a critical kernel:

1. Any complex X collapses onto its critical kernel.
2. If an essential subcomplex $Y \subset X$ includes the critical kernel of X, then X collapses onto Y.
3. If a subcomplex of Y includes the critical kernel of X and Z is an essential subcomplex of X such that $Y \subset Z$, then Z collapses onto Y.

Bertrand and Couprie [171] proved that the critical kernel is a unifying framework that encompasses parallel thinning approaches using both minimal nonsimple set and P-simple points. A notion of *constraint set* is introduced into this framework to capture the medial axis during the collapsing process. For example, the constraint set may include the centers of maximal included balls.

1.7 APPLICATIONS

Skeletonization in 2-D and 3-D has been popularly used in many image processing and computer vision applications, including shape recognition and analysis, shape decomposition, character recognition, analysis, animation, motion tracking, registration, interpolation, path-tracking, a large number of biometrical and medical imaging applications, etc. In the following, we list only a few examples.

Cornea et al. [172] reviewed several applications of curve skeletonization including computer graphics, virtual navigation, segmentation, quantification, registration, matching, morphing, object decomposition, etc. Sundar et al. [173] developed an algorithm for *shape matching* and retrieval using the geometric and topological information embedded in skeletal graphs. Brennecke and Isenberg [174] used skeletal graphs for 3-D shape matching. Aslan et al. [78] presented a new skeletal representation and demonstrated its effectiveness in addressing various challenges of deformable *shape recognition*. Bai and Latecki [175] introduced a new skeletal graph matching algorithm by comparing the geodesic paths between skeleton endpoints to develop a robust shape recognition algorithm based on object silhouettes.

Skeletonization has been popularly used for *shape decomposition* into meaningful segments [176,79,177–180]. Malandain et al. [177] and Saha et al. [179] presented algorithms to automatically detect topological classes (e.g., surface or curve interiors, edges, or junctions) of individual voxels of a surface skeleton and formulated an "unglue" operation to separate different topological segments in a surface skeleton. Others have used curve skeletons and combined both geometric and topological features to decompose different segments of a volumetric shape [176,79,178,180]. Recently, Serino et al. [180] have presented an object decomposition method achieved through skeleton decomposition. The skeleton is polygonally approximated, taking into account spatial position and distance values of its voxels, by segments along which no significant curvature changes occur and distance values are either constant or linearly changing. Each segment is interpreted as the spine of a simple part.

Several researchers have used skeletonization for *data compression* [181,182].

Since the beginning of OCR, skeletonization has been widely used for off-line *printed character recognition* [183–186]. It is also used for *handwriting recognition* and analysis; see, e.g., [187–189].

Fujiyoshi et al. [190] applied skeletonization to *track human motion* in a video stream. Kahn et al. [191] used skeletonization of the human silhouette for *gait analysis* to recognize Parkinson's decease, whereas Arai and Asmara [192] used dynamic skeletons in 3-D data from a Kinect camera to identify walking style.

Gagvani and Silver [193] applied multiscale skeletonization to construct a compact volumetric model of a 3-D object and used the model to reconstruct volumetric expressions under different *animation* poses and motions. Wade and Parent [194] used a skeletal path-tree and path-smoothing to generate a 3-D model from a volumetric object and applied it for animation. Da Fontoura Costa and Cesar Jr. [195] discussed the usefulness of skeletal points together with their distance to the object boundary in object animation, smoothing, and matching.

Various structures in the human body are used in biometrics for identification and verification. Some of these structures are essentially linear and thus very suited for skeletonization. Since a long time, skeletons are popular for *fingerprint analysis* [196–199]. The blood vessel pattern above the human *retina* is used for identification and skeletonization is a good tool to simplify the vessel tree. Two recent examples using this approach are Ahmed et al. [200] and Lajevardi et al. [201]. A more recent trend in biometrics is to use the *vein pattern* in the hand in infrared imagery. Wang and Leedham [202] use skeletonization of the veins for identification.

Skeletonization has been very widely used in different *medical imaging applications*. Among these applications, two popular areas are tracking and analysis of elongated anatomic structures and quantitative characterization of anatomic object morphology.

Centerline approaches have been widely applied to analyze images of *tubular structures* in the human body, e.g., blood vessel analysis, stenosis detection, colonoscopy, bronchoscopy, pulmonary imaging, etc. Fridman et al. [203] used skeletonization for computation of cores to extract *blood vessel* trees in MR angiogram data. Selle et al. [204] applied skeletonization to assess the structural properties of the hepatic vessels and their relationship with tumors for liver surgical planning. Yim et al. [100] designed a gray-scale skeletonization method to interactively detect small vessel paths and for determination of branching patterns of vascular trees. Tom et al. [205] applied skeletonization to determine the correspondence among skeletal pixels in two consecutive angiographic image frames using a dynamic programming-based approach. Kim et al. [206] used skeletal bifurcations as landmarks to compute local deformations between two consecutive image frames.

Skeletonization has become an essential step in *stenosis detection* [207,92,208,209,93,210,211], where vessel diameters are analyzed along the centerline, and sudden depressions in diameter values along the centerline are detected as locations of stenosis. Nyström and Smedby [212] applied distance transform and centerline analysis techniques to generate enhanced visualization of stenoses in MR angiography.

Centerline detection has been popularly used for automated path finding in *virtual colonoscopy* [213–216,35] and *bronchoscopy* [217,218]. Using curve skeletonization on in vivo CT imaging, Tschirren et al. [219] developed a method for automated matching of corresponding branch points between two human airway trees and for assigning anatomical names or labels to different branch segments of the human airway tree. Chaturvedi and Lee [220] applied skeletonization for airway tree analysis in micro-CT lung imaging for small animals. Also, curve skeletonization has been applied on other applications of pulmonary imaging [88,221] and tree analysis in surgical planning [204].

Another popular area of application of skeletonization in medical imaging is the quantitative characterization of *object morphology,* where a skeleton is used as a compact representation of the object. Kobatake and Yoshinaga [222] applied the Hough transform to detect radiating line structures on skeletons, which are used to locate spicules in *mammograms* for malignant tumor identification. Zwiggelaar et al. [223] used skeletonization for detecting linear structures in mammograms and for classifying those into different anatomical types, e.g., vessels, spicules, ducts, etc. Näf et al. [13] applied 3-D Voronoi skeletonization for characterization and recognition of complex anatomic shapes. Saha and his colleagues developed digital topological analysis [102,179] of surface skeletons [20], which has been popularly applied to characterize *trabecular bone* as plate or rod microarchitecture to assess bone strength from in vivo imaging [224–227,33,228,229]. Recently, they further generalized their theory to characterize individual trabeculae on the continuum between a perfect plate and a perfect rod [34]. Others have used skeletonization as the first step to assess trabecular bone thickness, spacing, and various network properties [230–234].

Besides the above two major areas, Pizer et al. [36] demonstrated applications of the skeleton-based model of the *object shape* in medical image segmentation and registration. Chatzis and Pitas [235] demonstrated an application of skeletonization in image interpolation, where skeletons on two adjacent image slices are used to build their correspondence and to compute the regional transformation field for *interpolation*.

Finally, we mention a few other applications of skeletons, mostly for various manufactured structures. Because of their shape, *fibers* are well suited for skeletonization before further analysis. This is the case both for manufactured fibers such as nanofibers and textile fibers: see, e.g., [236,237]. D'Amore et al. [238] use skeletonization to characterize engineered tissue fiber network topology, where the network can be used both in vitro and in vivo. Foams are used both in manufacturing materials for many different purposes and in food products. Therefore, is important to characterize the *structure of foams*. 3-D skeletons have been successfully used; see, e.g., [239–241]. Finally, Thibault and Gold [242] applied Voronoi skeletonization algorithms to construct *terrain models* of valleys and ridges and their topological relationships from contour data.

1.8 PERFORMANCE EVALUATION

Performance evaluation of a skeletonization algorithm is a serious research challenge emerging from the lack of definition of the "true" skeleton of a digital object. Therefore, a widely accepted approach evaluating the performance of a skeletonization algorithm is yet absent in the literature. Different research groups have adopted different paths to evaluate the performance of their algorithms. Early skeletonization algorithms [23,16,21] were qualitatively evaluated through illustrations of results of example digital shapes, which were then considered to look more or less "good." Haralick [243] discussed the guidelines for evaluating skeletonization algorithms and outlined some criteria of an error function for quantitative performance analysis. Jaisimha et al. [244] evaluated 2-D skeletonization algorithms in terms of robustness under added noise. Saha et al. [20] evaluated 3-D skeletonization in terms of robustness under border noise and image rotation.

Sanniti di Baja [7] and Arcelli et al. [24] recommended the performance criterion of a skeletonization algorithm as the accuracy of the reconstruction of the original object from its skeleton. However, criteria concentrating on missing (or added) spels in the reconstructed shape remains blind to the performance of an algorithm in terms of the quality of the shape of the skeleton and also ignore the sensitivity of an algorithm under noise and rotation and repeat scan data acquisition. A shape distance that also takes the position of the pixels into account is better (see [245]) but still is not ideal.

The Rotterdam coronary axis tracking evaluation group [208] reported results of a comprehensive evaluation study for centerline detection algorithms for coronary vasculature using consensus centerline with multiple observers and different quantitative measures of accuracy on a database containing 32 cardiac CTA datasets. Greenspan et al. [92] reported an evaluation study for centerline extraction algorithms in quantitative coronary angiography using a simulated coronary artery with known centerline and diameter function.

Jin and Saha [101] reported a framework starting from a known skeleton and constructing corresponding objects at different conditions of imaging artifacts of noise and image resolution. Recently, Sobiecki et al. [30,31] have reported the results of performance evaluating studies for curve and surface skeletonization algorithms in voxel grids.

In general, the above performance evaluation frameworks report an error defined as some measure of disagreement between the expected and the computed skeletons. A concern with such approaches is that the average measure of errors does not reflect the performance of an algorithm in terms of generating spurious branches, missing true branches, preserving sharp corners, or digital smoothness over a contiguous surface or curve under different conditions of noise, resolution, and rotation. A comprehensive and consensus framework for evaluating skeletonization algorithms providing structured knowledge of understanding of their performance under various categories at different imaging conditions is yet to emerge. However, based on the principle used in a skeletonization algorithm, one may anticipate its expected behavior, which may be useful in selecting an appropriate category of skeletonization algorithm for a specific application.

Parallel skeletonization algorithms usually have the advantage in terms of computational efficiency. At the same time, such algorithms face additional challenges in maintaining the expected geometry of the skeleton due to the asynchronous sequence of simple point removal. Distance transform-based approaches are expected to show a better performance under rotation and at sharp corners, due to the underlying robust strategy of using the DT, or depth measure, in defining the peel sequence. However, it is difficult to make such a distinction between DT-based and iterative skeletonization algorithms in terms of performance under noise. In the context of eliminating spurious branches created by noise, an algorithm based on assessment of the global significance of a branch is expected to be superior as compared to local decision-based approaches during peeling. A comprehensive and consensus framework for evaluating the performance of skeletonization algorithms for various conditions will be helpful in selecting the right skeletonization algorithm for a given application.

1.9 CONCLUSIONS

A large number of computational approaches have been published for extracting the skeleton of an object, some of which are widely different in terms of their principles. Several methods focus on symmetry and other geometric properties of skeletons of an object, whereas others apply continuous curve propagation approaches. A large number of skeletonization algorithms are directly applicable to digital grids. These algorithms are primarily based on some iterative erosion scheme under certain topological and geometric schemes. The distance transform field is used to efficiently and effectively simulate the erosion process where the time of erosion of individual spels are given by their distance transform value. Significant research efforts have been dedicated toward enhancing the performance of skeletonization algorithms in terms of spurious branches, sharpness at corners, invariance under image transformation, reconstruction of original objects, etc. Skeletonization has been widely applied in many different applications, especially in biomedical imaging, including pulmonary, cardiac, mammographic, abdominal, retinal, bone imaging, etc. Validation of skeletonization algorithms is a challenging issue and a comprehensive and consensus framework for evaluation is yet to emerge.

REFERENCES

[1] H. Blum, A transformation for extracting new descriptors of shape, Model Percept. Speech Vis. Form 19 (5) (1967) 362–380.

[2] H. Blum, R. Nagel, Shape description using weighted symmetric axis features, Pattern Recognit. Lett. 10 (3) (1978) 167–180.

[3] P.J. Giblin, B.B. Kimia, A formal classification of 3D medial axis points and their local geometry, IEEE Trans. Pattern Anal. Mach. Intell. 26 (2) (2004) 238–251.

[4] B.B. Kimia, A. Tannenbaum, S.W. Zucker, Shape, shocks, and deformations I: The components of two-dimensional shape and the reaction–diffusion space, Int. J. Comput. Vis. 15 (1995) 189–224.

[5] R. Kimmel, D. Shaked, N. Kiryati, Skeletonization via distance maps and level sets, Comput. Vis. Image Underst. 62 (1995) 382–391.

[6] F. Leymarie, M.D. Levine, Simulating the grassfire transform using an active contour model, IEEE Trans. Pattern Anal. Mach. Intell. 14 (1) (1992) 56–75.

[7] G. Sanniti di Baja, Well-shaped, stable, and reversible skeletons from the (3, 4)-distance transform, J. Vis. Commun. Image Represent. 5 (1994) 107–115.

[8] K. Siddiqi, S. Bouix, A. Tannenbaum, S.W. Zucker, Hamilton–Jacobi skeletons, Int. J. Comput. Vis. 48 (3) (2002) 215–231.

[9] K. Siddiqi, S.M. Pizer, Medial Representations: Mathematics, Algorithms and Applications, vol. 37, Springer, 2008.

[10] M. Brady, H. Asada, Smoothed local symmetries and their implementation, Int. J. Robot. Res. 3 (3) (1984) 36–61.

[11] M. Leyton, Symmetry-curvature duality, Comput. Vis. Graph. Image Process. 38 (3) (1987) 327–341.

[12] J.W. Brandt, V.R. Algazi, Continuous skeleton computation by Voronoi diagram, CVGIP, Image Underst. 55 (3) (1992) 329–338.

[13] M. Näf, G. Székely, R. Kikinis, M.E. Shenton, O. Kübler, 3D Voronoi skeletons and their usage for the characterization and recognition of 3D organ shape, Comput. Vis. Image Underst. 66 (1997) 147–161.

[14] R. Ogniewicz, M. Ilg, Voronoi skeletons: theory and applications, in: Int. Conf. Comp. Vis. Patt. Recogn., IEEE, 1992, pp. 63–69.

[15] R.L. Ogniewicz, O. Kübler, Hierarchic Voronoi skeletons, Pattern Recognit. 28 (3) (1995) 343–359.

[16] L. Lam, S.-W. Lee, C.Y. Suen, Thinning methodologies – a comprehensive survey, IEEE Trans. Pattern Anal. Mach. Intell. 14 (9) (1992) 869–885.

[17] T.-C. Lee, R.L. Kashyap, C.-N. Chu, Building skeleton models via 3-D medial surface/axis thinning algorithm, CVGIP, Graph. Models Image Process. 56 (6) (1994) 462–478.

[18] G. Németh, P. Kardos, K. Palágyi, Thinning combined with iteration-by-iteration smoothing for 3D binary images, Graph. Models 73 (2011) 335–345.
[19] K. Palágyi, A. Kuba, A parallel 3D 12-subiteration thinning algorithm, Graph. Models Image Process. 61 (1999) 199–221.
[20] P.K. Saha, B.B. Chaudhuri, D.D. Majumder, A new shape preserving parallel thinning algorithm for 3D digital images, Pattern Recognit. 30 (1997) 1939–1955.
[21] Y.F. Tsao, K.S. Fu, A parallel thinning algorithm for 3D pictures, Comput. Graph. Image Process. 17 (1981) 315–331.
[22] G. Borgefors, Distance transformations in digital images, Comput. Vis. Graph. Image Process. 34 (1986) 344–371.
[23] C. Arcelli, G. Sanniti di Baja, A width-independent fast thinning algorithm, IEEE Trans. Pattern Anal. Mach. Intell. 7 (4) (1985) 463–474.
[24] C. Arcelli, G. Sanniti di Baja, L. Serino, Distance-driven skeletonization in voxel images, IEEE Trans. Pattern Anal. Mach. Intell. 33 (4) (2011) 709–720.
[25] I. Bitter, A.E. Kaufman, M. Sato, Penalized-distance volumetric skeleton algorithm, IEEE Trans. Vis. Comput. Graph. 7 (3) (2001) 195–206.
[26] G. Borgefors, I. Nyström, G. Sanniti di Baja, Computing skeletons in three dimensions, Pattern Recognit. 32 (7) (1999) 1225–1236.
[27] C. Pudney, Distance-ordered homotopic thinning: a skeletonization algorithm for 3D digital images, Comput. Vis. Image Underst. 72 (2) (1998) 404–413.
[28] D. Attali, J.-D. Boissonnat, H. Edelsbrunner, Stability and Computation of Medial Axes: A State-of-the-Art Report, Springer-Verlag, Berlin, Germany, 2009, pp. 109–125.
[29] G. Matheron, Examples of Topological Properties of Skeletons, Academic Press, 1988, pp. 217–238.
[30] A. Sobiecki, A. Jalba, A. Telea, Comparison of curve and surface skeletonization methods for voxel shapes, Pattern Recognit. Lett. 47 (10) (2014) 147–156.
[31] A. Sobiecki, H.C. Yasan, A.C. Jalba, A.C. Telea, Qualitative Comparison of Contraction-Based Curve Skeletonization Methods, Springer, 2013, pp. 425–439.
[32] P.K. Saha, F.W. Wehrli, Measurement of trabecular bone thickness in the limited resolution regime of in vivo MRI by fuzzy distance transform, IEEE Trans. Med. Imaging 23 (2004) 53–62.
[33] P.K. Saha, B.R. Gomberg, F.W. Wehrli, Three-dimensional digital topological characterization of cancellous bone architecture, Int. J. Imaging Syst. Technol. 11 (2000) 81–90.
[34] P.K. Saha, Y. Xu, H. Duan, A. Heiner, G. Liang, Volumetric topological analysis: a novel approach for trabecular bone classification on the continuum between plates and rods, IEEE Trans. Med. Imaging 29 (11) (2010) 1821–1838.
[35] M. Wan, Z. Liang, Q. Ke, L. Hong, I. Bitter, A. Kaufman, Automatic centerline extraction for virtual colonoscopy, IEEE Trans. Med. Imaging 21 (2002) 1450–1460.
[36] S.M. Pizer, D.S. Fritsch, P.A. Yushkevich, V.E. Johnson, E.L. Chaney, Segmentation, registration, and measurement of shape variation via image object shape, IEEE Trans. Med. Imaging 18 (10) (1999) 851–865.
[37] P.K. Saha, G. Borgefors, G. Sanniti di Baja, A survey on skeletonization algorithms and their applications, in: Special Issue on Skeletonization and Its Application, Pattern Recognit. Lett. 76 (June 2016) 3–12.
[38] A. Tagliasacchi, T. Delame, M. Spagnuolo, N. Amenta, A. Telea, 3D skeletons: a state-of-the-art report, Comput. Graph. Forum 35 (2) (2016) 573–597.
[39] J. Damon, Smoothness and geometry of boundaries associated to skeletal structures I: sufficient conditions for smoothness, Ann. Inst. Fourier 53 (6) (2003) 1941–1985.

[40] J. Damon, Smoothness and geometry of boundaries associated to skeletal structures, II: geometry in the Blum case, Compos. Math. 140 (06) (2004) 1657–1674.
[41] D.S. Fritsch, S.M. Pizer, B.S. Morse, D.H. Eberly, A. Liu, The multiscale medial axis and its applications in image registration, Pattern Recognit. Lett. 15 (5) (1994) 445–452.
[42] B.S. Morse, S.M. Pizer, A. Liu, Multiscale medial analysis of medical images, in: Int. Conf. Inf. Process. Med. Imag., Springer, 1993, pp. 112–131.
[43] S.M. Pizer, D. Eberly, D.S. Fritsch, B.S. Morse, Zoom-invariant vision of figural shape: the mathematics of core, Comput. Vis. Image Underst. 69 (1) (1998) 55–71.
[44] G. Matheron, Random Sets and Integral Geometry, John Wiley & Sons, New York, NY, 1975.
[45] J. Serra, Image Analysis and Mathematical Morphology, Academic Press, London, 1982.
[46] C. Lantuejoul, Skeletonization in quantitative metallography, in: R. Haralick, J.-C. Simon (Eds.), Issues in Digital Image Processing, in: NATO ASI Ser. E, vol. 34, Sijthoff and Noordhoff, The Netherlands, 1980.
[47] P.A. Maragos, R.W. Schafer, Morphological skeleton representation and coding of binary images, IEEE Trans. Acoust. Speech Signal Process. 34 (1986) 228–244.
[48] P.P. Jonker, Morphological operations on 3D and 4D images: from shape primitive detection to skeletonization, in: International Conference on Discrete Geometry for Computer Imagery, Springer, 2000, pp. 371–391.
[49] Skeletons in *N* dimensions using shape primitives, Pattern Recognit. Lett. 23 (6) (2002) 677–686.
[50] M.J. Cardoso, M.J. Clarkson, M. Modat, S. Ourselin, On the extraction of topologically correct thickness measurements using Khalimsky's cubic complex, in: G. Székely, H. Hahn (Eds.), Proc. of Info Proc. Med. Imag. (IPMI), in: Lect. Notes Comput. Sci., vol. 6801, Springer, 2011, pp. 159–170.
[51] Y. Cointepas, I. Bloch, L. Garnero, A cellular model for multi-objects multi-dimensional homotopic deformations, Pattern Recognit. 34 (9) (2001) 1785–1798.
[52] M. Couprie, Topological maps and robust hierarchical Euclidean skeletons in cubical complexes, Comput. Vis. Image Underst. 117 (4) (2013) 355–369.
[53] P. Dlotko, R. Specogna, Topology preserving thinning of cell complexes, IEEE Trans. Image Process. 23 (10) (2014) 4486–4495.
[54] T.Y. Kong, Topology-preserving deletion of 1's from 2-, 3- and 4-dimensional binary images, in: Proc. of Int. Work Discr. Geom. Comp. Imag., Springer, Montpellier, France, 1997, pp. 1–18.
[55] L. Liu, E.W. Chambers, D. Letscher, T. Ju, A simple and robust thinning algorithm on cell complexes, Comput. Graph. Forum 29 (7) (2010) 2253–2260.
[56] G. Borgefors, G.S. di Baja, Skeletonizing the distance transform on the hexagonal grid, in: 9th International Conference on Pattern Recognition, 1988, IEEE, 1988, pp. 504–507.
[57] R.C. Staunton, An analysis of hexagonal thinning algorithms and skeletal shape representation, Pattern Recognit. 29 (7) (1996) 1131–1146.
[58] R. Strand, Surface skeletons in grids with non-cubic voxels, in: Proceedings of the 17th International Conference on Pattern Recognition, ICPR 2004, vol. 1, IEEE, 2004, pp. 548–551.
[59] S.K. Pal, A. Rosenfeld, A fuzzy medial axis transformation based on fuzzy disks, Pattern Recognit. Lett. 12 (10) (1991) 585–590.
[60] P.K. Saha, F.W. Wehrli, B.R. Gomberg, Fuzzy distance transform: theory, algorithms, and applications, Comput. Vis. Image Underst. 86 (2002) 171–190.

[61] P.K. Saha, F.W. Wehrli, Fuzzy distance transform in general digital grids and its applications, in: Proc. of 7th Joint Conf. Info Sc., Research Triangular Park, NC, 2003, pp. 201–213.
[62] S. Svensson, Aspects on the reverse fuzzy distance transform, Pattern Recognit. Lett. 29 (2008) 888–896.
[63] J. Serra, Image Analysis and Mathematical Morphology, vol II: Theoretical Advances, Academic Press, New York, NY, 1988.
[64] I. Bloch, Fuzzy skeleton by influence zones—application to interpolation between fuzzy sets, Fuzzy Sets Syst. 159 (15) (2008) 1973–1990.
[65] I. Bloch, H. Maître, Fuzzy mathematical morphologies: a comparative study, Pattern Recognit. 28 (9) (1995) 1341–1387.
[66] N. Amenta, S. Choi, R.K. Kolluri, The power crust, unions of balls, and the medial axis transform, Comput. Geom. 19 (2–3) (2001) 127–153.
[67] G. Székely, Shape Characterization by Local Symmetries, Habilitation, Swiss Federal Institute of Technology, Zürich, 1996.
[68] M. Schmitt, Some examples of algorithms analysis in computational geometry by means of mathematical morphological techniques, in: J.-D. Boissonnat, J.-P. Laumond (Eds.), Proc. of Geom. Robot, in: Lect. Notes Comput. Sci., vol. 391, Springer, Toulouse, France, 1989, pp. 225–246.
[69] D. Attali, A. Montanvert, Computing and simplifying 2D and 3D continuous skeletons, Comput. Vis. Image Underst. 67 (3) (1997) 261–273.
[70] T.K. Dey, W. Zhao, Approximating the medial axis from the Voronoi diagram with a convergence guarantee, Algorithmica 38 (1) (2004) 179–200.
[71] E.C. Sherbrooke, N.M. Patrikalakis, E. Brisson, An algorithm for the medial axis transform of 3D polyhedral solids, IEEE Trans. Vis. Comput. Graph. 2 (1) (1996) 44–61.
[72] A.C. Jalba, J. Kustra, A.C. Telea, Surface and curve skeletonization of large 3D models on the GPU, IEEE Trans. Pattern Anal. Mach. Intell. 35 (6) (2013) 1495–1508.
[73] A. Bucksch, R. Lindenbergh, Campino—a skeletonization method for point cloud processing, ISPRS J. Photogramm. Remote Sens. 63 (1) (2008) 115–127.
[74] J. Ma, S.W. Bae, S. Choi, 3D medial axis point approximation using nearest neighbors and the normal field, Vis. Comput. 28 (1) (2012) 7–19.
[75] S.M. Pizer, K. Siddiqi, G. Székely, J.N. Damon, S.W. Zucker, Multiscale medial loci and their properties, Int. J. Comput. Vis. 55 (2–3) (2003) 155–179.
[76] G. Borgefors, Distance transform in arbitrary dimensions, Comput. Vis. Graph. Image Process. 27 (1984) 321–345.
[77] N. Ahuja, J.-H. Chuang, Shape representation using a generalized potential field model, IEEE Trans. Pattern Anal. Mach. Intell. 19 (2) (1997) 169–176.
[78] C. Aslan, A. Erdem, E. Erdem, S. Tari, Disconnected skeleton: shape at its absolute scale, IEEE Trans. Pattern Anal. Mach. Intell. 30 (12) (2008) 2188–2203.
[79] N.D. Cornea, D. Silver, X. Yuan, R. Balasubramanian, Computing hierarchical curve-skeletons of 3D objects, Vis. Comput. 21 (11) (2005) 945–955.
[80] M.S. Hassouna, A.A. Farag, Variational curve skeletons using gradient vector flow, IEEE Trans. Pattern Anal. Mach. Intell. 31 (12) (2009) 2257–2274.
[81] Z.S.G. Tari, J. Shah, H. Pien, Extraction of shape skeletons from grayscale images, Comput. Vis. Image Underst. 66 (1997) 133–146.
[82] C. Lohou, G. Bertrand, A 3D 12-subiteration thinning algorithm based on P-simple points, Discrete Appl. Math. 139 (1) (2004) 171–195.

[83] C. Lohou, G. Bertrand, A 3D 6-subiteration curve thinning algorithm based on P-simple points, Discrete Appl. Math. 151 (1) (2005) 198–228.
[84] G. Borgefors, On digital distance transformation in three dimensions, Comput. Vis. Graph. Image Process. 64 (1996) 368–376.
[85] G. Borgefors, Weighted digital distance transforms in four dimensions, Discrete Appl. Math. 125 (1) (2003) 161–176.
[86] G. Bertrand, A parallel thinning algorithm for medial surfaces, Pattern Recognit. Lett. 16 (9) (1995) 979–986.
[87] C.M. Ma, Connectivity preservation of 3D 6-subiteration thinning algorithms, Graph. Models Image Process. 58 (4) (1996) 382–386.
[88] C.M. Ma, M. Sonka, A fully parallel 3D thinning algorithm and its applications, Comput. Vis. Image Underst. 64 (1996) 420–433.
[89] C.M. Ma, S.Y. Wan, Parallel thinning algorithms on 3D (18, 6) binary images, Comput. Vis. Image Underst. 80 (2000) 364–378.
[90] C.M. Ma, S.Y. Wan, J.D. Lee, Three-dimensional topology preserving reduction on the 4-subfields, IEEE Trans. Pattern Anal. Mach. Intell. 24 (2002) 1594–1605.
[91] V. Chatzis, I. Pitas, A generalized fuzzy mathematical morphology and its application in robust 2-D and 3-D object representation, IEEE Trans. Image Process. 9 (2000) 1798–1810.
[92] H. Greenspan, M. Laifenfeld, S. Einav, O. Barnea, Evaluation of center-line extraction algorithms in quantitative coronary angiography, IEEE Trans. Med. Imaging 20 (2001) 928–941.
[93] M. Sonka, M.D. Winniford, X. Zhang, S.M. Collins, Lumen centerline detection in complex coronary angiograms, IEEE Trans. Biomed. Eng. 41 (6) (1994) 520–528.
[94] S. Wang, J. Wu, M. Wei, X. Ma, Robust curve skeleton extraction for vascular structures, Graph. Models 74 (2012) 109–120.
[95] O. Wink, W.J. Niessen, M.A. Viergever, Multiscale vessel tracking, IEEE Trans. Med. Imaging 23 (2004) 130–133.
[96] G. Borgefors, I. Nyström, G. Sanniti di Baja, Surface skeletonization of volume objects, in: P. Perner, P. Wang, A. Rosenfeld (Eds.), Proc. of Int. Work Adv. Struct. Synt. Pat. Recog., Leipzig, Germany, in: Lect. Notes Comput. Sci., vol. 1121, 1996, pp. 251–259.
[97] G. Borgefors, I. Nyström, G. Sanniti di Baja, S. Svensson, Simplification of 3D skeletons using distance information, in: Proc. of Int. Conf. Vis Geomet., San Diego, CA, in: Proc. SPIE, vol. 4117, 2000, pp. 300–309.
[98] T. Saito, J.-I. Toriwaki, A sequential thinning algorithm for three dimensional digital pictures using the Euclidean distance transformation, in: Proc. of 9th Scand. Conf. Imag. Anal. (SCIA'95), IAPR, Uppsala, Sweden, 1995, pp. 507–516.
[99] J. Shah, Extraction of shape skeletons from grayscale images, Comput. Vis. Image Underst. 99 (2005) 96–109.
[100] P.J. Yim, P.L. Choyke, R.M. Summers, Gray-scale skeletonization of small vessels in magnetic resonance angiography, IEEE Trans. Med. Imaging 19 (6) (2000) 568–576.
[101] D. Jin, P.K. Saha, A new fuzzy skeletonization algorithm and its applications to medical imaging, in: Proc. of 17th Int. Conf. Image Anal. Proc. (ICIAP), Naples, Italy, in: Lect. Notes Comput. Sci., vol. 8156, 2013, pp. 662–671.
[102] P.K. Saha, B.B. Chaudhuri, Detection of 3-D simple points for topology preserving transformations with application to thinning, IEEE Trans. Pattern Anal. Mach. Intell. 16 (1994) 1028–1032.

[103] C. Arcelli, G. Sanniti di Baja, Finding local maxima in a pseudo-Euclidean distance transform, Comput. Vis. Graph. Image Process. 43 (1988) 361–367.

[104] G. Borgefors, Centres of maximal discs in the 5-7-11 distance transform, in: Proc. of 8th Scand. Conf. Imag. Anal., Tromsø, Norway, 1993, pp. 105–111.

[105] E. Remy, E. Thiel, Exact medial axis with Euclidean distance, Image Vis. Comput. 23 (2) (2005) 167–175.

[106] G. Borgefors, I. Nyström, Efficient shape representation by minimizing the set of centres of maximal discs/spheres, Pattern Recognit. Lett. 18 (5) (1997) 465–471.

[107] S. Svensson, I. Nyström, G. Borgefors, On reversible skeletonization using anchor points from distance transforms, Int. J. Vis. Commun. Image Represent. 10 (1999) 379–397.

[108] D. Jin, Y. Liu, P.K. Saha, Application of fuzzy skeletonization of quantitatively assess trabecular bone micro-architecture, in: Proc. of 35th IEEE Conf. Eng. Med. Biol. Soc. (EMBC), IEEE, Osaka, Japan, 2013, pp. 3682–3685.

[109] D. Jin, C. Chen, P.K. Saha, Filtering non-significant quench points using collision impact in grassfire propagation, in: Proc. of 18th Int. Conf. Image Anal. Proc. (ICIAP), in: Lect. Notes Comput. Sci., vol. 9279, Springer, Genova, Italy, 2015, pp. 432–443.

[110] C.J. Hilditch, Linear Skeletons from Square Cupboards, vol. 4, Edinburgh Univ. Press, Edinburgh, U.K., 1969, pp. 403–420.

[111] J. Mylopoulos, T. Pavlidis, On the topological properties of quantized spaces I: the notion of dimension, J. Assoc. Comput. Mach. 18 (2) (1971) 239–246.

[112] A. Rosenfeld, Connectivity in digital pictures, J. Assoc. Comput. Mach. 17 (1970) 146–160.

[113] A. Rosenfeld, Arcs and curves in digital pictures, J. Assoc. Comput. Mach. 20 (1973) 81–87.

[114] G. Tourlakis, J. Mylopoulos, Some results on computational topology, J. Assoc. Comput. Mach. 20 (1973) 439–455.

[115] D.G. Morgenthaler, Three-Dimensional Simple Points: Serial Erosion, Parallel Thinning and Skeletonization, Tech. Rep., Comp. Vis. Lab., Univ. of Maryland, College Park, MD, 1981.

[116] S. Lobregt, P.W. Verbeek, F.C.A. Groen, Three-dimensional skeletonization, principle, and algorithm, IEEE Trans. Pattern Anal. Mach. Intell. 2 (1980) 75–77.

[117] P.K. Saha, B.B. Chaudhuri, B. Chanda, D.D. Majumder, Topology preservation in 3D digital space, Pattern Recognit. 27 (1994) 295–300.

[118] P.K. Saha, 2D thinning algorithms and 3D shrinking, INRIA, Sophia Antipolis Cedex, France, Seminar Talk, June 1991.

[119] P.K. Saha, B. Chanda, D.D. Majumder, Principles and Algorithms for 2-D and 3-D Shrinking, Indian Statistical Institute, Calcutta, India, 1991, Tech. Rep. TR/KBCS/2/91.

[120] P.K. Saha, A. Rosenfeld, Determining simplicity and computing topological change in strongly normal partial tilings of R^2 or R^3, Pattern Recognit. 33 (2000) 105–118.

[121] G. Bertrand, G. Malandain, A new characterization of three-dimensional simple points, Pattern Recognit. Lett. 15 (1994) 169–175.

[122] G. Malandain, G. Bertrand, Fast characterization of 3-D simple points, in: Proc. of 11th Int. Conf. Pat. Recog., 1992, pp. 232–235.

[123] P.K. Saha, A. Rosenfeld, The digital topology of sets of convex voxels, Graph. Models 62 (5) (2000) 343–352.

[124] T.Y. Kong, P.K. Saha, A. Rosenfeld, Strongly normal sets of contractible tiles in n dimensions, Pattern Recognit. 40 (2) (2007) 530–543.

[125] D. Shaked, A.M. Bruckstein, Pruning medial axes, Comput. Vis. Image Underst. 69 (2) (1998) 156–169.
[126] X. Bai, L.J. Latecki, W.-Y. Liu, Skeleton pruning by contour partitioning with discrete curve evolution, IEEE Trans. Pattern Anal. Mach. Intell. 29 (3) (2007) 449–462.
[127] W. Shen, X. Bai, R. Hu, H. Wang, L.J. Latecki, Skeleton growing and pruning with bending potential ratio, Pattern Recognit. 44 (2) (2011) 196–209.
[128] A.D. Ward, G. Hamarneh, The groupwise medial axis transform for fuzzy skeletonization and pruning, IEEE Trans. Pattern Anal. Mach. Intell. 32 (6) (2010) 1084–1096.
[129] D. Jin, K.S. Iyer, C. Chen, E.A. Hoffman, P.K. Saha, A robust and efficient curve skeletonization algorithm for tree-like objects using minimum cost paths, in: Special Issue on Skeletonization and Its Application, Pattern Recognit. Lett. 76 (June 2016) 32–40.
[130] A.R. Dill, M.D. Levine, P.B. Noble, Multiple resolution skeletons, IEEE Trans. Pattern Anal. Mach. Intell. 4 (1987) 495–504.
[131] S.M. Pizer, W.R. Oliver, S.H. Bloomberg, Hierarchical shape description via the multiresolution symmetric axis transform, IEEE Trans. Pattern Anal. Mach. Intell. 4 (1987) 505–511.
[132] L. Gorelick, M. Galun, E. Sharon, R. Basri, A. Brandt, Shape representation and classification using the Poisson equation, IEEE Trans. Pattern Anal. Mach. Intell. 28 (12) (2006) 1991–2005.
[133] K. Siddiqi, B.B. Kimia, A shock grammar for recognition, in: IEEE Conf. Comp. Vis. Patt. Recog., IEEE, 1996, pp. 507–513.
[134] J. Giesen, B. Miklos, M. Pauly, C. Wormser, The scale axis transform, in: Proc. 25th Annual Symp. Comp. Geomet., ACM, 2009, pp. 106–115.
[135] B. Miklos, J. Giesen, M. Pauly, Discrete scale axis representations for 3D geometry, ACM Trans. Graph. 29 (4) (2010) 101.
[136] M. Postolski, M. Couprie, M. Janaszewski, Scale filtered Euclidean medial axis and its hierarchy, Comput. Vis. Image Underst. 129 (2014) 89–102.
[137] G. Borgefors, G. Ramella, G. Sanniti di Baja, Hierarchical decomposition of multiscale skeletons, IEEE Trans. Pattern Anal. Mach. Intell. 23 (11) (2001) 1296–1312.
[138] D. Attali, G. Sanniti di Baja, E. Thiel, Skeleton simplification through non significant branch removal, Image Process. & Commun. 3 (3–4) (1997) 63–72.
[139] D. Rutovitz, Pattern recognition, J. R. Stat. Soc. 129 (4) (1966) 504–530.
[140] W. Gong, G. Bertrand, A simple parallel 3D thinning algorithm, in: Proc. of 10th Int. Conf. Pat. Recog. (ICPR), vol. 1, 1990, pp. 188–190.
[141] J. Mukherjee, P.P. Das, B. Chatterji, On connectivity issues of ESPTA, Pattern Recognit. Lett. 11 (9) (1990) 643–648.
[142] K. Palágyi, A. Kuba, A 3D 6-subiteration thinning algorithm for extracting medial lines, Pattern Recognit. Lett. 19 (7) (1998) 613–627.
[143] W. Xie, R.P. Thompson, R. Perucchio, A topology-preserving parallel 3D thinning algorithm for extracting the curve skeleton, Pattern Recognit. 36 (2003) 1529–1544.
[144] G. Bertrand, Z. Aktouf, Three-dimensional thinning algorithm using subfields, in: Int. Conf. Vis. Geomet., Boston, MA, in: Proc. SPIE, vol. 2356, 1995, pp. 113–124.
[145] M.J. Golay, Hexagonal parallel pattern transformations, IEEE Trans. Comput. 18 (8) (1969) 733–740.
[146] K. Palágyi, A. Kuba, A hybrid thinning algorithm for 3D medical images, J. Comput. Inf. Technol. 6 (2) (1998) 149–164.
[147] P.K. Saha, B.B. Chaudhuri, Simple Point Computation and 3-D Thinning with Parallel Implementation, Indian Statistical Institute, Calcutta, India, 1993, Tech. Rep. TR/KBCS/1/93.

[148] K. Palágyi, A. Kuba, Directional 3D thinning using 8 subiterations, in: G. Bertrand, M. Couprie, L. Perroton (Eds.), Proc. of Int. Conf. Discr. Geom. Comp. Imag., in: Lect. Notes Comput. Sci., vol. 1568, Springer, 1999, pp. 325–336.
[149] K. Palágyi, A 3-subiteration 3D thinning algorithm for extracting medial surfaces, Pattern Recognit. Lett. 23 (6) (2002) 663–675.
[150] K. Palágyi, A 3D fully parallel surface-thinning algorithm, Theor. Comput. Sci. 406 (1) (2008) 119–135.
[151] C. Arcelli, G. Sanniti di Baja, L. Serino, A parallel algorithm to skeletonize the distance transform of 3D objects, Image Vis. Comput. 27 (6) (2009) 666–672.
[152] C. Ronse, A topological characterization of thinning, Theor. Comput. Sci. 43 (1986) 31–41.
[153] C. Ronse, Minimal test patterns for connectivity preservation in parallel thinning algorithms for binary digital images, Discrete Appl. Math. 21 (1) (1988) 67–79.
[154] C.-J. Gau, T.Y. Kong, Minimal nonsimple sets of voxels in binary images on a face-centered cubic grid, Int. J. Pattern Recognit. Artif. Intell. 13 (04) (1999) 485–502.
[155] C.-J. Gau, T.Y. Kong, Minimal non-simple sets in 4D binary images, Graph. Models 65 (1) (2003) 112–130.
[156] R.W. Hall, Tests for connectivity preservation for parallel reduction operators, Topol. Appl. 46 (3) (1992) 199–217.
[157] C.M. Ma, On topology preservation in 3D thinning, CVGIP, Image Underst. 59 (3) (1994) 328–339.
[158] C. Lohou, Detection of the non-topology preservation of Ma and Sonka's algorithm, by the use of P-simple points, Comput. Vis. Image Underst. 114 (2010) 384–399.
[159] T. Wang, A. Basu, A note on 'A fully parallel 3D thinning algorithm and its applications', Pattern Recognit. Lett. 28 (4) (2007) 501–506.
[160] C. Lohou, Detection of the non-topology preservation of Ma's 3D surface-thinning algorithm, by the use of P-simple points, Pattern Recognit. Lett. 29 (6) (2008) 822–827.
[161] C. Lohou, J. Dehos, Automatic correction of Ma and Sonka's thinning algorithm using P-simple points, IEEE Trans. Pattern Anal. Mach. Intell. 32 (6) (2010) 1148–1152.
[162] T.Y. Kong, Minimal non-deletable sets and minimal non-codeletable sets in binary images, Theor. Comput. Sci. 406 (1) (2008) 97–118.
[163] N. Passat, M. Couprie, G. Bertrand, Minimal simple pairs in the 3-D cubic grid, J. Math. Imaging Vis. 32 (3) (2008) 239–249.
[164] G. Bertrand, On P-simple points, C. R. Acad. Sci., Ser. 1 Math. 321 (321) (1995) 1077–1084.
[165] G. Bertrand, P-simple points: A solution for parallel thinning, in: Proc. of 5th Conf. Discrete Geom., France, 1995, pp. 233–242.
[166] J. Burguet, R. Malgouyres, Strong thinning and polyhedric approximation of the surface of a voxel object, Discrete Appl. Math. 125 (1) (2003) 93–114.
[167] R. Malgouyres, S. Fourey, Strong surfaces, surface skeletons, and image superimposition, in: Int. Conf. Vis. Geomet., San Diego, CA, in: Proc. SPIE, vol. 3454, 1998, pp. 16–27.
[168] C. Lohou, G. Bertrand, Two symmetrical thinning algorithms for 3D binary images, based on P-simple points, Pattern Recognit. 40 (8) (2007) 2301–2314.
[169] G. Bertrand, M. Couprie, A new 3D parallel thinning scheme based on critical kernels, in: Discrete Geom. Comp. Imag., Springer, France, 2006, pp. 580–591.
[170] G. Bertrand, M. Couprie, Two-dimensional parallel thinning algorithms based on critical kernels, J. Math. Imaging Vis. 31 (1) (2008) 35–56.

[171] G. Bertrand, M. Couprie, On parallel thinning algorithms: minimal non-simple sets, P-simple points and critical kernels, J. Math. Imaging Vis. 35 (1) (2009) 23–35.
[172] N.D. Cornea, D. Silver, P. Min, Curve-skeleton properties, applications, and algorithms, IEEE Trans. Vis. Comput. Graph. 13 (3) (2007) 530–548.
[173] H. Sundar, D. Silver, N. Gagvani, S. Dickinson, Skeleton based shape matching and retrieval, in: Shape Modeling International, 2003, IEEE, Seoul, South Korea, 2003, pp. 130–139.
[174] A. Brennecke, T. Isenberg, 3D shape matching using skeleton graphs, in: Proc. Simu. Visual., 2004, pp. 299–310.
[175] X. Bai, L.J. Latecki, Path similarity skeleton graph matching, IEEE Trans. Pattern Anal. Mach. Intell. 30 (7) (2008) 1282–1292.
[176] P.-Y. Chiang, C.-C.J. Kuo, Voxel-based shape decomposition for feature-preserving 3D thumbnail creation, J. Vis. Commun. Image Represent. 23 (1) (2012) 1–11.
[177] G. Malandain, G. Bertrand, N. Ayache, Topological segmentation of discrete surfaces, Int. J. Comput. Vis. 10 (1993) 183–197.
[178] D. Reniers, A. Telea, Skeleton-based hierarchical shape segmentation, in: Conf. Shape Model Appl., IEEE, 2007, pp. 179–188.
[179] P.K. Saha, B.B. Chaudhuri, 3D digital topology under binary transformation with applications, Comput. Vis. Image Underst. 63 (1996) 418–429.
[180] L. Serino, C. Arcelli, G. Sanniti di Baja, Decomposing 3D objects in simple parts characterized by rectilinear spines, Int. J. Pattern Recognit. Artif. Intell. 28 (7) (2014).
[181] L. Huang, A. Bijaoui, Astronomical image data compression by morphological skeleton transformation, Exp. Astron. 1 (5) (1990) 311–327.
[182] J.-M. Lien, G. Kurillo, R. Bajcsy, Skeleton-based data compression for multi-camera tele-immersion system, in: Proc. of International Symposium on Visual Computing (ISVC), Lake Tahoe, NV, USA, in: Lect. Notes Comput. Sci., vol. 4841, 2007, pp. 714–723.
[183] M. Ahmed, R. Ward, A rotation invariant rule-based thinning algorithm for character recognition, IEEE Trans. Pattern Anal. Mach. Intell. 24 (12) (2002) 1672–1678.
[184] N. Arica, F.T. Yarman-Vural, An overview of character recognition focused on off-line handwriting, IEEE Trans. Syst. Man Cybern., Part C, Appl. Rev. 31 (2) (2001) 216–233.
[185] Ø. Due Trier, A.K. Jain, T. Taxt, Feature extraction methods for character recognition – a survey, Pattern Recognit. 29 (4) (1996) 641–662.
[186] L. Lam, C.Y. Suen, An evaluation of parallel thinning algorithms for character recognition, IEEE Trans. Pattern Anal. Mach. Intell. 17 (9) (1995) 914–919.
[187] G. Boccignone, A. Chianese, L.P. Cordella, A. Marcelli, Recovering dynamic information from static handwriting, Pattern Recognit. 26 (3) (1993) 409–418.
[188] H. Bunke, M. Roth, E.G. Schukat-Talamazzini, Off-line cursive handwriting recognition using hidden Markov models, Pattern Recognit. 28 (9) (1995) 1399–1413.
[189] V. Pervouchine, G. Leedham, Extraction and analysis of forensic document examiner features used for writer identification, Pattern Recognit. 40 (3) (2007) 1004–1013.
[190] H. Fujiyoshi, A.J. Lipton, T. Kanade, Real-time human motion analysis by image skeletonization, IEICE Trans. Inf. Syst. 87 (1) (2004) 113–120.
[191] T. Khan, J. Westin, M. Dougherty, Motion cue analysis for Parkinsonian gait recognition, Open Biomed. Eng. J. 7 (2013) 1.
[192] K. Arai, R.A. Asmara, 3D skeleton model derived from Kinect Depth Sensor Camera and its application to walking style quality evaluations, Int. J. Adv. Res. Artif. Intell. 2 (7) (2013) 24–28.

[193] N. Gagvani, D. Silver, Animating volumetric models, Graph. Models 63 (6) (2001) 443–458.
[194] L. Wade, R.E. Parent, Automated generation of control skeletons for use in animation, Vis. Comput. 18 (2) (2002) 97–110.
[195] L. da Fontoura Costa, R.M. Cesar Jr., Shape Analysis and Classification: Theory and Practice, CRC Press, 2000.
[196] A. Farina, Z.M. Kovacs-Vajna, A. Leone, Fingerprint minutiae extraction from skeletonized binary images, Pattern Recognit. 32 (5) (1999) 877–889.
[197] S. Greenberg, M. Aladjem, D. Kogan, I. Dimitrov, Fingerprint image enhancement using filtering techniques, in: 15th Int. Conf. Patt. Recog., vol. 3, IEEE, Barcelona, Spain, 2000, pp. 322–325.
[198] M. Rao, T. Ch, Feature extraction for fingerprint classification, Pattern Recognit. 8 (3) (1976) 181–192.
[199] F. Zhao, X. Tang, Preprocessing and postprocessing for skeleton-based fingerprint minutiae extraction, Pattern Recognit. 40 (4) (2007) 1270–1281.
[200] M.I. Ahmed, M.A. Amin, B. Poon, H. Yan, Retina based biometric authentication using phase congruency, Int. J. Mach. Learn. Cybern. 5 (6) (2014) 933–945.
[201] S.M. Lajevardi, A. Arakala, S.A. Davis, K.J. Horadam, Retina verification system based on biometric graph matching, IEEE Trans. Image Process. 22 (9) (2013) 3625–3635.
[202] L. Wang, G. Leedham, Near- and far-infrared imaging for vein pattern biometrics, in: 2006 IEEE International Conference on Video and Signal Based Surveillance, IEEE, 2006, pp. 52–57.
[203] Y. Fridman, S.M. Pizer, S. Aylward, E. Bullitt, Extracting branching tubular object geometry via cores, Med. Image Anal. 8 (3) (2004) 169–176.
[204] D. Selle, B. Preim, A. Schenk, H.-O. Peitgen, Analysis of vasculature for liver surgical planning, IEEE Trans. Med. Imaging 21 (2002) 1344–1357.
[205] B.S. Tom, S.N. Efstratiadis, A.K. Katsaggelos, Motion estimation of skeletonized angiographic images using elastic registration, IEEE Trans. Med. Imaging 13 (3) (1994) 450–460.
[206] H.C. Kim, B.G. Min, M.K. Lee, J.D. Seo, Y.W. Lee, M.C. Han, Estimation of local cardiac wall deformation and regional wall stress from biplane coronary cineangiograms, IEEE Trans. Biomed. Eng. 32 (7) (1985) 503–512.
[207] S.Y. Chen, J.D. Carroll, J.C. Messenger, Quantitative analysis of reconstructed 3-D coronary arterial tree and intracoronary devices, IEEE Trans. Med. Imaging 21 (7) (2002) 724–740.
[208] M. Schaap, C.T. Metz, T. van Walsum, A.G. van der Giessen, A.C. Weustink, N.R. Mollet, C. Bauer, H. Bogunovic, C. Castro, X. Deng, E. Dikici, T. O'Donnell, M. Frenay, O. Friman, M. Hernandez Hoyos, P.H. Kitslaar, K. Krissian, C. Kuhnel, M.A. Luengo-Oroz, M. Orkisz, O. Smedby, M. Styner, A. Szymczak, H. Tek, C. Wang, S.K. Warfield, S. Zambal, Y. Zhang, G.P. Krestin, W.J. Niessen, Standardized evaluation methodology and reference database for evaluating coronary artery centerline extraction algorithms, Med. Image Anal. 13 (5) (2009) 701–714.
[209] M. Sonka, M.D. Winniford, S.M. Collins, Robust simultaneous detection of coronary borders in complex images, IEEE Trans. Med. Imaging 14 (1) (1995) 151–161.
[210] E. Sorantin, C. Halmai, B. Erdohelyi, K. Palágyi, L.G. Nyúl, K. Olle, B. Geiger, F. Lindbichler, G. Friedrich, K. Kiesler, Spiral-CT-based assessment of tracheal stenoses using 3-D-skeletonization, IEEE Trans. Med. Imaging 21 (3) (2002) 263–273.

[211] Y. Xu, G. Liang, G. Hu, Y. Yang, J. Geng, P.K. Saha, Quantification of coronary arterial stenoses in CTA using fuzzy distance transform, Comput. Med. Imaging Graph. 36 (2012) 11–24.
[212] I. Nyström, O. Smedby, Skeletonization of volumetric vascular images—distance information utilized for visualization, J. Comb. Optim. 5 (1) (2001) 27–41.
[213] H. Frimmel, J. Näppi, H. Yoshida, Centerline-based colon segmentation for CT colonography, Med. Phys. 32 (2005) 2665–2672.
[214] T. He, L. Hong, D. Chen, Z. Liang, Reliable path for virtual endoscopy: ensuring complete examination of human organs, IEEE Trans. Vis. Comput. Graph. 7 (4) (2001) 333–342.
[215] D.G. Kang, J.B. Ra, A new path planning algorithm for maximizing visibility in computed tomography colonography, IEEE Trans. Med. Imaging 24 (8) (2005) 957–968.
[216] R. Sadleir, P. Whelan, Fast colon centreline calculation using optimised 3D topological thinning, Comput. Med. Imaging Graph. 29 (2005) 251–258.
[217] A.P. Kiraly, J.P. Helferty, E.A. Hoffman, G. McLennan, W.E. Higgins, Three-dimensional path planning for virtual bronchoscopy, IEEE Trans. Med. Imaging 23 (11) (2004) 1365–1379.
[218] K. Mori, J. Hasegawa, Y. Suenaga, J. Toriwaki, Automated anatomical labeling of the bronchial branch and its application to the virtual bronchoscopy system, IEEE Trans. Med. Imaging 19 (2) (2000) 103–114.
[219] J. Tschirren, G. McLennan, K. Palágyi, E.A. Hoffman, M. Sonka, Matching and anatomical labeling of human airway tree, IEEE Trans. Med. Imaging 24 (12) (2005) 1540–1547.
[220] A. Chaturvedi, Z. Lee, Three-dimensional segmentation and skeletonization to build an airway tree data structure for small animals, Phys. Med. Biol. 50 (7) (2005) 1405.
[221] K. Palágyi, J. Tschirren, E.A. Hoffman, M. Sonka, Quantitative analysis of pulmonary airway tree structures, Comput. Biol. Med. 36 (9) (2006) 974–996.
[222] H. Kobatake, Y. Yoshinaga, Detection of spicules on mammogram based on skeleton analysis, IEEE Trans. Med. Imaging 15 (3) (1996) 235–245.
[223] R. Zwiggelaar, S.M. Astley, C.R. Boggis, C.J. Taylor, Linear structures in mammographic images: detection and classification, IEEE Trans. Med. Imaging 23 (9) (2004) 1077–1086.
[224] G. Chang, S.K. Pakin, M.E. Schweitzer, P.K. Saha, R.R. Regatte, Adaptations in trabecular bone microarchitecture in olympic athletes determined by 7T MRI, J. Magn. Reson. Imaging 27 (5) (2008) 1089–1095.
[225] B.G. Gomberg, P.K. Saha, H.K. Song, S.N. Hwang, F.W. Wehrli, Topological analysis of trabecular bone MR images, IEEE Trans. Med. Imaging 19 (2000) 166–174.
[226] G.A. Ladinsky, B. Vasilic, A.M. Popescu, M. Wald, B.S. Zemel, P.J. Snyder, L. Loh, H.K. Song, P.K. Saha, A.C. Wright, F.W. Wehrli, Trabecular structure quantified with the MRI-based virtual bone biopsy in postmenopausal women contributes to vertebral deformity burden independent of areal vertebral BMD, J. Bone Miner. Res. 23 (1) (2008) 64–74.
[227] X.S. Liu, P. Sajda, P.K. Saha, F.W. Wehrli, G. Bevill, T.M. Keaveny, X.E. Guo, Complete volumetric decomposition of individual trabecular plates and rods and its morphological correlations with anisotropic elastic moduli in human trabecular bone, J. Bone Miner. Res. 23 (2) (2008) 223–235.
[228] M. Stauber, R. Muller, Volumetric spatial decomposition of trabecular bone into rods and plates—a new method for local bone morphometry, Bone 38 (4) (2006) 475–484.

[229] F.W. Wehrli, B.R. Gomberg, P.K. Saha, H.K. Song, S.N. Hwang, P.J. Snyder, Digital topological analysis of in vivo magnetic resonance microimages of trabecular bone reveals structural implications of osteoporosis, J. Bone Miner. Res. 16 (2001) 1520–1531.
[230] T. Hildebrand, P. Rüegsegger, A new method for the model independent assessment of thickness in three-dimensional images, J. Microsc. 185 (1997) 67–75.
[231] A. Laib, D.C. Newitt, Y. Lu, S. Majumdar, New model-independent measures of trabecular bone structure applied to in vivo high-resolution MR images, Osteoporos. Int. 13 (2002) 130–136.
[232] Y. Liu, D. Jin, C. Li, K.F. Janz, T.L. Burns, J.C. Torner, S.M. Levy, P.K. Saha, A robust algorithm for thickness computation at low resolution and its application to in vivo trabecular bone CT imaging, IEEE Trans. Biomed. Eng. 61 (7) (2014) 2057–2069.
[233] F. Peyrin, Z. Peter, A. Larrue, A. Bonnassie, D. Attali, Local geometrical analysis of 3D porous networks based on the medial axis: application to bone microarchitecture microtomography images, Image Anal. Stereol. 26 (3) (2011) 179–185.
[234] L. Pothuaud, A. Laib, P. Levitz, C.L. Benhamou, S. Majumdar, Three-dimensional-line skeleton graph analysis of high-resolution magnetic resonance images: a validation study from 34-microm-resolution microcomputed tomography, J. Bone Miner. Res. 17 (10) (2002) 1883–1895.
[235] V. Chatzis, I. Pitas, Interpolation of 3-D binary images based on morphological skeletonization, IEEE Trans. Med. Imaging 19 (7) (2000) 699–710.
[236] E.H. Shin, K.S. Cho, M.H. Seo, H. Kim, Determination of electrospun fiber diameter distributions using image analysis processing, Macromol. Res. 16 (4) (2008) 314–319.
[237] Y. Wan, L. Yao, P. Zeng, B. Xu, Shaped fiber identification using a distance-based skeletonization algorithm, Tex. Res. J. 80 (10) (2010) 958–968.
[238] A. D'Amore, J.A. Stella, W.R. Wagner, M.S. Sacks, Characterization of the complete fiber network topology of planar fibrous tissues and scaffolds, Biomaterials 31 (20) (2010) 5345–5354.
[239] A.M. Kraynik, Foam structure: from soap froth to solid foams, Mater. Res. Soc. Bull. 28 (04) (2003) 275–278.
[240] A.M. Kraynik, D.A. Reinelt, F. van Swol, Structure of random foam, Phys. Rev. Lett. 93 (20) (2004) 208301.
[241] D.R. Baker, F. Brun, C. O'Shaughnessy, L. Mancini, J.L. Fife, M. Rivers, A four-dimensional X-ray tomographic microscopy study of bubble growth in basaltic foam, Nat. Commun. 3 (2012) 1135.
[242] D. Thibault, C.M. Gold, Terrain reconstruction from contours by skeleton construction, GeoInformatica 4 (4) (2000) 349–373.
[243] R.M. Haralick, Performance characterization in image analysis: thinning, a case in point, Pattern Recognit. Lett. 13 (1) (1992) 5–12.
[244] M.Y. Jaisimha, R.M. Haralick, D. Dori, Quantitative performance evaluation of thinning algorithms under noisy conditions, in: Int. Conf. Comp. Vis. Patt. Recog., Seattle, WA, 1994, pp. 678–683.
[245] V. Ćurić, J. Lindblad, N. Sladoje, H. Sarve, G. Borgefors, A new set distance and its application to shape registration, Pattern Anal. Appl. 17 (1) (2014) 141–152.

CHAPTER

Multiscale 2D medial axes and 3D surface skeletons by the image foresting transform

2

Alexandre Falcão*, **Cong Feng**†, **Jacek Kustra**‡, **Alexandru Telea**†

Department of Information Systems, Institute of Computing, University of Campinas (UNICAMP), Campinas, SP, Brazil Institute Johann Bernoulli, University of Groningen, Groningen, The Netherlands*† *Philips Research, Eindhoven, The Netherlands*‡

Contents

2.1 Introduction 43
2.2 Related Work 44
2.2.1 Definitions 45
2.2.2 Skeleton Regularization 45
2.3 Proposed Method 48
2.3.1 Multiscale Regularization—Strengths and Weaknesses 49
2.3.2 Image Foresting Transform 50
2.3.2.1 Single-Point Feature Transform 54
2.3.2.2 Shortest-Path Length Computation 55
2.3.3 Multiscale Skeletonization—Putting It All Together 56
2.4 Comparative Analysis 60
2.4.1 2D Medial Axes 60
2.4.2 3D Medial Surfaces 60
2.4.2.1 Global Comparison 60
2.4.2.2 Detailed Comparison 63
2.5 Conclusion 67
References 67

2.1 INTRODUCTION

Medial descriptors, or skeletons, are used in many applications such as path planning, shape retrieval and matching, computer animation, medical visualization, and shape processing [45,50]. In two dimensions, such descriptors are typically called medial axes. Three-dimensional shapes admit two types of skeletons, *surface* skeletons, which are sets of manifolds with boundaries that meet along a set of so-called

Skeletonization. DOI: 10.1016/B978-0-08-101291-8.00003-1

Y-intersection curves [12,30,6], and *curve* skeletons, which are 1D structures locally centered in the shape [8].

A fundamental and well-known problem of skeleton computation is that skeletons are inherently unstable to small perturbations of the input shape [50]. This leads to the appearance of so-called spurious branches, which have little or no application value, and considerably complicate the analysis and usage of skeletons. To alleviate this, various simplification methods have been proposed to eliminate such branches. In this context, *multiscale* methods are particularly interesting. They compute a so-called importance metric for skeletal points, which encodes the scale of the shape details, and next offer a continuous way for simplifying skeletons by simply thresholding that metric.

Several robust, simple to implement, and efficient methods exist for computing 2D multiscale skeletons; see, e.g., [35,17,53]. For surface skeletons, the situation is very different: Only a few such methods exist, and these are either computationally expensive [14,38,39], complex [26], or sensitive to numerical discretization [27].

In this chapter, we address the joint problem of computing multiscale 2D medial axes and 3D surface skeletons by a new method. For this, we cast the problem of computing the importance metrics proposed in [35,17,53] (for 2D skeletons) and in [39,26] (for 3D skeletons) as the search for an optimal path forest using the Image Foresting Transform framework [19]. In 2D, the skeletons are one-pixel wide and connected in all scales for genus-0 shapes. In 3D, the surface skeletons are one-voxel wide and can be connected in all scales for genus-0 shapes if the curve skeleton points are detected [39,26], which we do not address here. Next, we provide simple and efficient algorithms to compute these metrics for both 2D and 3D binary images. Compared to 3D techniques that use the same multiscale importance metric, our method is faster [14,38,39] or alternatively considerably simpler to implement [26]. Compared to other 3D multiscale techniques, our method is far less sensitive to numerical noise [27]. Compared to all above techniques, our method yields the same quality level in terms of centeredness, smoothness, thinness, and ease to simplify the skeleton.

This chapter is structured as follows. In Section 2.2, we overview multiscale skeletonization solutions and challenges. Section 2.3 introduces the Image Foresting Transform and its adaptations required for multiscale skeletonization. Section 2.3.3 details our multiscale skeletonization algorithms for 2D and 3D shapes. Section 2.4 compares our method with its multiscale competitors on a wide set of 2D and 3D real-world shapes. Section 2.5 concludes the chapter, summarizing our main contributions and outlining directions for future work.

2.2 RELATED WORK

In this section, we provide the basic definitions related to skeletonization and discuss skeleton regularization based on local and global measures.

2.2.1 DEFINITIONS

Given a shape $\Omega \subset \mathbb{R}^d$, $d \in \{2, 3\}$ with boundary $\partial\Omega$, we first define its Euclidean distance transform $\mathcal{D} : \mathbb{R}^d \to \mathbb{R}^+$ as

$$\mathcal{D}(\mathbf{x} \in \Omega) = \min_{\mathbf{y} \in \partial\Omega} \|\mathbf{x} - \mathbf{y}\|. \tag{2.1}$$

The Euclidean skeleton of Ω is next defined as

$$\mathcal{S} = \{\mathbf{x} \in \Omega | \exists \mathbf{f}_1, \mathbf{f}_2 \in \partial\Omega, \mathbf{f}_1 \neq \mathbf{f}_2, \|\mathbf{x} - \mathbf{f}_1\| = \|\mathbf{x} - \mathbf{f}_2\| = \mathcal{D}(\mathbf{x})\}, \tag{2.2}$$

where $\mathbf{f}_1$ and $\mathbf{f}_2$ are the contact points with $\partial\Omega$ of the maximally inscribed ball in Ω centered at $\mathbf{x}$ [23,39], also called *feature points* of $\mathbf{x}$ [25], where the feature transform $\mathcal{F} : \mathbb{R}^d \to \mathcal{P}(\partial\Omega)$ is defined as

$$\mathcal{F}(\mathbf{x} \in \Omega) = \arg\min_{\mathbf{y} \in \partial\Omega} \|\mathbf{x} - \mathbf{y}\|. \tag{2.3}$$

The vectors $\mathbf{f} - \mathbf{x}$ are called *spoke vectors* [47]. By definition (Eq. (2.2)), for any $\mathbf{x} \notin \mathcal{S}$, $\mathcal{F}(\mathbf{x})$ yields a single point, i.e., $|\mathcal{F}(\mathbf{x})| = 1$, whereas for any $\mathbf{x} \in \mathcal{S}$, $|\mathcal{F}(\mathbf{x})| \geq 2$. In practice, computing $\mathcal{F}$ can be quite expensive and/or complicated due to its multivalued nature. As such, many applications (see, e.g., [25,21,39]) use the so-called single-value feature transform $F : \mathbb{R}^d \to \partial\Omega$, defined as

$$F(\mathbf{x} \in \Omega) = \mathbf{y} \in \partial\Omega \quad \text{so that} \quad \|\mathbf{x} - \mathbf{y}\| = \mathcal{D}(\mathbf{x}). \tag{2.4}$$

For $d = 2$, $\mathcal{S}$ is a set of curves that meet at the so-called skeleton junction points [17]. For $d = 3$, $\mathcal{S}$ is a set of manifolds with boundaries that meet along a set of so-called Y-intersection curves [12,30,6]. The pair $MAT = (\mathcal{S}, \mathcal{D})$ defines the medial axis transform (MAT) of Ω, which is a dual representation of Ω, i.e., allows reconstructing the shape $\Omega_{\mathcal{S},\mathcal{D}} = \bigcup_{\mathbf{x} \in \mathcal{S}} B_{\mathcal{D}(\mathbf{x})}(\mathbf{x})$ as the union of balls $B_{\mathcal{D}(\mathbf{x})}(\mathbf{x})$ centered at $\mathbf{x} \in \mathcal{S}$ and with radii $\mathcal{D}(\mathbf{x})$.

Except when it is explicitly mentioned, we consider only the case where $\Omega \setminus \partial\Omega$ and $\partial\Omega$ have the same number of components.

2.2.2 SKELETON REGULARIZATION

In practice, skeletons are extracted from discretized (sampled) versions of Ω, using either an implicit (boundary mesh) representation [35,51,32,26] or an explicit (volumetric) representation [17,53,14,39]. In this chapter, we focus on the latter case. Here, the d-dimensional space is discretized in a uniform grid of so-called *spels* (space elements) having integer coordinates, i.e., pixels for $d = 2$ and voxels for $d = 3$. Due to this discretization, solving Eq. (2.2) on $\mathbb{Z}^d$ rather than on $\mathbb{R}^d$ yields skeletons that are not perfectly centered, not necessarily one-spel thin, and not necessarily homotopic to the input shape. Discretization also causes skeletons to have a large amount of spurious manifolds (branches). Formally, this means that skeletonization is not a Cauchy

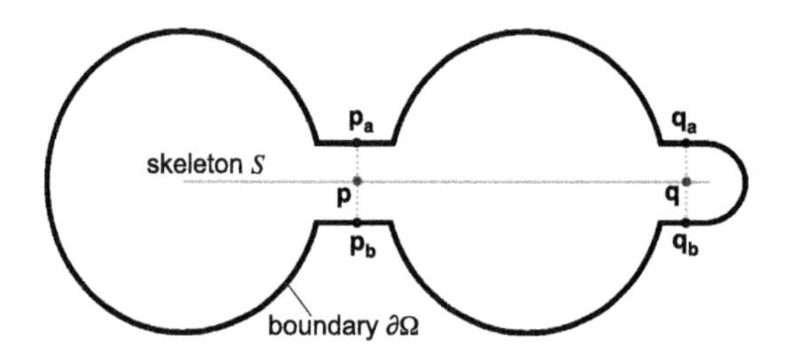

FIGURE 2.1

Problems of local regularization. For the shown shape, all local regularization metrics yield the same values for points $\mathbf{p}$ and $\mathbf{q}$. However, $\mathbf{p}$ is globally more important for the shape description than $\mathbf{q}$.

or Lipschitz continuous operation with respect to the Hausdorff distance between two shapes, but a semicontinuous operation [50]. Informally put, small variations of a shape can cause arbitrarily large variations of its skeleton.

To achieve Cauchy or Lipschitz continuity, desirable for most practical applications (which should not be sensitive to discretization issues), a *regularization* process is typically used. For this, we define the so-called *importance metric* $\rho : \Omega \to \mathbb{R}^+$, whose upper thresholding by some desired value $\tau > 0$ removes, or prunes, branches caused by small-scale details or noise on $\partial\Omega$ [43,12]. The regularized skeleton is defined as $\mathcal{S}_\tau = \{\mathbf{x} \in \mathcal{S} | \rho(\mathbf{x}) \geq \tau\}$. We distinguish between local and global importance measures, in line with [39,32,26,27], as follows.

Local measures essentially consider, for a skeletal point $\mathbf{x} \in \mathcal{S}$, only a small neighborhood of $\mathbf{x}$ to compute $\rho(\mathbf{x})$. The main advantage of these measures is that they are simple to implement and fast to compute. Local measures include the angle between the feature points and distance-to-boundary [1,22,49,25], divergence metrics [44,5], first-order moments [42], and detecting the multivalued points of ∇D [47,48]. Local measures are also, historically speaking, the first proposed skeleton regularization techniques. Good surveys of local methods are given in [45,50]. However, local measures have a fundamental issue: They cannot discriminate between locally identical, yet globally different, shape contexts. Fig. 2.1 illustrates this for a synthetic case: For the 2D shape with boundary $\partial\Omega$, the central skeletal point $\mathbf{p}$ is clearly more important to the shape description than the peripheral point $\mathbf{q}$ that corresponds to the right local protrusion. However, any local importance metric will rank $\mathbf{p}$ as important as $\mathbf{q}$ since their surroundings, including the placement of their feature points (shown in the figure), are identical. Similar situations can be easily found for 3D shapes.

Given the above, thresholding local measures can (and typically will) disconnect skeletons. Reconnection needs extra work [44,37,31,49] and makes skeleton pruning less intuitive and harder to implement [43]. Without this kind of work, no local measure can yield connected skeletons for all shapes. Note that this is a fundamental issue related to the local nature of these metrics; see also the discussion in [27]. Secondly, local metrics do not support the notion of a *multiscale* skeleton: Such skeletons $\mathcal{S}_\tau$ should ensure a continuous simplification (in the Cauchy or Lipschitz sense mentioned before) of the input shape Ω in terms of its reconstruction $\Omega_{\mathcal{S}_\tau,\mathcal{D}}$ as a function of the simplification parameter τ [35,50]. As such, although local measures can be "fixed" by reconnection work to yield connected skeletons, they still cannot provide an intuitive and easy-to-use way to simplify skeletons according to a user-prescribed threshold τ.

Global measures monotonically increase as one walks along $\mathcal{S}$ from the skeleton boundary $\partial\mathcal{S}$ inwards, toward points increasingly further away from $\partial\mathcal{S}$. For genus-0 shapes, such measures can be informally thought of as giving the removal order of skeletal points in a homotopy-preserving erosion process that starts at $\partial\mathcal{S}$ and ends when the entire skeleton has been eroded away. Given this property, thresholding them always yields connected skeletons, which also capture shape details at a user-given scale. For shapes with genus greater than 0, like having holes (in 2D) or cavities (in 3D), simple suitable postprocessing of the importance metric guarantees the joint connectivity and multiscale properties. For 2D shapes, a well-known global measure is the so-called boundary-collapse metric used to extract multiscale 2D skeletons, and proposed by various authors in different contexts [35,17,53,52]. For 3D shapes, the union-of-balls (UoB) approximation uses morphological dilation and erosion to define the scale of shape details captured by the regularized skeleton [24,32]. Dey and Sun propose as a regularization metric the medial geodesic function (MGF), equal to the length of the shortest-path between feature points [14,38] and use this metric to compute regularized 3D curve skeletons. Reniers et al. [39] extend the MGF for both surface and curve skeletons using geodesic lengths and surface areas between geodesics, respectively. A fast GPU implementation of this extended MGF for meshed shapes is given in [26].

The 3D MGF and its 2D boundary-collapse metric counterpart have an intuitive geometric meaning: They assign to a skeleton point $\mathbf{x} \in \mathcal{S}$ the amount of shape boundary that corresponds, or "collapses" to, $\mathbf{x}$ by some kind of advective boundary-to-skeleton transport. As such, skeleton points $\mathbf{x}$ with low importance values correspond to small-scale shape details or noise; points $\mathbf{x}$ with large importance values correspond to large-scale shape parts. Fig. 2.2 illustrates the boundary-collapse principle for both 2D medial axes (A) and 3D medial surfaces (B). In both cases, for a skeletal point $\mathbf{x}$, the importance is equal to the length of the shortest path $\gamma_\mathbf{x}$ that goes on $\partial\Omega$ between the feature points $\mathbf{f}_1^\mathbf{x}$ and $\mathbf{f}_2^\mathbf{x}$. This allows an intuitive and controllable skeleton simplification: Thresholding the MGF by a value τ eliminates all skeleton points that encode less than τ boundary length or area units. Since all the above-mentioned collapse metrics monotonically increase from the skeleton boundary $\partial\mathcal{S}$ to its center, thresholding them delivers a set of nested skeleton approximations, also called a multiscale skeleton. Importantly, these progressively simplified skeletons correspond, via the MAT (Section 2.2.1), to progressively simplified versions of Ω. More precisely, for all above collapse metrics, the reconstruction of a shape Ω from its simplified skeleton $\mathcal{S}_\tau$ yields a shape where all details of Ω of size smaller than τ have been replaced by circle arcs (in 2D) or spherical caps (in 3D) [53,39,26].

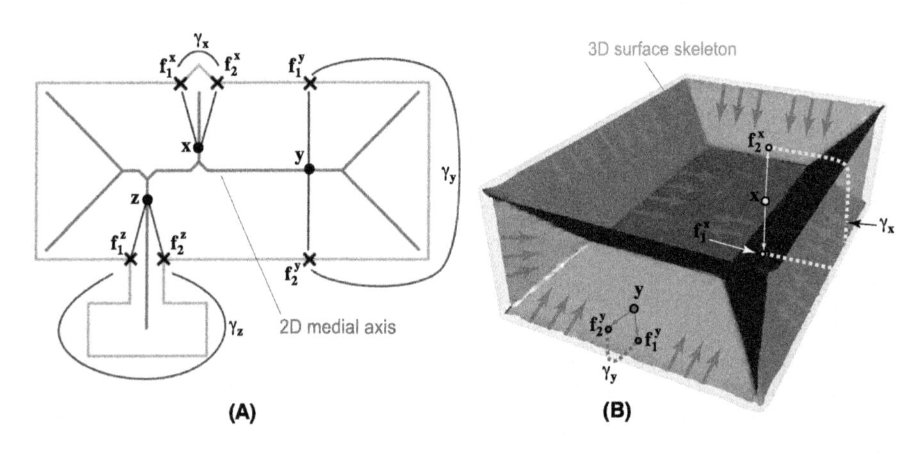

FIGURE 2.2

Multiscale collapse metric for 2D shapes (A) and 3D shapes (B).

The idea of a mass collapse process from $\partial\Omega$ to $\mathcal{S}$ was also used by other works to define multiscale skeletons. Couprie [9] proposed a discrete framework for computing 2D skeletons and 3D curve skeletons by a guided thinning process, which, for the 2D case, yields very similar results to [53,17]. However, 3D surface skeletons are not covered by this approach. Torsello et al. propose a conservative mass advection $\partial\Omega$ onto $\mathcal{S}$ to define ρ in 2D [4], which was next extended to 3D [41]. However, the numerical computation of this process is affected by serious stability issues. Recently, Jalba et al. extended this advection model to compute multiscale 2D skeletons and 3D surface and curve skeletons in a unified formulation [27]. Although this method delivers convincing results, it still suffers from numerical stability problems and is also relatively complex to implement.

Summarizing the above, we argue that multiscale regularization metrics are net superior, both in theory and practice, to local regularization metrics. However, as outlined and discussed next in more detail in Section 2.3.1, multiscale regularization is far from being simple and cheap. Our proposal, presented next, aims at solving these problems.

2.3 PROPOSED METHOD

To compute multiscale 2D and 3D skeletons of binary shapes efficiently and robustly, we propose to use the *Image Foresting Transform* (IFT) methodology [19]. We start, in Section 2.3.1, by introducing our general idea, which details the strengths and weaknesses of the MGF and advection-based regularization techniques introduced in Section 2.2.2. Next, we detail the use of the IFT to compute skeletons that combine the identified strengths of these two approaches (Section 2.3.2). Our final proposed skeletonization algorithms are detailed in Section 2.3.3.

2.3.1 MULTISCALE REGULARIZATION—STRENGTHS AND WEAKNESSES

Analyzing all multiscale skeletonization methods surveyed in Section 2.2 [35,17,53, 52,39,4,41,26,27], we notice that their various importance-metric definitions can be all explained, at a high level, by introducing a vector field $\mathbf{v} : \Omega \rightarrow \mathbb{R}^d$ as follows: If we imagine that the input shape surface $\partial\Omega$ is covered by uniformly distributed mass with density $\rho(\mathbf{x} \in \partial\Omega) = 1$, then all above methods explain the importance $\rho(\mathbf{x} \in \Omega)$ of a spel $\mathbf{x}$ as the amount of mass transported, or advected, by $\mathbf{v}$ from $\partial\Omega$ to $\mathbf{x}$. Studying the properties of $\mathbf{v}$ brings several insights into multiscale regularization as follows.

First, we note that the importance values of nonskeletal spels $\mathbf{x} \notin \mathcal{S}$ should be low and nearly locally constant, so that upper-thresholding ρ by this value allows us to reliably separate $\mathcal{S}$. All above-mentioned methods define $\rho(\mathbf{x} \notin \mathcal{S})$ to be equal to the importance of the single feature point of $\mathbf{x}$, i.e., $\rho(\mathbf{x} \notin \mathcal{S}) = \rho(\mathcal{F}(\mathbf{x}))$. This property is realized if we define $\mathbf{v} = \nabla\mathcal{D}$ for all $\mathbf{x} \notin \mathcal{S}$, as it is well known that gradient lines of the distance transform only intersect at skeletal points [44,5].

To fully define the multiscale importance ρ over Ω in terms of an advection process, it thus remains to define $\mathbf{v}$ on $\mathcal{S}$. Studying the above-mentioned multiscale methods and considering for now the case of connected genus-0 shapes, we see here that all such methods aim to define a field ρ that monotonically increases from $\partial\mathcal{S}$ to its center, or root $\mathbf{r} \in \mathcal{S}$. As explained earlier, this allows easy computing of multiscale skeletons $\mathcal{S}_\tau$ by simply upper-thresholding ρ with desired values τ. In advection terms, this is equivalent to defining a field $\mathbf{v}$ that transports mass along $\mathcal{S}$ from its boundary $\partial\mathcal{S}$ to $\mathbf{r}$, along paths that finally meet at $\mathbf{r}$. For $d = 2$, mass flows from $\partial\Omega$ to the one-dimensional medial axis $\mathcal{S}$ and then along the branches of $\mathcal{S}$ to its center $\mathbf{r}$. For $d = 3$, mass flows from $\partial\Omega$ to the two-dimensional surface-skeleton $\mathcal{S}$, then along $\mathcal{S}$ toward its local center (which is the curve skeleton of Ω), and then along the curve-skeleton branches toward the center $\mathbf{r}$ thereof.

The different multiscale importance listed above can be explained in terms of different definitions of $\mathbf{v}$ over $\mathcal{S}$ as follows. In 2D, for genus-0 shapes, all surveyed methods essentially reduce to defining $\mathbf{v}$ as being locally tangent to $\mathcal{S}$ and pointing toward the root of the skeleton (which in this case is a tree) [35,17,53,52]. In 3D, explaining multiscale importance in terms of a field $\mathbf{v}$ defined on $\mathcal{S}$ is more complicated. We distinguish here two classes of methods as follows.

MGF methods: For genus-0 shapes, MGF-based methods do not actually give a formal definition of $\mathbf{v}$, but compute $\rho(\mathbf{x} \in \mathcal{S})$ as the length of the longest shortest-path on $\partial\Omega$ between any two feature points $\mathbf{f}$ and $\mathbf{g}$ of $\mathbf{x}$, i.e.,

$$\rho(\mathbf{x} \in \mathcal{S}) = \max_{\mathbf{f}\in\mathcal{F}(\mathbf{x}),\mathbf{g}\in\mathcal{F}(\mathbf{x})} GL(\mathbf{f}, \mathbf{g}), \tag{2.5}$$

where $GL(\mathbf{x}, \mathbf{y})$ is the length of the shortest path on $\partial\Omega$ between two points $\mathbf{x} \in \partial\Omega$ and $\mathbf{y} \in \partial\Omega$. This is based on the empirical observation that Eq. (2.5) defines a field ρ that smoothly and monotonically increases from $\partial\mathcal{S}$ to its center [14,39]. This allows us to compute regularized surface skeletons even for noisy and complex 3D shapes. Another advantage of the MGF is that it makes simplification intuitive to understand: thresholding ρ at a value τ eliminates all spels from $\mathcal{S}$ where the local thickness of the shape is larger than τ. However, a formal justification of the MGF in terms of an advection process on $\mathcal{S}$ has not yet been given. A second limitation of the MGF model is that is becomes very expensive to compute for large 3D shapes, where we need to trace (at least one) shortest path on $\partial\Omega$ for each point $\mathbf{x} \in \mathcal{S}$, and so its cost is $O(|\mathcal{S}| \cdot |\partial\Omega| \log |\partial\Omega|)$. Using GPU techniques can accelerate this process [26], but also massively complicates the implementation of the method.

Advection methods: Aiming to solve the above issues with the MGF, Jalba et al. define $\mathbf{v}$ on $\mathcal{S}$ as the result of a mass-conserving advection model, with topological constraints used to ensure skeleton homotopy with the input shape [27]. In contrast to the MGF, computation of all types of skeletons (2D medial axes, 3D surface skeletons, and 3D curve skeletons) is now fully captured by a single unified advection model. Also, the method in [27] is considerably faster than MGF techniques, its cost being $O(|\Omega|)$, and is also simpler to implement. However, setting the simplification threshold τ is now less intuitive than for MGF techniques since this value does not have an immediate geometric interpretation. Also, all 3D advection methods we are aware of [4,41,27] suffer various amounts of from numerical diffusion, which means that the computed importance ρ will deliver regularized skeletons $\mathcal{S}_\tau$ having jagged boundaries.

The method that we propose in the next section aims to combine the strengths and limit the disadvantages of the MGF and advection methods outlined above. In detail, we use an incremental advection-like computation of the intuitive MGF metric, which makes our method considerably faster than existing MGF methods. We provide an implementation that is simple and does not suffer from numerical diffusion issues. Our proposal delivers smooth-boundary regularized, centered, one-spel-thin, multiscale skeletons of the same overall quality as skeletons delivered by existing state-of-the-art methods.

2.3.2 IMAGE FORESTING TRANSFORM

The Image Foresting Transform (IFT) interprets d-dimensional images as graphs and reduces image operators to the computation of an *optimum-path forest* followed by a local processing of its attributes. In essence, the IFT is Dijkstra's shortest-path algorithm modified to use multiple sources and more general connectivity (path-value) functions. In our context, the IFT propagates, from $\partial\Omega$ to $\mathbf{x} \in \Omega \setminus \partial\Omega$, both the feature points $\mathcal{F}(\mathbf{x})$ and the advection field $\mathbf{v}(\mathbf{x})$. Fig. 2.3 shows in yellow the spels s with undefined $\nabla\mathcal{D}(\mathbf{x})$. These are the leaves of the optimum-path forest whose paths follow the direction of $\mathbf{v}$.

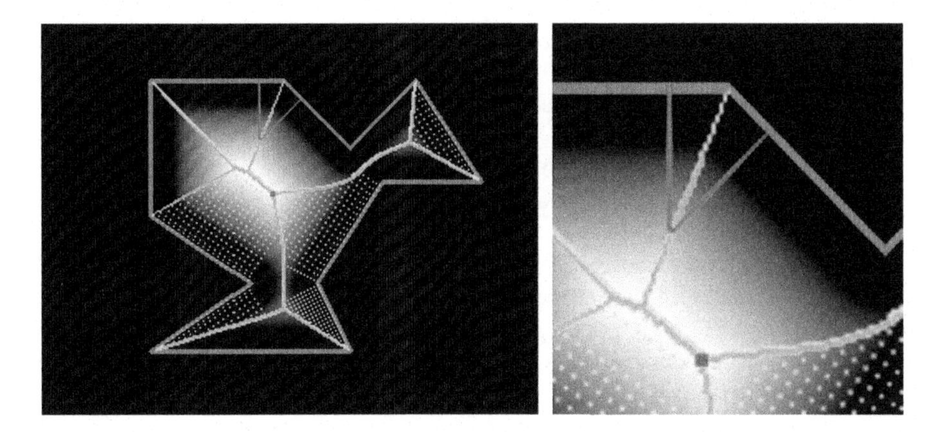

FIGURE 2.3

A polygon Ω with orange boundary $\partial\Omega$, its distance transform D (gray values), a plausible skeleton $\mathcal{S}$ (cyan), the root point r (red), the points with undefined $\nabla D(s)$ (yellow), and magenta lines connecting a given spel s with its feature points in $\partial\Omega$.

To design an image operator based on the IFT, we need to specify which image elements (points, edges, regions) are the nodes of the graph, an *adjacency relation* between them, and a *connectivity function* that assigns a value (e.g., strength, cost, distance) to any path in the graph. This methodology has been successfully used for boundary-based [20,18,33], region-based [16,13], and hybrid image segmentation [46,7]; connected filtering [15]; shape representation and description [17,11,10, 3]; and unsupervised [40], semisupervised [2], and supervised data classification [36]. In this section, we show how the IFT can be adapted to propagate the single-point feature transform $F(\mathbf{x}) \in \partial\Omega, \forall \mathbf{x} \in \Omega \setminus \partial\Omega$ (Eq. (2.4)) and also to compute the length $GL(\mathbf{x}, \mathbf{y})$ of the shortest path on $\partial\Omega$ between two feature points $\mathbf{x}, \mathbf{y} \in \partial\Omega$.

Graph definition: In the discrete space, the shape Ω is provided as a binary image $\mathbf{I} = (\mathcal{I}, I)$, where $\mathcal{I} \subset \mathbb{Z}^d$ is the image domain, and each spel $\mathbf{x} \in \mathcal{I}$ has a value $I(\mathbf{x}) \in \{0, 1\}$. For instance, $\Omega \subset \mathcal{I}$ may be the set of spels with value 1 and its complement $\overline{\Omega} = \mathcal{I} \setminus \Omega$ be defined by the spels with value 0. The image $\mathbf{I}$ can be interpreted as a graph whose nodes are the spels in $\mathcal{I}$ and arcs are defined by the adjacency relation

$$\mathcal{A}_{\mathcal{I},\delta} = \{(\mathbf{x}, \mathbf{y}) \in \mathcal{I} \times \mathcal{I} \mid \|\mathbf{x} - \mathbf{y}\| \leq \delta\} \tag{2.6}$$

for a given value $\delta \in \mathbb{R}^+$. Let also $\mathcal{A}_{\mathcal{I},\delta}(\mathbf{x})$ be the set of spels adjacent to a spel $\mathbf{x}$. The shape boundary $\partial\Omega$ is then defined by

$$\partial\Omega = \{\mathbf{x} \in \Omega | \overline{\Omega} \cap \mathcal{A}_{\mathcal{I},1}(\mathbf{x}) \neq \emptyset\}. \tag{2.7}$$

To compute the single-point feature transform F and the length $GL(\mathbf{f}, \mathbf{g})$ of shortest paths between point-pairs $(\mathbf{f}, \mathbf{g}) \in \partial\Omega \times \partial\Omega$, we constrain the adjacency relation $\mathcal{A}_{\mathcal{I},\delta}$ to it subsets $\mathcal{A}_{\Omega,\delta}$ and $\mathcal{A}_{\partial\Omega,\delta}$ defined as

$$\mathcal{A}_{\Omega,\delta} = \{(\mathbf{x}, \mathbf{y}) \in \Omega \times \Omega \mid \|\mathbf{x} - \mathbf{y}\| \leq \delta\}, \tag{2.8}$$

$$\mathcal{A}_{\partial\Omega,\delta} = \{(\mathbf{x}, \mathbf{y}) \in \partial\Omega \times \partial\Omega \mid \|\mathbf{x} - \mathbf{y}\| \leq \delta\}, \tag{2.9}$$

respectively. Given the above, we are next interested in the graphs $(\Omega, \mathcal{A}_{\Omega,\delta})$ and $(\partial\Omega, \mathcal{A}_{\partial\Omega,\delta})$ that describe the shape's interior and boundary, respectively.

For the graph $(\Omega, \mathcal{A}_{\Omega,\delta})$, let a *path* $\pi_{\mathbf{t}}$ be a spel-sequence $\langle \mathbf{x}_1, \mathbf{x}_2, \ldots, \mathbf{x}_n = \mathbf{t} \rangle$ with terminal spel $\mathbf{t}$ such that $(\mathbf{x}_i, \mathbf{x}_{i+1}) \in \mathcal{A}_{\Omega,\delta}$, $1 \leq i < n$. Let $\Pi\left(\Omega, \mathcal{A}_{\Omega,\delta}\right)$ be the set of all paths in $(\Omega, \mathcal{A}_{\Omega,\delta})$. A pair of spels is called *connected* in $(\Omega, \mathcal{A}_{\Omega,\delta})$ if there exists a path in $\Pi\left(\Omega, \mathcal{A}_{\Omega,\delta}\right)$ between them. A *connected component* in $(\Omega, \mathcal{A}_{\Omega,\delta})$ is a subgraph, maximal for the inclusion, therein all pairs of spels are connected. The same definitions apply to the graph $(\partial\Omega, \mathcal{A}_{\partial\Omega,\delta})$.

We next assume, with no generality loss, that the interior $\Omega \setminus \partial\Omega$ of Ω defines a single connected component in $(\Omega, \mathcal{A}_{\Omega,1})$. When this is not the case, we simply treat each such connected component separately to yield a separate skeleton. To eliminate cases where Ω and $\Omega \setminus \partial\Omega$ have different numbers of components, i.e., the removal of $\partial\Omega$ would disconnect the shape, we can preprocess Ω by, e.g., morphological dilation with the distance of the size of one spel.

Objects with holes (in 2D) can be easily treated by merging the internal skeletons derived from each component in $(\partial\Omega, \mathcal{A}_{\partial\Omega,\sqrt{d}})$ into a single one through the *skeleton by influence zones* (SKIZ) [17]. We will illustrate that for the 2D case, but since our examples of 3D objects do not present cavities, we assume for simplicity (in 3D) that $\partial\Omega$ has a single connected component in $(\partial\Omega, \mathcal{A}_{\partial\Omega,\sqrt{d}})$. Note also that the algorithm we next propose to compute the so-called interior skeleton of Ω can be also used to compute the so-called external skeleton of the complement $\overline{\Omega}$. For presentation simplicity, we focus here on the interior skeleton.

Distance and feature transforms: Using the graph $(\Omega, \mathcal{A}_{\Omega,\sqrt{d}})$, we can propagate from $\partial\Omega$ the distance value $\mathcal{D}(\mathbf{x})$ to every interior spel $\mathbf{x} \in \Omega \setminus \partial\Omega$ by using the connectivity function

$$\psi_{edt}(\langle \mathbf{t} \rangle) = \begin{cases} 0 & \text{if } \mathbf{t} \in \partial\Omega, \text{ and} \\ +\infty & \text{otherwise,} \end{cases}$$

$$\psi_{edt}(\pi_{\mathbf{s}} \cdot \langle \mathbf{s}, \mathbf{t} \rangle) = \|\mathbf{s} - F(\mathbf{s})\|, \tag{2.10}$$

where $F(\mathbf{s}) \in \partial\Omega$ is the single-point feature transform of $\mathbf{s}$ (Eq. (2.4)), and $\pi_{\mathbf{s}} \cdot \langle \mathbf{s}, \mathbf{t} \rangle$ is the extension of the path $\pi_{\mathbf{s}}$ by the arc $(\mathbf{s}, \mathbf{t}) \in \mathcal{A}_{\Omega,\sqrt{d}}$. The Euclidean distance transform $\mathcal{D}$ thus becomes

$$\mathcal{D}(\mathbf{t} \in \Omega) = \min_{\forall \pi_{\mathbf{t}} \in \Pi(\Omega, \mathcal{A}_{\Omega,\sqrt{d}})} \{\psi_{edt}(\pi_{\mathbf{t}})\}. \tag{2.11}$$

Besides propagating the distance transform $\mathcal{D}$, the IFT method also propagates the closest point $F(\mathbf{x}) \in \partial\Omega$ to any $\mathbf{x} \in \Omega$. This yields the single-point feature transform F for all spels in Ω, which we will heavily use, as outlined next.

MGF importance: To compute ρ (Eq. (2.5)), we need the complete feature transform $\mathcal{F}$. As explained in Section 2.2.1, computing $\mathcal{F}$ is difficult, especially for discrete-grid representations. In practice, this is often replaced by computing the so-called extended single-point feature transform $\mathcal{F}_{ext}$ [39,21] defined as

$$\mathcal{F}_{ext}(\mathbf{s}) = \{F(\mathbf{t}) \in \partial\Omega \mid \mathbf{t} \in \mathcal{A}_{\Omega,1}(\mathbf{s})\}, \tag{2.12}$$

which gathers the single-point feature transforms $F(\mathbf{t})$ of all adjacent spels $\mathbf{t} \in \mathcal{A}_{\Omega,1}(\mathbf{s})$. Having $\mathcal{F}_{ext}$, we can now immediately write a simpler version of Eq. (2.5) as

$$\rho(\mathbf{s} \in \Omega \setminus \partial\Omega) = \max_{\mathbf{f}=F(\mathbf{s}),\mathbf{g}\in\mathcal{F}_{ext}(\mathbf{s})} \{GL(\mathbf{f}, \mathbf{g})\}. \tag{2.13}$$

This simplification applies to multiscale planar and surface skeletons, leading to less shortest-path length computations, but Eq. (2.5) is still important if one desires to merge 3D multiscale curve and surface skeletons because the union of geodesic paths between all pairs of feature points of a spel $\mathbf{s}$ on the curve skeleton draws in $\partial\Omega$ a closed contour, splitting $\partial\Omega$ into two parts such that the geodesic surface area between them can be used as importance $\rho(\mathbf{s})$ [39].

To evaluate Eq. (2.13), we compute the shortest-path length GL from $\mathbf{f} = F(\mathbf{s})$ to $\mathbf{g} \in \mathcal{F}_{ext}(\mathbf{s})$ on the boundary-graph $(\partial\Omega, \mathcal{A}_{\partial\Omega,\sqrt{d}})$ by using the connectivity function ψ_{geo} defined as

$$\begin{aligned}\psi_{geo}(\langle \mathbf{w} \rangle) &= \begin{cases} 0 & \text{if } \mathbf{w} = \mathbf{f}, \text{ and} \\ +\infty & \text{otherwise,} \end{cases} \\ \psi_{geo}(\pi_{\mathbf{w}} \cdot \langle \mathbf{w}, \mathbf{h} \rangle) &= \psi_{geo}(\pi_{\mathbf{w}}) + \|\mathbf{h} - \mathbf{w}\| + \|\mathbf{g} - \mathbf{h}\|,\end{aligned} \tag{2.14}$$

where the term $\|\mathbf{g} - \mathbf{h}\|$ is the A^* heuristic optimization [34] used to reach $\mathbf{g}$ faster. Summarizing, we compute the shortest-path length $GL(\mathbf{f}, \mathbf{g})$ as

$$GL(\mathbf{f}, \mathbf{g}) = \min_{\pi_{\mathbf{g}} \in \Pi(\Omega, \mathcal{A}_{\Omega,\sqrt{d}})} \{\psi_{geo}(\pi_{\mathbf{g}})\}. \tag{2.15}$$

It is important to note that the IFT propagation for ψ_{edt} can also output an *optimum-path forest* P, i.e., a map that assigns a so-called predecessor $\mathbf{s} = P(\mathbf{t}) \in \Omega$ to every spel $\mathbf{t} \in \Omega \setminus \partial\Omega$, and a marker $P(\mathbf{t}) = nil \notin \Omega$ to spels $\mathbf{t} \in \partial\Omega$, respectively [19]. This defines a vector field $\mathbf{v}(\mathbf{t}) = \mathbf{t} - P(\mathbf{t})$ for every interior spel $\mathbf{t} \in \Omega \setminus \partial\Omega$. The forest P provides the direction of $\mathbf{v}$ for nonskeletal spels. The skeleton S is contained in the set of P's leaves (yellow lines in Fig. 2.3). The vector field $\mathbf{v}$ describes how all information computed by the IFT—that is, $\mathcal{D}$, F, and ρ—is iteratively propagated, or *advected*, from $\partial\Omega$ to all spels in the shape's interior $\Omega \setminus \partial\Omega$. This fundamentally links the MGF importance model and the advection importance model. As explained in Sections 2.2 and 2.3.1, these two importance models are typically used independently in the literature. The only work that we are aware of where an MGF model is linked with an advection model is [9]. However, our work here stands apart from [9] in terms of algorithmic model, and also by the fact that, for the 3D case, we compute multiscale *surface* skeletons (and not curve skeletons, whereas [9] approaches precisely the opposite).

Summarizing the above: To compute ρ (Eq. (2.13)), we need to compute the single-point distance transform F and the shortest-path length between feature points in the extended feature transform $\mathcal{F}_{ext}$. The algorithms for both these operations are described in Sections 2.3.2.1 and 2.3.3, respectively.

2.3.2.1 Single-Point Feature Transform

The single-point feature transform F is computed by the same algorithm used for computing the Euclidean distance transform $\mathcal{D}$, but returns F rather than $\mathcal{D}$. The full algorithm we use for computing F is listed below. It also returns $\partial\Omega$, which we next need to compute shortest-paths between feature points, and a component label map $L_C: \mathbf{s} \in \Omega \rightarrow \lambda(\mathbf{s}) \in \{1, 2, \ldots, c\}$ that assigns a subsequent integer number to each component of $\partial\Omega$ in $(\partial\Omega, \mathcal{A}_{\partial\Omega,\sqrt{d}})$ and its closest spels in Ω. The map L_C is used for SKIZ computation in 2D. Indeed, the component label propagation to every spel $\mathbf{s} \in \Omega \setminus \partial\Omega$ is not needed, but it helps to illustrate the location of the SKIZ (Section 2.3.3).

Algorithm 1 (SINGLE-POINT FEATURE TRANSFORM).

INPUT: An object Ω in dimension d represented on a uniform $\mathbb{Z}^d$ grid.
OUTPUT: The single-point feature transform F, object boundary $\partial\Omega$, and component label map L_C.
AUXILIARY: Priority queue Q, distance transform $\mathcal{D}$, and variable $tmp \in \mathbb{R}$.

1. *Compute* $\partial\Omega$ *of* Ω *by Eq.* (2.7).
2. **For each** $\mathbf{s} \in \Omega \setminus \partial\Omega$, $\mathcal{D}(\mathbf{s}) \leftarrow +\infty$.
3. **For each** $\mathbf{s} \in \partial\Omega$, **do**
4. $\quad$ $\mathcal{D}(\mathbf{s}) \leftarrow 0$; $F(\mathbf{s}) \leftarrow \mathbf{s}$; $L_C(\mathbf{s}) \leftarrow \lambda(\mathbf{s}) \in \{1, 2, \ldots, c\}$, *according to*
5. $\quad$ *its component in* $(\partial\Omega, \mathcal{A}_{\partial\Omega,\sqrt{d}})$; *and insert* $\mathbf{s}$ *in* Q.
6. **While** $Q \neq \emptyset$, **do**
7. $\quad$ *Remove* $\mathbf{s}$ *from* Q, *where* $\mathcal{D}(\mathbf{s})$ *is minimal over* Q.
8. $\quad$ **For each** $\mathbf{t} \in \mathcal{A}_{\Omega,\sqrt{d}}(\mathbf{s})$ *such that* $\mathcal{D}(\mathbf{t}) > \mathcal{D}(\mathbf{s})$, **do**
9. $\quad\quad$ $tmp \leftarrow \|\mathbf{t} - F(\mathbf{s})\|^2$.
10. $\quad\quad$ **If** $tmp < \mathcal{D}(\mathbf{t})$, **then**
11. $\quad\quad\quad$ $\mathcal{D}(\mathbf{t}) \leftarrow tmp$; $F(\mathbf{t}) \leftarrow F(\mathbf{s})$; $L_C(\mathbf{t}) \leftarrow L_C(\mathbf{s})$.
12. $\quad\quad\quad$ **If** $\mathcal{D}(\mathbf{t}) \neq +\infty$, **then**
13. $\quad\quad\quad\quad$ *Update position of* $\mathbf{t}$ *in* Q.
14. $\quad\quad\quad$ **Else**
15. $\quad\quad\quad\quad$ *Insert* $\mathbf{t}$ *in* Q.
16. **Return** $(F, \partial\Omega, L_C)$

Lines 1–5 essentially extract the object boundary $\partial\Omega$, initialize the trivial-path values of Eq. (2.10) in $\mathcal{D}(\mathbf{s})$ for all $\mathbf{s} \in \Omega$, set $F(\mathbf{s})$ for $\mathbf{s} \in \partial\Omega$, assign a distinct integer to each component of $\partial\Omega$ in $(\partial\Omega, \mathcal{A}_{\partial\Omega,\sqrt{d}})$, and insert $\mathbf{s} \in \partial\Omega$ in the priority queue Q. The main loop in Lines 6–15 propagates to every spel $\mathbf{t} \in \Omega \setminus \partial\Omega$ its single-point feature $F(\mathbf{t}) \in \partial\Omega$ in a nondecreasing order of distance values $\mathcal{D}(\mathbf{t})$ between $\mathbf{t}$ and $\partial\Omega$. In Line 7, when a spel $\mathbf{s}$ is removed from Q, $\mathcal{D}(\mathbf{s})$ stores the closest squared distance between $\mathbf{s}$ and $\partial\Omega$, $F(\mathbf{s})$ stores its single-point feature, and $L_c(\mathbf{s})$ indicates its closest component in $(\partial\Omega, \mathcal{A}_{\partial\Omega,\sqrt{d}})$. The loop in Lines 8–15 evaluates if $\mathbf{s}$ can offer a lower squared distance value $\|\mathbf{t} - F(\mathbf{s})\|^2$ (value of an extended path $\pi_{\mathbf{s}} \cdot \langle \mathbf{s}, \mathbf{t} \rangle$ in Eq. (2.10)) to the current value assigned to an adjacent spel $\mathbf{t}$ in $\mathcal{D}(\mathbf{t})$ (Lines 9–10). If this is the case, then Line 11 updates distance, single-point feature of $\mathbf{t}$ with respect to $\partial\Omega$, its closest component in $(\partial\Omega, \mathcal{A}_{\partial\Omega,\sqrt{d}})$, and Lines 12–15 update the status of $\mathbf{t}$ in Q.

Note that the use of the squared Euclidean distance $\|\mathbf{t} - F(\mathbf{s})\|^2$ in Line 9 allows us to implement Q by bucket sorting [20] since all distances are integers on a pixel/voxel grid representation. As such, Algorithm 1 has average complexity $O(|\Omega|)$.

2.3.2.2 Shortest-Path Length Computation

As mentioned earlier in Section 2.3.1, computing the multiscale regularization metric ρ for $d = 3$ heavily depends, cost-wise, on the rapid computation of shortest-path lengths on $\partial\Omega$ between single-point features. Accelerating these shortest-path computations is key to accelerating multiscale 3D skeletonization. To achieve this, we maintain, for each $\mathbf{f} \in \partial\Omega$, a set $\mathcal{C}(\mathbf{f}) = \{\mathbf{s} \in \Omega \setminus \partial\Omega | F(\mathbf{s}) = \mathbf{f}\}$. We use $\mathcal{C}$ to *incrementally* compute all shortest-path lengths between $\mathbf{f}$ and other single-point features $\mathbf{g} \neq \mathbf{f}, \mathbf{g} \in \partial\Omega$ as follows: The IFT algorithm returns $GL(\mathbf{f}, \mathbf{g})$ whenever $\mathbf{g}$ is reached; for any $\mathbf{h} \in \partial\Omega$ for which $GL(\mathbf{f}, \mathbf{h}) \leq GL(\mathbf{f}, \mathbf{g})$, we return immediately the already computed path length $GL(\mathbf{f}, \mathbf{h})$ and thus only continue computation for points $\mathbf{h} \in \partial\Omega$ where $GL(\mathbf{f}, \mathbf{h}) \geq GL(\mathbf{f}, \mathbf{g})$. To do the above, we store the computed shortest-path lengths $GL(\mathbf{f}, \mathbf{g})$ (Eq. (2.15)) between a given feature point $\mathbf{f} \in \partial\Omega$ and all other spels $\mathbf{g} \in \partial\Omega$ into a map $L_{\mathbf{f}} : \partial\Omega \rightarrow \mathbb{R}^+$, $L_{\mathbf{f}}(\mathbf{g}) = GL(\mathbf{f}, \mathbf{g})$. The computation of the shortest-path length between a given spel $\mathbf{f} \in \partial\Omega$ and all other spels $\mathbf{g} \in \partial\Omega$ is presented in Algorithm 4, Section 2.3.3.

For $d = 2$, the problem is trivial since $\partial\Omega$ may consist of closed *one-dimensional* contours: For an arbitrary spel $\mathbf{f}_0$ in each contour $C \subset \partial\Omega$, we first compute in $L_{\mathbf{f}}(\mathbf{g})$ the path length from $\mathbf{f}_0$ to each spel $\mathbf{g} \in C$ while circumscribing C from $\mathbf{f}_0$ in a single orientation (clockwise or anticlockwise). Now, for any two spels $\mathbf{f}, \mathbf{g} \in C$, let $\Delta(\mathbf{f}, \mathbf{g}) = |L_{\mathbf{f}}(\mathbf{g}) - L_{\mathbf{f}}(\mathbf{f})|$. The geodesic length $GL(\mathbf{f}, \mathbf{g})$ between $\mathbf{f}$ and $\mathbf{g}$ is then given by

$$GL(\mathbf{f}, \mathbf{g}) = \min\{|C| - \Delta(\mathbf{f}, \mathbf{g}), \Delta(\mathbf{f}, \mathbf{g})\}, \tag{2.16}$$

where $|C|$ is the perimeter length of the contour C. However, for the purpose of finding one-spel-wide skeletons by Eq. (2.13), we can drop the absolute difference and redefine $\Delta(\mathbf{f}, \mathbf{g}) = L_\mathbf{f}(\mathbf{g}) - L_\mathbf{f}(\mathbf{f})$ for $\mathbf{f} = F(\mathbf{s})$ and $\mathbf{g} \in \mathcal{F}_{ext}(\mathbf{s})$ on the boundary graph $(\partial\Omega, \mathcal{A}_{\partial\Omega,\sqrt{d}})$.

The next section presents the IFT-based multiscale skeletonization algorithms for $d = 2$ and $d = 3$, respectively.

2.3.3 MULTISCALE SKELETONIZATION—PUTTING IT ALL TOGETHER

The complete 2D multiscale skeletonization algorithm is presented below.

Algorithm 2 (MULTISCALE SKELETON COMPUTATION IN 2D).

INPUT:	An object Ω in dimension $d = 2$.
OUTPUT:	Multiscale skeleton importance ρ.
AUXILIARY:	Boundary $\partial\Omega$ with perimeter-length $\lvert\partial\Omega\rvert$; path length map $L_\mathbf{f}$; single-point feature transform F; component label map L_c; variable $tmp \in \mathbb{R}$.

1. $(F, \partial\Omega, L_c) \leftarrow$ Algorithm 1(Ω).
2. **For each** *component* $C \in (\partial\Omega, \mathcal{A}_{\partial\Omega,\sqrt{d}})$, **do**
3.   *Select an arbitrary point* $\mathbf{f}_0 \in C$.
4.   **For each** $\mathbf{g} \in C$ *found by circumscribing* C *from* $\mathbf{f}_0$, **do**
5. $L_\mathbf{f}(\mathbf{g}) \leftarrow$ *path length from* $\mathbf{f}_0$ *to* $\mathbf{g}$ *on* C.
6. **For each** $\mathbf{s} \in \Omega \setminus \partial\Omega$, **do**
7. $\rho(\mathbf{s}) \leftarrow 0$.
8.   *Compute* $\mathcal{F}_{ext}(\mathbf{s})$ *by Eq.* (2.12).
9.   **For each** $\mathbf{g} \in \mathcal{F}_{ext}(\mathbf{s})$, **do**
10.   **If** $L_c(\mathbf{g}) > L_c(F(\mathbf{s}))$, **then** *set* $\rho(\mathbf{s}) \leftarrow +\infty$ *and return to 6.*
11. $tmp \leftarrow L_\mathbf{f}(\mathbf{g}) - L_\mathbf{f}(F(\mathbf{s}))$.
12.   **If** $tmp > \lvert\partial\Omega\rvert - tmp$, **then** $tmp \leftarrow \lvert\partial\Omega\rvert - tmp$.
13.   **If** $tmp > \rho(\mathbf{s})$, **then** $\rho(\mathbf{s}) \leftarrow tmp$.
14. **Return** ρ.

Line 1 finds the object boundary $\partial\Omega$, the single-point feature transform F, and the component label map L_c by Algorithm 1. The remaining lines follow the procedure described in Section 2.3.2.2 for $d = 2$. Lines 2–5 compute in $L_\mathbf{f}(\mathbf{g})$ the path length from an arbitrary spel $\mathbf{f}_0 \in C$, selected for each component $C \in (\partial\Omega, \mathcal{A}_{\partial\Omega,\sqrt{d}})$, to every spel $\mathbf{g} \in C$ while circumscribing the contour C. The main loop in Lines 6–13 computes for each spel $\mathbf{s} \in \Omega \setminus \partial\Omega$ the shortest-path length by Eq. (2.16) between point feature $F(\mathbf{s})$ and each $\mathbf{g} \in \mathcal{F}_{ext}(\mathbf{s})$ (Lines 11–12), and use them to update the MGF $\rho(\mathbf{s})$ in Line 13, as proposed in Eq. (2.13). The SKIZ is detected whenever a point feature $\mathbf{g} \in \mathcal{F}_{ext}(\mathbf{s})$ comes from a distinct component than $F(\mathbf{s})$. For one-pixel-wide connected SKIZ, $\mathbf{s}$ is selected as belonging to the SKIZ whenever $L_c(\mathbf{g}) > L_c(F(\mathbf{s}))$. In this case, $\rho(\mathbf{s})$ is set to the maximum possible value, and the algorithm returns to Line 6 (see example in Fig. 2.4). It should be clear that Algorithm 2 has complexity $O(\lvert\Omega\rvert)$.

A comment regarding the multiscale skeleton homotopy with the input shape is needed here. As visible from Fig. 2.4, the importance ρ has now a different varia-

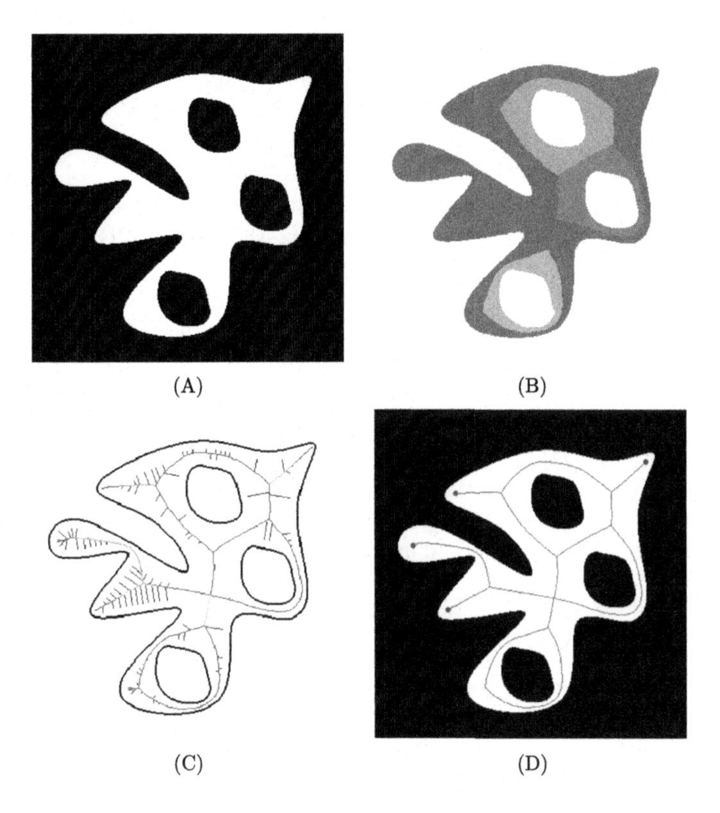

FIGURE 2.4

(A) A 2D object Ω with three holes. (B) The component label map L_c as computed by Algorithm 1. (C) The color-coded multiscale skeleton ρ of Ω, using the rainbow color map. The SKIZ is shown in red since its spels are assigned to the maximum importance in ρ. (D) A connected one-pixel-wide skeleton for a given scale of ρ, with its terminal points shown in blue.

tion across $\mathcal{S}$ than for genus-0 shapes (see, e.g., Fig. 2.5). Clearly, for sufficiently high thresholds, the skeleton in Fig. 2.4 will get disconnected, i.e., the three loops surrounding the holes in Ω will get separated from the central skeletal branch. Note that this *also* happens when using all other definitions of the same importance metric proposed by [35,17,53]. The root of the problem is that the collapsed-boundary importance metric used in all above works (and ours too) makes sense, in a multiscale way, only for genus-0 shapes whose skeleton is a tree. In other words, we know how to gradually simplify a tree (by removing its leafs), but we do not know how to do the same for a graph having loops. Issues here are how to assign an importance value to a loop (based on which geometric and/or topological criterion); and should the simplification of a loop remove it all at once, or should it allow its gradual disconnection. All these are (valid) questions that, however, go beyond our scope here.

For completeness, we note that disconnection of 2D nongenus-0 figures during simplification can be easily achieved, if this is a key issue. To do this, we can simply postprocess the computed skeleton $\mathcal{S}$: Trace all shortest paths in $\mathcal{S}$ linking each pair of loop components in $\mathcal{S}$ and assign spels along a value $+\infty$. This will only allow next the multiscale simplification of the tree parts of $\mathcal{S}$.

Essentially, Algorithm 2 is identical to the methods presented in [35,17,53]. As such, its main added-value is of theoretical nature—showing that 2D multiscale skeletonization can be easily cast in the IFT framework.

The situation in 3D ($d = 3$) is however very different: Here, our proposed multiscale skeletonization is *both* conceptually similar to the 2D case *and* very computationally efficient. This is in stark contrast with existing methods that are either similar in 2D and 3D but quite complex and do not provide an *explicit* definition of the regularization metric [27], or with existing methods that provide strongly related metrics in 2D [53,17,35] and 3D [14,39] but show a massive performance drop in the 3D case. The algorithm listed next shows our 3D multiscale skeletonization method. In contrast to the 2D proposal (Algorithm 2), we now use the efficient incremental shortest-path computation proposed in Algorithm 4.

Algorithm 3 (MULTISCALE SKELETON COMPUTATION IN 3D).

INPUT: An object Ω in dimension $d = 3$.
OUTPUT: Multiscale skeleton importance ρ.
AUXILIARY: Priority queue Q; list $\mathcal{V}$ of boundary points that have been inserted in Q; A^* path-cost map G; boundary $\partial\Omega$; sets $\mathcal{C}(\mathbf{s})$, $\forall \mathbf{s} \in \partial\Omega$; shortest-path length map $L_\mathbf{f}$; single-point feature transform F.

```
1.  (F, ∂Ω) ← Algorithm 1(Ω).
2.  For each s ∈ Ω \ ∂Ω, do
3.      └ Insert s in C(F(s)).
4.  For each f ∈ ∂Ω, do
5.      └ L_f(f) ← +∞; G(f) ← +∞.
6.  Q ← ∅; V ← ∅.
7.  For each f ∈ ∂Ω, do
8.      | L_f(f) ← 0; G(f) ← 0; insert f in Q; insert f in V.
9.      | While there exists s ∈ C(f), do
10.     |     | Remove s from C(f).
11.     |     | ρ(s) ← 0; compute F_ext(s) by Eq. (2.12).
12.     |     | For each g ∈ F_ext(s), do
13.     |     |     | L_f(g) ← Algorithm 4(∂Ω, g, Q, V, G, L_f).
14.     |     └     └ If L_f(g) > ρ(s), then ρ(s) ← L_f(g).
15.     | For each g ∈ V
16.     |     └ L_f(g) ← +∞; G(g) ← +∞.
17.     └ Q ← ∅; V ← ∅.
18. Return ρ.
```

Line 1 finds the object boundary $\partial\Omega$ and the single-point feature transform F by Algorithm 1. We are not interested in L_c since Algorithm 3 assumes that $\partial\Omega$ is a single surface. Lines 2–3 compute the sets $\mathcal{C}(\mathbf{f}) = \{\mathbf{s} \in \Omega \setminus \partial\Omega | F(\mathbf{s}) = \mathbf{f}\}$ that speed up shortest-path length computations, as described in Section 2.3.2.2 for $d = 3$. Note that G stores the A^* path costs, whereas $L_\mathbf{f}$ stores the desired path lengths, in Algorithm 4. The shortest-path lengths from each boundary point $f \in \partial\Omega$ to other boundary points $g \in \partial\Omega$ are *incrementally* computed in Algorithm 4 (Eq. (2.15)). Therefore, the trivial-path value initialization of ψ_{geo} (Eq. (2.14)) must be performed outside Algorithm 4 (Lines 4–5 before the main loop of Lines 7–17, and Lines 15–16 and 8 to restart computation for every initial boundary point $f \in \partial\Omega$). Lines 4–5 execute for the entire boundary $\partial\Omega$, so the purpose of set $\mathcal{V}$ is to revisit only the boundary points used in Algorithm 4, when reinitializing $L_\mathbf{f}$ and G. Line 8 initializes the priority queue Q and sets $\mathcal{V}$ with one initial boundary point f for Algorithm 4. The loop of Lines 9–14 computes the 3D MGF $\rho(\mathbf{s})$ by Eq. (2.13) for each spel $\mathbf{s}$ whose the single-point feature is the current point $\mathbf{f} \in \partial\Omega$. Line 10 removes a spel $\mathbf{s}$ from $\mathcal{C}(\mathbf{f})$, Line 11 initializes $\rho(\mathbf{s})$ and finds $\mathcal{F}_{ext}(\mathbf{s})$ by Eq. (2.12). For each point feature $\mathbf{g} \in \mathcal{F}_{ext}(\mathbf{s})$, Line 13 finds $GL(\mathbf{f}, \mathbf{g})$ (Eq. (2.15)) and stores it in $L_\mathbf{f}(\mathbf{g})$, and Line 14 updates $\rho(\mathbf{s})$ according to Eq. (2.13). Algorithm 4 is presented next.

Algorithm 4 (INCREMENTAL SHORTEST-PATH LENGTH COMPUTATION).

INPUT: Boundary $\partial\Omega$; terminal node $\mathbf{g} \in \partial\Omega$; priority queue Q; boundary points $\mathcal{V}$ that have been inserted in Q; A^* cost map G; shortest-path-length map $L_\mathbf{f}$.

OUTPUT: Shortest-path length $L_\mathbf{f}(\mathbf{g})$ at the terminal node with respect to the current starting node $\mathbf{f}$ chosen in Algorithm 3.

AUXILIARY: Variable $tmp \in \mathbb{R}$.

1. **If** $L_\mathbf{f}(\mathbf{g}) \neq +\infty$, **then** *return* $L_\mathbf{f}(\mathbf{g})$.
2. **While** $Q \neq \emptyset$ **do**
3. *Remove* $\mathbf{w}$ *from* Q, *where* $G(\mathbf{w})$ *is minimal over* Q.
4. **If** $\mathbf{w} = \mathbf{g}$, **then** *return* $L_\mathbf{f}(\mathbf{g})$.
5. **For each** $\mathbf{h} \in \mathcal{A}_{\partial\Omega,\sqrt{d}}(\mathbf{w})$ *such that* $G(\mathbf{h}) > G(\mathbf{w})$, **do**
6. $tmp \leftarrow L_\mathbf{f}(\mathbf{w}) + \|\mathbf{h} - \mathbf{w}\| + \|\mathbf{g} - \mathbf{h}\|$.
7. **If** $tmp < G(\mathbf{h})$, **then**
8. $G(\mathbf{h}) \leftarrow tmp$; $L_\mathbf{f}(\mathbf{h}) \leftarrow L_\mathbf{f}(\mathbf{w}) + \|\mathbf{h} - \mathbf{w}\|$.
9. **If** $G(\mathbf{h}) \neq +\infty$, **then**
10. *Update position of* $\mathbf{h}$ *in* Q.
11. **Else**
12. *Insert* $\mathbf{h}$ *in* Q *and in* $\mathcal{V}$.

Line 1 halts computation whenever the shortest-path length from $\mathbf{f}$ to $\mathbf{g}$ on $\partial\Omega$ has already been computed in a previous execution of Algorithm 4. The main loop of Lines 2–12 computes the shortest-path length to every boundary point $\mathbf{w} \in \partial\Omega$ in a nondecreasing order of the cost values in G until it finds the terminal point $\mathbf{g}$ in Line 4. In Line 3, when a point $\mathbf{w} \in \partial\Omega$ is removed from Q, $G(\mathbf{w})$ stores the minimum A^* path cost, and $L_\mathbf{f}(\mathbf{w})$ stores the shortest-path length from $\mathbf{f}$ to $\mathbf{w}$ on $\partial\Omega$, which may be used for early termination in Line 1 in a next execution of Algorithm 4. The loop in Lines 5–12 evaluates if $\mathbf{w}$ can offer a lower path cost $L_{\mathbf{f}}(\mathbf{w}) + \|\mathbf{h} - \mathbf{w}\| + \|\mathbf{g} - \mathbf{h}\|$ (value of an extended path $\pi_{\mathbf{w}} \cdot \langle \mathbf{w}, \mathbf{h} \rangle$ in Eq. (2.14)) to the current value assigned to an adjacent point $\mathbf{h} \in \partial\Omega$ (Lines 6–7). If this is the case, then Lines 8–12 update $G(\mathbf{h})$, $L_f(\mathbf{h})$, and the status of $\mathbf{h}$ in Q, accordingly.

The complexity of Algorithm 3 would be $O(|\Omega \setminus \partial\Omega||\partial\Omega| \log |\partial\Omega|)$ with a naive implementation of shortest-path length computation. In practice, however, Algorithm 4 finishes in Lines 1 or 4 much earlier than visiting all boundary points. This makes a considerable reduction in the processing time of Algorithm 3, as we will see next.

2.4 COMPARATIVE ANALYSIS

We next present and discuss our results as compared to other state-of-the-art multiscale skeletonization methods.

2.4.1 2D MEDIAL AXES

We first consider medial axes of 2D objects. Here, we compare our IFT method with its two main competitors, the augmented fast-marching method (AFMM) [53] (basically identical to [35,17]) and the more recent advection-based method (AS) in [27]. We compared the above three methods on a set of over 30 2D shapes, taken from relevant papers in the field [44,5,35,53]. Fig. 2.5 shows three such shapes with their progressively simplified skeletons. It is clearly visible that all three methods yield nearly identical skeletons, both in terms of location *and* importance values. In other words, our IFT-based method can compute multiscale 2D skeletons, which are nearly identical to those computed by existing methods. As visible, our method handles complex, noisy, and variable-scale shapes with the same ease as the other two analyzed methods.

2.4.2 3D MEDIAL SURFACES

2.4.2.1 *Global Comparison*

For 3D shapes, we compared our IFT-based methods with two classes of competing techniques. First, and most interesting, we considered all techniques that we are aware of that produce multiscale skeletons, in the sense described in Section 2.2.2. These are the multiscale MGF-based method in [39] (MS), the advection-based method in [27] (AS), and the multiscale ball-shrinking method that implements the MGF metric in [39] for mesh models [26] (MBS). Secondly, to illustrate the advantage of multiscale regularization, we compare our method with three local regularization nonmultiscale methods: Hamilton–Jacobi skeletons (HJ [44]), the Integer Medial Axis (IMA [25]), and Iterative Thinning Process (ITP [28]). We have chosen these methods since they are well known in the 3D skeletonization arena, are relatively efficient, produce good-quality 3D surface skeletons, and have public implementations.

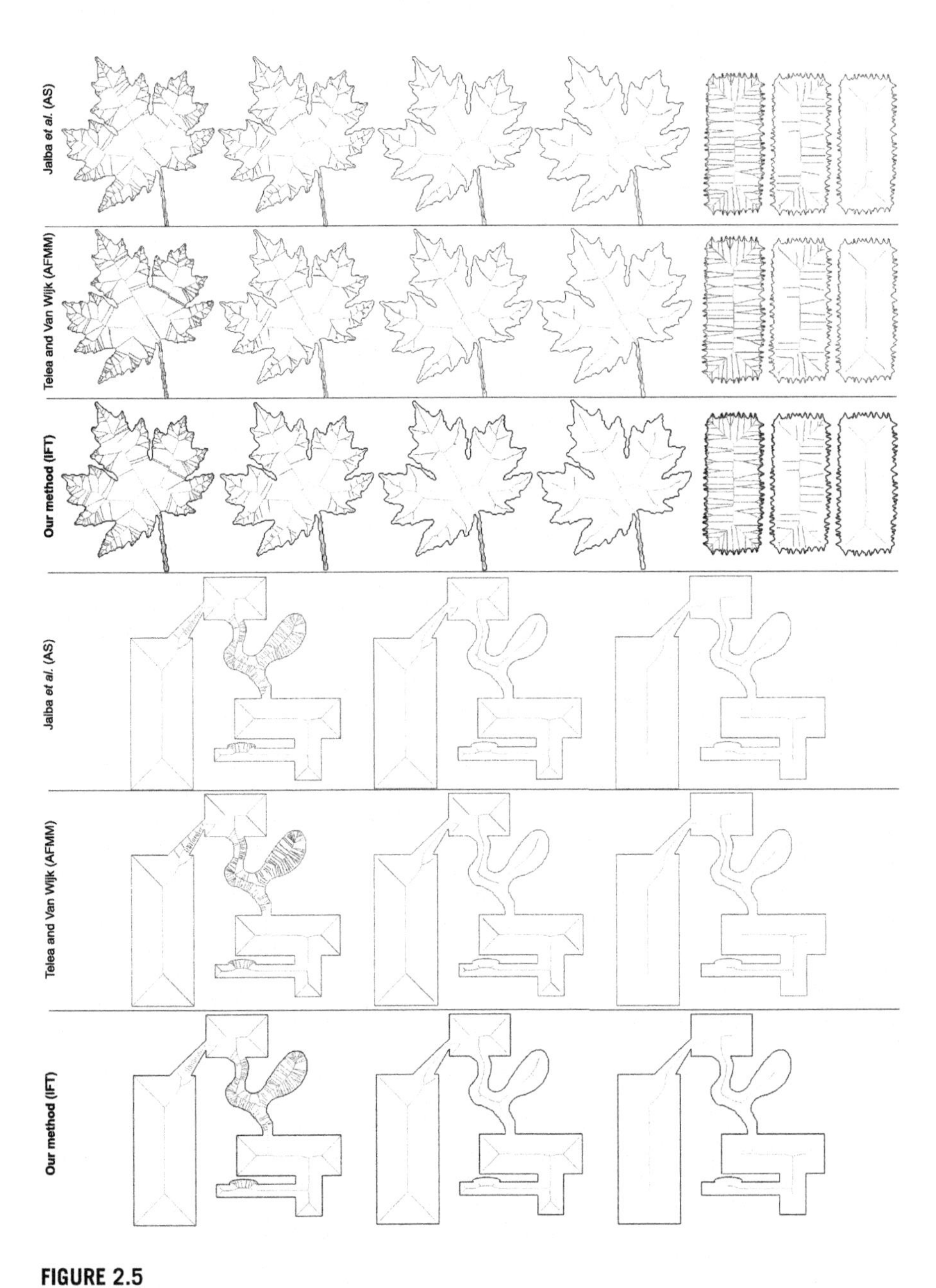

FIGURE 2.5

Our multiscale 2D skeletons compared with AFMM and AS.

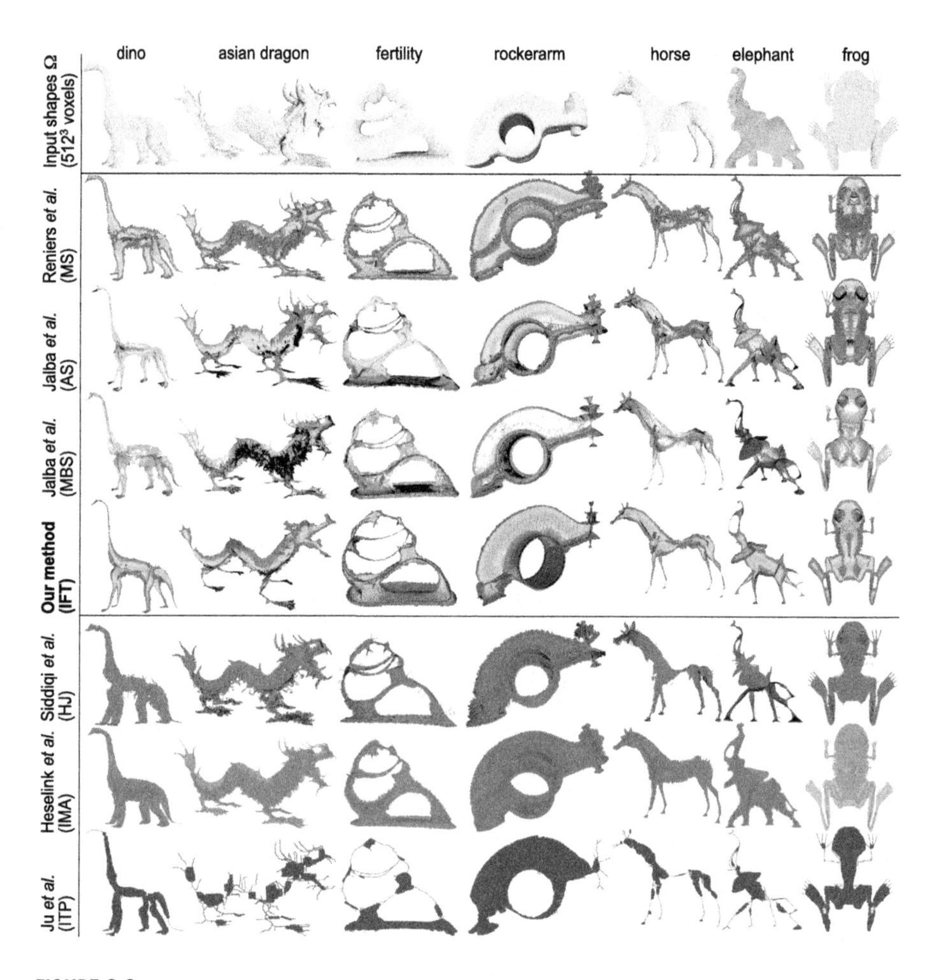

FIGURE 2.6

Global comparison of our 3D skeletonization method (IFT) with three multiscale skeletonization methods (MS, AS, MBS) and with three additional nonmultiscale methods (HJ, IMA, ITP). See Section 2.4.2.1.

Fig. 2.6 shows the results of the above-mentioned comparisons for seven shapes, processed by seven skeletonization methods. The multiscale skeletons computed by MS and AS are color-coded to reflect the importance metric, using a rainbow colormap, just as in Fig. 2.5. Multiscale skeletons computed by MBS are not importance color-coded in Fig. 2.6; the MBS importance is discussed separately in more detail in Section 2.4.2.2.

Quality-wise, our 3D surface skeletons are voxel-thin, centered within the shape (within the margin allowed by the voxel resolution), and have the same number of connected components and loops as the input shape by construction. These are key properties required by any skeletonization method [8]. For example, IMA yields centered and voxel-thin skeletons, but these can get disconnected when simplified too much since this method essentially uses the local angle-and-distance-based simplification criterion of the θ-SMA method of Foskey et al. [22]. Note that the disconnection implied above is not due to the existence of loops in the skeleton: IMA can easily disconnect also skeletons of genus-0 shapes. This does not happen with our method. Conversely, HJ yields connected skeletons, but for this, the method uses a thinning process ordered by the divergence of the distance transform gradient, which must explicitly checked to preserve homotopy [44]. The ITP method computes skeletons that are voxel-thin and homotopic to the input shape but not well centered in the shape, as seen by the various zig-zag branches of the *dragon* model (Fig. 2.6, bottom row).

The sensitivity of the skeletons shown in Fig. 2.6 to noise or small-scale details on the input shape surface varies quite a lot. As known, local regularization methods such as HJ, IMA, and ITP are more noise-sensitive than global regularization methods such as MS, AS, and MBS [50]. Our method (IFT) falls in the latter class of global methods, so it is less sensitive to noise and produces smoother surface skeletons, as visible in Fig. 2.6, fourth row from bottom.

2.4.2.2 Detailed Comparison

To gain more insight, we next compare our IFT method with several methods we are aware of that compute *multiscale* 3D surface skeletons (AS, MS, and MBS). The first two methods (AS, MS) are voxel-based, whereas the last one (MBS) is mesh-based. Figs. 2.7 and 2.8 show results for a selected set of shapes. Since all the above-mentioned methods produce multiscale skeletons, we regularized these by removing very low importance (spurious) skeleton points to yield comparably simplified skeletons. Several observations can be made when studying the compared methods as follows.

Regularization: Figs. 2.7 and 2.8 show that IFT delivers 3D surface skeletons that are, geometrically speaking, very similar to the ones produced by AS, MS, and MBS. This, in itself, is a good indication of quality of IFT. Indeed, surface-skeletonization methods should deliver similar results, given that they all aim to approximate the *same* surface skeleton definition (Eq. (2.2)). Secondly, we see that the IFT delivers the same degree of small-scale noise removal to create smooth and clean skeletal manifolds as AS, MS, and MBS, so it can be used for robust skeleton regularization. The IFT regularization is as easy to use as the one proposed by the other methods, the setting of a single importance thresholding parameter τ. Note that this is far simpler than the regularization proposed by local methods, e.g., HJ, IMA, or ITP, which require the careful setting of one or several parameters to obtain comparable results.

A more subtle insight regards the gradient of the importance metric ρ from the skeleton boundary $\partial\mathcal{S}$ to its center, visible in Figs. 2.7 and 2.8 in terms of the blue-to-red color change. All tested methods (IFT, AS, MS, MBF) yield a ρ that increases monotonically from $\partial\mathcal{S}$ to the center of $\mathcal{S}$. Separately, we see that ρ for IFT, MBS,

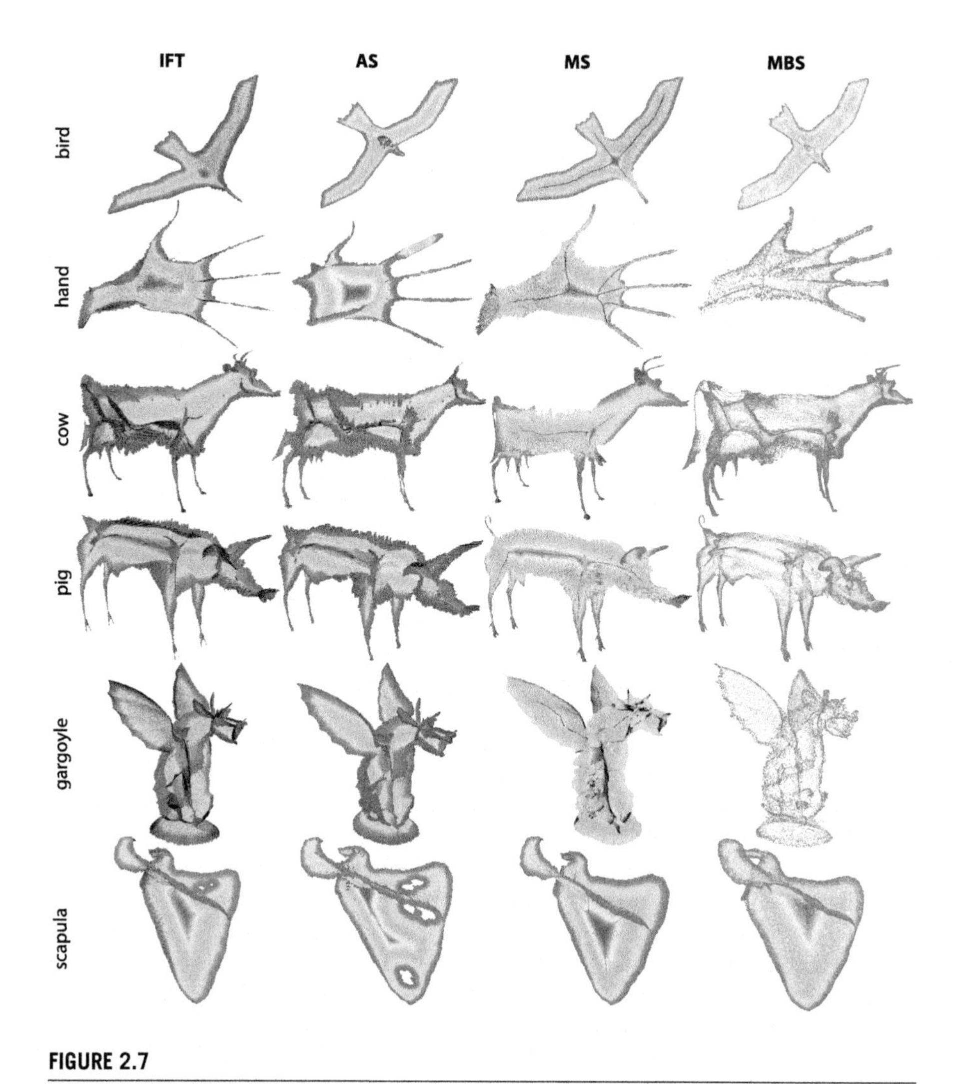

FIGURE 2.7

Detail comparison of 3D surface skeletons computed by our method (IFT) and other multiscale methods (MS, AS, and MBS). See Section 2.4.2.2.

and MS is not just increasing from $\partial\mathcal{S}$ to the center of $\mathcal{S}$, but has a very similar gradient. This implies that our method (IFT) delivers an importance metric ρ that is very similar to the ones delivered by MS and MBS. Since MS and MBS compute the medial geodesic function (MGF) metric, it follows that IFT also computes a very similar metric. This is indeed the expected outcome given the IFT algorithm (see Section 2.3.3). In contrast, the gradient of ρ delivered by AS is quite different. This is explained by the fact that AS is the only multiscale skeletonization method in the studied set that does not explicitly use the MGF metric.

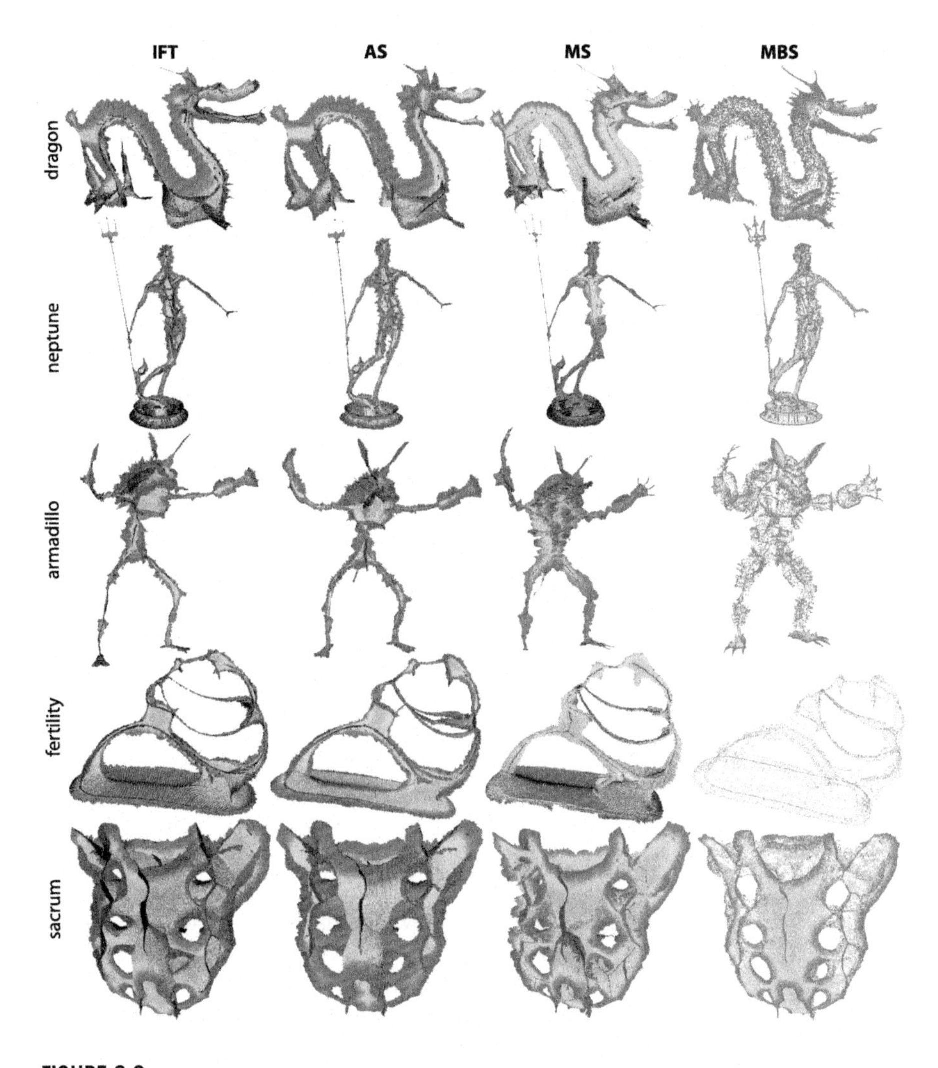

FIGURE 2.8

Additional examples of 3D multiscale skeletons computed by our method (IFT) and other multiscale methods (MS, AS, and MBS). See Section 2.4.2.2.

Connectivity: IFT, AS, and MS deliver a compact surface skeleton, whereas MBS delivers only a disconnected point cloud. This makes IFT (and AS and MS) more interesting than MBS for practical applications where one requires a compact surface skeleton. Indeed, point-cloud skeletons require complex post-processing methods for reconstructing a compact representation [26,29]. Voxel skeletons do not have this problem.

Table 2.1 Timings (seconds) for the compared surface skeletonization methods

Dataset	$\|\Omega\|$	MS [39]	AS [27]	IFT	$\|\partial\Omega\|^{mesh}$	MBS [26]
Bird	445,690	64.21	13.86	8.64	46,866	18.69
Hand	776,869	62.94	2.07	4.36	49,374	15.5
Cow	6,343,478	177.80	39.86	17.73	181,823	96.54
Pig	5,496,145	181.95	34.84	23.89	225,282	142.02
Gargoyle	6,614,154	566.52	25.66	79.43	25,002	7.54
Scapula	2,394,694	1717.37	29.99	609.33	116,930	102.57
Dragon	7,017,452	322.81	39.3	32.86	100,250	49.01
Neptune	2,870,546	322.75	47.25	68.72	28,052	5.85
Armadillo	1,854,858	45.43	7.2	4.25	172,952	104.65
Fertility	1,264,132	99.62	6.15	8.46	24,994	6.15
Sacrum	12,637,931	2015.59	39.83	417.54	204,710	213.49

Scalability: We implemented all tested methods in C++ and ran them on an Intel 3.5 GHz 8-core 32 MB RAM PC. The methods MBS, AS, and MS use CPU multithreading parallelization, as described in the respective papers. No GPU parallelization was used for MBS. Our method (IFT) is purely serial. Table 2.1 shows the timings for the compared methods for the shapes depicted in Figs. 2.7 and 2.8. Column $|\Omega|$ gives the number of foreground voxels of the tested models with MS, AS, and IFT. For MBS, the comparable metric, the number of sample points of the input mesh, is given in column $|\partial\Omega|^{mesh}$.

When testing scalability on large voxel volumes, we found that the MS implementation from [39] encountered problems: For the *scapula* shape (Fig. 2.7, bottom row), MS could not handle the 512^3-voxel resolution of our model, so we reduced the resolution to 370^3. At this resolution, the shape shows visible holes due to the very thin wall thickness (a few voxels). In contrast, IFT and AS (which are both voxel-based methods) could handle 512^3-voxel volumes without problems.

Performance-wise, Table 2.1 shows that IFT is roughly 3 to 10 times faster than MS, which is the only voxel-based method that implements the same MGF importance metric. This is an important result, as it tells us that the IFT algorithm produces significant speed-ups for the geodesic length evaluation, which was one of its main goals. Compared to AS, IFT is faster on some models but considerably slower on the *sacrum* and *scapula* models. This is explained by the fact that the complexity of AS is roughly $O(K|\partial\Omega|\log|\partial\Omega|)$, where K is $\max_{\mathbf{x}\in\Omega}\mathcal{D}(\mathbf{x})$, that is, the shape thickness. In contrast, the complexity of MS is roughly $O(L|\partial\Omega|)$, where K is the average geodesic-path length between two feature points on $\partial\Omega$. For large and locally tubular shapes, such as *cow* or *pig*, IFT is thus faster. For relatively thin and large-surface shapes, such as *scapula* and *sacrum*, the geodesic computation cost becomes very high, so IFT is slower than AS. However, as outlined earlier, this extra price of IFT delivers a higher-quality regularization in terms of smoothness of the importance metric. We note a similar effect when comparing IFT with MBS: for locally tubular shapes, IFT is faster than MBS, especially when the latter considers high-resolution mesh models. For locally thin and large-surface shapes, IFT becomes slower than MBS. Again, this extra price of IFT is counterbalanced by the higher-quality regularization metric it delivers and also by the fact that IFT delivers connected skeletons, whereas MBS delivers only a skeletal point-cloud. All in all, we argue that the performance of IFT compares favorably with methods using the same importance metric (MS) but also with other multiscale skeletonization methods (AS, MBS). This is especially salient when considering that we implemented IFT as a purely *serial* algorithm, whereas MS, AS, and MBS all use CPU-side 8-core multithreading parallelization.

2.5 CONCLUSION

In this chapter, we have presented a novel way of computing multiscale 2D medial axes and 3D surface skeletons of image, respectively voxel datasets. For this, we cast the problem of computing the medial geodesic function (MGF) regularization metric, known for its ability to deliver high-quality multiscale skeletons in the computation of optimal path forests with the Image Foresting Transform (IFT) framework. We show that the delivered 2D and 3D skeletons compare very favorably from the perspective of similarity and regularization with several other known multiscale skeletonization methods. Our IFT-based implementation is very simple and delivers good performance. To our knowledge, or method is the second one (aside [27]) that can compute both 2D and 3D multiscale medial skeletons with a unified formulation.

Several extensions of this work are possible. Performance-wise, extending IFT to use multithreaded parallelization has the potential to make this method the fastest multiscale skeletonization technique for 2D skeletons and 3D surface skeletons on the CPU in existence. Application-wise, the IFT framework allows one to easily change the cost function, thereby enabling one to design a whole family of multiscale regularization metrics beyond the MGF metric. Such metrics could, in turn, support various types of applications, such as feature-sensitive regularization. Finally, an interesting extension regards the computation of multiscale 3D curve skeletons.

REFERENCES

[1] N. Amenta, S. Choi, R. Kolluri, The power crust, in: Proc. SMA, ACM, 2001, pp. 65–73.

[2] W.P. Amorim, A.X. Falcão, M.H. de Carvalho, Semi-supervised pattern classification using optimum-path forest, in: 2014 27th SIBGRAPI Conference on Graphics, Patterns and Images, 2014, pp. 111–118.

[3] F. Andaló, P. Miranda, R. da Silva Torres, A. Falcão, Shape feature extraction and description based on tensor scale, Pattern Recognit. 43 (1) (2010) 26–36.

[4] A. Torsello, E. Hancock, Correcting curvature-density effects in the Hamilton–Jacobi skeleton, IEEE Trans. Image Process. 15 (4) (2006) 877–891.

[5] S. Bouix, K. Siddiqi, A. Tannenbaum, Flux driven automatic centerline extraction, Med. Image Anal. 9 (3) (2005) 209–221.
[6] M. Chang, F. Leymarie, B. Kimia, Surface reconstruction from point clouds by transforming the medial scaffold, Comput. Vis. Image Underst. 113 (11) (2009) 1130–1146.
[7] K. Ciesielski, P. Miranda, A. Falcão, J. Udupa, Joint graph cut and relative fuzzy connectedness image segmentation algorithm, Med. Image Anal. 17 (8) (2013) 1046–1057.
[8] N. Cornea, D. Silver, X. Yuan, R. Balasubramanian, Computing hierarchical curve-skeletons of 3D objects, Vis. Comput. 21 (11) (2005) 945–955.
[9] M. Couprie, Topological maps and robust hierarchical Euclidean skeletons in cubical complexes, Comput. Vis. Image Underst. 117 (4) (2013) 355–369.
[10] R. da, S. Torres, A. Falcão, Contour salience descriptors for effective image retrieval and analysis, Image Vis. Comput. 25 (1) (2007) 3–13.
[11] R. da, S. Torres, A. Falcão, L. da, F. Costa, A graph-based approach for multiscale shape analysis, Pattern Recognit. 37 (6) (2004) 1163–1174.
[12] J. Damon, Global medial structure of regions in $\mathbb{R}^3$, Geom. Topol. 10 (2006) 2385–2429.
[13] P. de Miranda, A. Falcão, J. Udupa, Synergistic arc-weight estimation for interactive image segmentation using graphs, Comput. Vis. Image Underst. 114 (1) (2010) 85–99.
[14] T. Dey, J. Sun, Defining and computing curve skeletons with medial geodesic functions, in: Proc. SGP, IEEE, 2006, pp. 143–152.
[15] A. Falcão, B. da Cunha, R. Lotufo, Design of connected operators using the image foresting transform, in: Proc. SPIE, vol. 4322, 2001, pp. 468–479.
[16] A. Falcão, F. Bergo, Interactive volume segmentation with differential image foresting transforms, IEEE Trans. Med. Imaging 23 (9) (2004) 1100–1108.
[17] A. Falcão, L. da, F. Costa, B. da Cunha, Multiscale skeletons by image foresting transform and its applications to neuromorphometry, Pattern Recognit. 35 (7) (2002) 1571–1582.
[18] A. Falcão, J. Udupa, A 3D generalization of user-steered live-wire segmentation, Med. Image Anal. 4 (4) (2000) 389–402.
[19] A.X. Falcão, J. Stolfi, R.A. Lotufo, The image foresting transform: theory, algorithms, and applications, IEEE Trans. Pattern Anal. Mach. Intell. 26 (1) (2004) 19–29.
[20] A.X. Falcão, J.K. Udupa, F.K. Miyazawa, An ultra-fast user-steered image segmentation paradigm: live wire on the fly, IEEE Trans. Med. Imaging 19 (1) (2000) 55–62.
[21] C. Feng, A. Jalba, A. Telea, Improved part-based segmentation of voxel shapes by skeleton cut spaces, Math. Morph. Theory Appl. 1 (1) (2016) 60–78.
[22] M. Foskey, M. Lin, D. Manocha, Efficient computation of a simplified medial axis, in: Proc. Shape Modeling, 2003, pp. 135–142.
[23] P. Giblin, B. Kimia, A formal classification of 3D medial axis points and their local geometry, IEEE Trans. Pattern Anal. Mach. Intell. 26 (2) (2004) 238–251.
[24] J. Giesen, B. Miklos, M. Pauly, C. Wormser, The scale axis transform, in: Proc. Annual Symp. Comp. Geom, 2009, pp. 106–115.
[25] W. Hesselink, J. Roerdink, Euclidean skeletons of digital image and volume data in linear time by the integer medial axis transform, IEEE Trans. Pattern Anal. Mach. Intell. 30 (12) (2008) 2204–2217.
[26] A. Jalba, J. Kustra, A. Telea, Surface and curve skeletonization of large 3D models on the GPU, IEEE Trans. Pattern Anal. Mach. Intell. 35 (6) (2013) 1495–1508.
[27] A. Jalba, A. Sobiecki, A. Telea, An unified multiscale framework for planar, surface, and curve skeletonization, IEEE Trans. Pattern Anal. Mach. Intell. 38 (1) (2015) 30–45.
[28] T. Ju, M. Baker, W. Chiu, Computing a family of skeletons of volumetric models for shape description, Comput. Aided Des. 39 (5) (2007) 352–360.

[29] J. Kustra, A. Jalba, A. Telea, Robust segmentation of multiple manifolds from unoriented noisy point clouds, Comput. Graph. Forum 33 (1) (2014) 73–87.

[30] F. Leymarie, B. Kimia, The medial scaffold of 3D unorganized point clouds, IEEE Trans. Vis. Comput. Graph. 29 (2) (2007) 313–330.

[31] G. Malandain, S. Fernandez-Vidal, Euclidean skeletons, Image Vis. Comput. 16 (5) (1998) 317–327.

[32] B. Miklos, J. Giesen, M. Pauly, Discrete scale axis representations for 3D geometry, in: Proc. ACM SIGGRAPH, 2010, pp. 394–493.

[33] P.A.V. Miranda, A.X. Falcão, T.V. Spina, Riverbed: a novel user-steered image segmentation method based on optimum boundary tracking, IEEE Trans. Image Process. 21 (6) (2012) 3042–3052.

[34] N.J. Nilsson, Artificial Intelligence: A New Synthesis, 1st edition, Morgan Kaufmann Publishers, Inc., 1998.

[35] R.L. Ogniewicz, O. Kubler, Hierarchic Voronoi skeletons, Pattern Recognit. 28 (3) (1995) 343–359.

[36] J. Papa, A. Falcão, V. de Albuquerque, J. Tavares, Efficient supervised optimum-path forest classification for large datasets, Pattern Recognit. 45 (1) (2012) 512–520.

[37] S. Pizer, K. Siddiqi, G. Szekely, J. Damon, S. Zucker, Multiscale medial loci and their properties, Int. J. Comput. Vis. 55 (2–3) (2003) 155–179.

[38] S. Prohaska, H.C. Hege, Fast visualization of plane-like structures in voxel data, in: Proc. IEEE Visualization, 2002, pp. 29–36.

[39] D. Reniers, J.J. van Wijk, A. Telea, Computing multiscale skeletons of genus 0 objects using a global importance measure, IEEE Trans. Vis. Comput. Graph. 14 (2) (2008) 355–368.

[40] L. Rocha, F. Cappabianco, A. Falcão, Data clustering as an optimum-path forest problem with applications in image analysis, Int. J. Imaging Syst. Technol. 19 (2) (2009) 50–68.

[41] L. Rossi, A. Torsello, An adaptive hierarchical approach to the extraction of high resolution medial surfaces, in: Proc. 3DIMPVT, 2012, pp. 371–378.

[42] M. Rumpf, A. Telea, A continuous skeletonization method based on level sets, in: Proc. VisSym, 2002, pp. 151–158.

[43] D. Shaked, A. Bruckstein, Pruning medial axes, Comput. Vis. Image Underst. 69 (2) (1998) 156–169.

[44] K. Siddiqi, S. Bouix, A. Tannenbaum, S. Zucker, Hamilton–Jacobi skeletons, Int. J. Comput. Vis. 48 (3) (2002) 215–231.

[45] K. Siddiqi, S. Pizer, Medial Representations: Mathematics, Algorithms and Applications, Springer, 2009.

[46] T. Spina, P. de Miranda, A. Falcão, Hybrid approaches for interactive image segmentation using the live markers paradigm, IEEE Trans. Image Process. 23 (12) (2014) 5756–5769.

[47] S. Stolpner, S. Whitesides, K. Siddiqi, Sampled medial loci and boundary differential geometry, in: Proc. IEEE 3DIM, 2009, pp. 87–95.

[48] S. Stolpner, S. Whitesides, K. Siddiqi, Sampled medial loci for 3D shape representation, Comput. Vis. Image Underst. 115 (5) (2011) 695–706.

[49] A. Sud, M. Foskey, D. Manocha, Homotopy-preserving medial axis simplification, in: Proc. SPM, 2005, pp. 103–110.

[50] A. Tagliasacchi, T. Delame, M. Spagnuolo, N. Amenta, A. Telea, 3D skeletons: a state-of-the-art report, Comput. Graph. Forum 35 (2) (2016) 573–597.

[51] A. Tagliasacchi, H. Zhang, D. Cohen-Or, Curve skeleton extraction from incomplete point cloud, in: Proc. SIGGRAPH, 2009, pp. 541–550.

[52] A. Telea, Feature preserving smoothing of shapes using saliency skeletons, in: Visualization in Medicine and Life Sciences, Springer, 2012, pp. 153–170.
[53] A. Telea, J.J. van Wijk, An augmented fast marching method for computing skeletons and centerlines, in: Proc. VisSym, 2002, pp. 251–259.

CHAPTER

Fuzzy skeleton and skeleton by influence zones: a review

3

Isabelle Bloch

LTCI, Télécom ParisTech, Université Paris-Saclay, Paris, France

Contents

3.1 Introduction 71
3.2 Distance-Based Approaches 73
3.3 Morphological Approaches to Compute the Centers of Maximal Balls 75
3.4 Morphological Thinning 76
3.5 Fuzzy Skeleton of Influence Zones 78
3.5.1 Definition Based on Fuzzy Dilations 79
3.5.2 Definitions Based on Distances 80
3.5.3 Illustrative Example (Reproduced from [8]) 82
3.6 Conclusion 84
References 85

3.1 INTRODUCTION

Representing an object by its skeleton is a widely addressed topic. It allows simplifying a shape and its description while keeping its most relevant features. When shapes and objects are imprecisely defined, it is convenient to represent them as fuzzy sets, instead of crisp ones. This may represent different types of imprecision, either on the boundary of the objects (due for instance to partial volume effect, to the spatial resolution, or to imperfect segmentation), on the variability of these objects, on the potential ambiguity between classes, etc. In a spatial domain $\mathcal{S}$ (typically $\mathbb{R}^n$ in the continuous case or $\mathbb{Z}^n$ in the discrete case), a fuzzy set is defined via its membership function μ, a function from $\mathcal{S}$ into $[0, 1]$, providing at each point the degree to which this point belongs to a fuzzy set. In this chapter, we will use equivalently fuzzy sets and membership functions. Basics on fuzzy sets can be found e.g. in [16], and a recent review of their use in image processing and understanding in [10]. Simplifying fuzzy objects then requires to extend definitions designed for the crisp case to the fuzzy case. Similarly, skeleton by influence zones (SKIZ), which structures the background of a set of objects, has to be extended when objects are imprecisely defined. In this paper, we review the main approaches for defining skeleton and skeleton by influence zones of fuzzy sets.

Skeletonization. DOI: 10.1016/B978-0-08-101291-8.00004-3

One of the main approaches to extend an operation to fuzzy sets is to apply this operation to α-cuts[1] and then reconstruct the result by a combination of these α-cuts. This approach is well suited for increasing operations, assuring that their application on α-cuts provides sets that can be considered as α-cuts of a resulting fuzzy set. Unfortunately, skeletons and SKIZs are not increasing, and therefore this approach cannot be applied directly. Other approaches have thus been developed.

Several problems have been addressed in the literature, exploiting fuzzy representations:

- define the skeleton (as a crisp set) of a fuzzy set;
- define the fuzzy skeleton (i.e. a fuzzy set) of a fuzzy set;
- define the fuzzy SKIZ of a set of fuzzy sets.

They will be reviewed in this paper and categorized according to the type of approach they use. Note that fuzzy sets can also be used to represent some additional information on a classical skeleton. This will not be addressed here since we focus only on fuzzy inputs, but as an example let us mention the work in [45], where a fuzzy measure of significance is computed on each branch of a classical skeleton, with applications to skeleton pruning.

Defining and computing the skeleton in the crisp case are a widely addressed topic. Whereas there is a consensus on continuous approaches that provide the required properties (in particular, representing semicontinuously a shape using thin and centered structures, respecting the topology of the initial shape, and allowing for its reconstruction),[2] several approaches have been defined in the discrete case, based on different principles, in order to address important issues directly caused by the digitization [26]. It is out of the scope of this paper to review them, and the reader may refer to [39] and in the recent work [33] for details. The first chapters of this book also provide an overview of the field, including multiscale aspects, and parallel implementations. The particular case of 1D skeletons of 3D objects is also reviewed in one chapter. Several applications are also described in some chapters, such as bone or vessel structure characterization. The main approaches can be grouped as follows:

- distance-based approaches, with the important notion of center of maximal balls (CMB), and related methods such as wavefront (or grass fire) propagation, ridges of the distance function, and minimal paths;
- morphological approaches for the computation of centers of maximal balls;
- morphological approaches based on homotopic thinning.

The two first approaches correspond to the intuitive idea proposed initially by Blum [12],[3] where the grass-fire principle led to the definition of the skeleton as

[1] An α-cut of a fuzzy set μ is a crisp set defined as $\mu_\alpha = \{x \in S \mid \mu(x) \geq \alpha\}$ for $\alpha \in [0, 1]$.
[2] A detailed analysis of the properties of the skeleton of an open set in $\mathbb{R}^n$ can be found in [25].
[3] Blum's work was first presented in a conference in the 1960s and formalized by Calabi also in the 1960s.

shock points at which wavefronts intersect. These points are those that are equidistant to at least two boundary points or, equivalently, the centers of maximal disks. The extensions to the fuzzy case of these two approaches are then equivalent to Blum's approach when restricted to crisp sets.

Approximate approaches in the continuous case rely on the Voronoï points sampled on the object contour or surface. The quality of this approximation can be estimated in the case of regular objects in the sense of mathematical morphology (i.e. sets equal to their opening and closing by a disk or sphere of given radius [37]); see e.g. [14]. Regularization methods [28] and surface reconstruction methods [1] have been proposed based on this approach. In this paper, we do not consider such approximations. However, the extended Voronoï shapes will be considered for defining fuzzy SKIZ.

In the fuzzy case, we also expect the skeleton to well represent the input fuzzy shape, to be centered in the shape, and to be thin enough (this notion being more complicated to define than in the crisp case). The function associating a skeleton to a fuzzy set should also be antiextensive and idempotent as in the crisp case. The question of homotopy is also a difficult one in the case of fuzzy sets. Finally as an extension, it is expected that the fuzzy definitions boil down to the crisp ones in the particular case where the membership functions take only values 0 and 1 (i.e. defining crisp sets).

The review proposed in this chapter is organized according to the type of approach used to define the skeleton or SKIZ of fuzzy sets, extending the above mentioned approaches. In Section 3.2, distance-based approaches are reviewed, with results being either crisp sets or fuzzy ones. Whereas several approaches propose to define a crisp skeleton of a fuzzy object, fuzzy versions are interesting too since if an object is imprecisely defined, we can expect its skeleton be imprecise too. Approaches based on mathematical morphology are presented in the next sections, for defining centers of maximal balls in Section 3.3, for extending the notion of thinning in Section 3.4, mostly based on extension of thinning to gray-scale images, and finally for defining fuzzy SKIZ in Section 3.5. Surprisingly enough, the question of thinning of gray-scale images has been addressed a long time ago, in the 1970s (maybe one of the first papers is [23]), with applications to fuzzy skeleton in the early eighties. Very recently, a renewed interest for fuzzy skeleton using various approaches can be observed.

3.2 DISTANCE-BASED APPROACHES

In the crisp case, this approach leads to several almost equivalent definitions. The idea is to define a distance function inside the shape, the ridges of which build the skeleton. The skeleton points are also defined as the points through which no minimal path from any point to the shape boundary goes. According to the underlying distance on the spatial domain, maximal balls included in the shape are defined (i.e. which cannot be included in any other ball included in the shape), the centers of which build the skeleton (also called medial axis in this case). The set of centers of maximal balls is denoted $CMB(\cdot)$ in the following. This approach can be computed efficiently according to various discrete distances [2,13].

The ridge approach was used for fuzzy sets in [20], where the fuzzy sets were defined using an S-shape transformation of the original image. Note that this implicitly assumes that higher gray levels correspond to the expected skeleton. A ridge following algorithm was proposed, using weighted neighborhood, along with a post-processing to get a one pixel width skeleton. This approach provides a crisp result, which is thin, connected, and passes though the center of the regions with the highest membership values (note that this is not necessarily equivalent to the center of the support or of any α-cut).

The notion of maximal fuzzy disks (or more generally fuzzy balls of a distance) was proposed by several authors. In [29], the space $\mathcal{S}$ is endowed with a metric d. A fuzzy disk included in a fuzzy set μ and centered at x is defined as the following fuzzy set:

$$\forall y \in \mathcal{S}, g_x^\mu(y) = \inf_{z | d(x,z)=d(x,y)} \mu(z). \tag{3.1}$$

Note that the standard inclusion of fuzzy sets is used, expressed here as $g_x^\mu(y) \leq \mu(y)$ for all $y \in \mathcal{S}$. The medial axis of μ is then defined as the set $CMB(\mu)$ of local maxima of g_x^μ (hence a crisp set), and the fuzzy medial axis transform is the set of fuzzy sets $\{g_x^\mu \mid x \in CMB(\mu)\}$.

Another definition of centers of fuzzy maximal balls was given in [24], based on the distance of a point to the thresholded fuzzy boundary of the initial fuzzy set.

An extension of the distance transform for fuzzy sets was defined in [36]. The idea is to weight the local distance between points by the membership of these points to the considered fuzzy set. For instance, the distance between two neighbor points x and y can be defined as $\max(\mu(x), \mu(y))||x - y||$ or $\frac{\mu(x)+\mu(y)}{2}||x - y||$. The length of a path is then classically defined by summing local weighted distances along the path, and the distance between two points is the length of a shortest path between these two points. Axial points, i.e. centers of maximal balls, are then derived from this weighted distance in a usual way [35]. Fuzziness is then taken into account only as weighting factors, leading to distance values defined as classical (crisp) numbers, and then a classical approach is used for the next steps. A similar approach using a discrete distance was developed in [41]. Then, based on the reverse distance transform, the centers of maximal fuzzy balls can be obtained and define the skeleton [42]. A crisp skeleton is then obtained. A similar approach was developed in [21,22] with further selection, filtering and refinement steps to obtain a thin crisp skeleton.

Another weighted approach was proposed in the early work published in [23], for gray-level images (but could be used for fuzzy sets as well). The length of an arc is defined by a weighted sum (or integral in the continuous case). Minimum paths from each point to the contour of the support of the initial function are then computed. The skeleton is then formed by the points that do not belong to the minimum path of another point.

All these approaches are equivalent to the crisp CMB one if the input is crisp. However, the final result ignores the imprecision on the input set, providing a crisp result. We may consider that an important information is then lost and that the result is an oversimplified representation. When transposed to the discrete case, CMB approaches suffer from the same limitations in terms of topology preservation as their crisp versions.

3.3 MORPHOLOGICAL APPROACHES TO COMPUTE THE CENTERS OF MAXIMAL BALLS

Centers of maximal balls can be obtained using mathematical morphology operations [37,38], in particular, erosion (denoted by ε) and opening (denoted by γ). The idea is that the center of a maximal ball of radius ρ is given by the set difference between the erosion of the set by a structuring element of size ρ and the opening (by the smallest possible structuring element) of this erosion. Formally, for a crisp set A, the skeleton is given by

$$r(A) = \bigcup_{\rho>0} s_\rho(A), \tag{3.2}$$

where $s_\rho(A)$ is the set of centers of maximal balls of radius ρ included in A, given by

$$s_\rho(A) = \bigcap_{\mu>0} [\varepsilon_{B_\rho}(A) \setminus \gamma_{\bar{B}_\mu}(\varepsilon_{B_\rho}(A))]$$

with B_ρ (respectively $\bar{B}_\rho$) denoting the open (respectively closed) ball of radius ρ. This definition is shown to have good properties for open sets [25]. In particular, the original set can be reconstructed from the skeleton by dilating all the s_ρ:

$$A = \bigcup_{\rho>0} \delta_{B_\rho}(s_\rho(A)).$$

In the discrete case, this definition of the skeleton is transposed as

$$S(A) = \bigcup_{n\in\mathbb{N}} [\varepsilon_{B_n}(A) \setminus \gamma_B(\varepsilon_{B_n}(A))], \tag{3.3}$$

where B is the elementary structuring element on the digital grid (i.e. of radius one), and ε_{B_0} (erosion by a structuring element of radius 0) is the identity mapping. Whereas the reconstruction property is preserved, the homotopy property is not, and we may obtain a disconnected skeleton of a connected object [26].

This type of construction was used in [30] for gray-scale images and could also be used for fuzzy sets. The idea of shrinking and expanding used in the construction

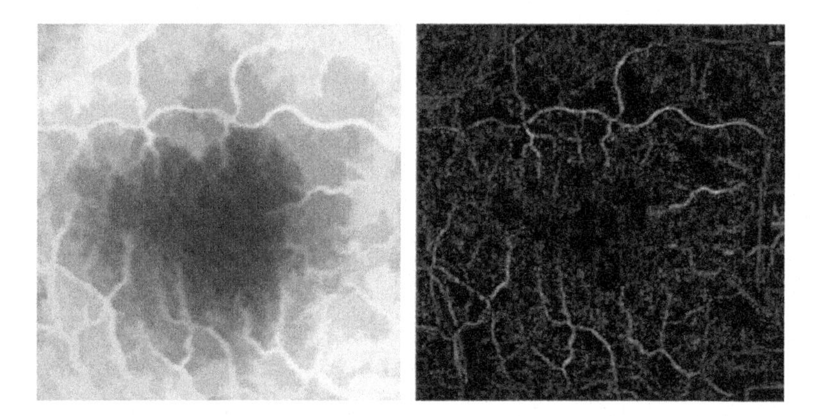

FIGURE 3.1

Left: original fuzzy set, derived from an eye vessel images. High gray levels indicate high membership degrees to the blood vessels. Right: fuzzy skeleton obtained using Eq. (3.4), where membership degrees range from 0 (black) to 1 (white).

of the skeleton is replaced by local min and max operators (hence corresponding to erosions and dilations). This was used in [43] for approximating an image.

Based on a similar idea, we propose to extend Eq. (3.3) to fuzzy sets using fuzzy mathematical morphology [9,11]. The fuzzy dilation is defined as a degree of conjunction between the translation of the (fuzzy) structuring element and the initial fuzzy set, whereas the fuzzy erosion is defined as a degree of implication. Opening is defined as usual as the combination of an erosion and the corresponding adjoint dilation. We propose to define the fuzzy skeleton of a fuzzy set in the discrete case as

$$FS(\mu) = \bigvee_{n \in \mathbb{N}} [\varepsilon_{B_n}(\mu) \setminus \gamma_B(\varepsilon_{B_n}(\mu))], \tag{3.4}$$

where the supremum $\bigvee$ is the fuzzy union. The continuous formulation extends in a similar way. This definition applies whatever the dimension of the space, as in the crisp case. Note that instead of a crisp structuring element, a fuzzy one could be used, for instance, representing the smallest spatial unit, given the imprecision in the image. An example illustrating this new definition is given in Fig. 3.1 and shows that the skeleton is fuzzy, representing the fact that areas, even on crest lines of the membership function, may belong to the fuzzy set with a low degree, and so do the corresponding parts of the skeleton.

3.4 MORPHOLOGICAL THINNING

In the crisp case, a common way to overcome the homotopy problems raised by a direct computation of centers of maximal balls on digital images consists in applying

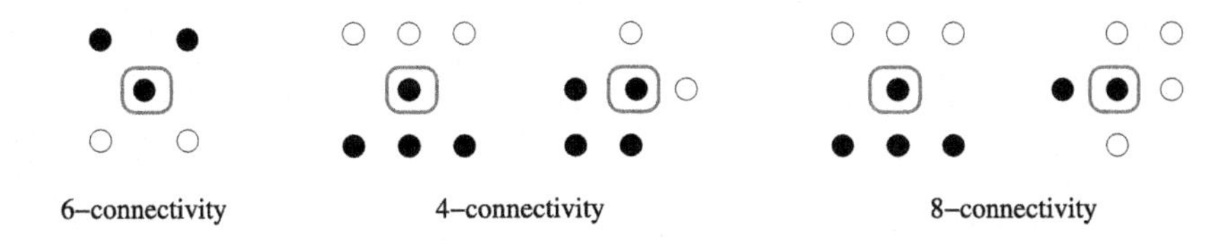

FIGURE 3.2

Structuring elements used for homotopic thinning on a hexagonal grid (6-connectivity) and on a square grid (4-connectivity or 8-connectivity). Black points represent object points, and white points correspond to background points. The center of the structuring elements is circled. All rotations of these structuring elements have to be used sequentially and iteratively to get the skeleton.

iterative thinning to the input shape, while guaranteeing the preservation of topology at each step, until convergence, leading to a homotopic skeleton [37]. The structuring elements used in this process to delete so-called simple points (i.e. whose deletion does not change the topology) are illustrated in Fig. 3.2. The 3D case was also addressed, e.g. in [4,34]. See also the corresponding chapter on 2D, 3D, and 4D thinning in this book.

This approach was extended to gray-level images in several ways. Note that these extensions directly apply to fuzzy sets if gray levels are bounded and can be matched monotonically to membership degrees such that high gray levels correspond to high membership degrees (or the reverse). Historically, a first approach was introduced in [18], based on a weighted connectedness: two points x and y are connected if there exists a path from x to y that does not go through a point with higher value. Then thinning of a function f is performed by replacing $f(x)$ by $\min_{y \in V(x)} f(x)$, where $V(x)$ denotes a neighborhood of x (i.e. this corresponds to an erosion), only if this change does not disconnect any pair of points in $V(x)$.

Still among the early works, a similar approach was proposed in [19], were local min and max operators are used, referring to fuzzy logic, with structuring elements adapted to the type of thinning (for instance a 3×3 neighborhood except the center to suppress of pepper and salt noise, or oriented neighborhoods to compute the skeleton, similarly as in Fig. 3.2).

Whereas thinning is usually defined as the set difference between the original image and the result of a hit-or-miss transformation with appropriate structuring elements, an interesting formulation in [39] allows for a direct extension to functions. Let B_1 be the part of the structuring element composed of object points (the black points in Fig. 3.2), and B_2 the other part of the structuring element. For thinning, the origin belongs to B_1. The thinning of a function f by $B = (B_1, B_2)$ is expressed as

$$\forall x \in S, (f \circ B)(x) = \begin{cases} \delta_{B_2}(f)(x) & \text{if } \varepsilon_{B_1}(f)(x) = f(x) \text{ and } \delta_{B_2}(f)(x) < f(x), \\ f(x) & \text{otherwise.} \end{cases} \tag{3.5}$$

If f is a binary image, then this definition reduces to the classical one.

Another extension was proposed in [15], where topological operators for gray-level images were defined. Thinning and skeleton are based on the notion of destructible point: a point x is destructible for a function f if it is simple (according to the binary case) for the threshold $f_k = \{y \mid f(y) \geq k\}$ with $k = f(x)$. Then the topology is not modified if $f(x)$ is replaced by $f(x) - 1$ (assuming that f takes values in a subset of $\mathbb{N}$). The skeleton of f is then obtained by reducing iteratively the value of destructible points that are not end-points (an end-point is a point x such that $CC(V(x) \setminus \{x\} \cap f_k) = 1$, with $k = f(x)$ and $CC(A)$ the number of connected components of A for a given discrete connectivity).

Since thinning is often derived from hit-or-miss transformation (HMT), let us finally mention some extensions of this transformation to gray-scale images. In [31], a function is expressed as the supremum of impulse functions below it, and a similar construction is used to define HMT, based on interval operators (an interval operator by (A, B) is equivalent to the HMT by (A, B^C) in the crisp case). A related approach, although somewhat different, is based on probing, as introduced in [3]. In [40], the cardinality of the set of gray levels such that a point belongs to the HMT of thresholds is considered. All these works have been nicely unified in [27] in the framework of interval operators. The idea is to decompose the HMT process into a fitting step and an evaluation step (which can be done using a sup as in [31], a sum as in [40], or any other set measure or using a newly proposed binary valuation). The authors show the links between the different approaches and the power of the unified framework. All definitions reduce to the classical ones if the functions are binary (i.e. sets).

3.5 FUZZY SKELETON OF INFLUENCE ZONES

The notions of Voronoï diagram, generalized Voronoï diagram, or skeleton by influence zones (SKIZ) define regions of space that are closer to a region or object than to another one, and have important properties and applications [37,39]. If knowledge or information is modeled using fuzzy sets, it is natural to see the influence zones of these sets as fuzzy sets too. The extension of these notions to the fuzzy case is therefore important for applications such as partitioning the space where fuzzy sets are defined, implementing the notion of separation, and reasoning on fuzzy sets (fusion, interpolation, negotiations, spatial reasoning on fuzzy regions of space, etc.). Despite their interest, surprisingly, such an extension has been little developed until now. In this section, we summarize the approach in [8], based on mathematical morphology.

In the crisp case, for a set X composed of several connected components ($X = \bigcup_i X_i$ with $X_i \cap X_j = \emptyset$ for $i \neq j$), the influence zone of X_i, denoted by $IZ(X_i)$, is defined as the set of points that are strictly closer to X_i than to X_j for $j \neq i$, according to a distance d defined on $\mathcal{S}$ (usually the Euclidean distance or a discrete version of it on digital spaces):

$$IZ(X_i) = \{x \in \mathcal{S} \mid d(x, X_i) < d(x, X \setminus X_i)\}. \tag{3.6}$$

The SKIZ of X, denoted by $SKIZ(X)$, is the set of points that belong to none of the influence zones, i.e. that are equidistant of at least two components X_i:

$$SKIZ(X) = (\bigcup_i IZ(X_i))^c. \tag{3.7}$$

The SKIZ is also called a generalized Voronoï diagram. Note that the SKIZ is a subset of the morphological skeleton of X^c (i.e. the set of centers of maximal balls included in X^c, where X^c denotes the complement of X in $\mathcal{S}$) [37,39]. It is not necessarily connected and contains in general less branches than the skeleton of X^c (this may be exploited in a number of applications).

An important and interesting property of this definition based on distance (Eq. (3.6)) is that it can be expressed in terms of morphological operations as well. Let us denote by δ_λ the dilation by a ball of radius λ and by ε_λ the erosion by a ball of radius λ. Then the influence zones can be expressed as [5]

$$IZ(X_i) = \bigcup_\lambda \left(\delta_\lambda(X_i) \cap \varepsilon_\lambda((\cup_{j\neq i} X_j)^c)\right) = \bigcup_\lambda \left(\delta_\lambda(X_i) \setminus \delta_\lambda(\cup_{j\neq i} X_j)\right). \tag{3.8}$$

Another link between SKIZ and distance can be expressed by involving the watersheds (WS) [39]:

$$SKIZ(X) = WS(d(y, X), y \in X^c). \tag{3.9}$$

Let us now extend these definitions and properties to fuzzy sets. For the sake of clarity, we assume two fuzzy sets with membership functions μ_1 and μ_2 defined on $\mathcal{S}$. The extension to an arbitrary number of fuzzy sets is straightforward. Fuzzy dilations and erosions are defined based on degrees of intersection (using a t-norm $\top$) and degrees of implication (using a fuzzy implication I), respectively:

$$\forall x \in \mathcal{S},\ \delta_\nu(\mu)(x) = \sup_{y\in\mathcal{S}} \top(\mu(y), \nu(x-y)), \tag{3.10}$$

$$\forall x \in \mathcal{S},\ \varepsilon_\nu(\mu)(x) = \inf_{y\in\mathcal{S}} I(\nu(y-x), \mu(y)), \tag{3.11}$$

and we choose here dual definitions of these operations with respect to a complementation c, using suitable $\top$ and I [9,11].

3.5.1 DEFINITION BASED ON FUZZY DILATIONS

Let us first consider the expression of influence zone using morphological dilations (Eq. (3.8)). This expression can be extended to fuzzy sets by using fuzzy intersection and union, and fuzzy mathematical morphology. For a given structuring element ν, the influence zone of μ_1 is defined as

$$IZ_{dil}(\mu_1) = \bigcup_\lambda \left(\delta_{\lambda\nu}(\mu_1) \cap \varepsilon_{\lambda\nu}(\mu_2^c)\right) = \bigcup_\lambda (\delta_{\lambda\nu}(\mu_1) \setminus \delta_{\lambda\nu}(\mu_2)). \tag{3.12}$$

The influence zone for μ_2 is defined in a similar way. The extension to any number of fuzzy sets μ_i is straightforward:

$$IZ_{dil}(\mu_i) = \bigcup_{\lambda}(\delta_{\lambda\nu}(\mu_i) \cap \varepsilon_{\lambda\nu}((\bigcup_{j\neq i}\mu_j)^c)). \tag{3.13}$$

In these equations, intersection and union of fuzzy sets are implemented as t-norms $\top$ and t-conorms $\perp$ (min and max for instance). The fuzzy complementation used in the following is always $c(a) = 1 - a$, but other forms could be employed as well. Eq. (3.12) then becomes

$$IZ_{dil}(\mu_1) = \sup_{\lambda} \top[\delta_{\lambda\nu}(\mu_1), 1 - \delta_{\lambda\nu}(\mu_2)]. \tag{3.14}$$

In the continuous case, if ν denotes the elementary structuring element of size 1, then $\lambda\nu$ denotes the corresponding structuring element of size λ (for instance, if ν is a ball of some distance of radius 1, then $\lambda\nu$ is the ball of radius λ). In the digital case, the operations performed using $\lambda\nu$ as structuring elements (λ being an integer in this case) are simply the iterations of λ operations performed with ν (iterativity property of fuzzy erosion and dilation [11]). Note that the number of dilations to be performed to compute influence zones in a digital bounded space $\mathcal{S}$ is always finite (and bounded by the length of the largest diagonal of $\mathcal{S}$).

The fuzzy SKIZ is then defined as

$$\text{SKIZ}(\bigcup_i \mu_i) = (\bigcup_i IZ(\mu_i))^c. \tag{3.15}$$

This expression also defines a fuzzy (generalized) Voronoï diagram.

3.5.2 DEFINITIONS BASED ON DISTANCES

Another approach consists in extending the definition in terms of distances (Eq. (3.6)) and defining a degree to which the distance to one of the sets is lower than the distance to the other sets. Several definitions of the distance of a point to a fuzzy set have been proposed in the literature. Some of them provide real numbers, and Eq. (3.6) can then be applied directly. But then the imprecision in the object definition is lost (the problem is the same as when using a weighted distance for computing CMB, as mentioned in Section 3.2). Definitions providing fuzzy numbers are therefore more interesting since if the sets are imprecise, we may expect that distances are imprecise too, as also underlined e.g. in [6,17,32]. In particular, as will be seen next, it may be interesting to use the distance proposed in [6], based on fuzzy dilation:

$$d(x, \mu)(n) = \top[\delta_{n\nu}(\mu)(x), 1 - \delta_{(n-1)\nu}(\mu)(x)]. \tag{3.16}$$

It expresses, in the digital case, the degree to which x is at a distance n of μ ($\top$ is a t-norm, and $n \in \mathbb{N}^*$). For $n = 0$, the degree becomes $d(x, \mu)(0) = \mu(x)$. This expression can be generalized to the continuous case as

$$d(x, \mu)(\lambda) = \inf_{\lambda' < \lambda} \top[\delta_{\lambda\nu}(\mu)(x), 1 - \delta_{\lambda'\nu}(\mu)(x)], \tag{3.17}$$

where $\lambda \in \mathbb{R}^{+*}$, and $d(x, \mu)(0) = \mu(x)$.

When distances are fuzzy numbers, the fact that $d(x, \mu_1)$ is lower than $d(x, \mu_2)$ becomes a matter of degree. The degree to which this relation is satisfied can be performed using methods for comparing fuzzy numbers (see e.g. [44]). Let us consider the definition in [16], which expresses the degree $\mu(d_1 < d_2)$ to which $d_1 < d_2$, d_1 and d_2 being two fuzzy numbers, using the extension principle [46]:

$$\mu(d_1 < d_2) = \sup_{a<b} \min(d_1(a), d_2(b)). \tag{3.18}$$

The influence zone of μ_1 based on the comparison of fuzzy numbers (using Eq. (3.18)) is defined as

$$\begin{aligned} IZ_{dist1}(\mu_1)(x) &= \mu(d(x, \mu_1) < d(x, \mu_2)) \\ &= \sup_{n<n'} \min[d(x, \mu_1)(n), d(x, \mu_2)(n')]. \end{aligned} \tag{3.19}$$

Note that this approach can be applied whatever the chosen definition of fuzzy distances.

When distances are more specifically derived from a dilation, as the ones in Eqs. (3.16) and (3.17), a more direct approach can be proposed, taking into account explicitly this link between distances and dilations. Indeed, in the binary case, the following equivalences hold:

$$\begin{aligned} (d(x, X_1) \le d(x, X_2)) &\Leftrightarrow (\forall \lambda, x \in \delta_\lambda(X_2) \Rightarrow x \in \delta_\lambda(X_1)) \\ &\Leftrightarrow (\forall \lambda, x \in \delta_\lambda(X_1) \vee x \notin \delta_\lambda(X_2)). \end{aligned} \tag{3.20}$$

This means that if x is closer to X_1 than to X_2, x is reached faster by dilating X_1 than by dilating X_2. This expression extends to the fuzzy case as follows. The degree $\mu(d(x, \mu_1) \le d(x, \mu_2))$ to which $d(x, \mu_1)$ is less than $d(x, \mu_2)$ is defined as

$$\mu(d(x, \mu_1) \le d(x, \mu_2)) = \inf_{\lambda} \bot(\delta_{\lambda\nu}(\mu_1)(x), 1 - \delta_{\lambda\nu}(\mu_2)(x)), \tag{3.21}$$

where $\bot$ is a t-conorm (fuzzy disjunction). This equation also defines a way to compare fuzzy numbers representing distances.

Defining influence zones requires a strict inequality between distances, which is derived by complementation:

$$\mu(d(x, \mu_1) < d(x, \mu_2)) = 1 - \mu(d(x, \mu_2) \le d(x, \mu_1)). \tag{3.22}$$

The influence zone of μ_1 is then defined as

$$IZ_{dist2}(\mu_1)(x) = 1 - \inf_{\lambda} \bot(\delta_{\lambda\nu}(\mu_2)(x), 1 - \delta_{\lambda\nu}(\mu_1)(x)). \tag{3.23}$$

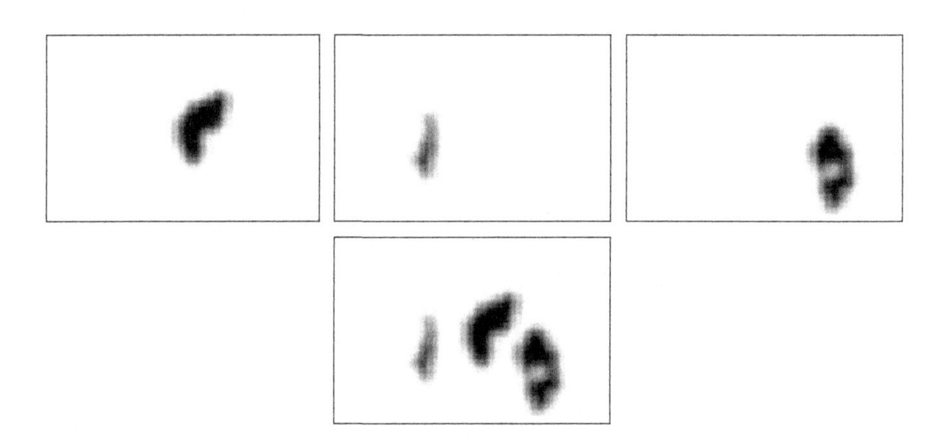

FIGURE 3.3

Three fuzzy objects and their union. Membership degrees range from 0 (white) to 1 (black).

Whatever the chosen definition of IZ, the SKIZ is always defined by Eq. (3.15).

Comparison and properties are detailed in [8]. In particular, the definitions derived from the dilation approach and from the direct distance approach are equivalent: $IZ_{dil}(\mu_1) = IZ_{dist2}(\mu_1)$. However, the two distance-based approaches are not equivalent since they rely on different orderings between fuzzy sets. Actually, the direct approach always provides a larger result: $IZ_{dist1}(\mu_1)(x) \leq IZ_{dist2}(\mu_1)(x)$ for all $x \in \mathcal{S}$. In terms of complexity, the direct approach is computationally less expensive. Another important property is the consistency with the crisp case, as generally required when extending an operation on crisp sets to fuzzy sets. Finally, the SKIZ is symmetric with respect to the μ_i and hence independent of their order.

3.5.3 ILLUSTRATIVE EXAMPLE (REPRODUCED FROM [8])

The notion of fuzzy SKIZ is illustrated on the three objects of Fig. 3.3. The structuring element ν is a crisp 3×3 square in Fig. 3.4 and a fuzzy set of paraboloid shape in Fig. 3.5. The influence zones of each object are displayed, as well as the SKIZ. These results are obtained with the dilation-based definition. Each influence zone is characterized by high membership values close to the corresponding object and decreasing when the distance to this object increases. The use of a fuzzy structuring element results in more fuzziness in the influence zones and SKIZ.

A binary decision can be made to obtain a crisp SKIZ of fuzzy objects. An appropriate approach consists in computing the watershed lines of the fuzzy SKIZ: it provides spatially consistent lines, without holes, and going through the crest lines of the membership function of the SKIZ. A result is provided in Fig. 3.6. For a fuzzy structuring element ν, the lines can go through the objects (Fig. 3.6B). Although this is impossible in the binary case, in the fuzzy case, this is explained by the fact that an object can, to some degree, be built of several connected components, linked together

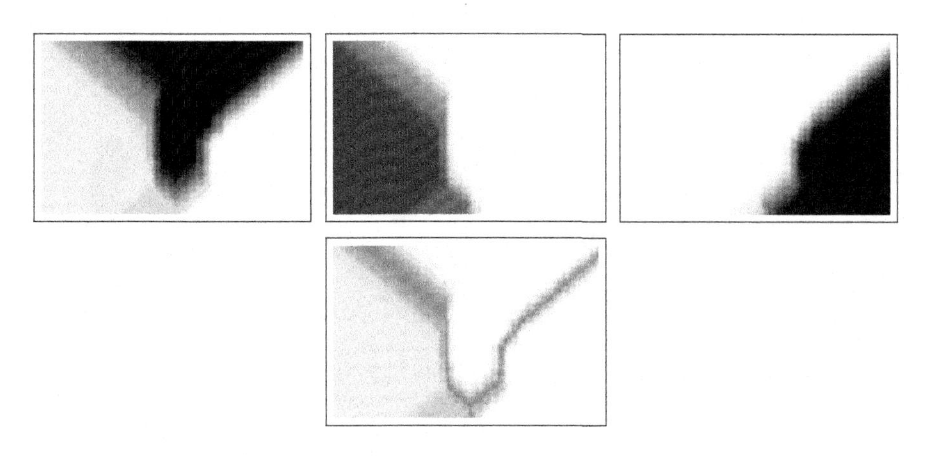

FIGURE 3.4

Influence zones of the three fuzzy objects displayed in Fig. 3.3 and resulting fuzzy SKIZ, obtained using a binary structuring element (3 × 3 square) and the dilation-based approach.

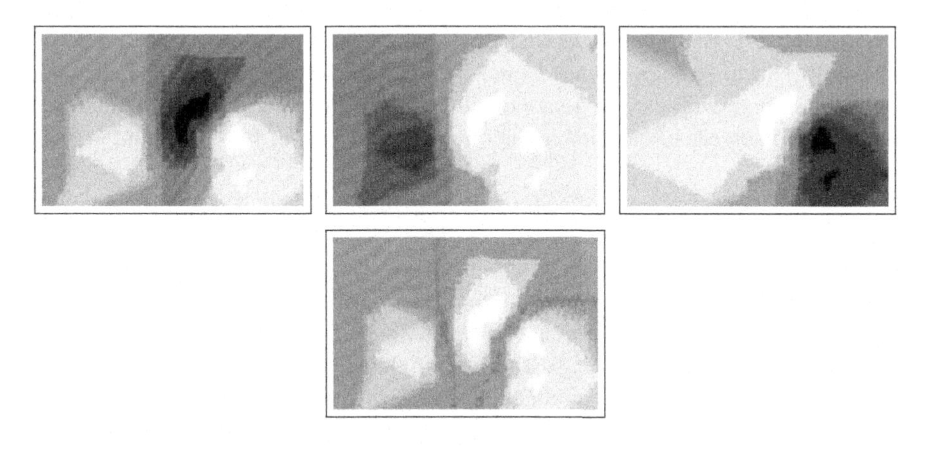

FIGURE 3.5

Influence zones of the three fuzzy objects displayed in Fig. 3.3 and resulting fuzzy SKIZ, obtained using a fuzzy structuring element (paraboloid-shaped) and the dilation-based approach.

by points with low membership degrees. The values of the SKIZ at those points are low too. This is the case for the third object in Fig. 3.3. The low values of the SKIZ along the line traversing this object are in accordance with the fact that the object has only one connected component with some low degree and two components with some higher degree. The line separating the third object can be suppressed by eliminating the parts of the watersheds having a very low degree in the fuzzy SKIZ (Fig. 3.6C). This requires to set a threshold value.

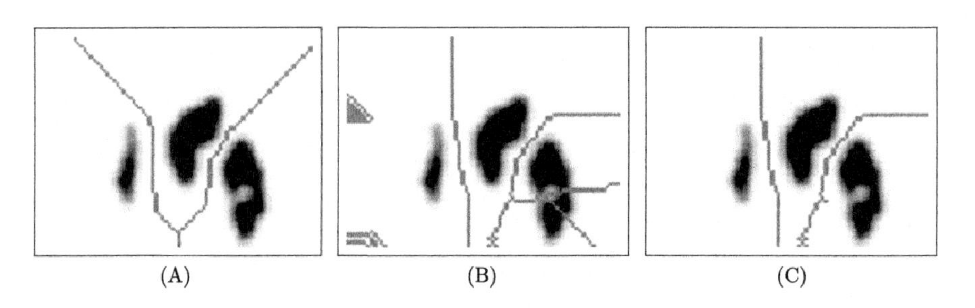

FIGURE 3.6

Binary decision using watershed for ν crisp (A) and fuzzy (B). Lines with a very low membership degree in the SKIZ of (B) have been suppressed in (C).

3.6 CONCLUSION

In this chapter, the main approaches for fuzzy skeleton and fuzzy SKIZ have been reviewed. Some of them were directly designed for fuzzy sets, whereas others were developed for gray-level images, but can be used for fuzzy sets as well, as soon as the gray-level scale is bounded (and then isomorphic to $[0, 1]$). Semantics has to be considered with care to guarantee that the transformation from gray levels to membership degrees has a suitable interpretation in terms of fuzzy sets and gradual membership.

Among these definitions, some provide crisp results, whereas other provide fuzzy results, which may be more convenient to keep track of the spatial imprecision of the input. This may be even more intuitive. For instance, if an object has imprecise boundaries, then we would expect that points at equal distance of two or more boundary points would not be precisely located either. Similarly, thinning methods that require to define both a shape and its complement have to account for the fact that the transition is not crisply defined. Concerning semantics, it may depend on the semantics of the fuzziness in the initial object, which may represent the observation of an intrinsically imprecise object, an imprecise object representation due to sensor limitations, an object suffering from imprecision during a detection or segmentation process, a preferred region of space, etc.

Therefore a hint for future work would be to extend distance-based approaches to provide fuzzy skeletons. This could be done by replacing the weighted distance transforms by distances taking values defined as fuzzy numbers (see e.g. [6,7] for such distances).

Another aspect that deserves to be further explored is a deeper analysis of the properties of the various definitions, in particular, fuzzy homotopy and its preservation (up to some degree if fuzziness if kept), measurement of how thin is a fuzzy skeleton and of how central it is in the input fuzzy set, and reconstruction capabilities. This would allow an easier comparison between different approaches at a theoretical level.

At a more practical level, besides efficient implementations, real-world applications would deserve to be more developed, to demonstrate the interest of fuzzy versions of skeleton and SKIZ.

REFERENCES

[1] N. Amenta, S. Choi, R.K. Kolluri, The power crust, unions of balls, and the medial axis transform, Comput. Geom. 19 (2) (2001) 127–153.

[2] C. Arcelli, G. Sanniti di Baja, Finding local maxima in a pseudo-Euclidean distance transform, Comput. Vis. Graph. Image Process. 43 (3) (1988) 361–367.

[3] C. Barat, C. Ducottet, M. Jourlin, Pattern matching using morphological probing, in: IEEE International Conference on Image Processing (ICIP), vol. 1, 2003, pp. 369–372.

[4] G. Bertrand, G. Malandain, A new characterization of three-dimensional simple points, Pattern Recognit. Lett. 15 (2) (1994) 169–175.

[5] S. Beucher, Sets, partitions and functions interpolations, in: H. Heijmans, J. Roerdink (Eds.), Mathematical Morphology and its Applications to Image Processing, ISMM'98, Kluwer Academic Publishers, Amsterdam, The Netherlands, 1998, pp. 307–314.

[6] I. Bloch, On fuzzy distances and their use in image processing under imprecision, Pattern Recognit. 32 (11) (1999) 1873–1895.

[7] I. Bloch, On fuzzy spatial distances, in: P. Hawkes (Ed.), Adv. Imaging Electron Phys., vol. 128, Elsevier, Amsterdam, 2003, pp. 51–122.

[8] I. Bloch, Fuzzy skeleton by influence zones – application to interpolation between fuzzy sets, Fuzzy Sets Syst. 159 (2008) 1973–1990.

[9] I. Bloch, Duality vs. adjunction for fuzzy mathematical morphology and general form of fuzzy erosions and dilations, Fuzzy Sets Syst. 160 (2009) 1858–1867.

[10] I. Bloch, Fuzzy sets for image processing and understanding, Fuzzy Sets Syst. 281 (2015) 280–291.

[11] I. Bloch, H. Maître, Fuzzy mathematical morphologies: a comparative study, Pattern Recognit. 28 (9) (1995) 1341–1387.

[12] H. Blum, Biological shape and visual science (Part I), J. Theor. Biol. 38 (2) (1973) 205–287.

[13] G. Borgefors, Centres of maximal discs in the 5-7-11 distance transform, in: Scandinavian Conference on Image Analysis, vol. 1, 1993, pp. 105–111.

[14] J.W. Brandt, V.R. Algazi, Continuous skeleton computation by Voronoi diagram, CVGIP, Image Underst. 55 (3) (1992) 329–338.

[15] M. Couprie, F.N. Bezerra, G. Bertrand, Topological operators for grayscale image processing, J. Electron. Imaging 10 (4) (2001) 1003–1015.

[16] D. Dubois, H. Prade, Fuzzy Sets and Systems: Theory and Applications, Academic Press, New-York, 1980.

[17] D. Dubois, H. Prade, On distance between fuzzy points and their use for plausible reasoning, in: International Conference on Systems, Man, and Cybernetics, 1983, pp. 300–303.

[18] C.R. Dyer, A. Rosenfeld, Thinning algorithms for gray-scale pictures, IEEE Trans. Pattern Anal. Mach. Intell. 1 (1) (1979) 88–89.

[19] V. Goetcherian, From binary to grey tone image processing using fuzzy logic concepts, Pattern Recognit. 12 (1) (1980) 7–15.

[20] M.E. Hoffman, E.K. Wong, A ridge-following algorithm for finding the skeleton of a fuzzy image, Inf. Sci. 105 (1) (1998) 227–238.

[21] D. Jin, C. Chen, P.K. Saha, Filtering non-significant quench points using collision impact in grassfire propagation, in: International Conference on Image Analysis and Processing, 2015, pp. 432–443.
[22] D. Jin, P.K. Saha, A new fuzzy skeletonization algorithm and its applications to medical imaging, in: International Conference on Image Analysis and Processing, 2013, pp. 662–671.
[23] G. Levi, U. Montanari, A grey-weighted skeleton, Inf. Control 17 (1) (1970) 62–91.
[24] M.C. Maccarone, M. Tripiciano, V. Di Gesu, An algorithm to compute medial axis of fuzzy images, in: Scandinavian Conference on Image Analysis, vol. 1, 1995, pp. 525–532.
[25] G. Matheron, Examples of topological properties of skeletons, in: J. Serra (Ed.), Image Analysis and Mathematical Morphology, Part II: Theoretical Advances, Academic Press, 1988, pp. 217–238, chapter 11.
[26] F. Meyer, Skeletons in digital spaces, in: J. Serra (Ed.), Image Analysis and Mathematical Morphology, Part II: Theoretical Advances, Academic Press, 1988, pp. 257–296, chapter 13.
[27] B. Naegel, N. Passat, C. Ronse, Grey-level hit-or-miss transforms, Part I: unified theory, Pattern Recognit. 40 (2) (2007) 635–647.
[28] R.L. Ogniewicz, O. Kübler, Hierarchic Voronoi skeletons, Pattern Recognit. 28 (3) (1995) 343–359.
[29] S.K. Pal, A. Rosenfeld, A fuzzy medial axis transformation based on fuzzy disks, Pattern Recognit. Lett. 12 (10) (1991) 585–590.
[30] S. Peleg, A. Rosenfeld, A min-max medial axis transformation, IEEE Trans. Pattern Anal. Mach. Intell. 2 (1981) 208–210.
[31] C. Ronse, A lattice-theoretical morphological view on template extraction in images, J. Vis. Commun. Image Represent. 7 (3) (1996) 273–295.
[32] A. Rosenfeld, Distances between fuzzy sets, Pattern Recognit. Lett. 3 (1985) 229–233.
[33] P.K. Saha, G. Borgefors, G. Sanniti di Baja, A survey on skeletonization algorithms and their applications, Pattern Recognit. Lett. 76 (2016) 3–12.
[34] P.K. Saha, B.B. Chaudhuri, Detection of 3-D simple points for topology preserving transformations with application to thinning, IEEE Trans. Pattern Anal. Mach. Intell. 16 (10) (1994) 1028–1032.
[35] P.K. Saha, F.W. Wehrli, Fuzzy distance transform in general digital grids and its applications, in: 7th Joint Conference on Information Sciences, Research Triangular Park, NC, 2003, pp. 201–213.
[36] P.K. Saha, F.W. Wehrli, B.R. Gomberg, Fuzzy distance transform: theory, algorithms, and applications, Comput. Vis. Image Underst. 86 (3) (2002) 171–190.
[37] J. Serra, Image Analysis and Mathematical Morphology, Academic Press, New-York, 1982.
[38] J. Serra (Ed.), Image Analysis and Mathematical Morphology, Part II: Theoretical Advances, Academic Press, London, 1988.
[39] P. Soille, Morphological Image Analysis: Principles and Applications, Springer Science, Berlin, 1999.
[40] P. Soille, Advances in the analysis of topographic features on discrete images, in: International Conference on Discrete Geometry for Computer Imagery (DGCI), 2002, pp. 175–186.
[41] S. Svensson, A decomposition scheme for 3D fuzzy objects based on fuzzy distance information, Pattern Recognit. Lett. 28 (2) (2007) 224–232.

[42] S. Svensson, Aspects on the reverse fuzzy distance transform, Pattern Recognit. Lett. 29 (7) (2008) 888–896.
[43] S. Wang, A.Y. Wu, A. Rosenfeld, Image approximation from gray scale "medial axes", IEEE Trans. Pattern Anal. Mach. Intell. 3 (6) (1981) 687–696.
[44] X. Wang, E. Kerre, Reasonable properties for the ordering of fuzzy quantities, Fuzzy Sets Syst. 118 (3) (2001) 375–405.
[45] A.D. Ward, G. Hamarneh, The groupwise medial axis transform for fuzzy skeletonization and pruning, IEEE Trans. Pattern Anal. Mach. Intell. 32 (6) (2010) 1084–1096.
[46] L.A. Zadeh, The concept of a linguistic variable and its application to approximate reasoning, Inf. Sci. 8 (1975) 199–249.

CHAPTER

Unified part-patch segmentation of mesh shapes using surface skeletons

4

Joost Koehoorn[*], Cong Feng[*], Jacek Kustra[†], Andrei Jalba[‡], Alexandru Telea[*]

Institute Johann Bernoulli, University of Groningen, Groningen, The Netherlands[*] Philips Research, Eindhoven, The Netherlands[†] Department of Mathematics and Computer Science, Technical University Eindhoven, The Netherlands[‡]

Contents

4.1 Introduction 90
4.2 Related Work 91
4.2.1 Skeletonization 91
4.2.2 Shape Segmentation 93
4.2.2.1 Part-Based Segmentation 93
4.2.2.2 Patch-Based Segmentation 95
4.2.3 Summary of Challenges 95
4.3 Method 96
4.3.1 Preliminaries 96
4.3.2 Regularized Surface Skeleton Computation 97
4.3.3 Cut-Space Computation 99
4.3.4 Cut-Space Partitioning 100
4.3.4.1 Histogram Valley Detection 101
4.3.4.2 Histogram-Based Cut Space Partitioning 102
4.3.5 Partitioning the Full Surface Skeleton 102
4.3.6 Partition Projection to Surface 104
4.3.7 Part-Based Partition Refinement 105
4.3.8 Unified (Part and Patch) Segmentation 107
4.3.8.1 Patch-Type Segmentation Using Surface Skeletons 108
4.3.8.2 Unification Desirable Properties 110
4.3.8.3 Unification Method 110
4.4 Results 112
4.5 Discussion 115
4.6 Conclusion 119
References 120

Skeletonization. DOI: 10.1016/B978-0-08-101291-8.00005-5

4.1 INTRODUCTION

Shape segmentation is an important problem in many application domains such as computer-aided design, computer graphics, scientific visualization, and medical imaging. Informally put, shape segmentation aims at partitioning a given shape into several components or segments that capture application-specific part–whole relations as well as possible. Segmentation enables several shape processing and shape analysis tasks such as editing and content creation, identifying important features that occur in a large dataset, and shape matching, retrieval, and registration [46,3]. Segmentation methods can be roughly classified into *part-based* methods, which aim to split articulated shapes into the components that would be perceived as naturally distinct by humans [40], and *patch-based* methods, which aim to find quasi-flat components separated by edges on synthetic faceted models [41].

Shapes are typically encoded following a boundary representation or a volume representation. Boundary (explicit) representations capture the surface that partitions the shape interior from the surrounding exterior space, using various sampling and reconstruction schemes, e.g. polygonal meshes or point clouds [7]. Volume (implicit) representations, such as voxel models, store a densely sampled labeling of the space in which shapes are embedded, marking points as interior vs exterior [5]. Both representations have their advantages and limitations: Voxel volumes are easy to create and manipulate, but can be very expensive when high resolution is needed; surface meshes can efficiently model high-resolution shapes, but mainly support operations focusing on the shape surface, rather than its volumetric structure.

Skeletal representations are a third way to represent shapes. Informally put, skeletons jointly capture the geometry, symmetry, and topology of a shape in compact ways. 3D shapes admit two types of skeletons: *Surface* skeletons are the locus of centers of maximally inscribed spheres in the shape and as such capture geometry, symmetry, and topology. They generalize to 3D the well-known concept of a 2D symmetry axis [6]. *Curve* skeletons are one-dimensional structures locally centered in the shape that mainly capture a shape's part–whole structure. Skeletons combine the compactness of boundary representations with the ability to model and reason about volumetric properties. Additionally, they provide explicit and efficient access to a shape's symmetry structure and part–whole properties. As such, skeletal representations are a valuable tool to design shape segmentation methods [52].

Many skeleton-based segmentation methods have been proposed [52]. However, most of these methods use only the topological information encoded by *curve* skeletons. As such, they target mostly part-based segmentation of organic articulated shapes. Producing patch-based segmentations of faceted synthetic shapes or, more generally, mixed part-patch segmentations of shapes that fall between the two categories (articulated, faceted) is rarely handled [41,9,26].

Recently, surface skeletons have been used to produce part-based segmentations of 3D shapes [17,16]. A key to this idea is the construction of a so-called *cut space* containing a large number of well-designed cuts that partition the shape in ways similar to how a human would cut it. By analyzing this space a small number of cuts is retained to yield the final shape segmentation. Although the method was shown to deliver good results, it has several limitations. First, it only produces part-based segmentations although it uses the surface skeleton that fully describes any type of shape (articulated or faceted). Secondly, it only handles voxel representations and as such is very expensive, or even prohibitive to use, for high-resolution models.

In this chapter, we extend the skeleton cut-space segmentation proposal in [16] with the following main contributions:

- We show how to compute part-based, patch-based, and mixed part-patch segmentations that cover a wide range of 3D shapes using only surface skeletons; as such, we show that surface skeletons are *effective* tools for handling complex 3D shape segmentation problems;
- We propose a fully mesh-based implementation of our method that can efficiently handle very large high-resolution mesh models with low computational and memory costs; as such, we show that surface skeletons are *efficient* tools for handling complex 3D shape segmentation problems.

The goals of this chapter are twofold: On the practical side, we show how we can improve on existing part- and patch-based segmentation methods. On the theoretical side, we show that 3D surface skeletons, so far used only rarely in practice, can effectively and efficiently be used to support such applications.

The remainder of this chapter is structured as follows. Section 4.2 reviews related work in part-based and patch-based shape segmentation, with a focus on skeleton-based methods. It also introduces the cut-space idea in [16]. Section 4.3 details our method, explaining the changes and enhancements proposed to the original cut-space segmentation. Section 4.4 presents several results of our method and compares these with related skeleton-based segmentation methods. Section 4.5 discusses our proposal. Section 4.6 concludes the chapter.

4.2 RELATED WORK

Given that our focus is skeleton-based segmentation methods, we proceed by introducing necessary background on skeletonization (Section 4.2.1). Next, we overview several part- and patch-based skeleton-based segmentation methods, outlining their advantages and limitations (Section 4.2.2).

4.2.1 SKELETONIZATION

To define skeletons, we first introduce the Euclidean *distance transform* $DT_{\partial\Omega} : \Omega \to \mathbb{R}^+$ that associates to every point in the shape the distance to the closest boundary point

$$DT_{\partial\Omega}(\mathbf{x} \in \Omega) = \min_{\mathbf{y} \in \partial\Omega} \|\mathbf{x} - \mathbf{y}\|, \tag{4.1}$$

where $\|\cdot\|$ denotes the Euclidean distance. The so-called medial skeleton, or surface skeleton, $S_{\partial\Omega} \subset \Omega$ is next defined as

$$S_{\partial\Omega} = \{\mathbf{x} \in \Omega \mid \exists \mathbf{f}_1 \in \partial\Omega, \mathbf{f}_2 \in \partial\Omega, \mathbf{f}_1 \neq \mathbf{f}_2, \|\mathbf{x} - \mathbf{f}_1\| = \|\mathbf{x} - \mathbf{f}_2\| = DT_{\partial\Omega}(\mathbf{x})\}. \quad (4.2)$$

In the above, $\mathbf{f}_1$ and $\mathbf{f}_2$ are two of the contact points with $\partial\Omega$ of the maximally inscribed sphere in Ω centered at $\mathbf{x}$ and of radius $DT(\mathbf{x})$. Such points are also known as *feature points* [21,52], whereas the vectors $\mathbf{f}_i - \mathbf{x}$ are known as *spoke vectors* [49]. These definitions are captured by the so-called *feature transform*

$$FT_{\partial\Omega}(\mathbf{x} \in \Omega) = \arg\min_{\mathbf{y} \in \partial\Omega} \|\mathbf{x} - \mathbf{y}\|, \quad (4.3)$$

which associates to any point inside the shape all its feature points on the shape boundary.

Surface skeletons of 3D shapes implied by the definition in Eq. (4.2) consist of a set manifolds with boundaries, or skeletal sheets, that meet along so-called Y-intersection curves [14]. The pair $(S_{\partial\Omega}, DT_{\partial\Omega}|S_{\partial\Omega})$ is called the medial axis transform (MAT) of Ω [49]. The MAT can be used to fully reconstruct a shape, e.g., by computing the union of balls centered at points on $S_{\partial\Omega}$ and having as radii the values of $DT_{\partial\Omega}$ at the respective points. As such, the MAT provides a third type of shape representation, along boundary and volumetric ones.

Skeleton points can be classified by the order of tangency of their maximally inscribed spheres with $\partial\Omega$ [19]. This classification enables several applications such as robust edge detection on 3D surfaces [39,26], finding Y-intersection curves for patch-based segmentation [30,41], and surface reconstruction from point clouds [9]. Until recently, surface skeletons have been hard to compute for large and complex shapes [43,21]. Recent methods significantly alleviated such issues for both voxel [1, 24] and mesh [34,23] representations.

Besides surface skeletons, 3D shapes admit also *curve* skeletons. These are generically defined as one-dimensional (curve) structures that are locally centered within the shape [12,52]. Their lower dimensionality implies a simpler structure (branches meeting at junctions), which makes them easier to use for part-based segmentation [15,40,4,47]. Also, curve skeletons are easier to compute for large and complex 3D shapes as compared to surface skeletons, both for mesh representations [4,51] and voxel representations [1,24]. However, in contrast to the MAT, curve skeletons do not fully represent shapes, except when these have locally circular symmetry. In the following, we will focus exclusively on surface skeletons since the use or curve skeletons in shape segmentation is well covered in the literature (see also Section 4.2.2.1).

Skeletons are well known to be unstable to small-scale noise on the input shape surface $\partial\Omega$ [49,52]. As such, several so-called regularization methods have been proposed. *Local* methods use information such as the angle between feature vectors [18, 21], distance transform values, or divergence of the distance transform gradient [48] to prune skeletal points caused by noise. *Global* measures approximate the amount of boundary that "collapses" to or corresponds to a skeletal point. This approximation can be done by computing the length of the geodesic path between the feature points of a skeleton point (the so-called medial geodesic function or MGF [15,43,23]) or by explicitly simulating the advection of mass from $\partial\Omega$ onto S along the gradient of the distance transform [24]. Important skeleton points are next defined to correspond to larger parts of the input boundary. Thresholding global importance measures can deliver a so-called multiscale skeleton, which reflects the input shape at a user-chosen level of detail [52].

4.2.2 SHAPE SEGMENTATION

Let $\Omega \in \mathbb{R}^3$ be a three-dimensional shape with boundary $\partial\Omega$. Segmenting Ω typically amounts to computing a so-called partition $\mathcal{C}$ of Ω into components C_i so that $\bigcup_i C_i = \Omega$ and $C_i \cap C_j = \varnothing, \forall i, j, i \neq j$. In other words, the set $\mathcal{C} = \{C_i\}$ consists of disjoint components that fully cover Ω. We next denote the borders of these segments by ∂C_i.

As mentioned in Section 4.1, segmentation methods can be classified into part-based and patch-based. The difference between the two classes amounts to different constraints put on the segments C_i. Regardless of the method type, however, segmentation can be seen as a combination of two key decisions: (1) finding *where* to cut a shape or where to place the segment borders ∂C_i; and (2) finding *how* to cut, or which properties the segment borders should respect. We next discuss part- and patch-based segmentation methods using skeletons from this perspective. In the following, we denote by C_i^{part} segments resulting from a part-based segmentation $\mathcal{C}^{part}$ and by C_i^{patch} segments resulting from a patch-based segmentation $\mathcal{C}^{patch}$. The notations $\mathcal{C}$ and C_i are used for the unified part-patch segmentation and respectively its segments that we aim to compute.

4.2.2.1 Part-Based Segmentation

As already stated, most part-based segmentations focus on natural articulated shapes, such as humanoids, animals, plants, or other objects showing a clear part–whole hierarchical structure. Parts are typically defined as elongated regions of a shape that significantly "stick out" of the shape rump. Such parts are separated from the rump by a negative curvature region, a principle also known as the "minima rule" in cognitive theory [22,8]. Such parts are easily detectable using curve skeletons since they correspond roughly one-to-one to the curve skeleton terminal branches [13]. Separately from the above heuristic that tells where to place cuts, part-based segmentation methods exploit other perceptual principles to constrain the cut shapes, such as the "short cut rule," which states that a cut should be as short (and wiggle free) as possible [50]. Lee et al. segment mesh models using the minima rule and optimizing for short cuts using snake models [28]. Additionally, a "part salience rule," which captures how much a part sticks out of the shape rump, can be used to limit oversegmentation [29].

Several part-based segmentation methods use curve skeletons in their design. Li et al. sweep the curve skeleton with a plane to cut the shape and keep those cuts that have important geometric and topological changes indicating a part joining the shape rump [31]. Au et al. compute curve skeletons by iteratively contracting 3D meshes [4]. This enables them to backproject each skeletal point to one or several surface points. Hence, segmenting the curve skeleton into separate branches and next backprojecting each branch enable part-based segmentation. Along the same line, Reniers et al. first define curve skeletons as those points in Ω having at least two equal shortest-paths between their feature points [43] and then use closed loops (cuts) formed by such shortest paths placed at the curve-skeleton junctions to segment a shape [42]. Using shortest paths (geodesics) to construct cuts guarantees that these obey the desirable short cut rule. The method was refined to discard cuts that are far from planar, which reduces unneeded oversegmentation [40]. Serino et al. refine this idea by detecting three kinds of skeletal parts (simple curves, complex sets, and single points), which, by backprojection to the input shape, partition it in parts that protrude from a rump (called simple regions and bumps) and the rump itself (called a kernel) [45]. In comparison to [40], this method can yield better segment boundaries and suffers less from oversegmentation. The method in [45] was subsequently refined to use a computationally efficient and simple to implement curve-skeleton extraction based on the selection of a small number of centers of maximally inscribed balls (so-called anchor points) that guide an iterative voxel removal or thinning process. Apart from shape segmentation, this method can be also used in other contexts where a 3D curve skeleton is required. The problem of oversegmentation due to the potentially large number of curve-skeleton junctions, which lead to skeleton branches that do not map to salient shape parts, also discussed in [40], is elegantly addressed in [44] by so-called "zones of influence," which compare the local shape thickness at junctions with interjunction distances.

Separately, part-based segmentation and skeletonization have been shown to be related operations, which allow computing the latter from the former [32]. Conversely, Shapira et al. [47] note the same relationship but use it to segment a shape by computing a shape-diameter function (SDF) based on the boundary-to-curve-skeleton distance and finding cuts in places where the SDF has sharp variations. Finally, Tierny et al. [55] analyze the Reeb graphs computed for scalar functions defined on the shape surface to yield a hierarchical shape segmentation. Although Reeb graphs are not identical to Euclidean (curve) skeletons, they share their ability to capture topology, and thus such methods can be seen as skeleton-based segmentation techniques.

Overall, curve skeletons have established themselves as good descriptors for producing part-based segmentations for articulated shapes whose parts have a (near) tubular local geometry. Multiscale or hierarchical segmentations can also be easily obtained by considering the curve skeleton hierarchical structure or by pruning less important curve skeleton branches [42]. Since curve skeletons are locally centered in a shape, they yield largely pose-invariant segmentations. Finally, curve skeletons are easy and fast to compute for both voxel and mesh representations (Section 4.2.1).

Recently, *surface* skeletons of voxel models have also been used for part-based segmentation [17]. The key idea is to construct a *cut space* $CS = \{c_i\}$ that contains a large set of cuts that have suitable properties to act as segment boundaries. This solves the problem of *how* to cut. Next, a subset of these cuts is selected to become segment borders, thereby solving the problem of *where* to cut. Cut spaces are constructed by building closed loops formed by shortest paths on $\partial\Omega$ between the two feature points of each surface skeleton point. Next, cuts that represent suitable segment boundaries are found by analyzing a histogram of the cut lengths [17] or, alternatively, clustering cuts in terms of length [16]. Related methods of analyzing cut spaces for segmenting shapes have also been proposed, though not using skeletons to construct the cuts; see, e.g., [25,20].

4.2.2.2 Patch-Based Segmentation

In contrast to part-based methods, patch-based segmentation methods focus mainly on synthetic faceted objects such as those produced by CAD applications. A segment, or patch, is here defined as a region of the shape surface that is relatively flat and is separated from it surroundings by a high-curvature area, such as edges or creases. Like for part-based segmentation, patch boundaries should usually be wiggle free; however, they need not be tight.

Most patch-based methods work directly on the shape surface by unsupervised clustering, or grouping, mesh facets found to be similar [35,38,33,10]. Although such methods can produce very good results, they generally are parameter-sensitive. Closer to our interest, surface skeletons have been used for patch-based segmentation. The first such method we are aware of backprojects the (voxel-based) surface skeleton boundaries to the input surface, using the inverse feature transform (Eq. (4.3)) to yield segment boundaries [41]. A different approach is to segment the surface skeleton manifolds using Giblin's skeletal point classification [19] and backproject these to the input surface [9]. All such methods produce good patch segmentations even for shapes having soft edges. Such methods require high-throughput skeletonization tools and a delicate analysis of the resulting surface skeletons to detect boundaries and isolate manifolds.

A simpler patch-based segmentation method using surface skeletons was recently presented in [26]. Briefly put, this method implements the same strategy as [41], i.e., finding segment borders by backprojecting the boundary of the surface skeleton. However, in contrast to [41], this method handles point-cloud skeletons and has a much simpler implementation than [9]. We will use this method further in our unified segmentation pipeline (Section 4.3.8.1) and refer to it as the skeleton boundary backprojection (SBB) method.

Other methods: Outside the class of skeleton-based methods, many other methods exist for shape segmentation. Given our focus on using skeletons for this task, these methods are of lesser interest. For a general survey on 3D shape segmentation, we refer the interested reader to [46,3].

4.2.3 SUMMARY OF CHALLENGES

Summarizing the above discussion on skeleton-based segmentation methods, we identify the following requirements and challenges. A good such method should

1. be able to efficiently and directly handle high-resolution mesh models since computing skeletons of high-resolution voxel models is prohibitive in terms of memory and time;
2. guarantee smoothness of the resulting segment borders;
3. be able to produce patch-based, part-based, and mixed-type segmentations, thereby handling all types of shapes.

No existing skeleton-based segmentation method complies with all the above. In the remainder of this chapter, we present such a method based on a complete recasting of the cut-space idea in [16] to use mesh models and to handle patch, part, and mixed segmentations.

4.3 METHOD

4.3.1 PRELIMINARIES

To start with, we briefly outline the idea of the cut-space part-based segmentation in [17,16], which we next refer to as the "voxel cut-space segmentation" (VCS) (see also Fig. 4.1, top). Given a voxel shape $\Omega \subset \mathbb{Z}^3$, its surface skeleton S is first computed using the method in [24]. Next, a simplified skeleton S_τ is extracted from S by removing voxels having an importance lower than a small predefined value τ. These are essentially voxels close to the boundary ∂S that correspond to small-scale details on the surface $\partial\Omega$ (Fig. 4.1B, top). This regularization makes sure that cuts (to be computed next) are only created from stable important skeleton parts. In this step, for each voxel $\mathbf{x} \in S_\tau$, a cut $c(\mathbf{x}) \in \partial\Omega$ is computed by tracing three shortest paths $\gamma_1, \gamma_2, \gamma_3$ on $\partial\Omega$ between the feature points $\mathbf{f}_1$ and $\mathbf{f}_2$ of $\mathbf{x}$ (γ_1), $\mathbf{f}_1$ and $\mathbf{m}$ (γ_2), and $\mathbf{f}_2$ and $\mathbf{m}$ (γ_3), where $\mathbf{m}$ is the reflection on $\partial\Omega$ of the midpoint of γ_1 with respect to $\mathbf{x}$; see red curve in Fig. 4.1C, top. Cuts are piecewise-geodesic (thus, smooth and tightly wrapping around Ω), closed, and locally orthogonal to the object symmetry axis (see again Fig. 4.1C, top). These properties ensure that cuts form good candidates for segment boundaries. The cut space $\mathcal{CS} = \{c(\mathbf{x}) | \mathbf{x} \in S_\tau\}$ is next partitioned into several cut-sets $\mathcal{K}_i$ containing similar-length cuts, using either an analysis of the cut-length histogram [17] or hierarchical clustering [16] (Fig. 4.1D, top). Next, the borders ∂C_i^{part} of the segments are computed by searching for cuts that separate the cut-sets $\mathcal{K}_i$ (Fig. 4.1E, top). Finally, the actual segments C_i^{part} are computed by searching for connected components on $\partial\Omega$ separated by the borders ∂C_i^{part} (Fig. 4.1F, top). For full implementation details, we refer to [17,16].

In the remainder of this section, we outline how we adapt the above voxel-based pipeline to work on mesh-based shapes admitting a point-cloud skeleton and also handle mixed part-patch segmentations. Sections 4.3.2–4.3.7 describe our part-based pipeline (steps in Fig. 4.1B–G, bottom). Section 4.3.8 describes the patch-based pipeline and how this is unified with the part-based one (Fig. 4.1H–L, bottom).

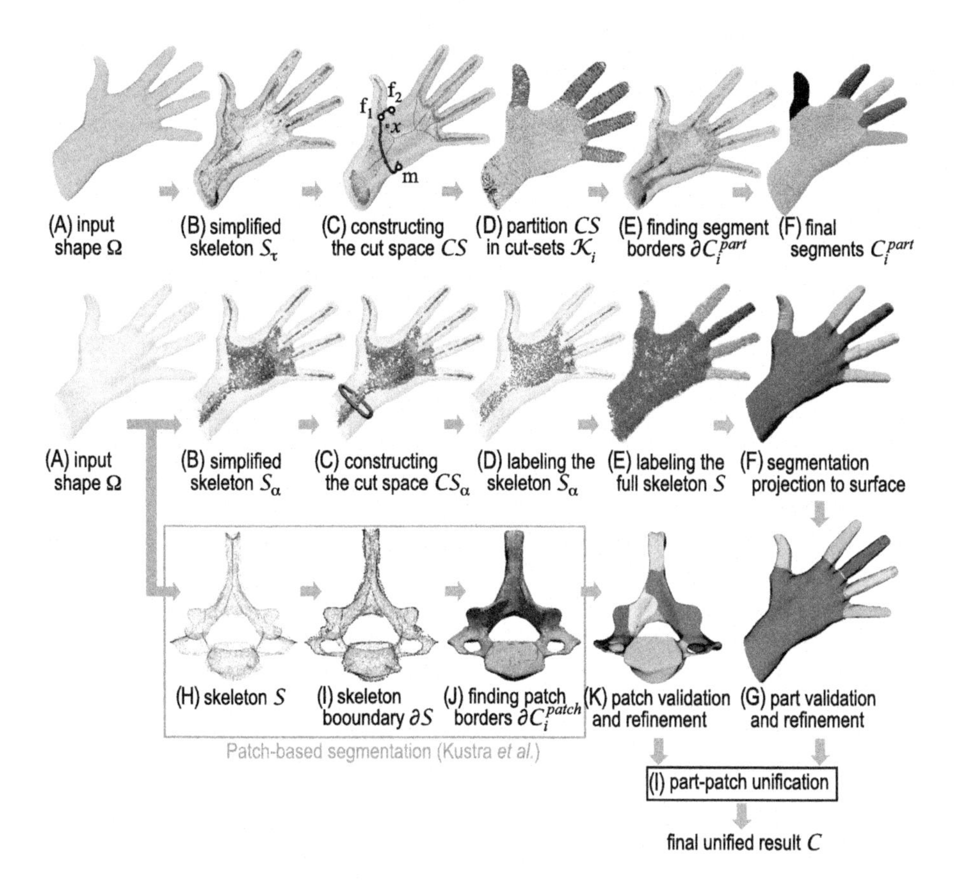

FIGURE 4.1

Top: cut-space segmentation pipeline from [17,16]. Bottom: our proposed pipeline.

4.3.2 REGULARIZED SURFACE SKELETON COMPUTATION

Step 1 of our proposed method parallels step 1 of VCS: We compute a simplified surface skeleton of the input shape. However, we have to skeletonize mesh shapes rather than voxel volumes. For this, we use the technique in [23], which is, to our knowledge, the fastest method to compute surface skeletons of mesh shapes to date. This method outputs a *point-cloud* representation of the skeleton S. Also, per skeleton point, the method computes two feature points, i.e., $\mathbf{f}_1$ and $\mathbf{f}_2$ in Eq. (4.2). In contrast, the voxel skeletonization method in [24] used by VCS outputs a voxel-based skeleton. As we shall see, point-cloud skeleton representations introduce several challenges to be addressed.

Similarly to VCS, we want to regularize S to avoid creating cuts from unimportant skeletal points that are caused by small-scale noise on $\partial\Omega$. At first sight, we could use for this the MGF importance metric delivered by the underlying skeletonization method [23] (see also Section 4.2.1). The rationale for this would be that the MGF metric is analogous to the collapse metric provided by [24] and used by VCS.

However, upon careful examination, we note several issues. If we compare the MGF and collapse metrics for the same shape (Figs. 4.2A, B), then we see that they are similar, except for points close to the shape curve skeleton, where the collapse metric attains much higher values than over the surrounding surface skeleton (red curves in Fig. 4.2B). Hence, thresholding the collapse metric, as done in VCS, eliminates noisy skeleton-points *and* also preserves the important curve-skeleton points, which capture the shape part–whole structure (Section 4.2.2.1). In contrast, thresholding the MGF metric eliminates noise, but also all skeleton points in thin shape areas, which is undesired. Fig. 4.2C illustrates this: At the current threshold level, the shown surface skeleton contains both important points (such as the one generating the well-oriented red cut) and unimportant points (such as the one generating the green cut). If we thresholded the MGF metric with values larger than the one corresponding to the green cut, i.e., around the value τ shown in the color legend, then we would loose about 70% of the entire skeleton, including the complete legs and ears of the horse model. Even a very conservative setting of the threshold $\tau = 0.01$ immediately eliminates detail parts such as the ears (see Fig. 4.2E). Hence, we cannot use the MGF metric to reliably keep skeletal points that correspond to small shape parts *and* in the same time eliminate spurious skeleton points that generate badly oriented cuts.

To solve this problem, we note that surface-skeleton points close to the curve-skeleton have large angles between their feature vectors [43]. Separately, skeleton points created by small-scale noise on the input surface have low MGF values and thus close feature vectors. Hence, we regularize the point-cloud skeleton using the angle between the feature vectors $\mathbf{f}_1$ and $\mathbf{f}_2$ of a skeletal point, which is a well-known local importance metric [18,21]. We define the simplified skeleton as

$$S_\alpha = \{\mathbf{x} \in S | \angle(\mathbf{f}_1, \mathbf{f}_2) > \alpha\} \tag{4.4}$$

where $\alpha = 120°$ is a fixed preset that delivered good results for all our tested shapes. Fig. 4.2D shows the result: The angle metric consistently gets high values close to the curve-skeleton branches of *all* shape parts (rump, legs, muzzle, ears) and gets low values close to the noisy boundary of the surface skeleton. Hence, if we threshold the surface skeleton above the value α indicated in the figure, then we robustly eliminate spurious cuts like the green one and keep cuts that wrap around the shape local symmetry axis (curve skeleton) like the red one. The resulting simplified skeleton S_α is shown in Fig. 4.2F. Note that this skeleton is not connected, first because it is just a point cloud, and next due to the removal of low-importance points. However, this does not pose any problem further on since we use it only to construct our segmentation cuts.

The proposed angle-based regularization of the surface skeleton is simple to implement (involves a dot product of the normalized feature vectors). Additionally, it eliminates the need to compute the MGF metric for surface skeletons, which is the most expensive part of the skeletonization method for mesh shapes in [23].

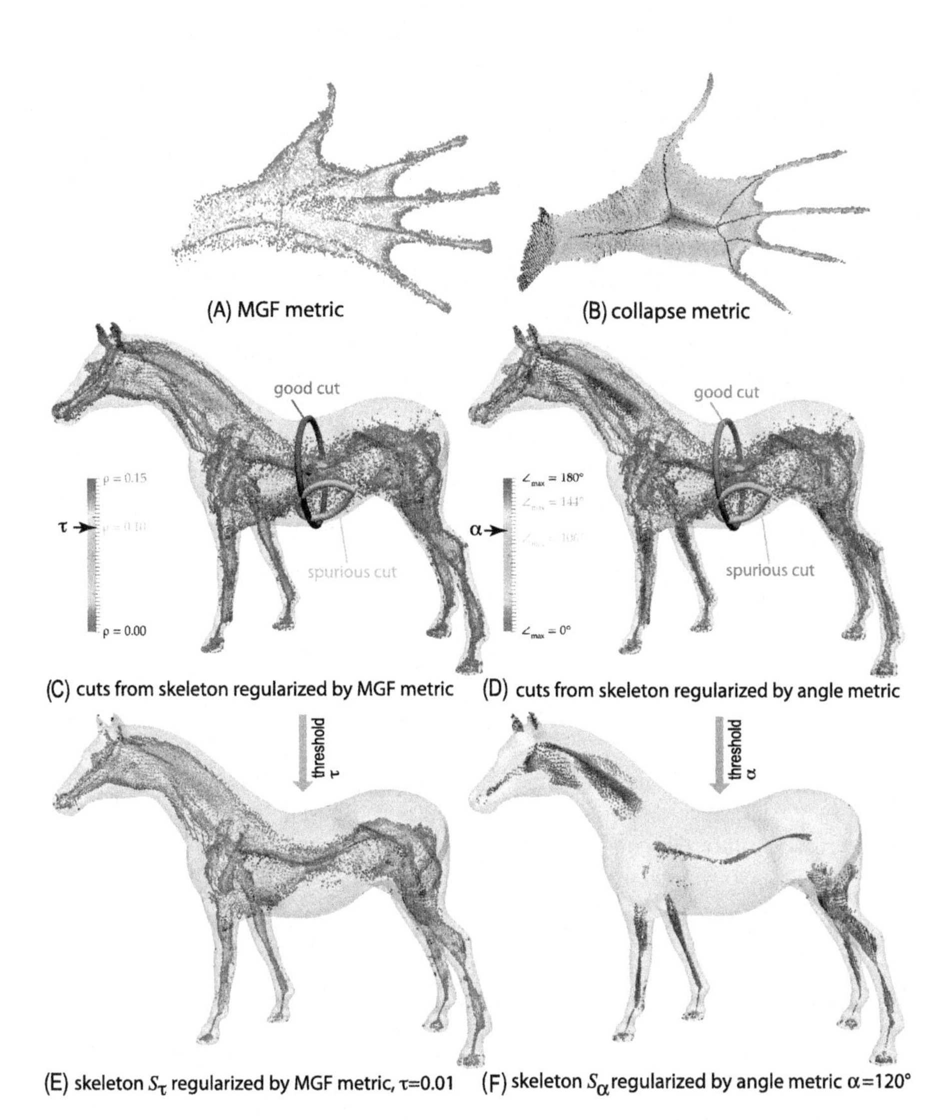

FIGURE 4.2

(A, B) Comparison of MGF and collapse importance metrics. (C) Thresholding the MGF metric keeps skeleton points that generate spurious cuts. (D) Thresholding the angle metric allows robust separating good from spurious cuts. All models are color-coded by the respective importance metrics.

4.3.3 CUT-SPACE COMPUTATION

Step 2 of our method computes the cut-space $\mathcal{CS}_\alpha = \{c(\mathbf{x}) | \mathbf{x} \in S_\alpha\}$ from the regularized skeleton S_α. To construct cuts, we cannot use the VCS approach, which employs

Dijkstra's algorithm to connect the feature points $\mathbf{f}_1$ and $\mathbf{f}_2$ of a skeleton point $\mathbf{x}$ by three shortest paths in the adjacency graph implied by the voxel surface $\partial\Omega$ (see again Fig. 4.1C and related explanation). If we did so, e.g., by using the adjacency graph implied by the surface mesh vertices and triangle edges, then we would obtain heavily zig-zagging cuts, which would be useless for our task of inferring segment borders. To create cuts, we propose here to use the geodesic-tracing technique introduced in [23] for the different purpose of evaluating the MGF skeleton-importance metric (see a discussion in Section 4.3.2).

In detail, we proceed as follows. A straightest geodesic (SG) $\gamma_S : \mathbb{R}^+ \to \partial\Omega$ is defined as the unique solution of the initial-value problem $\gamma_S(0) = \mathbf{p}$, $\gamma_S'(0) = \mathbf{v}$ with $\mathbf{p} \in \partial\Omega$ being a point on the shape boundary having tangent vector $\mathbf{v} \in T_\mathbf{p}$, where $T_\mathbf{p}$ is the plane tangent to $\partial\Omega$ at $\mathbf{p}$. Jalba et al. [23] proposed an extension to define shortest-and-straightest geodesics (SSGs) γ_{se} between two points $\mathbf{s} \in \partial\Omega$ and $\mathbf{e} \in \partial\Omega$ to be an accurate approximation of the SG from $\mathbf{s}$ to $\mathbf{e}$. SSGs are computed by tracing multiple straightest geodesics over tangent vectors $\mathbf{v}_i \in T_\mathbf{s}$ and then selecting the one with shortest length $\|\gamma_{S,i}\|$, i.e.,

$$
\begin{aligned}
\gamma_{S,i}(0) = \mathbf{s}, \quad \gamma_{S,i}'(0) = \mathbf{v}_i,\\
\gamma_{S,i}(\|\gamma_{S,i}\|) = \mathbf{e},\\
\gamma_{se} = \arg\min_i \left\|\gamma_{S,i}\right\|.
\end{aligned}
\tag{4.5}
$$

For computing the MGF metric, Jalba et al. computed the SSG between feature points $\mathbf{f}_1$ and $\mathbf{f}_2$ of each skeleton point $\mathbf{x} \in S$. For our purpose of computing shortest cuts, however, we require the geodesic to start and end in $\mathbf{f}_1$ and pass $\mathbf{f}_2$ somewhere in-between. As such, we redefine $\gamma_{S,i}$ as

$$
\gamma_{S,i}(0) = \mathbf{f}_1, \qquad \exists t \in \mathbb{R}^+ : \gamma_{S,i}(t) = \mathbf{f}_2, \qquad \gamma_{S,i}(\|\gamma_{S,i}\|) = \mathbf{f}_1. \tag{4.6}
$$

To compute γ_{se} using Eqs. (4.5) and (4.6), we proceed similarly to [23]: For each skeleton point $\mathbf{x} \in S$, we trace $M = 30$ straightest geodesics $\gamma_{S,i}$, $1 \leq i \leq M$, with starting directions $\mathbf{v}_i$ uniformly distributed in $T_\mathbf{s}$ and retain finally the one with minimal length. For each direction $\mathbf{v}_i$, we compute $\gamma_{S,i}$ by intersecting the mesh $\partial\Omega$ with the plane with normal $\mathbf{n}_i = \mathbf{f}_1 \times \mathbf{v}_i$ that passes through $\mathbf{s}$. In contrast to [23], we continue tracing until having found an intersection with both $\mathbf{f}_2$ and finally arrive back at $\mathbf{f}_1$. Finally, we gather all such SSGs to construct the cut-space

$$
\mathcal{CS}_\alpha = \{\gamma_{\mathbf{f}_1\mathbf{f}_2} | (\mathbf{f}_1, \mathbf{f}_2) \in FT_{\partial\Omega}|_{S_\alpha}\}, \tag{4.7}
$$

i.e., all cuts generated by points of the simplified skeleton S_α.

4.3.4 CUT-SPACE PARTITIONING

In step 3, we identify how to partition the cut-space $\mathcal{CS}_\alpha$ into cut-sets that correspond to the shape segments. We employ a similar approach to the VCS method, i.e., use

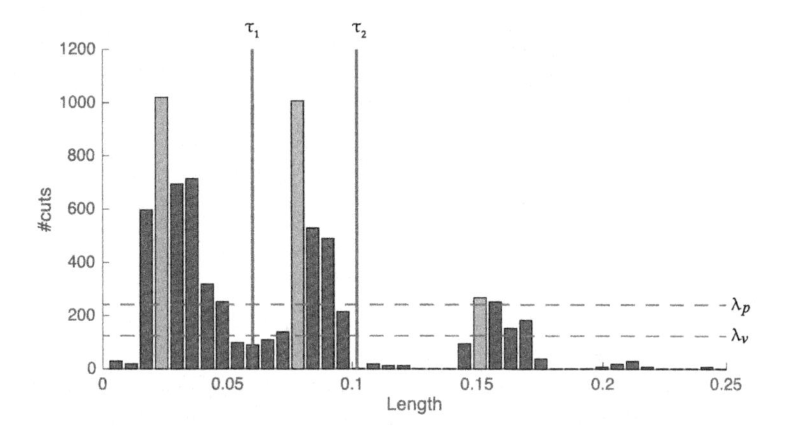

FIGURE 4.3

Histogram of cut-lengths of a horse shape from which two thresholds have been detected. The cuts in range $[0, 0.06)$ represent the horse legs, the range $[0.06, 0.1)$ corresponds with the neck and longer cuts correspond with the torso.

a histogram of cut lengths, in which peaks indicate many cuts with similar lengths, which likely belong to similar shape parts. With this assumption, we can find valleys separating the histogram peaks (Section 4.3.4.1), which in turn provide us with length thresholds to partition $\mathcal{CS}_\alpha$ (Section 4.3.4.2).

4.3.4.1 Histogram Valley Detection

To automatically and robustly detect histogram peaks and valleys, we need to analyze the histogram bins and their interrelationships. For this, VCS first searches for a bin high enough to be considered as peak, then continues searching for the next bin smaller than a certain quantity, which is considered a valley. Although this method does give an indication of where peaks and valleys are located, it suffers from not finding the smallest valley because of the greedy search for valleys. Moreover, what one would consider a valley depends on the neighboring bins. Hence, we propose to make the search for valleys taking into account their surroundings in the histogram.

To do this, we proceed as follows (see also Fig. 4.3). First, we use the mean shift algorithm [11] on the histogram bin heights to sharpen the differences between peaks and valleys. Next, we start scanning the histogram bins, left to right, for a peak that contains at least $\lambda_p = H_{\text{peak}} \cdot \|\mathcal{CS}_\alpha\|$ cuts. When found, we remember its height h_p and continue scanning and updating h_p as long as higher peaks are found. In the same time, we search also for valleys, i.e., for bins having fewer than $\lambda_v = H_{\text{decrease}} \cdot h_p$ cuts. This way, detection of a valley depends on the height of the peak (h_p) it follows. Just as for peak detection, when a bin having fewer than λ_v cuts is found, we remember its height h_v and continue scanning and updating h_v as long as lower peaks are found. If, during this valley-scan, we find a bin taller than λ_p, we have found a new peak. We have now two peaks and a valley of height h_v

in-between. We store the valley height $\theta_0 = h_v$ and continue scanning the histogram as above. After scanning the entire histogram this way, we obtain a set of valley heights $\Theta = \{\theta_i\}$. These will deliver our thresholds used to partition the cut space, as explained next in Section 4.3.4.2.

We established that good parameter choices are $H_{\text{peak}} = 0.01$ and $H_{\text{decrease}} = 0.25$, so that a peak should represent at least 1% of all cuts and a valley is smaller than a quarter of its accompanying peak.

4.3.4.2 Histogram-Based Cut Space Partitioning

Using the set of valley thresholds Θ, we can now partition $\mathcal{CS}_\alpha$. This can be done in linear time in the cut-space size $\|\mathcal{CS}_\alpha\|$ by iterating over histogram bins left-to-right and assigning all cuts between two consecutive thresholds θ_i, θ_{i+1} to the cut-set $\mathcal{K}_i$.

As for the VCS method, a cut-set $\mathcal{K}_i$ does not necessarily represent a compact shape segment, but could contain several such segments; for example, the five fingers of a hand have similar thickness, so they end in the same cut-set. To separate segments from cut-sets, we partition each cut-set into connected components, based on the connectivity of the skeleton points that generate the cuts. This was trivial to accomplish for VCS, given the explicit connectivity of skeleton voxels. In our case, the skeleton is an unorganized point-cloud lacking connectivity, as explained in Section 4.3.2. To address this, we define connectivity of skeletal points in terms of the neighborhood relation $\mathcal{N}_S(\mathbf{x})$ of a skeleton point $\mathbf{x} \in S$ to other skeleton points within a distance r from $\mathbf{x}$ as

$$\mathcal{N}_S(\mathbf{x}) = \{\mathbf{y} \in S \mid \mathbf{y} \neq \mathbf{x} \wedge \|\mathbf{x} - \mathbf{y}\| < r\}, \tag{4.8}$$

where setting r to 1% of the diameter of $\partial\Omega$ gives good results, following similar approaches in, e.g., [23,26].

The result of the cut-space partitioning is a labeling $\mathcal{L}_\alpha : S_\alpha \rightarrow \mathbb{N}$ that associates with each point in the simplified skeleton (or, alternatively, each cut in $\mathcal{CS}_\alpha$) an integer ID that tells the segment $C_i^{part} \subset \mathcal{C}^{part}$ these are part of. Given the nature of the neighborhood relation $\mathcal{N}_S$ and the imprecise nature of determining the thresholds in Θ, the labeling can be unstable for cuts whose skeleton points are very close to each other. To eliminate such variations, we apply a mode filter over all label values for points in the same neighborhood $\mathcal{N}_S(\mathbf{x})$, i.e., we assign to $\mathcal{L}_\alpha(\mathbf{x})$ the most frequently occurring label over $\mathcal{N}_S(\mathbf{x})$.

4.3.5 PARTITIONING THE FULL SURFACE SKELETON

The labeling $\mathcal{L}$ computed as outcome of the cut-space partitioning (Section 4.3.4.2) is not our final desired result, i.e., the segmentation $\mathcal{C}^{part}$ of $\partial\Omega$. Indeed, $\mathcal{L}$ only assigns segment IDs to a *subset* S_α of the entire surface skeleton S. To obtain the final part-based segmentation $\mathcal{C}^{part}$, we proceed in two steps. First, we extend the labeling $\mathcal{L}_\alpha$ from the simplified skeleton S_α to a labeling $\mathcal{L}$ of the full surface skeleton S. This is described in this section. Next, we project the full labeling $\mathcal{L}$ from S to the shape boundary $\partial\Omega$. This is described in Section 4.3.6.

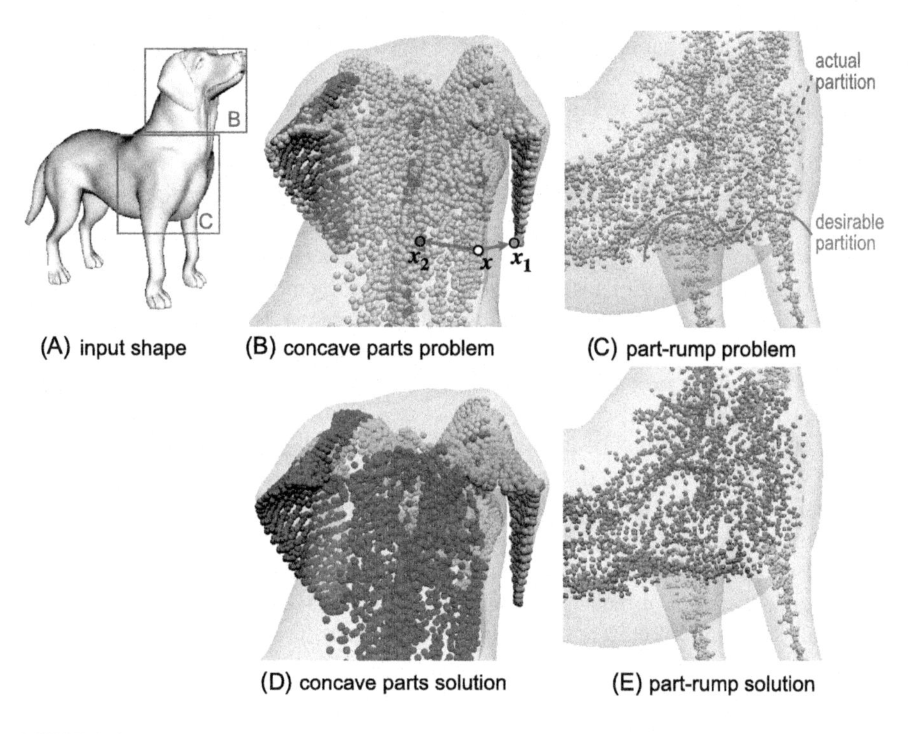

FIGURE 4.4

Label interpolation over the full skeleton. (A) Input shape. (B, C) Problems caused by naive nearest-neighbors interpolation. (D, E) Effective solution using the distance transform of the skeleton points.

Naive solution: To interpolate $\mathcal{L}_\alpha$ over the entire point-cloud skeleton S, one can use several strategies. A simple (but naive) one is to use a nearest-neighbor scheme where $\mathcal{L}(\mathbf{x} \in S \setminus S_\alpha) = \mathcal{L}_\alpha(\arg\min_{\mathbf{y} \in S_\alpha} \|\mathbf{x} - \mathbf{y}\|)$. This approach has several problems. First, using the $\mathbb{R}^3$ Euclidean distance metric to determine nearest neighbors will not work for nonconvex shapes that have nonconvex surface skeletons. Fig. 4.4B illustrates this for a dog model shown in Fig. 4.4A. Here, colored points are labeled points in S_α, and gray points are points to be labeled in $S \setminus S_\alpha$. The marked (red) point $\mathbf{x}$, which is located on the neck skeleton, is closer to point $\mathbf{x}_1$ (on the dog's ear) than to $\mathbf{x}_2$ (on the neck skeleton), so it will wrongly get the label of $\mathbf{x}_1$, i.e., be assigned as part of the ear. The second problem of this nearest-neighbor interpolation is that labels will always meet halfway between labeled points in S_α. This causes problems at junctions of parts having widely different thicknesses. Fig. 4.4C illustrates this. We see here a detail of the dog model having two label values over S_α, dark blue for rump points and cyan for points in the legs. The area where the two label values would meet, when using the simple nearest-neighbor interpolation, is shown in red, and it is actually the border of a generalized Voronoi diagram having all blue, respective cyan, points as two sites. This partition is undesirable since it would assign a large part of the rump skeleton to the segments corresponding to the legs. The desirable leg-rump partition is shown by the blue line in Fig. 4.4C.

Effective solution: To correct the first problem, one way would be to consider the geodesic distance along the S manifold rather than Euclidean distance in $\mathbb{R}^3$. Doing so is however not possible for skeletons represented as point clouds, and reconstructing smooth manifolds from such skeletons is a highly complex and expensive process [27]. Also, this would not correct the second problem. We next propose a much simpler and faster solution that addresses both problems. A key to this idea is the observation that the local shape thickness around a skeletal point should determine how far a label should be propagated. This local thickness is precisely represented by the distance transform $DT_{\partial\Omega}$ values on S. Hence, for a labeled point $\mathbf{x} \in S_\alpha$, we search all its neighbors $\mathbf{y}$ in a ball of radius $DT_{\partial\Omega}(\mathbf{x})$ and if $DT_{\partial\Omega}(\mathbf{y}) < DT_{\partial\Omega}(\mathbf{x})$, then assign $\mathcal{L}(\mathbf{y}) \leftarrow \mathcal{L}(\mathbf{x})$. The last distance condition is required to have labels of skeleton points in thick regions dominate those of nearby skeleton points in thin regions and also to ensure that the labeling does not depend on the order of visiting of the skeleton points. Finally, the (few) skeleton points that are still unlabeled after this operation are assigned the label of their nearest neighbor. Figs. 4.4D, E show the results of our proposal. As visible, both concave-part and part-rump problems mentioned earlier are now solved as desired.

4.3.6 PARTITION PROJECTION TO SURFACE

Having now a part labeling defined on all skeleton points, we map, or project, this labeling from S to $\partial\Omega$. For this, the original VCS proposal proceeds as follows. All cuts are tested for being borders of the final part segments C_i^{part}. A cut $c(\mathbf{x})$ from a cut-set $\mathcal{K}_i$ is deemed to be a border candidate if at least one of the 26 neighbors of the skeleton voxel $\mathbf{x}$ has a cut that belongs to a different cut-set $\mathcal{K}_j$, $j \neq i$. Next, a single border is picked from a set of border candidates that separates two cut-sets $\mathcal{K}_i$ and $\mathcal{K}_j$, based on heuristics involving the cut length.

In our context, there are several issues with using this method to project the labeling from the skeleton to the surface. First, our point-cloud skeleton does not admit a direct equivalent of the 26-neighbors relationship used for voxel shapes. The nearest-neighbor relation $\mathcal{N}_S$ (Eq. (4.8)) is too coarse to capture such fine details. Much more critically, though, is the fact that our cuts are *planar* SSGs, whereas the ones used by VCS consist of three geodesic curves, not necessarily located in a plane (see Section 4.3.3). As such, VCS cuts have a much larger freedom to model flexible segment boundaries than our cuts. However, despite their flexibility, the VCS cuts are still constrained by their construction process to consist of three geodesic curves (see Section 4.3.1). Although this appears to handle quite well part-based segmentation, it is arguably too rigid for producing good patch-based and mixed segmentations, which are our ultimate goal.

We propose a different way of projecting the segmentation information from the skeleton to the shape surface, which addresses all above issues. For each skeleton point $\mathbf{x} \in S$ having label $\mathcal{L}(\mathbf{x})$, we assign the label to all closest surface points to $\mathbf{x}$, i.e., set $\mathcal{L}(\mathbf{y} \in FT_{\partial\Omega}(\mathbf{x}) \leftarrow \mathcal{L}(\mathbf{x})$. Note that this may not assign labels to boundary points in convex surface regions since the skeletonization algorithm we use [23] only computes two feature points per skeleton point, rather than the full feature transform (Section 4.3.2). We compensate this by assigning labels to all yet unlabeled surface points by simple nearest-neighbor interpolation. Finally, we derive the part segments $C_i^{part} \subset \partial\Omega$ as the connected-components on $\partial\Omega$ that have the same label values.

4.3.7 PART-BASED PARTITION REFINEMENT

The previous step delivered a partition $\bigcup_i C_i^{part} = \partial\Omega$ of the input surface into segments. The final step of our part-based segmentation pipeline takes this partition and refines it to yield the final part-based segmentation. Three operations are performed in this refinement as follows.

Segment validation: First, segments C_i^{part} are checked as to their validity, in terms of what a human observer would qualify as being a "part" or not. Recalling the assumptions of part-based segmentation [20,40,17,16], we see that actual parts in a good part-based segmentation should be covered well by their associated cuts in the cut-space. Intuitively put, if we cut through a part with tight cuts orthogonal to the shape local symmetry axis, then the cuts should stay in the part. Conversely, segments whose cuts cover far more than the segment itself do not coincide with what is typically perceived as a part. To measure this coverage, let $K(C_i^{part}) = \{c \in \mathcal{CS} | c \cap C_i^{part} \neq \varnothing\}$ be the set of cuts in our cut-space that intersects segments C_i^{part}. We then define the coverage of C_i^{part} by cuts as

$$\nu(C_i^{part}) = \frac{\|\{c \in K(C_i^{part}) \mid c \setminus C_i^{part} = \varnothing\}\|}{\|K(C_i^{part})\|}, \tag{4.9}$$

i.e., the fraction of cuts passing through a part that are fully confined to that part. High values of ν indicate segments that are well covered by their cuts and thus that are plausible parts. We keep these in the final segmentation. Lower values of ν—below an empirically determined threshold of 0.8—indicate poor part properties. We remove such segments from the final result by merging them with one of their neighbor segments on $\partial\Omega$.

The coverage test effectively filters most, but not all, segments that are not part-like. A salient exception are corners of faceted objects such as the box in Fig. 4.5A. These segments are covered by cuts that nearly all stay inside them, as shown by the cut-space of the shape colored by the labels of the respective cuts (Fig. 4.5B). Hence, such segments will not be marked as invalid. A similar problem (without a solution) was highlighted by the curve-skeleton-based part segmentation technique in [40]. To discard such segments, we use a second heuristic: Parts should "stick out" of the shape as much as possible [29], or, in other words, they should have as many points with opposite normals. To test this, we compute angles between all surface

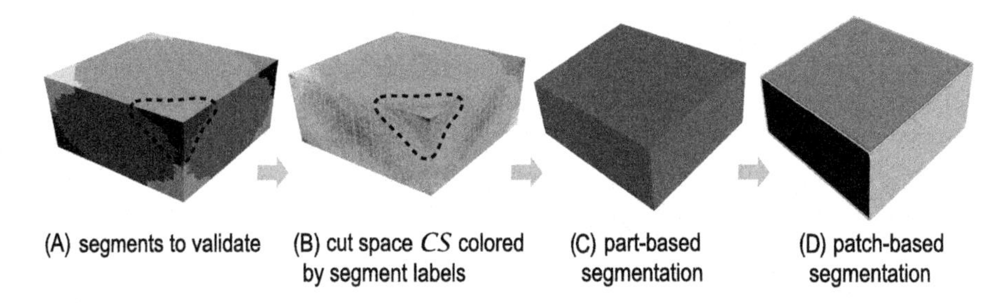

FIGURE 4.5

Part validation for a box shape. The produced segments (A) are validated by computing their cut-coverages (B). This results in a trivial part-based segmentation (C), which finally determines that a patch-based segmentation will handle this shape (D).

normals to points in a segment C_i^{part} and require that the median of these angles be at least 90°. For the box shape, this invalidates the corner segments, yielding a single segment for the entire shape in terms of part-based segmentation (Fig. 4.5C). As such, the patch-based segmentation of the same shape, which we next discuss in Section 4.3.8, gets full freedom to process the shape, leading to the expected result (Fig. 4.5D).

For completeness, we should note that our part validation heuristic is not the only possible one. For instance, Serino et al. propose a so-called "visibility criterion" [45], which aims to detect whether a peripheral part, far from the object main rump, actively contributes to a meaningful segmentation. This heuristic uses only the curve skeleton in its computation in a voxel-based setting. Since our pipeline only uses the *surface* skeleton and since computing the curve skeleton of mesh-based shapes is more complex when considering the technique in [23], which forms the backbone of our approach, the possibility of integrating this criterion in our framework is a topic of further investigation.

Segment border smoothing: Recall that segments C_i^{part} on $\partial\Omega$ are essentially backprojections of skeleton fragments sharing the same label values via the feature transform (Section 4.3.6). In the ideal continuous case, equivalent to an infinitely dense sampling of a smooth $\partial\Omega$, and a similar sampling of S, the segments would have smooth continuous borders. However, real-world meshes have a limited *and* often highly nonuniform sampling resolution. The used skeletonization method [23] essentially copies this sampling resolution to S. More critically, this method only uses the mesh vertices of $\partial\Omega$ as feature points for estimating the feature transform (Eq. (4.3)). As such, backprojecting labeled skeleton points to $\partial\Omega$ do not result in smooth segment borders. Fig. 4.6A shows this for a densely and uniformly sampled hand model of 197K vertices. As visible, segment borders exhibit problematic small-scale fractal-like noise. We solve this issue by computing smooth segment borders as follows. First, we create borders ∂C_i^{part} of the segments C_i^{part} by connecting the

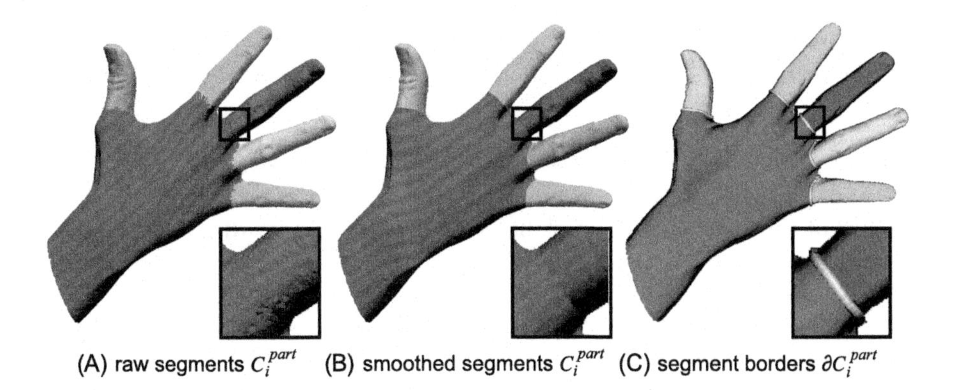

FIGURE 4.6

Computing part-segment borders. (A) Raw borders produced by skeleton-labeling backprojection. (B) Borders after Laplacian smoothing. (C) Visually emphasized borders.

midpoints of surface triangle-cell edges whose vertices have different labels (thus fall into different segments). Next, we apply classical Laplacian smoothing [54] to remove small-scale wiggles along these boundaries. After each smoothing pass (10 in total), we reproject the smoothed boundary ∂C_i^{part} to $\partial\Omega$ since unconstrained 3D smoothing makes the boundary curve leave the shape surface. The overall effect is identical to performing Laplacian smoothing *constrained* on the surface $\partial\Omega$. As the smoothed segment border moves on $\partial\Omega$, we update the labels of the vertices to ensure consistency. Fig. 4.6B shows the result: The smoothed boundaries stay roughly in the same position as the initial ones but are considerably tighter and smoother, thus similar in terms of desirable requirements to the original piecewise-geodesic boundaries of the VCS method. Finally, we draw these smooth boundaries as thick 3D tubes for ease of perception (Fig. 4.6C).

Visual representation: Our segmentation $\mathcal{C}^{part}$ describes parts on $\partial\Omega$ in terms of *vertices* having the same label. For all practical purposes, such as visualization or further geometric processing, we need a *cell*-based description. For this, we split the triangle cells in $\partial\Omega$ in a Voronoi-like fashion to interpolate, in the nearest-neighbor sense, the categorical vertex label values.

4.3.8 UNIFIED (PART AND PATCH) SEGMENTATION

So far, we described how to produce *part-based* segmentations $\mathcal{C}^{part}$ of mesh shapes. However, as outlined already in Section 4.1, our goal is to segment any shape and thus to propose a way to combine part-based and patch-based segmentations in a flexible way. We present here a way to combine the two segmentation types in a new segmentation model, which we refer to as *unified* (part-and-patch) segmentation $\mathcal{C}$. We first introduce the patch-based segmentation method we use (Section 4.3.8.1). Next, we discuss the desirable properties of a good unification method and possible strategies to implement it (Section 4.3.8.2). Finally, the unification technique we propose is presented in Section 4.3.8.3.

4.3.8.1 Patch-Type Segmentation Using Surface Skeletons

For patch-based segmentation, we use the skeleton boundary backprojection (SBB) method of Kustra et al. [26], which has several desirable properties in our context. First, this method treats high-resolution mesh shapes, which is our application target. Secondly, the method is also based on surface skeletons, which matches our toplevel goal of showing how such skeletons can effectively support shape segmentation. Thirdly, the method uses the same skeletonization technique [23] as our part-based segmentation, which makes easy the technical combination of part and patch segmentation. We next briefly explain this method and also outline its limitations relevant to our unified segmentation goal.

As all other patch segmentation methods, Kustra et al. define patches C_i^{patch} implicitly by requiring that borders separating them should occur in high surface-curvature areas. To do this, they proceed as follows. First, the point-cloud surface skeleton S of the input shape Ω is computed using the technique in [23]. Next, points on the boundary ∂S of the surface skeleton, or so-called A_3 points [19], are detected as those skeletal points whose images on $\partial\Omega$, via the feature transform, contain a single compact cluster. However, as noted earlier in this paper, the underlying skeletonization method [23] does not compute the full feature transform, but only two feature points per skeleton point. Kustra et al. note that computing the exact (full) feature transform is sensitive, so they propose instead the so-called extended feature transform

$$FT^{\tau}_{\partial\Omega}(\mathbf{x} \in S) = \{\mathbf{f} \in \partial\Omega \| \|\mathbf{x} - \mathbf{f}\| \leq DT_{\partial\Omega}(\mathbf{x}) + \tau\}, \tag{4.10}$$

which gathers, for each skeletal point $\mathbf{x}$, all boundary points within a range $DT_{\partial\Omega}(\mathbf{x}) + \tau$, where τ is a small positive value. The extended feature transform provides a conservative approximation of the actual feature transform (Eq. (4.3)), and can be readily used to detect ∂S as outlined above. Similar conservative approximations of the feature transform are also proposed by the VCS method for voxel shapes [17,16].

After the skeleton boundary is detected, the method projects ∂S to the shape surface via the extended feature transform, i.e., compute the set

$$\Phi = \{\mathbf{f} \in FT^{\tau}_{\partial\Omega}(\mathbf{x}) \mid \mathbf{x} \in \partial S\}, \tag{4.11}$$

which conservatively captures convex edges on $\partial\Omega$. Patches C_i^{patch} are now easily found as the connected components of $\partial\Omega \setminus \Phi$. Finally, points in Φ get assigned to the closest established patch, thereby completing the patch-based segmentation $\mathcal{C}^{patch}$ of $\partial\Omega$.

Fig. 4.7 (top row) shows the results of the patch-based segmentation method described above on two shapes (*fandisk* and *horse*). For *fandisk*, which has clear

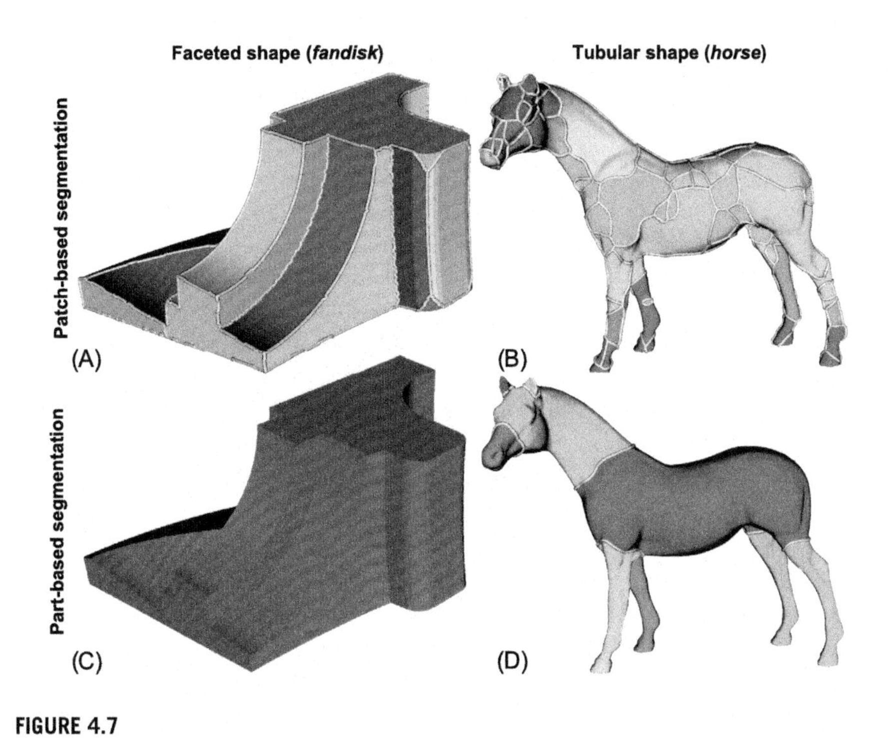

FIGURE 4.7

Two typical patch-type and part-type shapes, segmented by patch-based and part-based segmentation.

and salient edges, a very good patch segmentation is produced (Fig. 4.7A). The *horse* shape has much softer and fuzzier edges and as such yields a poor patch-type segmentation (Fig. 4.7B). This can be explained as follows: Finding a reliable set Φ that captures edges on the surface $\partial\Omega$ requires computing stable A_3 points to detect the skeleton boundary ∂S (Eq. (4.11)). However, the heuristic described earlier of finding A_3 points via the extended feature transform fails for cylinder-like parts of a shape, as discussed in [26]. Although fine-tuning various parameters of the method of Kustra et al. sometimes improve results, patch-type segmentation does not work well for shape regions having too soft edges. In contrast, using our part-based segmentation method described in Sections 4.3.2–4.3.7 yields very good results for tubular soft-edged shapes like *horse* (Fig. 4.7D). Conversely, part-based segmentation produces a poor result for the faceted shapes like *fandisk* (Fig. 4.7C). In this case, no part is identified, which is correct, given the part-validation criteria outlined in Section 4.3.7. Hence, to obtain the best segmentation results for all shape types, including those that have a mix of faceted and tubular parts, a unification of part- and patch-based segmentation is needed. Such a method is proposed next.

4.3.8.2 Unification Desirable Properties

Before designing a part-patch unification strategy, we must define its desirable properties. Based on our experience, we outline the following key properties:

1. *hybrid:* the strategy should handle shapes admitting a full part-based segmentation, shapes admitting a full patch-based segmentation, and also shapes admitting a mix of the two segmentation types in various areas;
2. *intuitive:* the unified segmentation should make sense according to human perception, that is, the produced parts, respectively patches, should visually make sense with respect to the part and patch properties established in Sections 4.2.2.1 and 4.2.2.2;
3. *balanced:* unified segmentations should not be oversegmented due to incorporating both segment types.

The simplest unification approach would be to compute separate part-based and patch-based segmentations and then choose the result that maximizes the respective quality properties of the two segmentations. However, this solution cannot handle shapes that require a hybrid segmentation. An alternative is to compute the two segmentation types separately and then *mix* their results in terms of selecting the optimal segments with respect to both local quality criteria (that decide which type of segmentation best fits a region of the shape) and global criteria (the user's preference for part- or patch-based segmentations). This approach can satisfy all above-mentioned requirements. As such, we next choose this way as follows. For part-based segmentation $\mathcal{C}^{part}$, we use our pipeline presented in Sections 4.3.2–4.3.7. For patch-based segmentation $\mathcal{C}^{patch}$, we use the method discussed in Section 4.3.8.1. Our unification method is discussed next.

4.3.8.3 Unification Method

As outlined in Section 4.3.8.2, our approach to a unified segmentation is to compute both part- and patch-based segmentations of the shape and next decide which of the produced segments are *valid* in terms of specific part and patch requirements. Valid segments are kept and finally merged to yield the unified segmentation. The process is detailed next.

Part validation: For part-based segmentation, we use the segment validation procedure described in Section 4.3.7, which consists of the cut-coverage and angle-coverage criteria. As such, part-based segmentation will only produce valid part-segments C_i^{part} in areas where such segments can be computed. When and where valid segments cannot be computed, the shape will be left unsegmented.

Patch validation: For patch-based segmentation, the method of Kustra et al. does not provide patch validation. Hence, this method may oversegment the shape or produce otherwise suboptimal patches, such as shown by the example in Fig. 4.7B. Merging such suboptimal patches with otherwise good parts will yield an overall poor unified segmentation. We address this by proposing a patch validation scheme, similar in spirit to the part validation discussed earlier.

As already outlined, a good (valid) patch C_i^{patch} should have its borders in high-curvature regions of the input surface $\partial\Omega$. Several methods exist for computing curvature on mesh surfaces, such as using the curvature tensor [36,53], using moment analysis [10], or estimating the shape thickness over the surface-skeleton boundary [26]. The first two methods are quite sensitive in terms of setting the scale at which edges are detected, as also noted in [26]. The last method, which is also used to detect patch borders in the segmentation technique described in Section 4.3.8.1, does not have this problem, but captures also soft edges, leading to issues such as the oversegmentation in Fig. 4.7B.

For our patch-segmentation context, we need to compute segment borders that follow as precisely as possible salient edges on $\partial\Omega$. We propose a simple but effective edge detector for this, as follows. Let $V = \{(\mathbf{x}, \mathbf{n}(\mathbf{x}))\}$ be the vertices and their normals of the surface mesh $\partial\Omega$, and let $E = \{(\mathbf{x}, \mathbf{y}) \mid \mathbf{x} \in V, \mathbf{y} \in V\}$ be the mesh edges. For a mesh point $\mathbf{x} \in V$, denote by

$$\mathcal{N}_V(\mathbf{x}) = \{\mathbf{y} \in V \mid \|\mathbf{x} - \mathbf{y}\| < r\} \tag{4.12}$$

the nearest neighbors of $\mathbf{x}$ in the mesh within a radius r. Next, let

$$\partial\mathcal{N}_V(\mathbf{x}) = \{\mathbf{y} \in V \setminus \mathcal{N}_V(\mathbf{x}) \mid (\mathbf{x}, \mathbf{y}) \in E\} \tag{4.13}$$

be the boundary of the neighborhood $\mathcal{N}_V(\mathbf{x})$. We approximate the curvature $\kappa(\mathbf{x})$ of the surface at point $\mathbf{x}$ by the mean-angle of all point combinations in $\partial\mathcal{N}_V(\mathbf{x})$, i.e.,

$$\kappa(\mathbf{x}) = \text{mean}\{\angle(\mathbf{n}(\mathbf{y}), \mathbf{n}(\mathbf{z})) \mid (\mathbf{y}, \mathbf{z} \neq \mathbf{y}) \subset \partial\mathcal{N}_V(\mathbf{x}) \times \partial\mathcal{N}_V(\mathbf{x})\}, \tag{4.14}$$

where $\times$ denotes Cartesian product.

Fig. 4.8 shows our curvature estimator κ for two shapes. As visible, κ captures all zones where salient edges exist, both for the *fandisk* model, which exhibits clear edges, and for the *frontal bone* model, which has much noisier edges. Here, the neighborhood size r (Eq. (4.12)) is set to roughly 2 to 5% of the diameter of $\partial\Omega$. This setting has given good results for all other tested shapes.

Using the curvature field κ, we can now find and remove invalid patches. For this, we consider each border fragment $\mathcal{B}_{ij}$ that separates patches C_i^{patch} and C_j^{patch}, which is defined as

$$\begin{aligned}\mathcal{B}_{ij} = &\left\{\mathbf{x} \in C_i^{patch} \,\middle|\, \exists \mathbf{y} \in C_j^{patch}, (\mathbf{x}, \mathbf{y}) \in E\right\} \\ &\cup \left\{\mathbf{x} \in C_j^{patch} \,\middle|\, \exists \mathbf{x} \in C_i^{patch}, (\mathbf{x}, \mathbf{y}) \in E\right\}.\end{aligned} \tag{4.15}$$

For each such border fragment, we compute the median curvature $\hat{\kappa}(\mathcal{B}_{ij})$ over all its vertices $\mathbf{x} \in \mathcal{B}_{ij}$. If $\hat{\kappa}(\mathcal{B}_{ij})$ is larger than a threshold $\beta = \pi/4$, determined empirically for our studied models, then $\mathcal{B}_{ij}$ is a valid (good) boundary of our patch-based

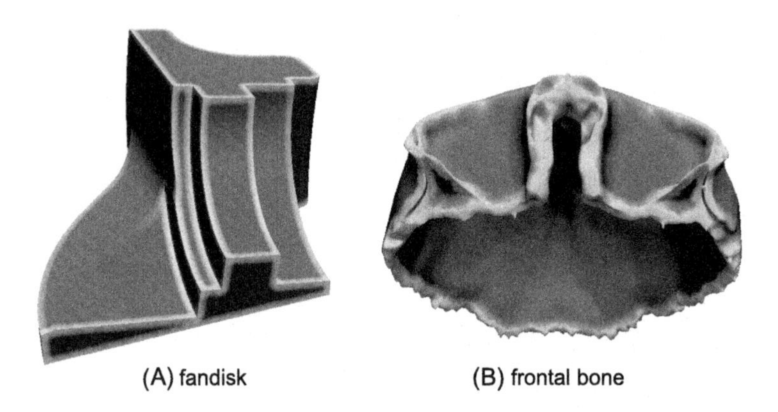

FIGURE 4.8

Proposed curvature detector κ for two shapes.

segmentation, so we keep it. If not, then we erase $\mathcal{B}_{ij}$ from the segmentation, i.e., merge patches C_i^{patch} and C_j^{patch}. To ensure a deterministic patch-validation result, we process patch border segments $\mathcal{B}_{ij}$ in increasing order of $\hat{\kappa}(\mathcal{B}_{ij})$.

Merging segmentations: After removing invalid patches, we have both a validated part-based ($\mathcal{C}^{part}$) and patch-based ($\mathcal{C}^{patch}$) segmentation. We now construct the unified segmentation $\mathcal{C}$ by merging $\mathcal{C}^{part}$ and $\mathcal{C}^{patch}$ as follows. We first initialize $\mathcal{C}$ with $\mathcal{C}^{part}$, i.e., keep all part segments that have been already validated (see Section 4.3.8.3). Next, we merge in $\mathcal{C}$ the patches C_i^{patch}. This essentially refines, or subdivides, those parts that have been found to consist of several patches in $\mathcal{C}^{patch}$.

Let us explain how the unified segmentation yields the desirable results for the shapes in Fig. 4.7. For the *fandisk* shape, $\mathcal{C}^{patch}$ contains patches having very high curvature along their borders (Fig. 4.7A). Hence, all borders shown in the figure will be validated and kept as shown. In contrast, its $\mathcal{C}^{part}$ contains a single part (Fig. 4.7C). The merging process outlined above will split this single part into precisely the patches of $\mathcal{C}^{patch}$. Hence, the unified result will be identical to $\mathcal{C}^{patch}$, as desired. For the *horse* shape, $\mathcal{C}^{patch}$ contains patches having (very) low curvature values along their borders (Fig. 4.7B). As such, all these patches will be invalidated, yielding $\mathcal{C}^{patch} = \varnothing$. Hence, the merging process will keep $\mathcal{C}^{part}$ unchanged. This is the desired result since $\mathcal{C}^{part}$ for this shape is of good quality (Fig. 4.7D).

4.4 RESULTS

We next present several segmentation results obtained with our unified method.

We start by comparing part-based segmentations. Fig. 4.9 compares our method with the original voxel-based cut-space segmentation (VCS) in [16]. As visible, our

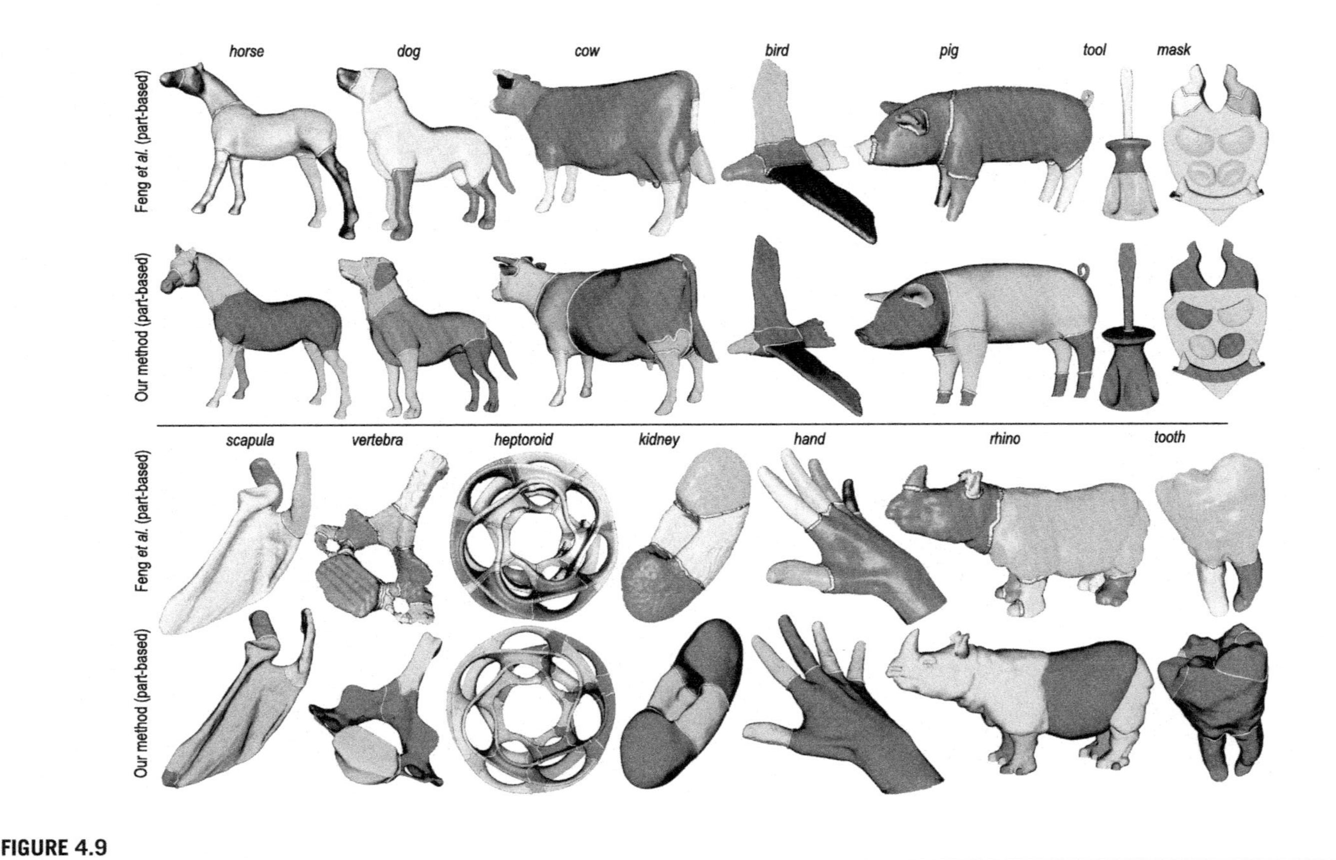

FIGURE 4.9

Comparison of segmentations between VCS (Feng et al. [16]) and our part-based method.

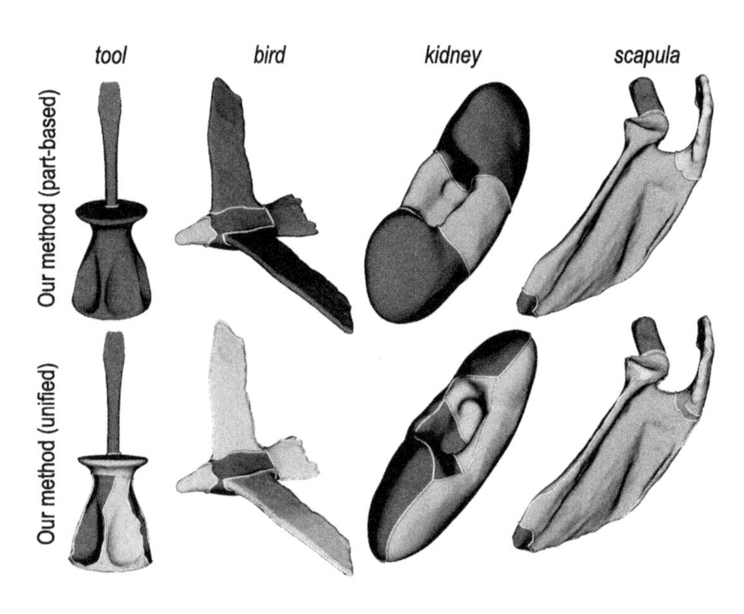

FIGURE 4.10

Soft-edged shapes segmented by part and unified techniques with our method.

results are very similar in terms of position, smoothness, and tightness of the delivered segments. We also see that our method succeeds in cases where VCS visibly undersegments the shape (*tool*, *mask*). Small-scale details may differ between the two segmentations, such as for the *rhino* model where VCS undersegments the toes, whereas our method undersegments the horn and ears. In this particular case (*rhino*), the VCS segmentation is likely better since separating the larger horns and ear details should be more important than separating the smaller toes. These variations are explainable by the two different parameter settings of the two methods.

Fig. 4.10 shows the added-value of the unified segmentation for a set of shapes having relatively soft (shallow) edges. We see how the unified segmentation (bottom row) refines the already-discussed part-based segmentation (top row) by splitting segments along lines of high curvature, e.g., the edge of the bird wing or of the scapular bone. This allows getting a mix of parts that capture the shape topology, and faces, or patches, that capture the shape geometry. The unified segmentation can thus be seen as a refinement of the part-based segmentation.

Fig. 4.11 compares our unified segmentation with the SBB method of Kustra et al. [26], which is also using surface skeletons to produce patch-based segmentations of 3D shapes. The figure contains the anatomic shapes used as benchmark in [26]. These are quite challenging to segment since they have complex geometries, a mix of sharp and soft tortuous edges, and consist of both parts and patches. As visible, our method is able to find all patch-like segments that the SBB method can. However,

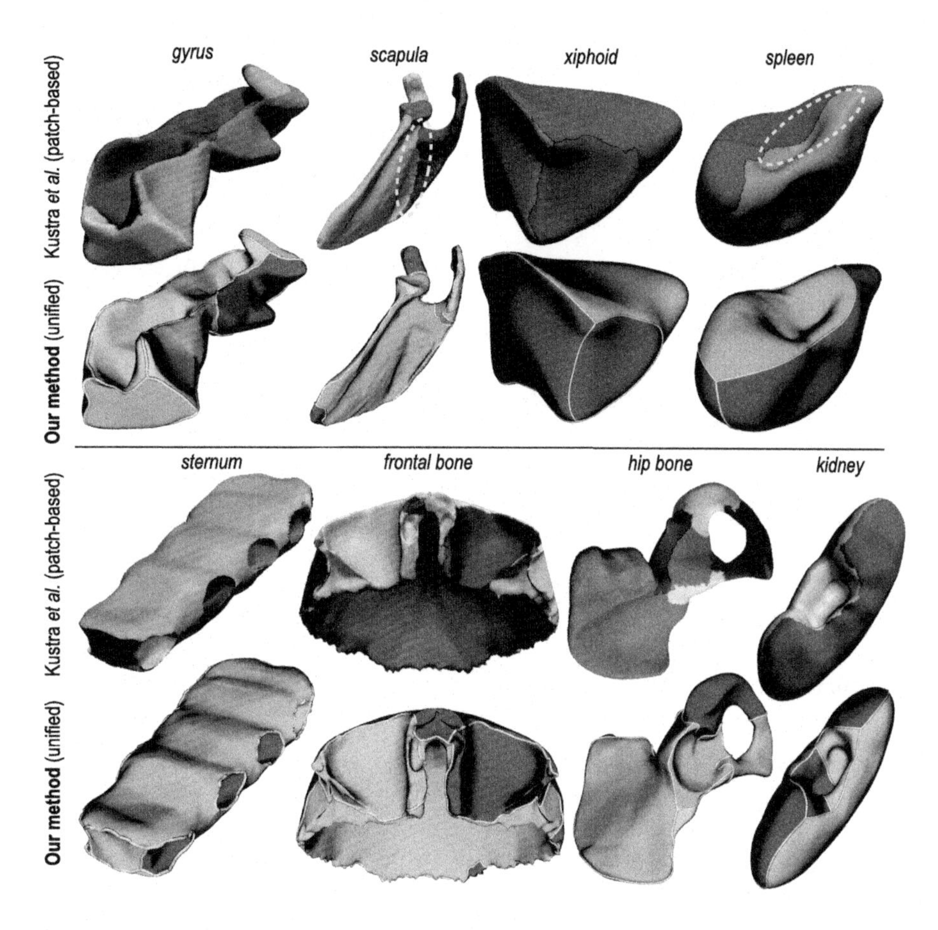

FIGURE 4.11

Anatomic shapes segmented by the SBB method of Kustra et al. [26] and our unified method.

our method generates visibly smoother segment borders, which also better follow the shape edges, even in the case of very complex geometries such as *frontal bone* and *gyrus*. We also see that SBB places two segment borders in areas where no apparent patch or part transition occurs (*scapula* and *spleen*, marked details). In contrast, our unified method does not allow such borders to exist due to its part and patch validation steps (Sections 4.3.7 and 4.3.8.3).

4.5 DISCUSSION

We next discuss several aspects of our unified segmentation method.

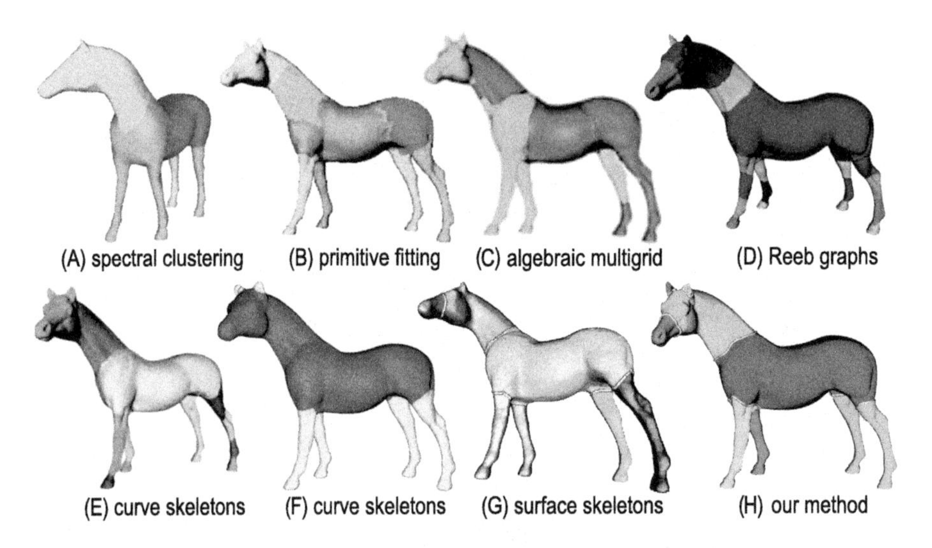

FIGURE 4.12

Comparison of eight part-based segmentation methods. (A) Liu and Zhang [33]; (B) Attene et al. [2]; (C) Clarenz et al. [10]; (D) Tierny et al. [55]; (E) Lien et al. [32]; (F) Reniers et al. [40]; (G) Feng et al. [16]; (H) our method.

Comparison: Section 4.4 has compared our method with the two main related methods, which use surface skeletons to segment shapes (VCS [16] and SBB [26]). It is also interesting to compare our method with the larger class of shape segmentation techniques out there. Although it is impossible to do so in the same detail as provided in Section 4.4, Fig. 4.12 shows a qualitative comparison of our method with seven well-known part-based segmentation methods. Images (A–D) correspond to methods that do not use skeletons. Images (E–F) correspond to methods that use either curve or surface skeletons. For an overview of these methods, we refer to Section 4.2.2.1.

The most salient observation on Fig. 4.12 is that skeleton-based methods tend, in general, to provide more "natural" part-based segmentations than the other studied methods in terms of positioning and smoothness of the segment borders. This is due to the inherent ability of skeletons to model a shape local axis of symmetry, across which segment borders can be fit. In contrast, non-skeleton-based methods cannot ensure this proper border orientation. Secondly, we see that the cut-space based methods (Figs. 4.12G, H) do not *oversegment*. This observation is also confirmed by all other earlier examples showing our method (Figs. 4.9, 4.10, 4.11). This is due to two design elements in our method: (1) the cut-space partitioning ensures that similar-length cuts will never yield different parts (Section 4.3.4), and (2) the part and patch validations ensure that superfluous parts and patches are automatically removed (Sections 4.3.7 and 4.3.8.3). This is in contrast to several of the other methods depicted here (Figs. 4.12B–E). All in all, the above observations strongly plead for the added-value of skeletons for shape segmentation.

Table 4.1 Performance of the proposed unified segmentation method

Shape	$\|\partial\Omega\|$	$\|CS_\alpha\|$	t_{skel}	t_{cuts}	t_{patch}	t_{unif}	$T_{unified}$
Horse	49,749	7453	5.6	63.2	5.2	6.4	5866
Hound	16,158	364	0.7	5.1	2.0	2.7	1153
Cow	137,862	16,447	23.0	242.3	10.9	63.0	9357
Bird	47,184	26,485	6.3	262.0	5.3	14.1	4769
Pig	4800	756	0.2	2.0	0.6	0.6	1814
Vertebra	22,789	7779	0.2	57.4	2.4	2.2	3088
Scapula	117,432	83,340	17.3	2325.7	11.0	26.6	4208
Heptoroid	79,056	63,876	5.0	539.1	6.3	5.8	9367
Kidney	30,389	9986	6.5	177.7	4.0	27.0	1707
Hand	49,546	2815	7.3	47.9	4.2	11.1	2911

The fact that there is quite some variation in the segmentations produced by different methods and, implicitly, in their perceived quality, should however be interpreted with care. The comparison presented here, as is far from exhaustive, cannot thus be used to derive generalizing value judgments of one method vs another. For instance, the examples in Fig. 4.12E, F show that curve-skeleton-based methods can sometimes oversegment (see horse legs, Fig. 4.12E) and sometimes undersegment (see horse neck and rump, Fig. 4.12F). The fact that surface-skeleton-based methods do not show such artifacts (Fig. 4.12G, H) should not be interpreted as pointing to a general superiority of such methods as opposed to curve-skeleton-based methods. The main conclusion from this comparison is limited to showing that segmentation methods using *surface* skeletons are an interesting and viable alternative for *part*-based shape segmentation.

Unification: If we study the state-of-the-art in shape segmentation methods [46,3], then we see that most such methods are, implicitly or explicitly, focused on handling either part-based or patch-based segmentations, but rarely both. This is easy to explain since we have seen that the criteria defining (good) parts and patches are very different, as the two have different natures. Indeed, parts are inherently volumetrically described; whereas patches are best described on the shape surface [40,41]. As such, one typically needs to know in practice what is the nature of shapes one wants to segment in order to choose the best segmentation method for that task; or else, one needs to manually run several such methods and hopes that one of them will be optimal for the shapes at hand. Our unified segmentation proposed here shows that one can use a single descriptor—surface skeletons—to automatically compute *and* combine both segmentation types. This allows users to simply "drop" their shapes in the tool and let it choose the optimal mix of parts and patches to segment with.

Performance: Table 4.1 shows the performance of our unified method, implemented in single-threaded C++, on an Intel Core i7 3.8 GHz PC with 32 GB RAM. The tested

Table 4.2 Performance of the VCS part-based segmentation method of Feng et al. [16]

Shape	$\|\partial\Omega\|$	$\|\mathcal{CS}\|$	t_{skel}	t_{cuts}	T_{VCS}
Horse	109,555	884	1.24	9.58	10,109
Hound	245,759	1530	1.51	23.24	16,179
Cow	143,938	1009	0.96	8.15	17,820
Bird	45,638	476	0.18	2.28	9527
Pig	145,215	959	1.51	10.97	12,694
Vertebra	68,632	683	0.37	8.56	5472
Scapula	285,854	4329	30.0	301.3	4106
Heptoroid	651,478	4873	3.36	400.5	7926
Kidney	31,874	403	0.16	3.91	3278
Hand	58,071	584	0.22	2.15	15,773

shapes vary considerably in terms of mesh resolution ($\|\partial\Omega\|$) and mesh-sampling uniformity, and also in terms of the number of cuts we compute ($\|\mathcal{CS}_\alpha\|$). We next see that the total cost is dominated by the cut-space computation. At first sight, the overall cost appears to be quite high. To better assess this cost, we compute the throughput of our method

$$T_{unified} = \frac{1}{1000} \frac{|\partial\Omega\| \cdot \|\mathcal{CS}_\alpha\|}{t_{cuts}}, \tag{4.16}$$

i.e., the number of cuts, for a given resolution of the mesh (in thousands of vertices) that the method can deliver per second. Using the mesh resolution in Eq. (4.16) accounts for the fact that the cut computation cost is proportional with the mesh resolution. Table 4.2 shows the throughput T_{VCS} for the original voxel-based VCS method. Here, $\|\partial\Omega\|$ denotes the number of voxels on the shape surface. The ratio of the average T_{VCS} to the average $T_{unified}$ is 2.32, i.e., the VCS method is 2.32 times faster than ours. However, the original VCS method does use CPU parallelism (8 threads) to compute cuts, whereas our method is purely sequential. Parallelizing our cut computation is very easy since cuts are traced completely independently. This makes our method potentially over three times faster than the VCS method. Further, using GPU parallelism to compute the cuts and following the technique originally proposed by [23] for computing surface-skeleton importance would make our method significantly faster than the VCS method, which lends itself far less to GPU parallelization. Interestingly, we see that the relative throughput of our method as compared to VCS increases significantly for shapes where many cuts are used, e.g., *scapula* and *heptoroid*—for the second shape, our method has an even higher throughput than VCS. Memory-wise, our method needs to store only the input mesh, point-cloud skeleton, and two feature points per skeleton point, its memory cost being $O(\|\partial\Omega\|)$. In contrast, VCS needs to store four full densely-sampled volumes, yielding a memory cost of $O(\|\partial\Omega\|^{3/2})$ (for details, see the underlying skeletonization method [24]). All in all, we see that our method scales far better than VCS with respect to shape size.

Implementation: Although at first sight complex, our unified method requires only a few ingredients to be implemented: the point-cloud skeletonization method for mesh models in [23], which delivers surface skeletons, distance transforms, and feature transforms; and a way to find the nearest neighbors in a point cloud (Eqs. (4.8) and (4.12)). The former is provided by the respective algorithm, and the latter is readily available via the ANN package [37].

Limitations: Although delivering good-quality segmentations of complex shapes, our unified method still has several limitations. First and foremost, its quality essentially depends on the underlying qualities of the parts and patches delivered by its two branches (Section 4.3). Although the part and patch validation steps we introduce considerably help delivering good quality parts and patches, the overall result still depends on the possibility of *independently* finding good parts and patches on a shape. There are, obviously, cases where this assumption does not hold. In such cases, designing a new "joint segment" model able to capture the full continuum between parts and patches is desirable, and is a challenge for future work.

Secondly, our unified method is technically more complex than other part-based and patch-based segmentation methods taken separately. Indeed, it comprises a full part-based segmentation pipeline (7 steps, Sections 4.3.2–4.3.7), a full patch-based segmentation pipeline (Section 4.3.8.1), and a unification pipeline (two steps, Section 4.3.8.3). However, this is justified by the fact that our method is, to our knowledge, the only existing one that can handle mixed part-patch-based segmentations with good results.

The comparison of our method with related segmentation methods [17,16,40,26, 33,2,10,55,32] is, of course, not covering the entire spectrum of segmentation methods out there. This would be highly challenging, if not impossible, to do given the available space and the availability of implementations of such methods. More effort (from the entire shape segmentation community) is required here. Yet, we argue that our main point, showing that our method can leverage surface skeletons to generate part-patch-based segmentations than other tools in the same class, has been well defended by the presented results.

4.6 CONCLUSION

In this chapter, we have presented a novel method to create segmentations of 3D mesh shapes. In contrast to most existing segmentation methods out there, we show that it is possible to design a method that effectively handles tubular (articulated) shapes, faceted shapes, and shapes containing a mix of the two. On a more fundamental level, we show that surface skeletons are an effective tool to support complex segmentation tasks. Thereby, we strengthen the existing evidence that this type of medial descriptors, so far sparsely used in actual applications, are effective tools with practical added value. We support our claims by comparing our method with the most relevant part-based and patch-based segmentation methods using 3D skeletons on a collection of shapes ranging from purely faceted to purely tubular.

Future work can handle several directions. First and foremost, the current results show that it is possible to create hybrid part-patch segmentations using a single descriptor type (surface skeletons). As such, it is interesting to explore refinements of the presented part and patch detection and merging heuristics to design methods where users can control the resulting segmentation more easily and intuitively. Secondly, low-hanging fruits include the GPU acceleration of all steps of the proposed pipeline to yield a single method able to compete speed-wise with current state-of-the-art segmentation methods.

REFERENCES

[1] C. Arcelli, G. Sanniti di Baja, L. Serino, Distance-driven skeletonization in voxel images, IEEE Trans. Pattern Anal. Mach. Intell. 33 (4) (2011) 709–720.
[2] M. Attene, B. Falcidieno, M. Spagnuolo, Hierarchical mesh segmentation based on fitting primitives, Vis. Comput. 22 (3) (2006).
[3] M. Attene, S. Katz, M. Mortara, G. Patané, M. Spagnuolo, A. Tal, Mesh segmentation – a comparative study, in: Proc. IEEE SMI, 2006, pp. 134–141.
[4] O.K.C. Au, C. Tai, H. Chu, D. Cohen-Or, T. Lee, Skeleton extraction by mesh contraction, in: Proc. ACM SIGGRAPH, 2008, pp. 441–449.
[5] J. Bloomenthal, C. Bajaj, J. Blinn, M.-P. Cani, A. Rockwood, B. Wyvill, G. Wyvill, Introduction to Implicit Surfaces, Morgan Kaufmann, 1997.
[6] H. Blum, A transformation for extracting new descriptors of shape, in: Models for the Perception of Speech and Visual Form, MIT Press, 1967, pp. 362–380.
[7] M. Botsch, L. Kobbelt, M. Pauly, P. Alliez, B. Lévy, Polygon Mesh Processing, A K Peters, 2010.
[8] M. Braunstein, D. Hoffman, A. Saidpour, Parts of visual objects: and experimental test of the minima rule, Perception 18 (1989) 817–826.
[9] M. Chang, F. Leymarie, B. Kimia, Surface reconstruction from point clouds by transforming the medial scaffold, Comput. Vis. Image Underst. 113 (11) (2009) 1130–1146.
[10] U. Clarenz, M. Griebel, M. Schewitzer, A. Telea, Feature sensitive multiscale editing on surfaces, Vis. Comput. 20 (5) (2004) 329–343.
[11] D. Comaniciu, P. Meer, Mean shift: a robust approach toward feature space analysis, IEEE Trans. Pattern Anal. Mach. Intell. 24 (5) (2002) 603–619.
[12] N. Cornea, D. Silver, P. Min, Curve-skeleton properties, applications, and algorithms, IEEE Trans. Vis. Comput. Graph. 13 (3) (2007) 87–95.
[13] N. Cornea, D. Silver, X. Yuan, R. Balasubramanian, Computing hierarchical curve-skeletons of 3D objects, Vis. Comput. 21 (11) (2005) 945–955.
[14] J. Damon, Global medial structure of regions in $\mathbb{R}^3$, Geom. Topol. 10 (2006) 2385–2429.
[15] T. Dey, J. Sun, Defining and computing curve skeletons with medial geodesic functions, in: Proc. IEEE SGP, 2006, pp. 143–152.
[16] C. Feng, A.C. Jalba, A.C. Telea, Improved part-based segmentation of voxel shapes by skeleton cut spaces, Math. Morphol. Theory Appl. 1 (1) (2015) 1–20.

[17] C. Feng, A.C. Jalba, A.C. Telea, Part-based segmentation by skeleton cut space analysis, in: Proc. ISMM, Springer, Jan. 2015, pp. 1–12.
[18] M. Foskey, M. Lin, D. Manocha, Efficient computation of a simplified medial axis, in: Proc. SMA, 2003, pp. 135–142.
[19] P. Giblin, B.B. Kimia, A formal classification of 3D medial axis points and their local geometry, IEEE Trans. Pattern Anal. Mach. Intell. 26 (2) (2004) 238–251.
[20] A. Golovinskiy, T. Funkhouser, Randomized cuts for 3D mesh analysis, ACM Trans. Graph. 27 (2008) 454–463.
[21] W. Hesselink, J. Roerdink, Euclidean skeletons of digital image and volume data in linear time by the integer medial axis transform, IEEE Trans. Pattern Anal. Mach. Intell. 30 (12) (2008) 2204–2217.
[22] D. Hoffman, W. Richards, Parts of recognition, Cognition 18 (1984) 65–96.
[23] A. Jalba, J. Kustra, A. Telea, Surface and curve skeletonization of large 3D models on the GPU, IEEE Trans. Pattern Anal. Mach. Intell. 35 (6) (2013) 1495–1508.
[24] A.C. Jalba, A. Sobiecki, A.C. Telea, An unified multiscale framework for planar, surface, and curve skeletonization, IEEE Trans. Vis. Comput. Graph. 38 (1) (2016) 30–45.
[25] S. Katz, A. Tal, Hierarchical mesh decomposition using fuzzy clustering and cuts, ACM Trans. Graph. 22 (2003) 954–961.
[26] J. Kustra, A. Jalba, A. Telea, Computing refined skeletal features from medial point clouds, Pattern Recognit. Lett. 76 (2014) 13–21.
[27] J. Kustra, A. Jalba, A. Telea, Robust segmentation of multiple intersecting manifolds from unoriented noisy point clouds, Comput. Graph. Forum 33 (1) (2014) 73–87.
[28] Y. Lee, S. Lee, A. Shamir, D. Cohen-Or, Intelligent mesh scissoring using 3D snakes, in: Proc. IEEE Pacific Graphics, 2004, pp. 279–287.
[29] Y. Lee, S. Lee, A. Shamir, D. Cohen-Or, H.P. Seidel, Mesh scissoring with minima rule and part salience, Comput. Aided Geom. Des. 22 (2005) 444–465.
[30] F. Leymarie, B. Kimia, The medial scaffold of 3D unorganized point clouds, IEEE Trans. Vis. Comput. Graph. 29 (2) (2007) 313–330.
[31] X. Li, T.W. Woon, T.S. Tan, Z. Huang, Decomposing polygon meshes for interactive applications, in: Proc. ACM I3D, 2001, pp. 35–42.
[32] J. Lien, J. Keyser, N. Amato, Simultaneous shape decomposition and skeletonization, in: Proc. ACM SPM, 2005, pp. 219–228.
[33] R. Liu, H. Zhang, Segmentation of 3D meshes through spectral clustering, in: Proc. Pacific Graphics, 2004, pp. 298–305.
[34] J. Ma, S. Bae, S. Choi, 3D medial axis point approximation using nearest neighbors and the normal field, Vis. Comput. 28 (1) (2012) 7–19.
[35] A. Mangan, R. Whitaker, Partitioning 3D surface meshes using watershed segmentation, IEEE Trans. Vis. Comput. Graph. 5 (4) (1999) 308–321.
[36] H. Moreton, C. Séquin, Functional optimization for fair surface design, in: Proc. ACM SIGGRAPH, 1992, pp. 167–176.
[37] D. Mount, S. Arya, Approximate nearest neighbor search software, www.cs.umd.edu/~mount/ANN, 2016.
[38] D. Page, A. Koschan, M. Abidi, Perception-based 3D triangle mesh segmentation using fast marching watersheds, in: Proc. IEEE CVPR, 2003, pp. 27–32.
[39] D. Reniers, A. Jalba, A. Telea, Robust classification and analysis of anatomical surfaces using 3D skeletons, in: Proc. VCBM, Eurographics, 2008, pp. 61–68.
[40] D. Reniers, A. Telea, Part-type segmentation of articulated voxel-shapes using the junction rule, Comput. Graph. Forum 27 (7) (2008) 1837–1844.

[41] D. Reniers, A. Telea, Patch-type segmentation of voxel shapes using simplified surface skeletons, Comput. Graph. Forum 27 (7) (2008) 1837–1844.
[42] D. Reniers, A.C. Telea, Skeleton-based hierarchical shape segmentation, in: Proc. IEEE SMA, 2007, pp. 179–188.
[43] D. Reniers, J.J. van Wijk, A. Telea, Computing multiscale skeletons of genus 0 objects using a global importance measure, IEEE Trans. Vis. Comput. Graph. 14 (2) (2008) 355–368.
[44] L. Serino, C. Arcelli, G.S. di Baja, From skeleton branches to object parts, Comput. Vis. Image Underst. 129C (2014) 42–51.
[45] L. Serino, G.S. di Baja, C. Arcelli, Using the skeleton for 3D object decomposition, in: Proc. SCIA, in: Lect. Notes Comput. Sci., Springer, 2011, pp. 447–456.
[46] A. Shamir, A survey on mesh segmentation techniques, Comput. Graph. Forum 27 (6) (2008) 1539–1556.
[47] L. Shapira, A. Shamir, D. Cohen-Or, Consistent mesh partitioning and skeletonisation using the shape diameter function, Vis. Comput. 24 (2008) 249–259.
[48] K. Siddiqi, S. Bouix, A. Tannenbaum, S. Zucker, Hamilton–Jacobi skeletons, Int. J. Comput. Vis. 48 (3) (2002) 215–231.
[49] K. Siddiqi, S. Pizer, Medial Representations: Mathematics, Algorithms and Applications, Springer, 2009.
[50] M. Singh, G. Seyranian, D. Hoffman, Parsing silhouettes: the short-cut rule, Percept. Psychophys. 61 (4) (1999) 636–660.
[51] A. Tagliasacchi, I. Alhashim, M. Olson, H. Zhang, Skeletonization by mean curvature flow, in: Proc. Symp. Geom. Proc., 2012, pp. 342–350.
[52] A. Tagliasacchi, T. Delame, M. Spagnuolo, N. Amenta, A. Telea, 3D skeletons: a state-of-the-art report, Comput. Graph. Forum (2016), http://dx.doi.org/10.1111/cgf.12865.
[53] G. Taubin, Estimating the tensor of curvature of a surface from a polyhedral approximation, in: Proc. ICCV, 1995, pp. 902–907.
[54] G. Taubin, Geometric signal processing on polygonal meshes, in: Proc. Eurographics – STARs, Eurographics Association, 2000.
[55] J. Tierny, J. Vandeborre, M. Daoudi, Topology driven 3D mesh hierarchical segmentation, in: Proc. SMI, 2007, pp. 215–220.

CHAPTER

Improving the visual aspect of the skeleton and simplifying its structure

5

Gabriella Sanniti di Baja, Luca Serino
Institute for High Performance Computing and Networking, CNR, Naples, Italy

Contents

5.1 Introduction 123
5.2 Preliminary Notions and Definitions 125
5.3 Tools Improving the Visual Aspect of the Skeleton 127
5.3.1 Zig-zag Straightening 127
5.3.2 Fusion of Close Branch Points 129
5.3.3 Pruning 131
5.4 Experimental Results on the Improvement of Skeleton Visual Aspect 135
5.5 Tools to Simplify Skeleton Structure 138
5.5.1 Polygonal Approximation 140
5.5.2 Building a Linearized Version of the Skeleton 144
5.6 Experimental Results on the Simplification of Skeleton Structure 145
5.7 Conclusion 147
References 149

5.1 INTRODUCTION

Since the introduction of the Medial Axis Transform MAT [1], the skeleton of 2D objects has been regarded as a powerful shape representation scheme, especially for objects that can be interpreted as consisting of ribbon-like parts. In fact, skeleton points are symmetrically placed within the object and can be assigned the value of their distance from the complement of the object. Thus, the object can be recovered by the envelope of the discs centered on the skeleton's points and having radii equal to the distance values of the skeleton's points.

Ideally, the skeleton of an object should be characterized by the following properties: i) Unit thickness. The skeleton should consist of the union of curves; ii) Centrality. The skeleton should be symmetrically placed within the object; iii) Homotopy. The skeleton should have the same number of connected components as the object, and each hole of the object should be surrounded by a loop of the skeleton; iv) Re-

Skeletonization. DOI: 10.1016/B978-0-08-101291-8.00006-7

coverability. The object should be recovered faithfully by the envelope of the discs centered on the skeleton points and with radii equal to the distance values of the corresponding skeleton points; and v) Significance. Skeleton branches should originate only in correspondence of significant limbs and strong boundary convexities of the object.

Actually, whichever skeletonization algorithm is applied to a digital object, the obtained skeleton S unavoidably differs in a more or less significant manner from the ideal skeleton. Some of the above properties cannot be simultaneously fully satisfied by S: if S is characterized by unit thickness, then complete object recovery cannot be guaranteed in presence of object parts with thickness given by an even number of pixels. Generally, the number of object pixels that are not recovered by the envelope of the discs centered on the pixels of the unit wide S constitutes a disregardable percentage of the total number of pixels of the input object. In addition, a skeleton with linear structure is more manageable to capture relevant shape information than a skeleton that is 2-pixel thick at parts. Thus, unit thickness is mostly preferred to full recoverability.

Also, branch significance and recoverability may not be completely satisfied at the same time by S. In fact, skeletonization may be strongly conditioned by small changes of curvature along the boundary of the digital object. As a consequence, noisy object's protrusions are likely to be mapped into scarcely significant peripheral branches of S that would be anyway necessary for a full recovery of the object. Resorting to polygonal approximation of the object's boundary to identify a number of vertices to be used for the computation of S based on the Voronoi diagram of the detected vertices does not prevent the creation of spurious noisy branches [2]. In fact, to approximate reasonably well the boundary of natural shapes, a large number of vertices is necessary, and, hence, a large number of peripheral branches is expected to characterize S since a peripheral branch is obtained in correspondence with each side of the approximating polygon. Thus, to keep in the skeleton only branches mapped into individually significant regions, pruning is necessary to identify and remove all noisy peripheral skeleton branches originating from weak convexities of the object's boundary or from scarcely relevant object's protrusions. Of course, pruning will unavoidably have an impact on recoverability.

Besides the impossibility to satisfy simultaneously all the above five properties desired for the skeleton, other differences between S and the ideal skeleton regard number and position of the branch points, and jaggedness of skeleton branches.

Schematically, a branch point of the ideal skeleton is understood as the meeting point of a number of skeleton branches associated with individually significant partially overlapping regions. However, often a number of branch points are present in S, where a unique meeting point would be expected for the ideal skeleton. These branch points can be connected to each other, forming a cluster consisting only of branch points, or can just be rather close to each other, forming a cluster that includes a few skeleton pixels linking pairs of branch points, besides the branch points themselves. In the latter case, "short" skeleton branches linking close branch points would exist in S, which are not mapped into individually significant object's parts. To establish a one-to-one correspondence between skeleton branches and individually significant object's regions, each cluster of branch points should be interpreted as a single branch point.

The jaggedness of skeleton branches is mainly caused by the process generally called "final thinning," which is accomplished once an at most 2-pixel-thick set of skeletal pixels is available to reduce it to unit thickness. Though zig-zags along the branches of S do not constitute a serious drawback and do not prevent to use S for shape analysis, their presence has an unpleasant impact on the visual aspect of S. Jaggedness may regard also the distance values associated with a sequence of skeletal pixels along a branch. In fact, erratic variations in the thickness of an object may cause a fluctuating trend of the distance values of skeletal pixels along the branch. Zig-zags of S, regarding both curvature and distance values, should be straightened without altering the topology of S, to obtain a more visually appealing skeleton.

Processes such as pruning scarcely significant peripheral branches, fusion of close branch points to get rid of "short" branches that do not individually correspond to significant parts of the object, and zig-zags straightening along skeleton branches improve the visual aspect of S making its structure resembling in a satisfactory manner that of the ideal skeleton, without significantly reducing the representative power of the skeleton. Moreover, also a further simplification of S by means of polygonal approximation can be useful in the framework of shape analysis. In fact, shape information can be conveyed into just a few points of any skeleton branch, which would significantly reduce the cardinality of the representation scheme still having a rather faithful representation of the object.

In this chapter, we describe possible ways to improve the visual aspect of S and to simplify its structure. In the following, we refer to the skeleton computed by the algorithm [3], which is based on the use of the $\langle 3, 4 \rangle$ distance transform of the object [4]. However, we point out that the suggested tools can be used with any other skeletonization algorithm that originates a distance labeled skeleton, provided that suitable modifications of some of the involved parameters are accomplished.

The rest of this chapter is organized as follows. Preliminary notions and definitions are given in Section 5.2. Tools for zig-zag straightening, fusion of close branch points, and pruning are described in Section 5.3. Some results obtained by using the above tools are discussed in Section 5.4. Then, simplification of skeleton structure by means of polygonal approximation and construction of a linearized version of the skeleton are treated in Section 5.5, and the corresponding results are discussed in Section 5.6. A brief conclusion is finally given in Section 5.7.

5.2 PRELIMINARY NOTIONS AND DEFINITIONS

We work with binary images, where the object is composed by snake-shaped parts and consists of the pixels with value 1, whereas the background consists of the pixels with value 0. The 3×3 neighborhood $N(p)$ of a pixel p includes, besides p, the four edge-neighbors and the four vertex-neighbors of p. The 8-connectedness and the 4-connectedness are used for the object and the background, respectively. Since the skeleton S is a subset of the object, the 8-connectedness is used also for S.

Given two pixels p and q, their $\langle 3,4\rangle$ distance $d(p,q)$ is the length of a minimal 8-connected path linking p to q, where unit moves toward an edge-neighbor and a vertex-neighbor along the path are respectively weighted 3 and 4 [4]. The $\langle 3,4\rangle$ distance provides a good approximation of the Euclidean distance and is characterized by the simplicity of path-based distances.

The $\langle 3,4\rangle$ distance transform of the object, DT, is a replica of the object where each pixel p is assigned the value $d(p)$ of its $\langle 3,4\rangle$ distance from the background. Any pixel p in DT can be interpreted as the center of a discrete disc with radius $d(p)$. The disc associated with p can be built by applying to p the $\langle 3,4\rangle$ reverse-distance transformation. If the disc associated with p is not included by any other single disc associated to any pixel in DT, then the disc is a maximal disc, and p is called the center of a maximal disc, CMD. Both the $\langle 3,4\rangle$ distance transform and the $\langle 3,4\rangle$ reverse-distance transform can be conveniently computed in two raster scans of the image [4].

The distance transform can be interpreted as the result of a local propagation of distance information during which any pixel p of the object receives distance information from some of its neighbors belonging to the background or closer to the background than p and propagates distance information to its object neighbors that are farther than p from the background. A CMD is a pixel that does not provide distance information to any neighbor. Thus, a pixel p is a CMD if for each of its n neighbors, it results in $d(n) - d(p) < w$, where w is equal to 3 or 4 depending on whether n is an edge-neighbor or a vertex-neighbor of p [5]. For completeness, we remark that since distance values in DT are linear combinations with positive coefficients of the two weights 3 and 4, obviously, values 1, 2, and 5 never exist in DT. On the other hand, if the $\langle 3,4\rangle$ reverse-distance transformation is applied to pixels with value 1 or 2 (with value 5), then a disc identical in size and shape to the disc with radius 3 (radius 6) is obtained. Thus, to avoid that the CMD selection criterion identifies pixels whose associated discs are not actually maximal discs, the two distance values 3 and 6 in DT should be replaced by their smallest equivalent values 1 and 5 when applying the above CMD detection criterion [5].

For simplicity, let us suppose that only one 8-connected object I exists in the image at hand, so that its skeleton S consists of exactly one 8-connected component. In the following, we refer to the skeleton S computed by the algorithm [3], whose pixels are labeled with their $\langle 3,4\rangle$ distance from the background. We point out that the skeletonization algorithm [3] includes three main phases: 1) extraction of an at most 2-pixel-thick set of skeletal pixels, 2) final thinning, and 3) postprocessing to slightly improve the quality of the resulting skeleton. Only the first two phases are used to obtain the skeleton S in this work. Thus, I can be almost fully reconstructed by applying to S the $\langle 3,4\rangle$ reverse-distance transformation. From now on, we will not consider the unavoidably existing, but generally small, difference between the number of pixels in I and the number of pixels in the object I_S recovered by S. We will regard the object I_S as a suitable replacement of I and, accordingly, will consider I_S instead of I as the input object.

Pixels of S are classified as end points, normal points, or branch points, depending on whether they have one, two, or more than two neighbors in S.

An object pixel p is a simple pixel if its removal from the object does not alter object's topology. If removal of p does not alter topology in $N(p)$, then removal of p does not alter object's topology. Thus, p is a simple pixel if the following local conditions are satisfied in $N(p)$: 1) p has at least one edge-neighbor in the complement of the object, and 2) the number of 8-connected components of object's pixels is equal to one.

Each region of the input object is mapped into a branch of S, which is a curve entirely consisting of normal points with the exception of the two pixels delimiting the curve, which can be end points or branch points. A branch is termed peripheral branch if at least one delimiting pixel is an end point. Internal branches of S are delimited by two branch points.

A concatenation in S is a sequence of skeleton branches where any two successive branches share a branch point. The length L of a concatenation of n branches is measured by the number of normal points in the concatenation plus $n + 1$ to account for the pixels delimiting the n branches.

5.3 TOOLS IMPROVING THE VISUAL ASPECT OF THE SKELETON

The tools described in this section are aimed at zig-zag straightening, fusion of close branch points, and pruning.

5.3.1 ZIG-ZAG STRAIGHTENING

The skeleton may be affected by jaggedness as regards both the spatial distribution and the distance values of its pixels. To improve the visual aspect of S, zig-zags should be straightened by suitably shifting some pixels of S without altering its topology. The only configurations of $N(p)$ where a skeletal pixel p can be shifted without causing topology changes are those shown in Fig. 5.1, where black squares are pixels of S, whereas white and gray squares denote pixels in the complement of S. If the gray edge-neighbor n of p is a pixel of the input object, then the skeletal pixel p can be replaced by n in S so as to straighten the zig-zag. As for the distance value of the shifted pixel, this is the distance value pertaining n in DT.

Zig-zag straightening may cause creation of superfluous pixels, i.e., pixels whose presence is not necessary in S for topology preservation. An example is shown in Fig. 5.2, left and middle. Once the pixel p is replaced by q in S, the pixel r becomes a simple pixel that is superfluous for connectedness preservation. Of course, also end points of S are simple pixels that, differently from r, should be kept in S to reflect the geometrical features of the object. Thus, after zig-zag straightening, removal of

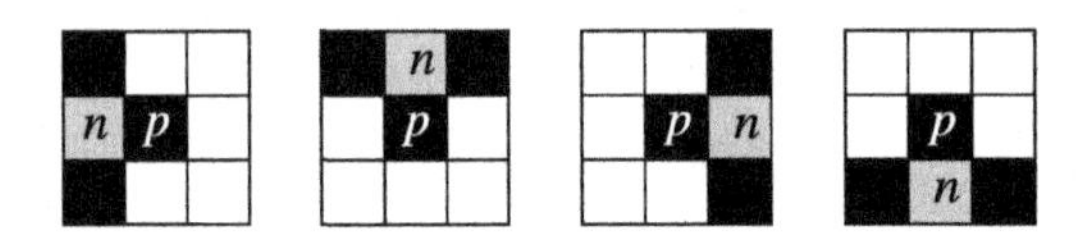

FIGURE 5.1

The four configurations where shifting the central pixel p does not alter topology.

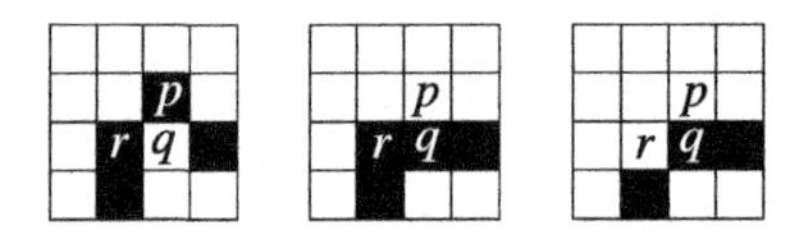

FIGURE 5.2

Pixel r becomes superfluous after p is replaced by q to straighten the zig-zag.

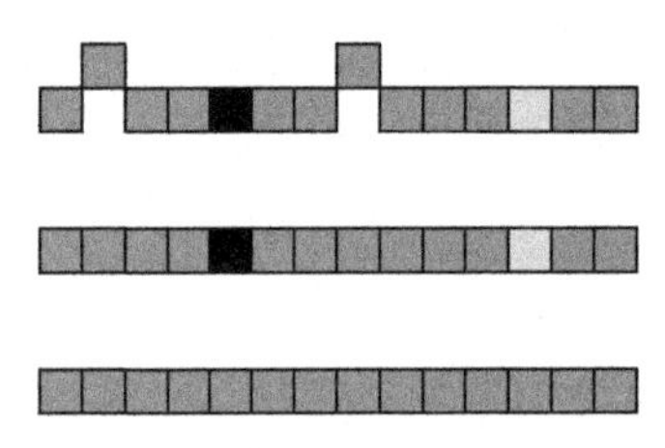

FIGURE 5.3

From top to bottom, the skeleton before zig-zag straightening, after shifting of position, and after distance value correction.

simple skeletal pixels with more than one neighbor in the skeleton is accomplished. The result is shown in Fig. 5.2, right.

Jaggedness can also regard the distribution of distance values among the skeletal pixels. If a pixel p with distance value $d(p)$ is a normal pixel and its two neighbors in S have distance values larger than $d(p)$, then the distance value of p is changed into the smallest distance value out of those of its neighbors, so as to remove the presence of a local minimum and have a distribution of distance values that is not fluctuating. Similarly, if the two neighbors of a normal pixel p have distance values smaller than $d(p)$, then p is a local maximum that can be can be removed by changing the distance value of p into the largest distance value out of those pertaining to its neighbors.

Shifting of position and distance value correction are applied to S one after the other. In both processes, pixels of S are sequentially checked. Fig. 5.3 shows the results of the two processes on a synthetically drawn example. The skeleton before zig-zag straightening is shown in Fig. 5.3, top, the skeleton after shifting of position in Fig. 5.3, middle, and the skeleton after distance value correction in Fig. 5.3, bottom. Different gray levels are used to denote different distance values (darker gray levels correspond to higher distance values).

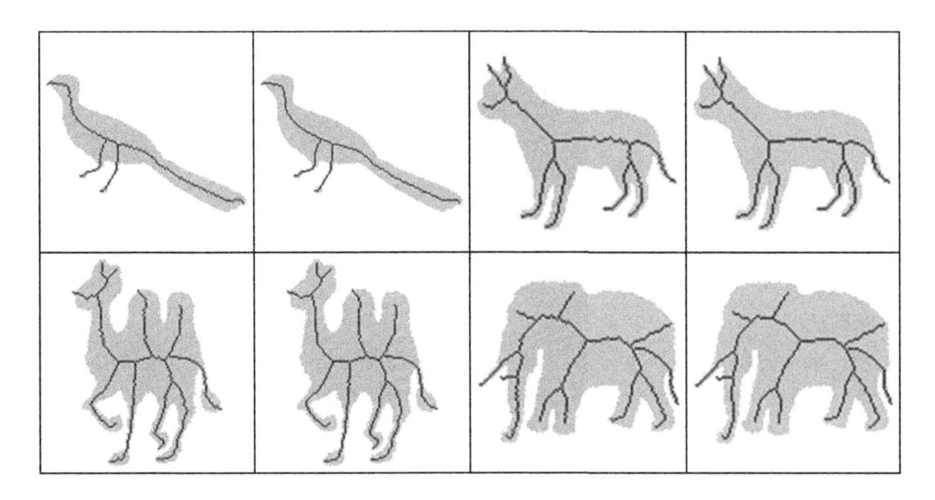

FIGURE 5.4

Some examples of skeletons before and after zig-zag straightening.

A few examples are given in Fig. 5.4, showing the skeletons of different objects before zig-zag straightening and after zig-zag straightening superimposed onto the objects.

5.3.2 FUSION OF CLOSE BRANCH POINTS

In the ideal skeleton, the branch points are the points where different branches meet. In a discrete skeleton, clusters of branch points are possible. A cluster consists either exclusively of branch points forming a connected component or of branch points rather close to each other and of the normal points linking them. In any case, each cluster of branch points should be replaced by a single meeting point, so that S can better reflect the structure of the ideal skeleton. We consider as belonging to the same cluster all the branch points that are closer to each other than the sum of their associated distance values. In such a case, the discs associated to the branch points partially overlap, and their envelope constitutes what we call the "zone of influence" of the branch points in the cluster [6]. See Fig. 5.5, left, showing a cluster consisting of four branch points. The envelope of the discs associated with the four branch points is shown in light gray.

We suggest to replace all the skeletal pixels included in the same zone of influence by their centroid. Then, we interpret the centroid as delimiting pixel for each skeleton subset outside the zone of influence. See Fig. 5.5, right. Of course, replacing a connected subset of the skeleton by its centroid creates disconnections in S. Thus, track is kept of the skeletal pixels outside each zone of influence with whom the centroid of that zone of influence is interpreted as linked. Alternatively, straight-line segments

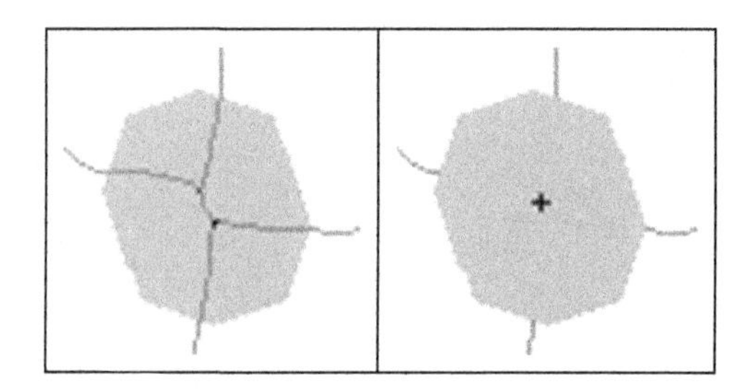

FIGURE 5.5

Branch points, black pixels, forming a cluster, left; the centroid of the cluster, marked by a cross, is taken as the meeting point replacing the cluster, right.

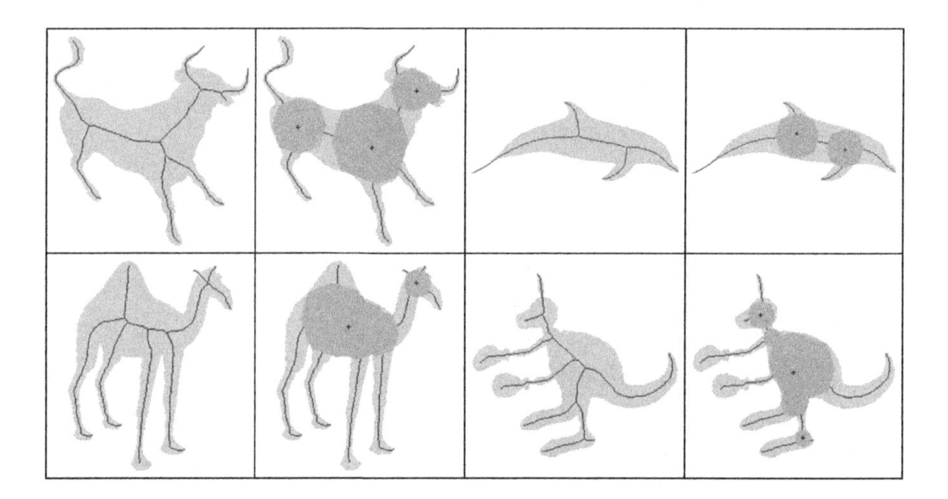

FIGURE 5.6

Skeletons before and after replacing all skeletal pixels within the zones of influence by their centroids, marked by crosses.

are built to link the centroid with the skeleton subsets outside the zone of influence. Details on how to build these segments can be found in Section 5.5.2.

Replacing the skeletal pixels in a zone of influence by their centroid is useful not only to associate a single meeting point to each cluster of branch points, but also to remove from S "short" internal branches, which do not correspond to individually meaningful object's parts. In this way a one-to-one correspondence between skeleton branches and individually meaningful object's parts can be established.

A few examples are shown in Fig. 5.6, where all skeletal pixels included in the zones of influence are replaced by their centroids.

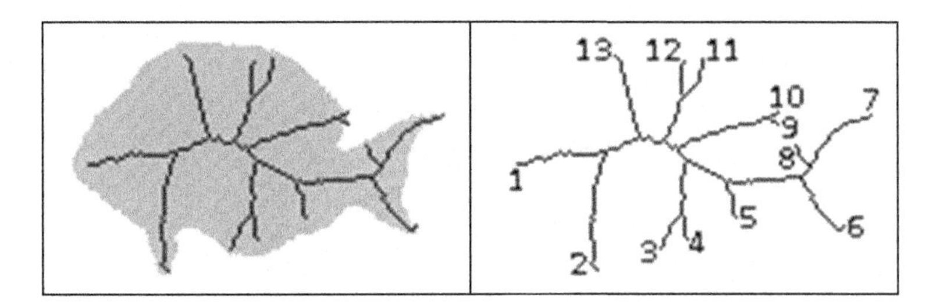

FIGURE 5.7

A skeleton with a number of peripheral branches larger than the number of object parts perceived as individually meaningful. End points delimiting peripheral branches are numbered.

5.3.3 PRUNING

Whereas straightening of zig-zags to remove jaggedness is mainly accomplished for aesthetical reasons, replacing the pixels inside a zone of influence by the centroid and pruning are important to have a skeleton resembling as much as possible the ideal skeleton. In particular, pruning is an indispensable step for any skeletonization algorithm. In fact, the skeleton is likely to include a number of peripheral branches not all obtained in correspondence of individually meaningful object protrusions or boundary's convexities. See, for example, Fig. 5.7. In particular, the end points delimiting the peripheral branches of S are numbered in Fig. 5.7, right.

Pruning is generally accomplished during a postprocessing phase. However, instead of pruning scarcely significant peripheral branches, their creation can be avoided by performing a preliminary filtering process with the aim of identifying in the object suitable anchor points to guarantee that during skeletonization peripheral branches would start only from the tips of object's convexities regarded as significant. See, for example, [7,8]. Different criteria have been suggested in the literature to distinguish peripheral branches corresponding to significant parts of the object from those removable without biasing the representative power of the skeleton. In [9], a comprehensive survey is presented of pruning criteria based on propagation velocity, maximal thickness, radius function, axis arc length, and boundary/axis length ratio. A pruning criterion based on the maximal thickness of slices of object pixels that will not be recovered starting from the pruned skeleton has been proposed in [3]. More recent pruning methods involve contour partitioning via discrete curve evolution and bending potential ratio [10,11], where pruning is either performed during a postprocessing step or is embedded into the skeleton computation process. In [12], the selection of sequential or parallel pruning strategies is discussed, and a hybrid approach is suggested. In this chapter, we describe in detail a recently suggested criterion to remove concatenations of branches without altering the topology of S or reducing its representative power [13]. Only concatenations including peripheral branches undergo pruning.

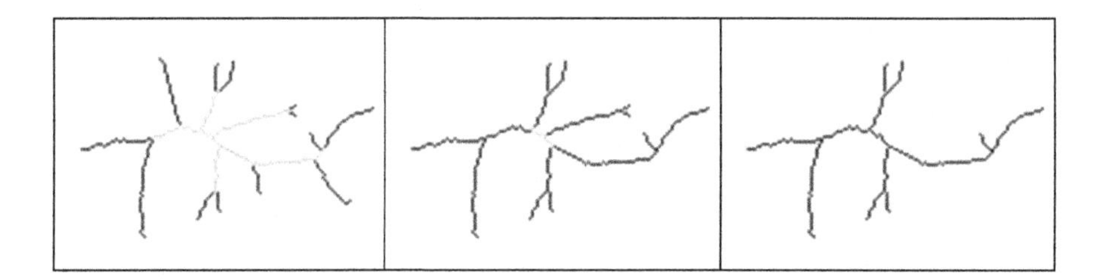

FIGURE 5.8

The hierarchical skeleton structure. Black pixels belong to concatenations of branches at level 1, left, concatenations of branches at levels 1 and 2, middle, and concatenations of branches at levels 1, 2, and 3, right.

To identify the concatenations, we build a hierarchical skeleton structure, where each level of the hierarchy includes the branches of S that can be simultaneously considered as peripheral branches. Let us consider as running example the skeleton in Fig. 5.7. Since S includes 13 end points, 13 concatenations are possible, each of which having one of the peripheral branches as starting branch.

The 13 branches of S that are initially detected as peripheral branches are assigned to the lowest level of the hierarchy, i.e., level 1. See Fig. 5.8, left, where the peripheral branches assigned to level 1 are shown in black, whereas more internal skeleton branches are shown in a lighter gray tone.

Then, all branches at level 1 are temporarily regarded as removed, which means that the skeleton temporarily consists only of the pixels in lighter gray tone of Fig. 5.8, left. By looking at the skeleton after removal of its peripheral branches, we note that new end points may have been originated, depending on the number of internal branches that meet in branch points delimiting peripheral branches. Branch points common to a number of peripheral branches and to only one internal branch of S are transformed into new end points after removal of the peripheral branches originated from end points 1, 2, 3, 4, 7, 8, 9, 10, 11, 12. On the contrary, if a branch point is common to more than one internal branch, then removal of peripheral branches will leave it still as a branch point or will at most transform it into a normal point after removal of peripheral branches originated from end points 5, 6, 13. The end points created by removal of peripheral branches of S are the delimiting pixels of new peripheral branches that are assigned to level 2 of the hierarchy. Not all 13 concatenations found at level 1 extend to level 2 of the hierarchy. In fact, only concatenations with branches of level 1 whose delimiting branch points have been transformed into new end points extend to level 2. The remaining concatenations terminate at level 1. See Fig. 5.8, middle, where concatenations of branches at levels 1 and 2 are shown in black, concatenations that do not extend to level 2 are not shown, and the remaining more internal branches are in lighter gray tone.

Level 3 of the hierarchy can be built by following the same scheme by regarding also all peripheral branches at level 2 as removed. Skeleton hierarchy construction and detection of all possible concatenations continue as far as new end points are generated at level n, after all branches detected as peripheral at level $n-1$ are regarded as removed. In the example shown in Fig. 5.8, it is $n = 3$. Thus, we can associate a concatenation with each of the 13 end points of S, and for each concatenation, we can identify the highest level of the hierarchy to which it extends. The root of each concatenation is the most internal branch point of S along the concatenation. For the running example, the eight concatenations originated from end points 1, 2, 3, 4, 7, 8, 11, 12 extend to level 3, the two concatenation originated from end points 9 and 10 include only branches at levels 1 and 2, and the three concatenations originated from end points 5, 6, 13 include only branches at level 1.

To evaluate the significance of a concatenation, we use the combination of a number of significance measures, which involve the computation of different parameters. Let p be the hth end point, $h = 1, 2, \ldots, k$, of S, and let q be the branch point delimiting the concatenation C_h originating from p. We compute the following parameters at each level i, $i = 1, 2, \ldots, n$, at which the concatenation C_h exists:

- The difference δ_h^i between the distance values of p and q;
- The $\langle 3, 4 \rangle$ distance $d(p, q)$ between p and q;
- The length L_h^i of C_h;
- The number $\#CMD_h^i$ of the centers of maximal discs along C_h;
- The ratio $R_h^i = \#CMD_h^i / L_h^i$;
- The value $\lambda_h^i = \delta_h^i + d(p, q)$.

Generally, distance values along a scarcely significant branch increase from the end point toward more internal pixels. Thus, the first parameter can be roughly used to check if the distance values along the branch have an increasing trend. The distance between the extremes of a concatenation, second parameter, gives an idea on whether the concatenation is short or long, especially if also the length of the concatenation, third parameter, is considered. The number of CMDs in the branch, fourth parameter, is important since the region associated with a branch of S is the envelope of the discs centered on the CMDs of the branch. Of course, the ratio between the number of CMDs and the length of a concatenation, fifth parameter, gives a clearer idea on the importance of a branch: the higher the ratio, the higher the importance of the branch. Finally, the sixth parameter roughly evaluates the thickness of the slices of the object that would not be recovered by the pruned skeleton, as shown in Fig. 5.9.

To decide whether the concatenation C_h can be removed at level i without affecting the representative power of S, we combine the above parameters into a pruning criterion according to which C_h is removable if the following logical operation returns TRUE:

$$\delta_h^i < \theta 1 \text{ AND } \lambda_h^i < \theta 2 \text{ AND } [L_h^i < \theta 3 \text{ OR } R_h^i < \theta 4].$$

Clearly, the values for the four thresholds depend on the specific application. We think that $\theta 1$ and $\theta 2$ should take into account the size of the input object, whereas $\theta 3$ and $\theta 4$ could be set independently of image resolution, but $\theta 3$ should take into

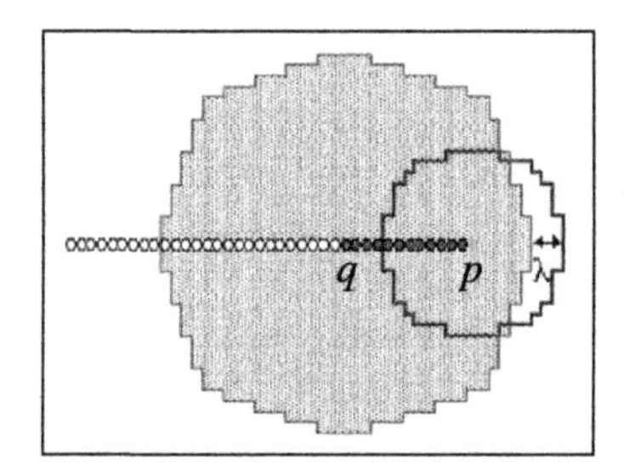

FIGURE 5.9

The thickness λ of the slice of object pixels lost if pruning the skeleton from p to q is a function of the distance between p and q and of the radii of the discs centered on p and q.

account the hierarchy level. To correlate $\theta 1$ and $\theta 2$ with the size of the object, we use the bounding box including the object itself. In particular, we set $\theta 1$ and $\theta 2$ to a suitable percentage of the length of the shorter side of the bounding box, *min*. We have experimentally found that in the average satisfactory results are obtained by setting $\theta 1 = 3\%\ min$, and $\theta 2 = 10\%\ min$. As for $\theta 3$, if the concatenation includes i levels, then we set $\theta 3$ equal to the arithmetic mean of the lengths of all concatenations of S that include i levels. Finally, we have fixed $\theta 4 = 0.5$ by considering scarcely significant any concatenation with less than half of pixels that are CMDs.

Our strategy to identify in correspondence with every end point p the longest concatenation whose removal does not significantly diminish the representative power of S is to check for removal first the concatenation with the deepest root. If such a concatenation is removable, then we mark it as removable. Otherwise, we move the root to a less internal branch point going toward the end point and check removability for the concatenation with one level less until a removable concatenation is found (which is marked for removal without checking shorter concatenations), or the logical operation always returns FALSE (meaning that no pruning at all can be done starting from the end point p since the concatenations starting from it resulted as significant at every level).

Some of the internal branches of the longest removable concatenation starting from the end point p may also be included in nonremovable portions of other concatenations starting from other end points. Thus, though the concatenation at hand is completely removable, pruning will start from its peripheral branch and will continue through its internal branches only as far as the current internal branch belongs exclusively to removable portions in all other concatenations including it. For example, let us consider two concatenations originating from two end points p and q and sharing their successive internal levels. Let us suppose that the longest removable concatenation starting from p includes two levels (the peripheral branch and the successive internal branch), whereas no concatenation at all is prunable starting from q. Thus, pruning the longest prunable concatenation originating from p will only remove the peripheral branch but will not remove the successive internal branch since removal

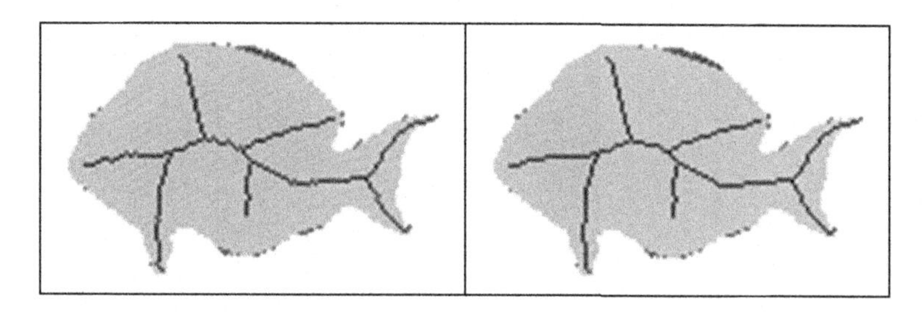

FIGURE 5.10

The skeleton after pruning, left, and after zig-zag straightening, right.

of such an internal branch would disconnect from the pruned S the peripheral branch in the unprunable concatenation starting from q.

For the running example, the shorter side of the bounding box measures 70 pixels. Thus, by taking the integer part of the computed percentages, it results in $\theta 1 = 2.1$ and $\theta 2 = 7$. The arithmetic mean of the lengths of the concatenations including 3, 2, and 1 successive branches is 50, 38.40, and 15.15, respectively. These are the values of $\theta 3$ for the concatenations including 3, 2, and 1 successive branches. Finally, $\theta 4 = 0.5$ has been used for all concatenations. The resulting pruned skeleton is shown in Fig. 5.10, left, whereas the pruned skeleton obtained after zig-zag straightening is shown in Fig. 5.10, right. The object pixels that would be recovered by the unpruned skeleton but are not recovered by the pruned skeleton are shown in darker gray tone.

By comparing the pruned skeleton with the unpruned skeleton in Fig. 5.7 we may observe that only pixels belonging to thin peripheral slices of the object are not recovered by the pruned skeleton, whereas the structure of the skeleton is remarkably more manageable.

A few more examples showing the effect of pruning are given in Fig. 5.11. Also, in these examples, zig-zag straightening has been performed after pruning.

5.4 EXPERIMENTAL RESULTS ON THE IMPROVEMENT OF SKELETON VISUAL ASPECT

We have applied the tools described in the previous section to a large set of binary images, most of which have been taken from publicly available shape repositories such as [14–17]. In this section, we mainly discuss the results obtained by applying pruning and in particular focus on the robustness of the suggested method as regards its application in the presence of noise and when the input object appears at a different scale or in a different pose. We point out that, for all the examples in this chapter, the values of $\theta 1$, $\theta 2$, $\theta 3$, and $\theta 4$ have been set as suggested in Section 5.3.3.

The results achieved in the presence of noise can be appreciated with reference to Fig. 5.12, where we show from left to right the unpruned skeleton, top, and the

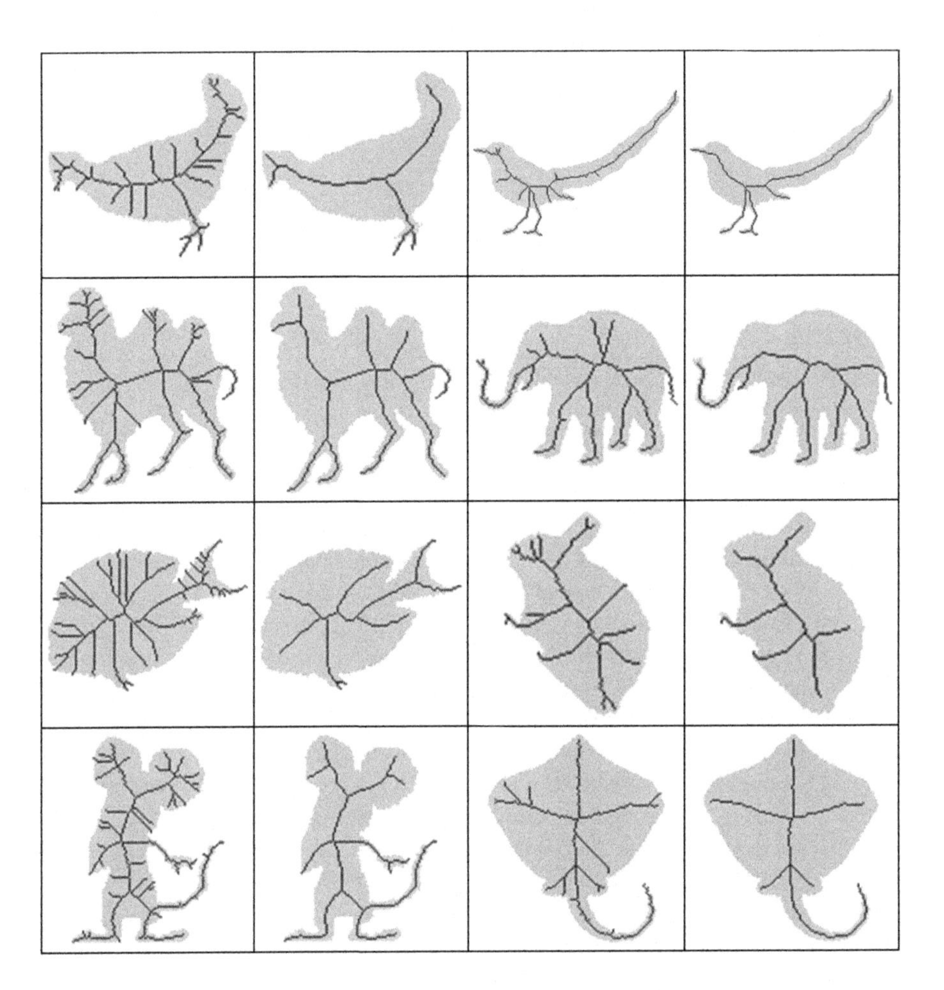

FIGURE 5.11

Skeletons of different objects before pruning and after pruning and zig-zag straightening.

pruned skeleton, bottom, of a clean object and of three noisy versions of the object. We observe that whereas the unpruned skeletons are rather different from each other as regards the number of branches, the pruned skeletons are characterized by a similar structure in all cases. Some differences may still be present, which are due to the relevance of boundary changes that are classified as noise. Obviously, if noisy protrusions have a size comparable with the size of other regions of the object that are regarded as significant regions, then pruning will either remove skeleton branches present in both noisy regions and significant regions with the same extension or will keep them in both types of regions.

The behavior of pruning in case of objects in different poses can be appreciated in Fig. 5.13, where the same two objects, "cat" and "hand," appear in four distinct poses.

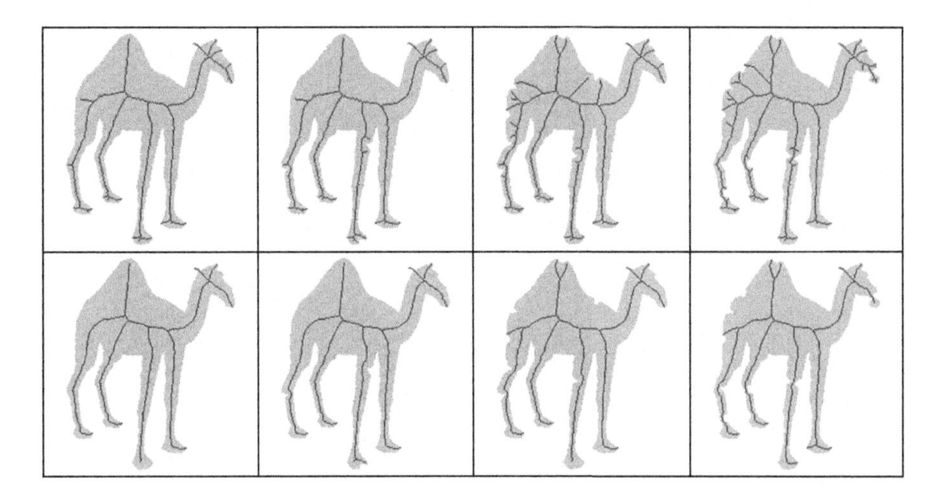

FIGURE 5.12

From left to right, unpruned skeletons, top, and the skeletons after pruning and zig-zag straightening, bottom, for a clean object and for three noisy versions of the object.

The skeletons before pruning are shown in the odd lines of Fig. 5.13, and those after pruning and zig-zag straightening in the even lines of Fig. 5.13. Also in this case, we note that pruning allows us to obtain skeletons with mostly the same structure independently of the pose. Of course, if some details can be seen only in a specific pose of the object, then unavoidable differences will characterize the skeletons. For example, in one of the poses, only three legs of "cat" are visible, which unavoidably produces a skeleton with a peripheral branch less.

We have also checked the performance of our pruning method in case of objects at different resolutions. In the previous section, we suggested how to correlate the two thresholds $\theta 1$ and $\theta 2$ with the size of the object through the bounding box including the object itself. The effect of pruning applied to an object in four resolutions (512×512, 256×256, 128×128 and 64×64) can be seen in Fig. 5.14, where for a better visualization, the true scale factor has not been used. It can be noted that the skeleton maintains almost the same structure at all resolutions.

Since the $\langle 3, 4 \rangle$ distance is a good approximation of the Euclidean distance, the labeled skeleton is expected to be not affected by object rotation. However, rotation may create boundary noisy configurations, which cause a larger number of peripheral skeleton branches with respect to the expected one. Also in this case, pruning can help to obtain skeletons with the same main structure even after object rotation. An example is given in Fig. 5.15, where from left to right we show the skeletons before, top, and after pruning and zig-zag straightening, bottom, of an object and of its rotated version for angles of 12°, 25°, 50°, and 75°, respectively.

Finally, to evaluate the performance of pruning, we also computed the ratio between the number of pixels recovered by the pruned skeleton and the number of

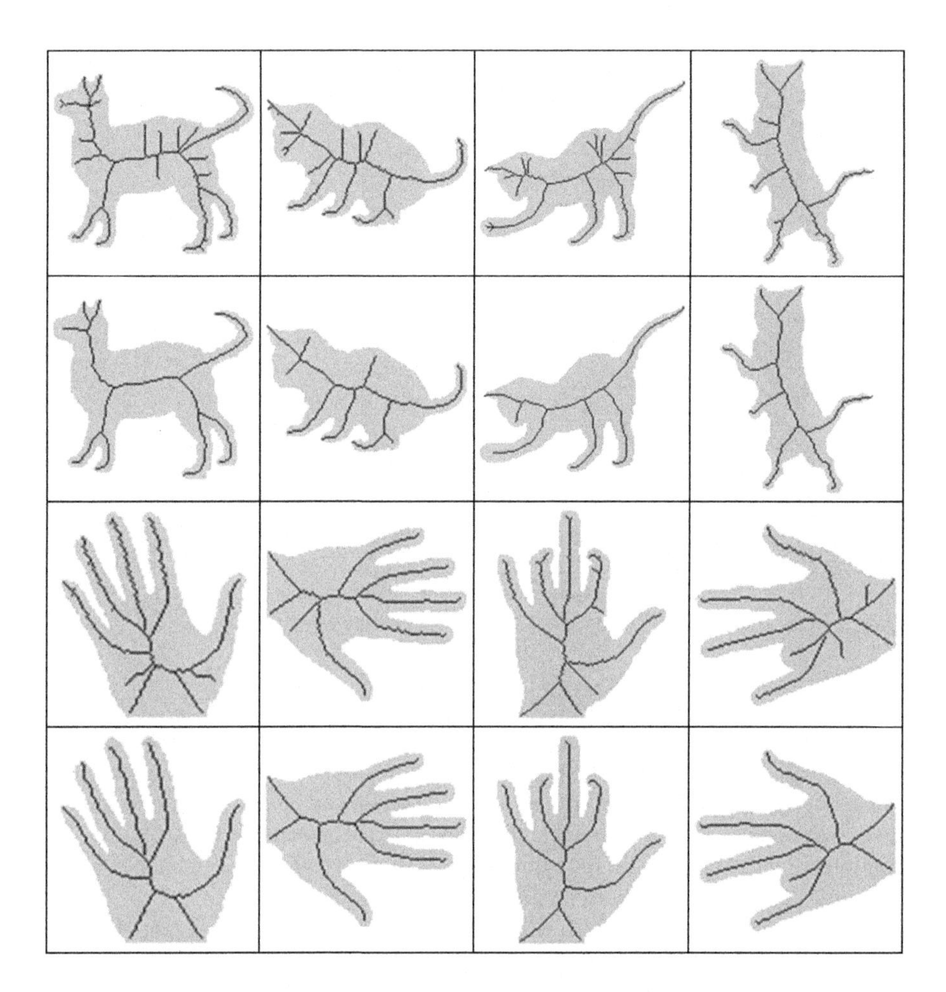

FIGURE 5.13

The unpruned skeletons, lines 1 and 3, and the skeletons after pruning and zig-zag straightening, lines 2 and 4, of two objects in four different poses.

pixels recovered by the unpruned skeleton. The average value of the ratio is above 98%, showing that the representative power of the pruned skeleton is not significantly modified with respect to that of the unpruned skeleton.

5.5 TOOLS TO SIMPLIFY SKELETON STRUCTURE

In this section, we suggest how to simplify the skeleton structure by means of polygonal approximation. The skeleton is replaced by the list of the spatial coordinates and

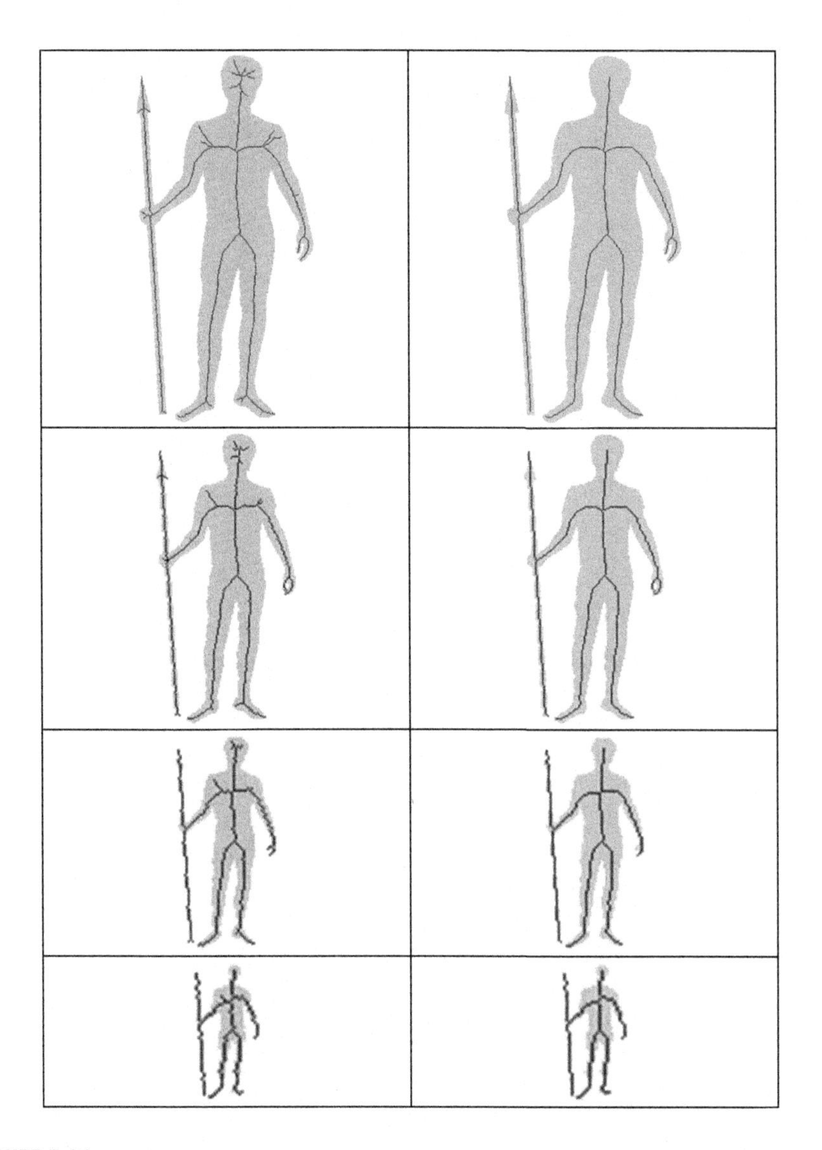

FIGURE 5.14

The skeleton of the same object at different resolutions. Unpruned skeleton, left, skeleton after pruning and zig-zag straightening, right.

distance values of the vertices found along its branches, which is enough to represent the object in a very compact manner. We also show how to build a linearized version of the skeleton starting from the list of spatial coordinates and distance values of the vertices.

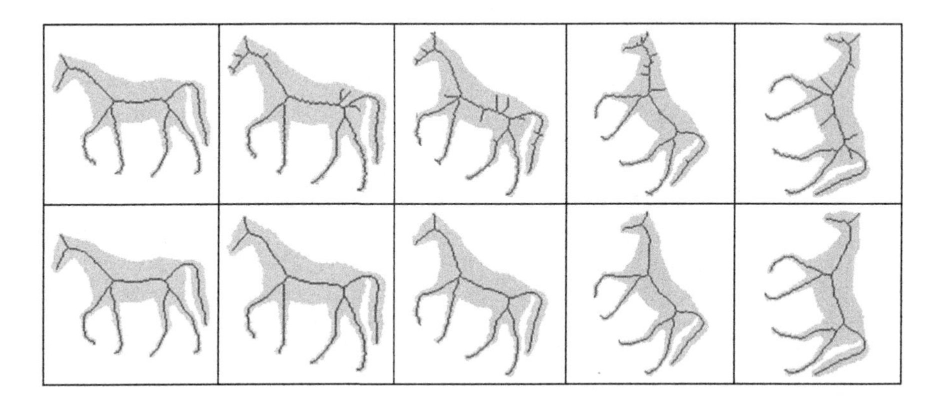

FIGURE 5.15

The skeleton of the same object in different orientations. Unpruned skeleton, top, skeleton after pruning and zig-zag straightening, bottom.

5.5.1 POLYGONAL APPROXIMATION

Resorting to shape representation at different scales is of interest for pattern recognition. In fact, a schematic representation where only the main features of the object are taken into account could be enough to identify the class to which the object belongs, whereas a more detailed representation could be employed to distinguish the objects in the same class. In this framework, the skeleton can be used for object recognition based on graph matching. In fact, S can be represented by a graph, where nodes are skeleton branches and edges account for adjacency relationships among branches. In turn, skeleton pruning can be used not only to remove scarcely significant skeleton branches, but also to provide representations characterized by different level of detail. In fact, if the thresholds on the pruning parameters are suitably modified, a number of skeletons are obtained for the same object, which differ from each other for the number of their constituting branches. The different skeletons represent different approximations of the object, each characterized with respect to the other approximations by the presence/absence of some limbs. This approach has been illustrated in an early paper [18]. Here we present a different strategy, originally introduced in the framework of object decomposition [19–21], to simplify the skeleton structure without changing the number of branches constituting S. The skeleton is first divided into its constituting branches. Then, polygonal approximation of each branch is done in the 3D space, where the three coordinates of each point are the two spatial coordinates and the distance value of the corresponding skeleton pixel. In this way, each branch is divided into a number of straight-line segments, which can be interpreted as the spines of the parts composing the region of the object mapped into that skeleton branch. In the limits of the tolerance adopted during polygonal approximation, the spines are straight-line segments along which distance values are either constant or change in a monotonically increasing/decreasing manner. Thus, the parts associ-

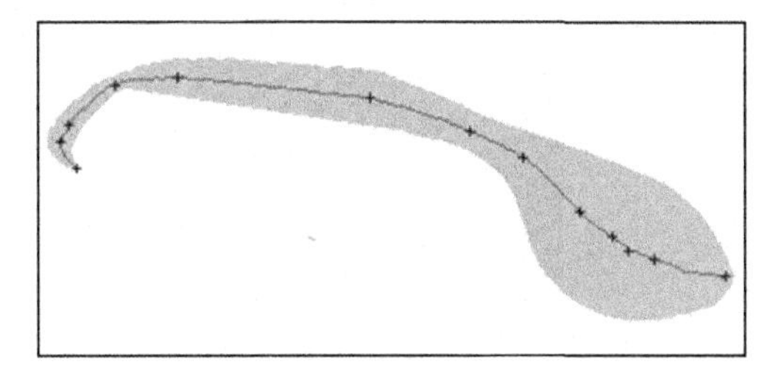

FIGURE 5.16

A simple snake-shaped object with its skeleton. Vertices found by the polygonal approximation in the 3D space, where the 3D coordinates of each point are the two spatial coordinates and the distance value of the corresponding skeleton pixel, are marked by crosses.

ated with these spines have boundaries rid of significant curvature changes and are characterized by either constant thickness or by monotonically increasing/decreasing thickness. Obviously, a different number of segments can be found along the same skeleton branch when polygonal approximation is accomplished with different tolerance values. Thus, S can be replaced by a structure at a coarser or finer level of detail.

We use the split-type polygonal approximation algorithm described in [22]. For each skeleton branch, we accept its delimiting pixels as vertices of the polygonal approximation and compute the distance of each other pixel in the branch from the straight-line joining the two vertices. If the largest distance computed in this way overcomes an a priori fixed threshold value, then the corresponding pixel is taken as a new vertex. The found vertex divides the skeleton pixels along the branch into two subsets, which undergo the same split-type approximation process. Splitting is repeated as far as skeleton pixels are found at a distance higher than the threshold from the straight-lines joining the extremes of the corresponding subsets of the skeleton branch. At the end of the process, the skeleton branch results to be approximated by a number of straight-line segments in the limit of the adopted tolerance. The skeleton branch can be recorded by the ordered sequence of the found vertices.

As an example, refer to Fig. 5.16, where a simple snake-shaped object whose skeleton consists of a single skeleton branch is shown. The skeleton is shown with only one gray tone for a better visualization, but we remark that different distance values characterize the skeletal pixels. Though simple enough to have a skeleton consisting of a single branch, the object is characterized by changes of curvature along its boundary, as well as by changes of local thickness. By taking into account the spatial coordinates (x_p, y_p) of each pixel p of the skeleton S and the associated distance label $d(p)$, S can be represented by an ordered, but nonnecessarily connected, set of points S_{3D} in the 3D space $(x_p, y_p, d(p))$.

Changes in curvature along the boundary of the object are reflected by changes of curvature along S_{3D}. Also the changes in local thickness of the object, corresponding

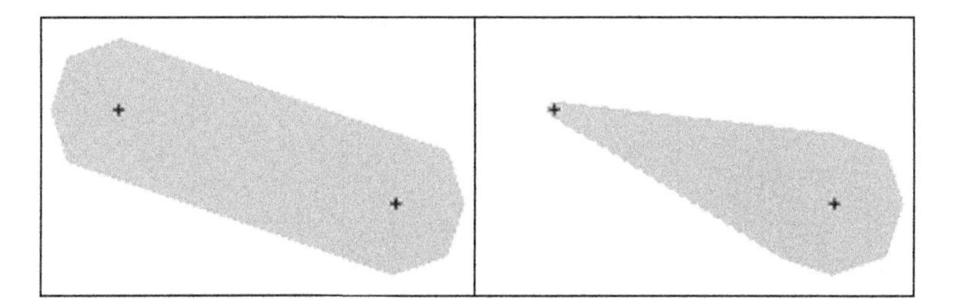

FIGURE 5.17

Rectangle-shaped and trapezium-shaped simple parts corresponding to skeleton segments delimited by two vertices of the polygonal approximation, marked by crosses.

to changes in the distance values of the skeletal pixels, are reflected by changes of curvature along S_{3D}. Thus, the polygonal approximation of S_{3D} takes into account simultaneously both the changes in curvature along the boundary and those in the distribution of distance values.

To identify the vertices for the polygonal approximation of S_{3D}, we compute the Euclidean distance $d_E(c)$ of each point c of S_{3D} from the straight-line joining the two delimiting points v and w. By denoting by $\|vw\|$ the norm of the vector vw and by P_{vwc} the scalar product between vectors vw and vc, the Euclidean distance $d_E(c)$ can be computed as follows:

$$d_E(c)^2 = \|vc\|^2 - \mathrm{P}_{vwc} * \mathrm{P}_{vwc}/\|vw\|^2.$$

Of course, the point c at which the Euclidean distance $d_E(c)$ assumes the maximal value is taken as a vertex only, provided that $d_E(c) > \tau$, where τ is a threshold whose value is fixed depending on the desired approximation.

The vertices found in S_{3D} are mapped back on the corresponding pixels of S, which is then split into a number of straight-line segments each of which is interpreted as the spine of a part of the 2D object, called simple part. A simple part is characterized by the following two properties, derived from the linearity of its spine in terms of curvature and distance values: 1) the thickness is either constant or changes in a monotonically increasing (decreasing) manner, and 2) the boundary is not affected by significant curvature changes. The vertices found for $\tau = 3.83$ are marked by crosses in Fig. 5.16.

We point out that object features such as shape, size, position, and orientation of a simple part can be approximately derived starting from the spatial coordinates and distance values of the two vertices delimiting its spine. In fact, if the two vertices delimiting the spine have the same distance value, then the simple part is rectangle-shaped, whereas in case of delimiting vertices with different distance values, the simple part is trapezium-shaped. In both cases, two half discs with radii equal to the distance values of the vertices complete the simple part. See Fig. 5.17.

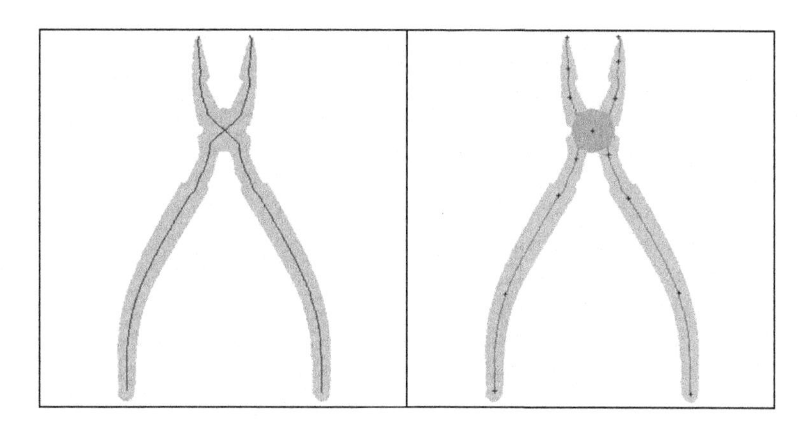

FIGURE 5.18

Left, original skeleton. Right, centroids and vertices are marked by crosses.

The area of a simple part can be evaluated by using elementary notions of Euclidean geometry and by taking into account the length of the associated spine and the distance values of the vertices delimiting the spine. Position and orientation of the simple part can be evaluated by referring to the spatial coordinates of the vertices. Of course, to evaluate the area of the whole object, it must be taken into account that simple parts whose spines share a common vertex partially overlap. Another interesting feature that can be evaluated is the existence of significant curvature or thickness changes where two simple parts corresponding to two spines sharing a vertex overlap, which can be measured by the angle between the two spines.

In a more general case, the skeleton consists of a number of branches; e.g., see Fig. 5.18, left. For a better visualization, though S is a distance-labeled skeleton, its pixels are all shown with the same gray tone. After clusters of branch points have been replaced by the centroids and once polygonal approximation has been done on each branch of S_{3D} and the found vertices have been mapped onto S, each skeleton branch can be represented simply by orderly recording the set of spatial coordinates and radii of the vertices detected along it. In Fig. 5.18, right, the subsets of S included in the zones of influence are not shown, whereas the centroids and the vertices found for $\tau = 7, 5$ are marked by crosses. Of course, for each centroid, track is kept of the linking relations of the centroid with the proper skeleton subsets outside the zone of influence.

Finally, if the skeleton consists exclusively of a loop, which happens for ring-shaped objects possibly having variable thickness, the approximation might be biased by a random selection of the two delimiting vertices. To avoid this, we select the skeleton pixel at maximal Euclidean distance from the barycenter of the skeleton as both starting and ending point for the approximation of the loop.

5.5.2 BUILDING A LINEARIZED VERSION OF THE SKELETON

After polygonal approximation in the 3D space, the skeleton S can be seen as consisting of segments. In the limits of the tolerance adopted for the approximation, each segment is a straight-line segment, and distance values along it either have always the same value or change in a monotonically increasing (decreasing) manner. These segments can be built to produce a linearized version SL of the skeleton.

To build a straight-line segment delimited by two given vertices, say v_i and v_{i+1}, we need to establish 1) the total number of moves along the segment linking the two vertices, 2) how many of these moves are toward edge-neighbors and how many toward vertex-neighbors, 3) the proper alternation of the two types of moves, and, finally, 4) which distance values should be distributed to the pixels of the built segment.

Let (x_i, y_i) and r_i be the spatial coordinates and distance value of v_i, and (x_{i+1}, y_{i+1}) and r_{i+1} those of v_{i+1}. Let Δx, Δy, and Δr denote the differences in absolute value between homologous coordinates and distance values of the two vertices.

The total number N of pixels in the segment can be immediately computed as the chessboard distance of the two vertices, $d_8(v_i, v_{i+1})$, which is equal to the maximum between Δx and Δy. For one move toward a vertex-neighbor, both spatial coordinates must increase/decrease by one unit. Thus, the total number of vertex-moves, n_v, is equal to the minimum between Δx and Δy. By subtracting the value n_v from the maximum between Δx and Δy the number of moves toward edge-neighbors, n_e, is computed.

It is well known that a number of different minimal length paths with the same numbers of moves n_v and n_e can be built to link v_i and v_{i+1}. Thus, the n_v and n_e moves should be properly alternated so as to identify the path that satisfies the chord property [23], and actually is a digital straight-line segment. If n_v is the minimum between n_e and n_v, the ratio n_e/n_v accounts for the number of moves toward edge-neighbors, after which one move toward a vertex-neighbor has to be accomplished. Of course, cases where the ratios are not integer numbers or the denominator is zero have to be properly treated. The first move from v_i will be toward an edge-neighbor since we are assuming that n_v is the minimum between n_e and n_v (it would be toward a vertex-neighbor if n_e was the minimum). The right edge-neighbor, out of the four possible ones, is the one at minimal distance from v_{i+1}. Then, only the neighbors of the current pixel that can be reached with the current move are considered, and the neighbor closer to v_{i+1} is selected for building the path.

The last task to obtain the linearized labeled skeleton SL is to assign distance values to the pixels of the path. Let us suppose that $r_i < r_{i+1}$. Then, the ratio $\Delta r/N$ denotes the increment in distance value that, starting from the value r_i pertaining v_i, has to be used to assign distance values to the remaining pixels in the segment until v_{i+1} is reached. Distance values are approximated by the closest integers in order S can be regarded as a digital skeleton.

Of course, different polygonal approximations of S and, hence, different linearized distance labeled skeletons SL are possible, depending on the value chosen

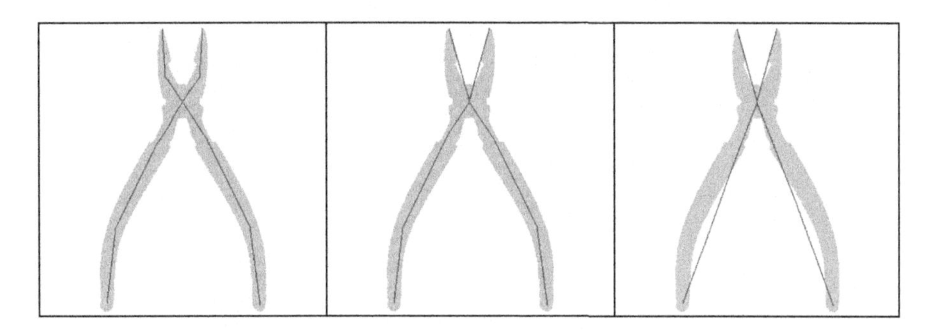

FIGURE 5.19

From left to right, the linearized distance labeled skeleton SL, obtained for $\tau = 13.5$, $\tau = 18.5$, and $\tau = 26.5$, respectively.

for the threshold τ. The smallest possible value is $\tau = 1$ since a digital curve is a digital straight-line if the chord property [23] is satisfied for all its pixels, meaning that all the pixels have distance at most equal to 1 from the straight-line joining the two extremes of the curve. In principle, the largest possible value for τ can be obtained by taking for each skeleton branch the maximal distance of its pixels from the straight-line joining the two extremes of the branch and by computing the maximum value out of the maxima detected for all the branches. Obviously, such a value would be definitely too high to obtain a reasonable approximation of the skeleton. Thus, a strategy to select a proper value for τ is necessary.

Some different linearized version SL obtained by polygonal approximation of S with different values for the threshold τ are shown in Fig. 5.19.

To obtain an optimal approximation, where the pixels in each single segment of SL almost coincide with the pixels of S that it replaces, the value for τ should be different for each skeleton branch. We suggest to use as threshold for a skeleton branch delimited by two pixels v_i and v_{i+1}, either the largest value between their associated radii r_i and r_{i+1}, or the arithmetic mean of r_i and r_{i+1}. In Fig. 5.20, the optimal approximations are shown that are obtained by using as threshold for each branch the largest value and the arithmetic mean of the distance values of the extremes, respectively.

A few more examples are given in Fig. 5.21. In all examples, for each skeleton, branch τ has been set equal to the arithmetic mean of the distance values of the extremes of the branch.

5.6 EXPERIMENTAL RESULTS ON THE SIMPLIFICATION OF SKELETON STRUCTURE

Polygonal approximation accomplished on the skeleton is particularly useful in the framework of the structural approach to shape analysis. According to such an ap-

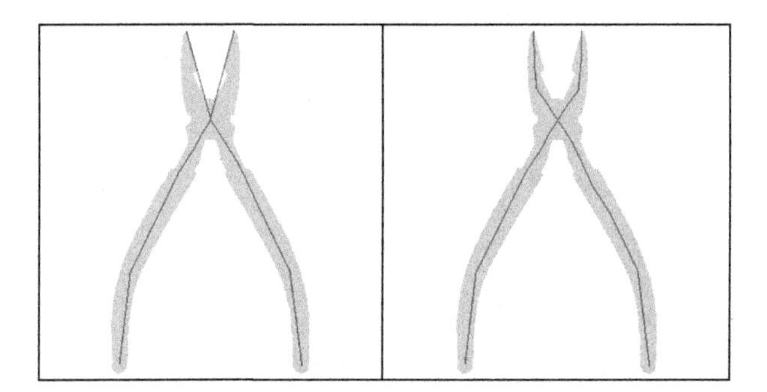

FIGURE 5.20

The optimal polygonal approximations obtained by selecting as threshold for each skeleton branch the largest distance value of the extremes, left, and the arithmetic mean of the distance values of the extremes, right.

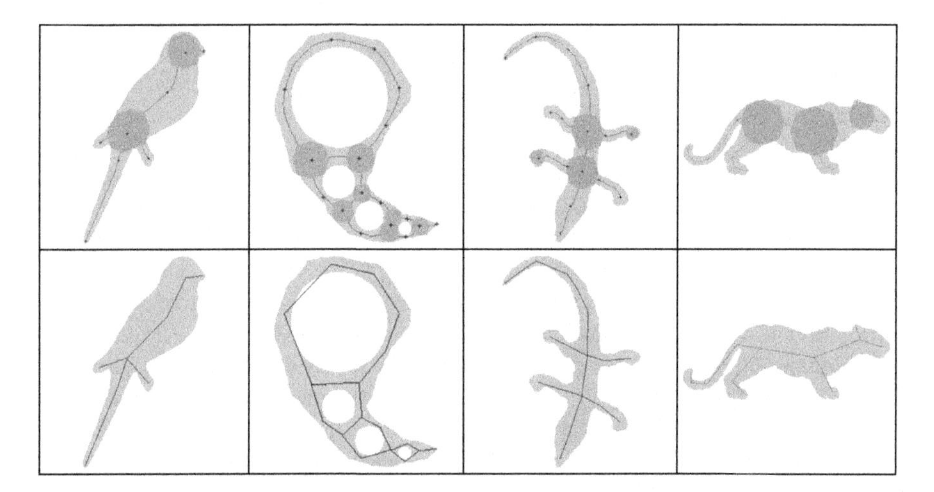

FIGURE 5.21

Skeletons after polygonal approximation, top, and linearized skeletons, bottom.

proach, an object characterized by complex shape can be divided into simpler parts. Then, the description of the object can be achieved in terms of the description of each part constituting it and of the spatial relationships among the parts. Since the skeleton of an object represents the object, the possibility of decomposing the skeleton into parts, each of which corresponding to a region of the object with simpler shape, is very important for analyzing the shape of articulated objects.

The notion of zone of influence introduced in Section 5.2, besides being important to detect the correct number of branch points that will allow us to establish a match between the ideal skeleton and the digital skeleton, is useful in the context of shape

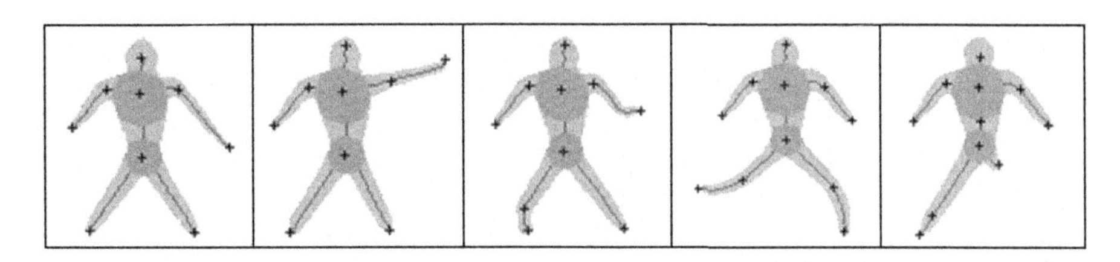

FIGURE 5.22

Skeletons of the same object in different poses, which can be distinguished due to the vertices of the polygonal approximation, crosses.

decomposition and object recognition. In fact, a partition of the skeleton into significant simple curves is immediately obtained once each subset of skeletal pixels inside any zone of influence is replaced by its centroid. Each simple curve of S represents a region of the object. These regions partially overlap since the centroids are common to a number of skeleton branches.

By performing the polygonal approximation of each simple curve in the 3D space, so as to take into account also the distance information attached to each skeletal pixel, we split the simple curve into a sequence of straight-line segments, which can be interpreted as the spines of simple parts. Of course, also the simple parts constituting the object are partially overlapping since two successive spines along the same skeleton branch share a common hinge.

The initial decomposition of S can be used to select the class to which the object at hand belongs by taking into account the number of branches into which the skeleton is decomposed, how many of them are peripheral branches and how many are internal branches, and the spatial relationships among the various branches. Then, the segments found by means of polygonal approximation can be used to distinguish the objects within the same class, which are characterized by different size or pose. As an example, refer to Fig. 5.22, where the skeleton of an object in different poses is shown. The subsets of pixels inside the zones of influence, which are actually replaced by the corresponding centroids, are not shown; centroids and vertices found by the 3D polygonal approximation of the skeleton are marked by crosses. We observe that the object is characterized by the same number of branches, independently of the pose. The vertices found by polygonal approximation can be used to distinguish the different poses.

5.7 CONCLUSION

The skeleton is a shape representation system that can be profitably used in the framework of shape analysis. Since the definition of Medial Axis Transform, a huge number of skeletonization algorithms have been suggested, which are all more or less affected by some drawbacks that limit the actual use of the skeleton for applications.

Aim of this contribution is to introduce a number of tools, to be used individually or suitably combined, that can improve the visual aspect of the skeleton and simplify its structure. In particular, we have shown 1) how to straighten zig-zags possibly present along skeleton branches; 2) replace each set of branch points sufficiently close to each other by the centroid in the zone of influence of the set of branch points; 3) prune branches originating from scarcely significant object's protrusions or arising in correspondence with weak convexities of the object's boundary; 4) polygonally approximate skeleton branches by taking into account both the changes of curvature along the skeleton branches and the variations in local thickness of the object's regions into which the skeleton branches are mapped; and 5) build a simplified version of the skeleton, where each skeleton branch consists of a sequence of hinged straight-line segments along each of which the distance values of the skeleton points either assume a unique value or assume values that are monotonically increasing or monotonically decreasing.

Removing zig-zags is done to reduce the jaggedness often affecting the skeleton as regards both the spatial distribution and the distance values of its pixels. Though the presence of zig-zags along skeleton branches does not prevent the use of S for shape analysis, the results obtained after zig-zags straightening are more visually pleasant.

Detecting the zones of influence and replacing the skeletal pixels included therein is crucial to associate a single meeting point with a cluster of branch points. Doing this, it is also possible to get rid of short branches connecting very close branch points, which do not correspond to individually meaningful object's parts, so that a one-to-one correspondence between skeleton branches and object's parts can be established. The zones of influence have also another important role in the framework of object analysis and recognition. In fact, once the zones of influence have been found and the subset of skeletal pixels included in each zone of influence is replaced by the centroid, an immediate decomposition of the skeleton into simple curves is obtained, where each simple curve is in correspondence with a part of the object. Thus, decomposing the skeleton simulates the decomposition of the object into a number of partially overlapping regions, each of which represented by the associated skeleton branch.

Pruning has a very important role. In fact, it is a fundamental step whichever skeletonization algorithm is used to have a skeleton that can really be used for practical applications. The method suggested in this chapter can treat objects with different resolution. Its more interesting feature is that it does not require to be iterated since the longest concatenation that can be pruned starting from every branch point is automatically determined. The suggested pruning method is robust, producing skeletons with mostly the same structure independently of whether the input object is clean or noisy, and independently on object orientation, scale or pose.

Polygonal approximation is useful to replace the skeleton by the list of the vertices found along its branches, so obtaining a very compact representation of the object. Since an object can be characterized by changes of curvature along its boundary and also by changes of local thickness, we perform the polygonal approximation in a 3D space, where the three coordinates of each point are the spatial coordinates and distance values of the skeletal pixels. In this way, each skeleton branch is decomposed into segments that are straight-line segments and the distance values of the points in each segment either are constant or change in a monotonically increasing or monotonically decreasing manner. Each segment can be interpreted as the spine of an object part that is characterized by simple shape. In fact, any such an object part is shaped as a rectangle or as a trapezium, depending on whether the distance values of the two vertices of the spine are equal or not.

Finally, a linearized version of the skeleton can be obtained starting from the spatial coordinates and distance values of the vertices found during polygonal approximation. Such a linearized version can be adopted to recover the object or its individual parts.

REFERENCES

[1] H. Blum, A transformation for extracting new descriptors of shape, in: W. Wathen-Dunn (Ed.), Models for the Perception of Speech and Visual Form, MIT, Cambridge, MA, 1967, pp. 362–380.

[2] R.L. Ogniewicz, O. Kubler, Hierarchic Voronoi skeletons, Pattern Recognit. 28 (3) (1995) 343–359.

[3] G. Sanniti di Baja, Well-shaped, stable and reversible skeletons from the (3, 4)-distance transform, J. Vis. Commun. Image Represent. 5 (1994) 107–115.

[4] G. Borgefors, Distance transformations in digital images, Comput. Vis. Graph. Image Process. 34 (3) (1986) 344–371.

[5] C. Arcelli, G. Sanniti di Baja, Finding local maxima in a pseudo Euclidean distance transform, Comput. Vis. Graph. Image Process. 43 (1988) 361–367.

[6] L. Serino, C. Arcelli, G. Sanniti di Baja, From skeleton branches to object parts, Comput. Vis. Image Underst. 129 (2014) 42–51.

[7] L. Serino, G. Sanniti di Baja, Selecting anchor points for 2D skeletonization, in: M. Kamel, A. Campilho (Eds.), ICIAR 2011, in: Lect. Notes Comput. Sci., vol. 6753, Springer-Verlag, 2011, pp. 344–353, Part I.

[8] M. Postolski, M. Couprie, M. Janaszewski, Scale filtered Euclidean medial axis and its hierarchy, Comput. Vis. Image Underst. 129 (2014) 89–102.

[9] D. Shaked, A.M. Bruckstein, Pruning medial axes, Comput. Vis. Image Underst. 69 (2) (1998) 156–169.

[10] X. Bai, L.J. Latecki, W-Y. Liu, Skeleton pruning by contour partitioning with discrete curve evolution, IEEE Trans. Pattern Anal. Mach. Intell. 29 (3) (2007) 449–462.

[11] W. Shen, X. Bai, R. Hu, H. Wang, L.J. Latecki, Skeleton growing and pruning with bending potential ratio, Pattern Recognit. 44 (2011) 196–209.

[12] M. Frucci, G. Sanniti di Baja, C. Arcelli, L.P. Cordella, On the strategy to follow for skeleton pruning, in: M. De Marsico, A. Fred (Eds.), ICPRAM 2013, SCITEPRESS, Lisboa, Portugal, 2013, pp. 263–266.

[13] L. Serino, G. Sanniti di Baja, A new strategy for skeleton pruning, Pattern Recognit. Lett. 76 (2016) 41–48.

[14] http://tosca.cs.technion.ac.il/book/resources_data.html.

[15] http://www.cs.toronto.edu/~dmac/ShapeMatcher/.
[16] http://vision.lems.brown.edu/faculty/kimia.
[17] http://www.cs.rug.nl/svcg/Shapes/SkelBenchmark.
[18] C. Arcelli, G. Sanniti di Baja, Medial lines and figure analysis, in: Proc. 5th ICPR, vol. 2, IEEE, 1980, pp. 1016–1018.
[19] G. Sanniti di Baja, E. Thiel, (3, 4)-weighted skeleton decomposition for pattern representation and description, Pattern Recognit. 27 (1994) 1039–1049.
[20] G. Sanniti di Baja, E. Thiel, Shape description via weighted skeleton partition, in: S. Impedovo (Ed.), Progress in Image Analysis and Processing III, World Scientific, Singapore, 1994, pp. 87–94.
[21] G. Sanniti di Baja, E. Thiel, The path-based distance skeleton: a flexible tool to analyse silhouette shape, in: Proc. 12th International Conference on Pattern Recognition, Jerusalem, 1994, pp. 570–572.
[22] U. Ramer, An iterative procedure for the polygonal approximation of plane curves, Comput. Vis. Graph. Image Process. 1 (1972) 244–256.
[23] A. Rosenfeld, Convex digital arcs, IEEE Trans. Comput. C-23 (1974) 1264–1269.

CHAPTER

Curve skeletonization using minimum-cost path

Dakai Jin*, **Cheng Chen***, **Eric A. Hoffman**†, **Punam K. Saha***,†

*Department of Electrical and Computer Engineering, University of Iowa, Iowa City, IA, USA**

Department of Radiology, University of Iowa, Iowa City, IA, USA†

Contents

6.1 Introduction 151
6.2 General Principles of Curve Skeletonization 153
6.2.1 Geodesic Distance Transform Based Methods 154
6.2.2 Direct Methods 155
6.2.3 Minimum-Cost Path Based Methods 156
6.3 Curve Skeletonization with Minimum-Cost Path 157
6.3.1 Overall Outline 158
6.3.2 Minimum-Cost Path 161
6.3.3 Centered Minimum-Cost Paths 162
6.3.4 Skeletal Branch Detection 164
6.3.5 Object Volume Marking 165
6.3.6 Skeletal Branch Significance 166
6.3.7 Termination Criterion 166
6.3.8 Algorithm Efficiency 167
6.4 Applications 169
6.4.1 Virtual Colonoscopy and Bronchoscopy 170
6.4.2 Anatomical Labeling of Human Airway Trees 172
6.4.3 Morphologic Assessment of Pulmonary Arteries 173
6.5 Conclusions 174
References 175

6.1 INTRODUCTION

Skeletonization provides a simple yet compact representation of an object while capturing its essential topologic and geometric features [1–3]. Over the last several decades, it has been widely used in various applications related to computer graphics and vision, image processing, image understanding, medical image analysis, etc. Popular applications of skeletonization include object description, retrieval, virtual

Skeletonization. DOI: 10.1016/B978-0-08-101291-8.00007-9

FIGURE 6.1

Illustrations of surface and curve skeletons for binary 3-D digital objects in a voxel grid. Voxels representing curve skeletons are shown in red, whereas those either in red or green constitute surface skeletons.

navigation, matching, recognition, animation, compression, registration, and tracking [2–4]. Blum [5] presented the foundation of skeletonization in the form of *medial loci* of an object in R^n that forms the skeleton of the object, which consists of structures of symmetry at lower dimensions. He described the skeletonization process using his seminal notion of grassfire transform. Specifically, the skeleton of a two-dimensional (2-D) object is formed by the *centers of maximally inscribed discs* (*CMIDs*), whereas *the centers of maximally inscribed balls* (*CMIBs*) constructs the skeleton of a 3-D object. Following Blum's principle of skeletonization, the skeleton of an object in R^n consists of $(n - 1)$ or lower dimensional structures. For example, in 2-D, it reduces an object to a skeleton consisting of one-dimensional (1-D) structures, i.e., curves, only. In three dimensions (3-D), it converts an object into a skeleton that consists of surfaces, i.e., 2-D structures and curves. This skeletonization process in 3-D is commonly referred to as *surface skeletonization*, as it allows surface structures to exist in a skeleton. In a different formulation of skeletonization, a 3-D object is reduced to a "curve skeleton," where a skeleton is formed strictly by 1-D elements, i.e., curves, and no surface elements may survive in the skeleton. Such a process is referred to as *curve skeletonization* and the skeleton as *curve skeleton*. A few examples of surface and curve skeletons for 3-D digital objects are illustrated in Fig. 6.1. For an object consisting of tubular structures, both surface and curve skeletons are identical, and they are formed by the principle of Blum's grassfire propagation. However, the same may not be true for other objects.

Most skeletonization algorithms available in the image processing, computer vision, and graphics literature use Blum's grassfire propagation as the underlying principle. However, a large number of computational and algorithmic approaches have been adopted by various research groups. A thorough discussion on different computational approaches of skeletonization is presented in Chapter 1. See Chapter 3 for discussion on fuzzy skeletonization [6–8]. A strategy of iterative constrained digital erosion of object boundary has been popularly used in many 3-D skeletonization algorithms [9–13]. In these algorithms, the constraint for erosion includes preservation of both topology [14,15] and geometric features. A few algorithms modified the geometric constraints only to preserve elongated features for directly computing the curve skeleton from a 3-D object [16–19].

Different principles, algorithms, and applications of curve skeletonization with primary focus on methods using centered minimum-cost path tools are discussed in this chapter. Specifically, general principles of different curve skeletonization approaches and their computational solutions are discussed in Section 6.2. A comprehensive solution for curve skeletonization of tree-like 3-D digital objects using minimum-cost paths and related topics are presented in Section 6.3. In this context, various important and related ideas including minimum-cost paths, their centeredness properties, volume marking, skeletal branch significance, termination criterion, etc. are described. An efficient algorithm with significantly improved computational complexity as compared to conventional approach of adding one skeletal branch per iteration is described. Also, several real-life applications of curve skeletons are discussed in Section 6.4. Finally, the conclusions are drawn in Section 6.5.

6.2 GENERAL PRINCIPLES OF CURVE SKELETONIZATION

Blum's medial loci or skeleton is defined using the process of grassfire propagation, where a 2-D object is treated as a field of dry grass, and the fire is simultaneously lit at the entire boundary. The fire burns the grass field, and fire-front propagates inside the object at a uniform speed; finally, the skeleton is formed at *quench points* where independent fire-fronts meet. In 3-D, this definition forms a surface skeleton or a medial surface representation, which consists of both 1-D and 2-D manifold structures. The fire propagation process can be simulated using distance transform (DT) [20,21] from the object boundary to its inner points, where the singularity points on the distance transform map coincide with the quench points of grassfire transform. Although the same process leads to the curve skeleton of a tubular object whose orthogonal cutting plane is a circle, it fails to generate curve skeletons for arbitrary objects in 3-D. For example, the surface skeleton of the tubular object in Fig. 6.2 (left) coincides with its curve skeleton, and both can be computed using Blum's grassfire transform. On the other hand, the curve skeleton of the object in Fig. 6.2 (right) is different from its surface skeleton. Also, for the object in Fig. 6.2 (right), Blum's grassfire transform fails to directly compute its curve skeleton. The major theoretical hurdle with curve skeletonization is that, in 3-D, it attempts to reduce the dimensionality of an object by two, whereas the premise of skeletonization is established on single dimensionality reduction only. It is easy to find examples in real-life applications, where the curve skeleton of a 3-D object fails to capture its geometric properties, or it does not even exist, e.g., a hollow ball.

Despite the theoretical hurdles, curve skeletonization has remained an active research topic due to its topologic and geometric simplicity, especially, in applications related to tubular or quasi-tubular objects. Also curve skeletonization is useful in hierarchical representation of shapes for 3-D animation, where motion-related transforms may be propagated from curve skeleton to surface skeleton, and then to the 3-D

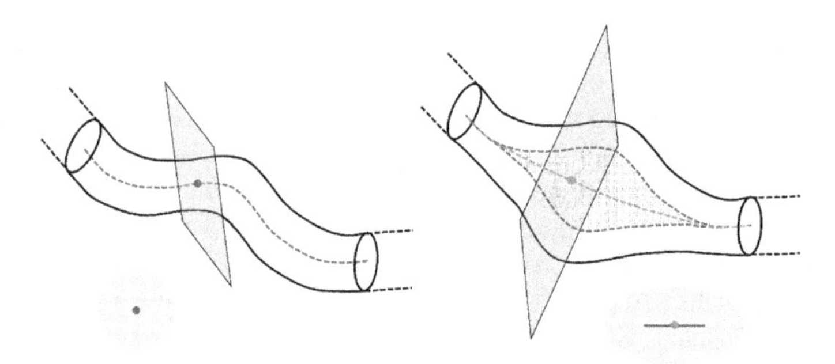

FIGURE 6.2

Illustrations of surface (red) and curve (green) skeletons for tubular (left) and nontubular (right) objects. For a tubular object, its surface and curve skeletons are identical, which is the loci of CMIBs. In contrast, the surface skeleton of a nontubular object is a surface. On an orthogonal cutting plane, a nontubular object creates an image of an elliptical region, whereas its surface skeleton forms the line joining the focal points of that region; finally, the curve skeleton intersects the plane at the center.

volume. Several research groups have made efforts to formulate a concise definition for curve skeletons of 3-D objects [22–24], whereas others attempted to approximate it using computational approaches [9–11,22–34]. Cornea et al. [4] discussed the desired properties for the curve skeletons in terms of topology preservation, invariance under isometric transformations, ability to reconstruct the original object, thinness, centeredness, reliability, i.e., visibility of every boundary point from of a curve skeleton point, junction detection and component-wise differentiation, smoothness, and robustness in terms of sensitivity to small changes on the object boundary. Also, they discussed the impact of various properties of curve skeletons in different real-life applications. Ward and Hamarneh [35] developed a group-wise skeletonization framework that yields a fuzzy significance measure for each skeletal branch, derived from information provided by a group of shapes.

In the rest of this section, principles and computational strategies of different curve skeletonization methods are discussed.

6.2.1 GEODESIC DISTANCE TRANSFORM BASED METHODS

A conventional approach of defining the curve skeleton of an object is based on imitating geodesic grassfire propagation on the surface skeleton of the object [9,11, 36–38]. Specifically, the surface skeleton is considered as a grass field, and fire is lit at all the surface edge points. The fire propagates inward at a uniform speed along geodesic paths on the surface skeleton, and the curve skeleton is formed as quench points on the surface skeleton, where independent geodesic fire-fronts meet. This definition leads to the algorithm for computing curve skeletons using geodesic distance transform (GDT) from the surface skeleton edge to inner surface points [38,39]. The singularities on the GDT map coincide with the quench points, which is also called the *centers of maximal geodesic balls* (CMGB) [38]. It should be clarified that topology preservation is imposed during transformation from a 3-D object to its surface skeleton and from the surface skeleton to its curve skeleton. In digital grids, the set of CMIBs or CMGBs are not topologically equivalent to the original object. Therefore, explicit constraints [14,15] are necessary to ensure topology preservation during these transformations.

Most algorithms under this category adopted a two-step approach. First, a surface skeleton is computed from a 3-D object, which is used as an input to compute the curve skeleton. Iterative surface erosion is used to simulate the process of geodesic fire propagation, and the curve skeleton is computed by detecting end points and, also, by imposing topology preservation [9,11]. Svensson et al. [37] used topological classification of boundary and junction voxels in surface skeleton [40,41] to guide curve skeletonization of surface-like objects in 3-D digital images. A different approach is to compute the GDT from the surface skeleton edges of an object followed by computation of the curve skeleton by locating CMGBs and removing surface skeleton voxels under topology preservation [36–38]. An important observation about GDT computation is that the geodesic distance propagation must be stopped at surface junctions [25,39]; otherwise, the singularity points on the GDT map may lead to topologically unstable situations. For example, if the geodesic distance is allowed to propagate topological junctions between planes, then undesired tunnels may be created in the curve skeleton of a surface structure consisting of a relatively narrow upright plane situated at the center of a wider horizontal plane. The methods currently available in literature under this principle were developed for binary images. However, this principle may be generalized for fuzzy objects using a surface skeletonization algorithm that is directly applicable to fuzzy objects; see Chapter 3 for a detailed discussion on skeletonization of fuzzy objects.

6.2.2 DIRECT METHODS

Several methods are designed to directly compute the curve skeleton of an object, which do not require the intermediate step of surface skeletonization. There are various methods under this category that use different mechanisms or functional fields to directly compute the curve skeleton of a 3-D object. These mechanisms or function fields include erosion or thinning [10,26], distance transform [27,28,42,43], potential field [29,31], feature distance or medial geodesic transform [22,23,33], wave front propagation [24,32], etc. Erosion-based thinning algorithms iteratively remove voxels under topology preservation and protect elongated structures by locating curve-end points [10,12,13,26,27]. A few erosion-based thinning methods [12,13,26] use a mask-based strategy, where both topology and curve-end preservation are combined within preset mask-rules. The basic principle of distance transform or DT-based methods are similar except that the curve-end points are located by analyzing the local DT map without performing the actual erosion process [27,28,42,43].

Potential field-based methods assume uniform distribution of electric charges along the object boundary and use the generated potential field to locate the curve skeleton. Specifically, the curve skeleton is computed by locating local extrema on the potential field and then by connecting them to preserve the object topology. A wave front propagation-based method first identifies an internal object point with the global maximum DT as the starting point and then simulates a wave propagation from the starting point to each object points with a speed that is proportional to local DT value. These waves propagate faster along the centerline of the object creating high curvature. Finally, the curve skeleton is computed by locating merging and extrema points and then by connecting them along the paths with high positive curvature of wave fronts.

A relatively recent approach is to locate singularity points of a feature-distance map to compute the curve skeleton in a 3-D object [22]. The outline of the method is as follows. For a given point p on the surface skeleton $S_{\mathcal{O}}$ of a 3-D object $\mathcal{O}$, the *feature-distance* [23,33,44–46] of p is the length of the geodesic path along object boundary connecting two farthest contact points of the maximally inscribed ball (MIB) centered at p. This measure is referred to as the medial geodesic function (MGF), and the curve skeleton is defined as the singularity points on the MGF map [22]. The outward flux of the MGF map is used to locate the singularity points similar to the method originally proposed in [47]. This framework is further generalized into the notion of density advection model from the surface of an object leading to a unified multiscale approach for computing both surface and curve skeletons [23,33]. In this approach, the *skeleton importance* measure λ is defined as the maximum density that reaches a given object point. Intuitively, this model defines a process where, "mass" is assumed to be uniformly distributed along the object boundary, which flows inward along the shortest path to the surface skeleton; then along the surface skeleton to the curve skeleton; and finally, along the curve skeleton until it reaches the shape center where flows from different parts of the object meet; see Fig. 1 in [33] for graphical illustration of the idea. Once, all masses reach the shape center, the simplified surface and curve skeleton is computed by applying a threshold on the skeleton importance measure λ.

Currently available algorithms under this category were developed for binary objects. However, the basic principle of the algorithms based on distance transform, potential field, feature distance transform, or wave front propagation may be generalized for fuzzy objects. It is relatively straightforward to generalize distance transform-based algorithms for fuzzy objects using fuzzy distance analysis instead of binary distance. The other approaches need additional research and development generalizing related functional formulations for fuzzy objects.

6.2.3 MINIMUM-COST PATH BASED METHODS

Minimum-cost path is a popular tool, which has been used for computing curve skeletons or tracing centerlines, especially, in those medical imaging applications where the underlying object is quasi-tubular such as colonoscopy [30,48–50], pulmonary imaging [51,52], coronary artery tracking [53,54], etc. Methods of centerline extraction using minimum-cost path approach can be classified into two major categories—(1) direct computation of centerlines of tubular objects from raw intensity image without prior object segmentation and (2) centerline computation for presegmented binary or fuzzy tubular objects. Often, the methods under the first category use multiscale Hessian operators on an intensity image for locating potential centerline segments. Finally, minimum-cost path techniques are applied to connect such centerline segments [55–59]. A few methods under this category require graphical user inputs to locate end points of a tubular structure, which are connected to the main centerline tree using minimum-cost paths to interactively grow the target tubular tree object. Antúnez and Guibas [60] presented a 3-D grayscale curve skeletonization algorithm that does not require image segmentation. Specifically, they used least-extremal paths to generate a topologically consistent skeleton representing curve-like features at different scales and intensities of the image These methods are applicable to original grayscale images requiring to presegmentation; a thorough survey on related topics is available in [61]. Lesage et al. [52] reviewed direct centerline tracking and minimal path techniques in the context of 3-D vessel lumen segmentation in medical imaging.

Methods under the second category use a prior segmentation of the volumetric tree object as an input and generate its curve skeleton [30,34,48–50,62,63]. These methods use a local cost function designed to encourage centered paths as compared to paths coming close to object boundaries. Often, distance transform map from the object boundary is used to define centeredness of a given point. A method under this category starts with a root as the initial centerline tree and iteratively augment the tree by adding new centerline branches. First, it identifies a significant extremum point, and then joins it with a new centerline branch computed as a centered minimum-cost path. This tree-augmentation process continues until all significant branches are added. A significant difference between a minimum-cost path approach and those discussed in previous subsections is that a minimum-cost path method sequentially adds branches in the order of their significance as compared to other methods, where the entire skeleton is generated at once. This unique property of a minimum-cost path approach offers a natural hierarchical representation of a centerline tree and provides a robust performance in terms of noisy branch detection. The methods presented in [30,48–50,62] are applicable for binary images, only, whereas the methods by Jin et al. [34,63] are applicable to both binary and fuzzy objects. However, both methods need either binary or fuzzy segmentation of target objects, and it is not obvious how to generalize the basic principle of these methods for gray-scale images without prior segmentation of the target object.

The primary focus in the rest of this chapter is curve skeletonization of tubular or quasi-tubular tree-like objects such as airway or vascular trees.

6.3 CURVE SKELETONIZATION WITH MINIMUM-COST PATH

In the context of image processing, *minimum-cost path* or *minimal path* is used as a popular tool for boundary or centerline-based image segmentation. In many image segmentation applications, it is relatively more efficient to track object boundaries (e.g., 2-D boundary tracing graphical tools) or the centerlines (e.g., tracing centerlines of vascular structures) instead of growing the object region from a seed. In such applications, often, minimum-cost paths are used as a tool for tracing object boundaries or centerlines [64–67]. Cohen and Kimmel [65] adapted the minimum-cost path approach to develop an active model with the global minimum property, which avoids active contours getting trapped at local minima. Although the minimum-cost path approach was initially introduced for image segmentation applications, later, it became popular for tracing centerlines of quasi-tubular objects, especially, in medical imaging applications [48–50,56–59,62,68]. The basic premise of such algorithms is based on computation of *centered minimum-cost paths* that correspond to branches of the curve skeleton of an object.

Sato et al. [49] proposed the first algorithm for computing a complete centerline tree of a presegmented quasi-tubular tree object. From the same research group, Bitter et al. [30] discussed the limitations of their earlier works and presented a stable framework for computing curve skeleton of quasi-tubular tree structures. Recently, Jin et al. [34] made fundamental changes in the framework of minimum-cost path-based curve skeletonization and presented a comprehensive and efficient solution. Some of the following key contributions of this recent work are discussed in this chapter—(1) volume marking represented by each skeletal tree branch leading significant improvements in terms of noisy branch detection and termination criterion, (2) scale-independent cost function for centered paths, (3) improvement of computational efficiency from $O(N)$ to $O(\log_2 N)$ for a balanced tree and $O(\sqrt{N})$ for a general tree, where N is the number of terminal nodes, and (4) generalization of the algorithm for fuzzy objects.

6.3.1 OVERALL OUTLINE

The overall outline of curve skeletonization using minimum-cost paths is graphically illustrated in Fig. 6.3. The method starts with a root point o as the initial skeleton, which is iteratively grown by finding the farthest CMIB and then connecting it to the current skeleton with a new skeletal branch. The iterative expansion of the skeleton continues until no new meaningful branch can be found. During the first iteration, the method finds the farthest CMIB (in 3-D) p_1 from the current skeleton o. Subsequently, the skeleton is expanded by adding a new skeletal branch joining p_1 to the current skeleton. This step is solved by finding a minimum-cost path from o to p_1 (see Fig. 6.3, left). Here, it is important that the cost function is chosen such that a minimum-cost path always runs along the centerline of an object and a high cost is applied when it attempts to deviate from the centerline. After the new skeletal branch op_1 is added, its representative object volume is filled using a local scale-adaptive dilation along the new skeletal branch and marked as shown in Fig. 6.3 (right). The volume marking is used to indicate the object region already assigned with skeletal

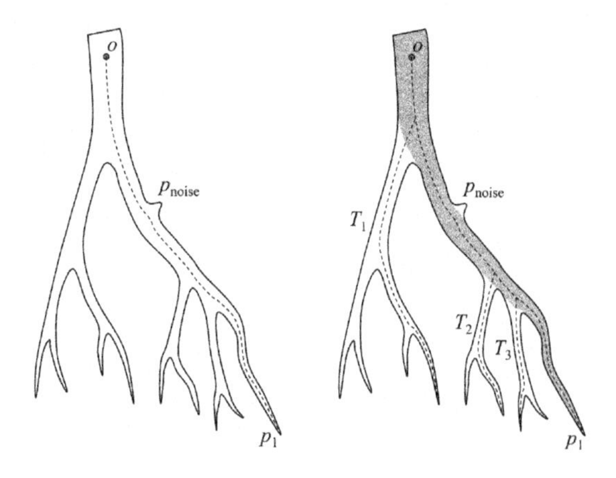

FIGURE 6.3

Overall outline of computing curve skeletons of tubular tree objects using minimum-cost paths. Here, 2-D shapes are used for simplicity; however, the principles are equally applicable to both 2-D and 3-D objects. (Left) The algorithm starts with a root point o as the initial curve skeleton and finds the farthest CMID (CMIB in 3-D) p_1. Next, p_1 is connected to the skeleton using a centered minimum-cost path op_1, which adds a new skeletal branch. (Right) The object area (volume in 3-D) corresponding to the new branch is marked as shown in gray. In the next iteration, three connected components T_1, T_2, and T_3 are identified in the unmarked object region, each representing an unmarked subtree. During this iteration, three skeletal branches are simultaneously added, where each branch connects the farthest CMID (CMIB in 3-D) in one of the three unmarked subtrees. Note that the noisy protrusion p_{noise} does not create a significant component in the unmarked region, and therefore, generates no noisy branch even after all meaningful branches are added.

branches. Thus, new skeletal branch searching in subsequent iterations is confined over unmarked regions, only.

In the second iteration, a connected component labeling algorithm is applied over unmarked object regions. For the specific example of Fig. 6.3, three meaningful components T_1, T_2, and T_3 of unmarked object regions are found. Each component represents an unmarked object subtree. Skeletal branches are simultaneously added for each of these unmarked object subtrees. Specifically, the farthest CMIB in each unmarked object subtree T_i from the currently marked object volume is identified, and a new skeletal branch is added connecting the farthest CMIB to the current skeleton. Note that the current marked object region is used to locate the farthest CMIB in each subtree, whereas the minimum-cost path from the current skeleton is used for computing the new skeletal branch. To complete the second iteration, the marked object volume is augmented using dilation along the three new skeletal branches in the subtrees T_1, T_2, and T_3. This iterative process continues until no new meaningful branch can be found. Fig. 6.4 presents a color-coded illustration of the marked object volume corresponding to the branches added after different iterations. This solution

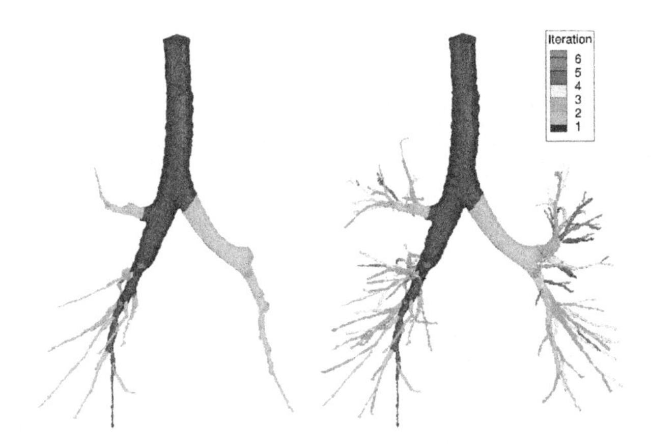

FIGURE 6.4

Results of minimum-cost path-based curve skeletonization after different iterations. (Left) The marked object volume on a CT-derived human intrathoracic airway tree corresponding to the skeletal branches computed after two iterations. (Right) Same as (left) but at the terminal iteration.

of curve skeletonization with minimum-cost path works for 2-D or 3-D tree-like objects (e.g., vascular or airway trees), which are simply connected objects without tunnels or cavities [14,69]. Major steps of this algorithm are summarized using the following high-level pseudo code.

Begin Algorithm: compute-curve-skeleton using minimum-cost path.

Input: the original object volume O

Output: curve skeleton S

Initialize a root voxel o as the current skeleton S and the current marked object volume O_{marked}

While new branches are found

  Detect disconnected subtrees $T_1, T_2, T_3, \cdots$ in the unmarked object volume $O - O_{\text{marked}}$

  For each subtree T_i

    Find the CMIB voxel $v_i \in T_i$ that is farthest from O_{marked}

    If the potential branch from v_i to O_{marked} is significant

      Add a new skeletal branch B_i joining v_i to the current skeleton S using a minimum-cost path

      Augment $S = S \cup B_i$

      Compute local scale-adaptive dilatation D_i along B_i

      Augment $O_{\text{marked}} = O_{\text{marked}} \cup D_i$

End Algorithm: compute-curve-skeleton using minimum-cost path

In the rest of this section, minimum-cost paths are discussed along with detail description of different steps in curve skeletonization. The following definitions and notations are used.

Three-dimensional digital images are popularly represented on a *cubic grid*, which is often denoted by Z^3, where Z is the set of integers. An element $p \in Z^3$ in the image grid is referred to as a *voxel*. A digital image is defined using an image intensity function $I : Z^3 \to [I_{\text{MIN}}, I_{\text{MAX}}]$, where I_{MIN} and I_{MAX} are the minimum and maximum image intensity values. A segmented digital object may be obtained from a digital image using a suitable segmentation algorithm [70]. A *fuzzy digital object* $\mathcal{O} = \{(p, \mu_{\mathcal{O}}(p)) \mid p \in Z^3\}$ is a fuzzy subset of Z^3, where $\mu_{\mathcal{O}} : Z^3 \to [0, 1]$ is the *membership function*. The *support* O of $\mathcal{O}$ is the set of voxels with nonzero membership values, i.e., $O = \{p \mid p \in Z^3 \wedge \mu_{\mathcal{O}}(p) \neq 0\}$. A voxel inside the support is referred to as an *object voxel*. Let $DT_{\mathcal{O}}(p)$ and $FDT_{\mathcal{O}}(p)$ denote the distance transform (DT) [20,21] and fuzzy distance transform (FDT) [71] at an object voxel $p \in O$, respectively. Note that $DT_{\mathcal{O}}(p)$ is computed using the support O as the object. Let $N(p)$ and $N^*(p)$ denote the $3 \times 3 \times 3$ neighbor voxels of p with and without p, respectively. In the rest of this chapter, a center of maximal inscribed ball will be referred to as a "center of maximal ball" in the context of digital images as popularly used in literature. A voxel p is a *center of maximal ball* (CMB) [72] in a binary digital object if $DT_{\mathcal{O}}(q) - DT_{\mathcal{O}}(p) < |p - q|$ holds for every neighbor voxel $q \in N^*(p)$. A voxel p is a *fuzzy center of maximal ball* (fCMB) [73,74] in a fuzzy digital object if $FDT_{\mathcal{O}}(q) - FDT_{\mathcal{O}}(p) < \frac{1}{2}(\mu_{\mathcal{O}}(p) + \mu_{\mathcal{O}}(q))|p - q|$ for every neighbor voxel $q \in N^*(p)$.

6.3.2 MINIMUM-COST PATH

Minimum-cost paths are formulated in a digital domain using a predefined step-cost function $\tau : Z^3 \times Z^3 \to R^+$, where R^+ is the set of all positive numbers. The function τ defines the cost of a step between two adjacent voxels; note that the step-cost is an infinitely large value when the two voxels of a step are nonadjacent. To ensure the metric property of the derived minimum-cost path function, a step-cost function must satisfy the following three properties:

1) $\tau(p, p) = 0$ for any $p \in Z^3$,
2) $\tau(p, q) > 0$ for any $p, q \in Z^3$ and $p \neq q$, and
3) $\tau(p, q) = \tau(q, p)$ for any $p, q \in Z^3$.

Note that a step-cost function is not a metric as it does not satisfy triangular property. The purpose of step-cost function is to define a distance between two adjacent voxels. A minimum-cost path metric is derived from such a step-cost function using the formulations presented in Eqs. (6.1)–(6.3). Further, note that when the second constraint is relaxed, the minimum-cost path function becomes a pseudo-metric. A *path* $\pi = \langle p_0, p_1, \ldots, p_{l-1} \rangle$ is an ordered sequence of voxels where every two successive voxels $p_{i-1}, p_i \in Z^3$, $i = 1, \ldots, l-1$, are 26-adjacent. Note that $\langle p_{i-1}, p_i \rangle$, $i = 1, \ldots, l-1$, are the steps on the path π. The cost of a path

$\pi = \langle p_0, p_1, \ldots, p_{l-1} \rangle$, denoted as $Cost(\pi)$, is computed as the sum of the costs of all individual steps-along the path, i.e.,

$$Cost(\pi) = \sum_{i=1}^{l-1} \tau(p_{i-1}, p_i). \tag{6.1}$$

In order to define the minimum-cost path between any two voxels $p, q \in Z^3$, not necessarily adjacent, we need to identify the set of all paths joining p, q. Let $\Pi_{p,q}$ denote the set of all paths between p, q. The *minimum-cost* between the two voxel p, q, denoted by $\mathrm{T}(p, q)$, is defined as follows:

$$\mathrm{T}(p, q) = \min_{\pi \in \Pi_{p,q}} Cost(\pi). \tag{6.2}$$

Finally, the *minimum-cost path* between two voxel p, q, denoted as $\pi^*_{p,q}$, is computed as follows:

$$\pi^*_{p,q} = \arg \min_{\pi \in \Pi_{p,q}} Cost(\pi). \tag{6.3}$$

The minimum-cost path between the starting point p and a given point q is derived from the minimum-cost map T using a gradient descent back-propagation starting at q until p is reached. The gradient descent back-propagation always converges to p from q since it has no local minima except at p. The fast marching algorithm [75], which uses a front-propagation approach similar to Dijkstra's algorithm [76], is used to efficiently compute the minimum-cost map. Deschamps and Cohen [56] proposed a simultaneous front propagation approach for minimum-cost path computation. The idea is to simultaneously propagate a front from each of the two voxels p and q and finding the voxel r where the two fronts meet. The minimum-cost path between p and q is computed by separately finding the minimum-cost path from r to each of p and q and then adjoining the two geodesics. This approach greatly reduces the region covered during front propagation as compared to the fast-marching algorithms, and, also, it allows parallel implementation of the two front propagations.

6.3.3 CENTERED MINIMUM-COST PATHS

The core principle of a minimum-cost path-based curve skeletonization algorithm is to iteratively add skeletal branches each connecting the current farthest CMB voxel to the main skeleton using a minimum-cost path. As discussed in Section 6.3.1, such a skeletal branch should follow the centerline of the corresponding volumetric brunch structure. To fulfill this objective, an appropriate step-cost function is used that encourages path propagation along the centerline while heavily penalizing any deviation from it. DT has been popularly used to impose the centeredness property of minimum-cost paths. Often, an inverse function of DT is used to encourage centered paths as central voxels possess larger DT value as compared to peripheral ones [50, 55,57,65]. Normalization of DT values by the global maximum of DT has been used

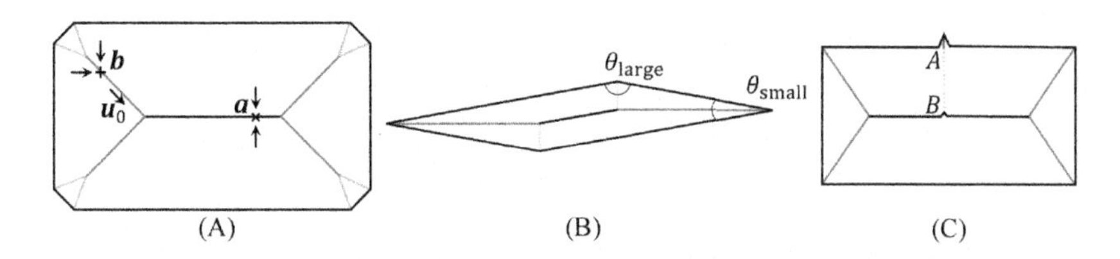

FIGURE 6.5

Collision impacts at different skeletal points during grassfire propagation on binary shapes. (A) The fire-fronts make head-on collision at the point $\boldsymbol{a}$ with the maximum collision impact of 1. At the point $\boldsymbol{b}$, the fire-fronts collide obliquely generating a weaker collision impact. (B) The collision impact along a skeletal branch, originated from a polygonal vertex with a small interior angle, e.g., θ_{small}, is higher than that along a skeletal branch generated from a vertex with a large interior angle, e.g., θ_{large}. (C) The collision impact along the skeletal branch-segment AB connecting a small protruding structure to the central skeleton, shown by the dotted line, is low.

to reduce the dependency on the global scale of a target object [30,48,49]. However, the normalization of DT value by the global maximum does not account for local scale dependency.

An improved performance of minimum-cost paths in terms of alignment with the object centerline is accomplished when a local scale independent measure of collision impact is used to impose centeredness of minimum-cost paths [34]. Collision impact is always "0" at non-CMB, whereas it has a positive value at CMB voxels. This property of collision impact ensures that minimum-cost paths are confined to the set of CMBs unless required for topological connectivity. The magnitude of the collision impact capture the strength of impact when Blum's colliding fire fronts meet [77]. As illustrated in Fig. 6.5, the collision impact reflects the angle between the boundary sections from which the colliding fronts had originated. Thus, it generates high values of collision impact at CMBs located around centerlines of elongated structures, whereas small values for CMBs are produced by small fluctuations at object boundaries. Collision impact at a voxel p in a fuzzy digital object $\mathcal{O}$, denoted by $\xi_{\mathcal{O}}(p)$, is defined as follows:

$$\xi_{\mathcal{O}}(p) = 1 - \max_{q \in N_{26}^*(p)} \frac{f_+(FDT_{\mathcal{O}}(q) - FDT_{\mathcal{O}}(p))}{\frac{1}{2}(\mu_{\mathcal{O}}(p) + \mu_{\mathcal{O}}(q))|p - q|}, \tag{6.4}$$

where the function $f_+(\cdot)$ returns the value of the parameter when it is positive and zero otherwise. Note that the value of collision impact at a voxel always lies in the normalized range of [0, 1]. Voxels within the central region of a structure will take the value close to 1, whereas it takes the value close to 0 at no-central voxels. A CMB with its collision impact greater than 0.5 may be considered as a *strong CMB* or a *strong quench voxel*. The step cost function τ ensuring centered minimum-cost path

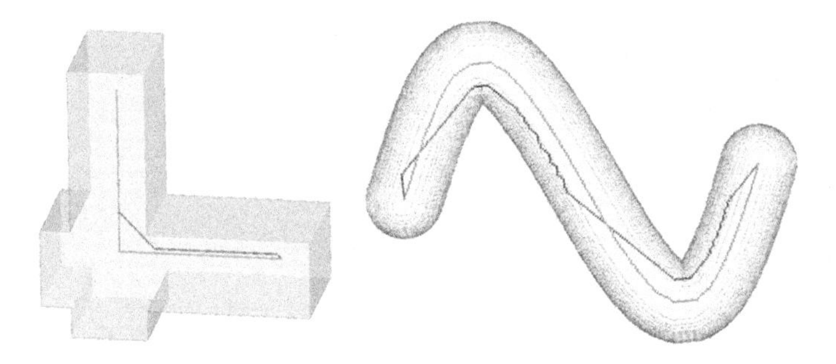

FIGURE 6.6

Centeredness performance of minimum-cost paths using different step cost functions. Minimum-cost paths using a DT-based step function (red) [30] often make corner cutting or sharp deviation from the centerline. On the other hand, a minimum-cost path using collision impact (green) [34] strictly adheres to the centerline and smoothly follows it.

is defined using collision impact as follows:

$$\tau(p,q) = \frac{|p-q|}{\epsilon + (\frac{\xi_D(p)+\xi_D(q)}{2})^2}, \tag{6.5}$$

where the parameter ϵ is a small number used to overcome problem related to numerical computation of division by zero. A constant value of 0.01 may be used for ϵ. A continuous formulation of above step-cost function for a compact object in R^3 ensures that a minimum-cost path always sticks to the Blum's skeleton. Performance of different step-cost functions is illustrated in Fig. 6.6.

6.3.4 SKELETAL BRANCH DETECTION

Skeletal branches are computed by locating the farthest CMB within the unmarked object region and then finding the centered minimum-cost path joining the farthest CMB to the current skeleton. As mentioned earlier, during an iteration, multiple skeletal branches may be simultaneously added, where each branch comes from a connected subtree T_i within the unmarked object region (see Fig. 6.3). Let O denote the support of the original object, and let O_{marked} denote the marked object region already represented by the current skeletal branches. Thus, $O_{\text{unmarked}} = O - O_{\text{marked}}$ is the unmarked object region where new skeletal branches are added. Let $Q_{\mathcal{O}} \subset O$ be the set of all strong CMBs in the original object $\mathcal{O}$; a voxel with its collision impact value greater than 0.5 is considered as a strong CMB. Note that the locations of CMBs and their collision impact are defined with respect to the original images, and these values are not changes during the iterative augmentation of skeletal branches or marking of object volume. To locate the branch-end voxel in an unmarked subtree T_i, the geodesic DT is computed from O_{marked} where paths are constrained within the support O of the original object. The branch-end point in T_i is determined as the voxel in $Q_{\mathcal{O}} \cap T_i$ with the largest geodesic DT from O_{marked}. This algorithm of branch-end detection ensures that a skeletal branch always ends at a CMB instead of the object boundary and thus complies with the principle of Blum's skeletonization. Let e_i denote the branch-end voxel located in T_i, and let S denote the current skeleton. The new skeletal branch B_i is computed using the following equation, and the skeleton is augmented to $S \cup B_i$:

$$B_i = \arg \min_{\pi \in \bigcup_{q \in S} \Pi_{e_i, q}} Cost(\pi). \tag{6.6}$$

It is reemphasized that geodesic binary DT from O_{marked} is used to locate the branch end point e_i, whereas a collision impact-based minimum-cost path from the current skeleton S is used to compute the new skeletal branch B_i.

6.3.5 OBJECT VOLUME MARKING

The last step in an iteration of minimum-cost path-based curve skeletonization algorithm is to mark the object volume represented by the newly added skeletal branches. The purpose of volume marking is to exclude the region already represented by current skeletal branches from subsequent searches for new skeletal branches. This simple idea significantly improves the performance of minimum-cost path-based curve skeletonization. First, it drastically reduces repetitious transverse skeletal branches within an object branch, which is a major source of noisy branches, especially, at large scale regions. Also, it improves the ability to distinguish between true and false skeletal branches because the volume assigned with a spurious branch will be small although the length of the branch may be relatively large; for example, consider the possible spurious branch in Fig. 6.3 from the point p_{noise}. Finally, the volume marking step allows a natural termination criterion for the iterative skeletal expansion process. Using volume marking, the iterative process can be terminated when the entire object region is marked, and it does not require any ad hoc thresholding on branch-length for termination.

The object volume marking step utilizes local scales along a skeletal branch. For a given point p on the skeleton S of an object $\mathcal{O}$, its *local scale* is $FDT_{\mathcal{O}}(p)$ ($DT_{\mathcal{O}}(p)$ for a binary object). For an object in the continuous domain, the dilation scale at p is exactly same as its local scale. However, due to various artifacts related to a digital space, we suggest the dilation scale as two times its local scale. Note that a little extra geodesic dilation does not create any risk of missing a true skeletal branch. On the other hand, it reduces false branches, adds stability in the termination criterion by filling small protrusions, and improves computational efficiency. Finally, during an iteration, a connected component of unmarked object region adding no new skeletal branch is assigned as marked volume.

Begin Algorithm: compute-local-scale-adaptive-dilation.
Input: support O of a fuzzy object
a new skeletal branch B_i
local dilation scale function $scale : O \rightarrow R^+$
Output: Dilated object volume O_{B_i}
$\forall p \in B_i$, initialize the local dilation scale $DS(p) = scale(p)$
$\forall p \in O - B_i$, initialize dilation scale $DS(p) = -max$
While the dilation scale map DS is changed
$\forall p \in O - B_i$, set $DS(p) = \max_{q \in N^*(p)} DS(q) - |p - q|$
Set the output $O_{B_i} = \{p \mid p \in O \wedge DS(p) \geq 0\}$
Augment the marked object volume $O_{\text{marked}} = O_{\text{marked}} \cup O_{B_i}$
End Algorithm: compute-local-scale-adaptive-dilation

6.3.6 SKELETAL BRANCH SIGNIFICANCE

The objective of computing skeletal branch significance is to determine the global shape information represented by a skeletal branch. During a specific iteration, let v_i be the CMB in a connected unmarked object region T_i that is farthest from the current marked object volume O_{marked}. The possible skeletal branch joining v_i to the current skeleton is computed using a centered minimum-cost path approach, as described in Section 6.3.4. However, before adding the new branch, it is necessary to determine its significance. Although the length of a skeletal branch had been earlier used as a measure of significance, a noticeably improved performance is obtained when a collision impact based measure of branch significance is used. It is even better than counting CMBs on the branch because, in a digital object, especially, for fuzzy digital objects, noisy CMBs with low significance are detected. The *significance* of a branch B_i joining an end voxel v_i to the current skeleton S is computed by adding collision impact values of along the skeletal branch path, i.e.,

$$significance(B_i) = \sum_{p \in B_i} \xi_D(p). \tag{6.7}$$

6.3.7 TERMINATION CRITERION

As described earlier, the algorithm iteratively adds skeletal branches, and it terminates when no more significant skeletal branches are found. Specifically, the termination is triggered by either of the following two different situations: (1) the marked volume O_{marked} fills the entire object region, or (2) none of the strong quench voxels in the unmarked object region $O - O_{\text{marked}}$ generates a significant branch. The first criterion characterizes the situation when the entire object region has already been represented with skeleton branches and no further branch is needed. The second situation occurs when there are small protrusions left in the unmarked region, however, none of those protrusions warrants a meaningful skeletal branch.

The measure of skeletal branch significance, described in Section 6.3.6, is used. A local scale-adaptive threshold is applied on the measure to determine the significance of a specific skeletal branch. Let $p_v \in S$ be the skeletal voxel where the branch B_i meets the current skeleton S. The scale-adaptive significance threshold for the selection of the new branch B_i is set at $\alpha + \beta DT_{\mathcal{O}}(p_v)$ in the voxel unit. The constant term α plays the major role in determining significance of branches originated from a small scale regions, mostly attributed with branches at a higher tree generation. The term $\beta DT_{\mathcal{O}}(p_v)$ becomes the determining factor for branches emanating from large-scale regions mostly true for branches at a lower tree generation. Although, the constant parameters α and β are application dependent, for most applications, the values $\alpha = 3$ and $\beta = 1$ seem to be acceptable. Thus, the length of the protruded part of a significant branch at large scale region should be at least as long as the radius of the immediate parent branch. Also, the parameter value $\alpha = 3$ implies that the length of a significant branch at a small scale region is at least 3 voxels. It should be noted that the parameter β is resolution independent, whereas α is linearly dependent on image resolution. In a specific application, the values of the parameters may be optimized using sensitivity and specificity analysis of true and false branches.

Two important features of the above termination criterion need to be highlighted. First, a measure of branch significance is used instead of simple path-length, which elegantly subtracts the portion of a path merely contributing to topological connectivity with little or no significance to object shape properties. Secondly, the termination criterion uses a scale-adaptive threshold for significance. At large-scale object regions, it is possible to visualize a situation where a branch is long enough while failing to become significant under the new criterion. Under such a situation, a large portion of the target branch falls inside the marked object volume resulting in a low significance measure. A scale-adaptive threshold of significance further ensures that such false branch at large-scale regions are arrested. Final results of skeletonization for a CT-based human intrathoracic airway tree are shown in Fig. 6.7B. As it appears visually, the described algorithm successfully traces all true branches without creating any false branches. In contrast, an algorithm using simple path-length-based measure of branch significance [30] creates several false branches while missing quite a few obvious ones (marked with red circles in Fig. 6.7A).

6.3.8 ALGORITHM EFFICIENCY

The computational bottleneck of the minimum-cost path-based curve skeletonization is that it requires recomputation of path-cost map over the entire object after each iteration. In earlier methods [30,63], only one branch is added in an iteration. Therefore, using the earlier algorithms, the computational complexity is determined by the number of terminal branches N in the skeleton. This is somewhat slow when it is applied for computing the skeleton of tree-like objects with a large number of terminal branches, e.g., airway or vascular trees. A recent algorithm has solved this

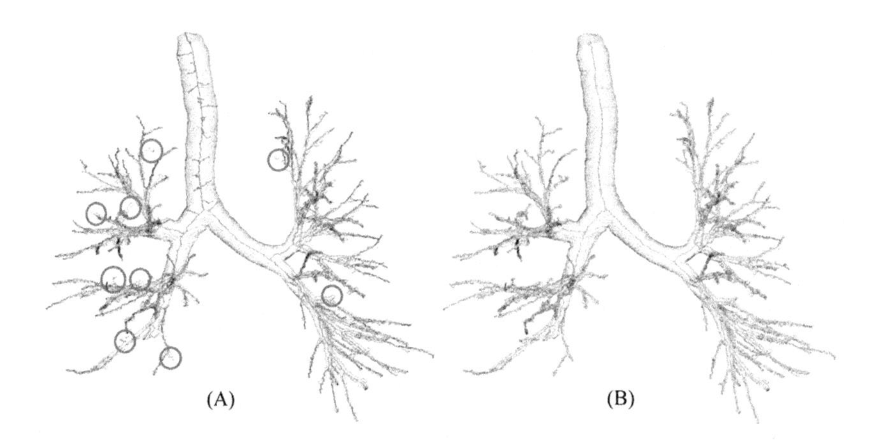

FIGURE 6.7

Comparison of the performance of skeletal branch selection using simple branch-length [30] (A) and collision-impact-based measure [34] of significance (B) on a human airway tree generated from pulmonary CT imaging. The missing branches are marked with red circles. The first algorithm, shown in (A), generates several noisy skeletal while already missing quite a few true branches. On the other hand, the method using collision impact, shown in (B), neither generates a noisy branch nor misses a visually obvious branch.

problem that makes a major improvement in its computational complexity [34]. The basic principle of the computational improvement is described using the illustration of Fig. 6.3 as follows.

After the first iteration, the skeletal branch op_1 is added, and the represented object volume along op_1 is marked as shown in the figure. At this stage, the unmarked object region generates three connected subtrees T_1, T_2, and T_3 (Fig. 6.3, right). These three subtrees represent three object regions for which skeletal branches are yet to be detected. Here, an important observation is that, since these subtrees are disconnected, their representative skeletal branches are independent. Therefore, new skeletal branches can be simultaneously computed in T_1, T_2, and T_3. In other words, in the second iteration, three branches can be simultaneously added where each branch connects to the farthest CMB within each subtree. After adding the three branches, the marked volume is augmented using local scale-adaptive dilation along the three new branches. This process continues until dilated skeletal branches mark the entire object region or all meaningful branches are found. For the example of Fig. 6.3, the algorithm terminates in four iterations although it has nine terminal branches.

This simple yet powerful observation reduces the computational complexity of the algorithm from the order of number of terminal branches to worst-case performance of the order of tree-depth. For a tree with N terminal nodes, the average depth of unbalanced bifurcating tree is $O(\sqrt{N})$ [78]. Thus, the average computation complexity of the improved algorithm is better than $O(\sqrt{N})$ as compared to

N using earlier algorithms adding only one branch per iteration. For a complete bifurcating tree, the average tree depth is $O(\log_2 N)$, which will be the number of iterations by the improved method. For example, the airway tree in Fig. 6.4 contains 118 terminal nodes, and the improved method adds twelve skeletal branches after two iterations (Fig. 6.4, left) and completes the skeletonization process in 6 iterations, only (Fig. 6.4, right). Tested on ten CT-derived human airway trees [34], the average number of terminal branches observed in the ten airway trees was 121.7, and the average number of iterations required by the new algorithm was 7.1, which is close to $\log_2 N = \log_2 121.7 = 6.93$ and better than the observed average tree-depth of 11.6, which is close to $\sqrt{N} = \sqrt{121.7} = 11.03$.

6.4 APPLICATIONS

Fig. 6.8 presents a few examples of curve skeleton for several 3-D objects using a state-of-the-art minimum-cost path algorithm [34]. For these examples, the minimum-cost path method generates a skeletal branch for every visually prominent geometric features while arresting noisy branches. It may be clarified that no post-processing steps was applied. Note that, although all examples in this figure are not "tree-like," these objects were carefully chosen such that they contain no cavities or tunnels. The basic principle of the minimum-cost path approach fails when an object contains cavities or tunnels. Also, as discussed in Section 6.2, there are many examples in 3-D, where the object contains no tunnel or cavity; however, it has no meaningful curve skeleton.

Fig. 6.9 compares the performance of a recent minimum-cost path-based method [34] with two popular conventional methods [10,79]. Computerized human airway tree phantoms were generated at different levels of noisy granules simulating protrusions and dents. The conventional methods failed to match the results of the minimum-cost path approach and produced several noisy branches. As observed in a quantitative evaluation using labeling of false and missing branches by mutually blinded experts, the minimum-cost path-based curve skeletonization method generated no false branches over the entire range of noise levels. On the other hand, one conventional method [10] produced 4.4, 9.2, and 12.8 false branches at low, medium, and high noise levels, respectively. The other conventional method [79] produced 3.8,14.2, and 17.4 false branches at three different noise levels. These experimental results suggest that a well-conceived minimum-cost path-based curve skeletonization method is superior to conventional methods in terms of noisy branch detection and the difference increases at higher noise levels.

Curve skeletonization of 3-D objects is popularly used in various applications in medical image analysis, including virtual endoscopy [43,56,80], image registration [81,82], morphological quantifications [83–86], etc. Besides medical imaging applications, curve skeletonization is also widely used in computer vision and graphics applications, such as 3-D animation [87,88], shape decomposition [89,90], object

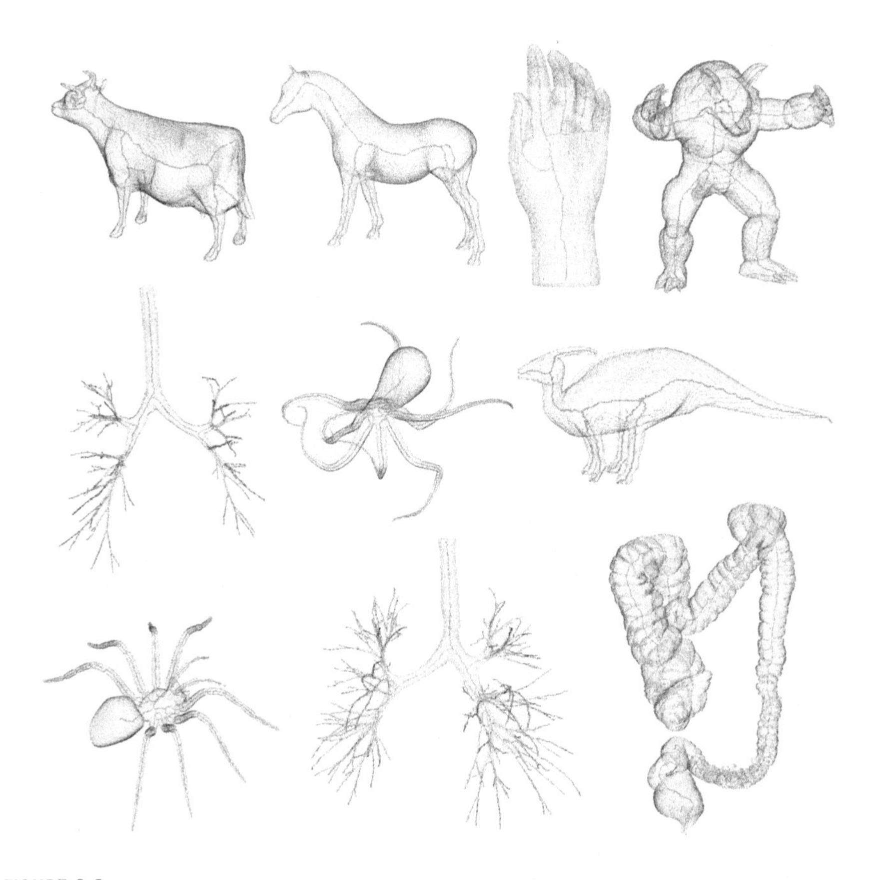

FIGURE 6.8

Examples of curve skeletons for several 3-D objects available online. These curve skeletons were computed using a recent minimum-cost path approach [34]. Object volumes are displayed with partial transparency, whereas computed curve skeletons are shown in green. As visually apparent in these results, the recent minimum-cost path method generates no noisy branch while producing branches for all prominent geometric features in these objects, which can be represented using a curve skeleton.

matching and retrieving [91], etc. Several applications of curve skeletonization are discussed in the following.

6.4.1 VIRTUAL COLONOSCOPY AND BRONCHOSCOPY

Minimum-cost path tools have been popularly used for centerline tracing in virtual endoscopy including virtual colonoscopy and bronchoscopy. Practical requirements of both applications perfectly matches with the principle of minimum-cost path methods. In these applications, often, a user specifies two end points of the target structure,

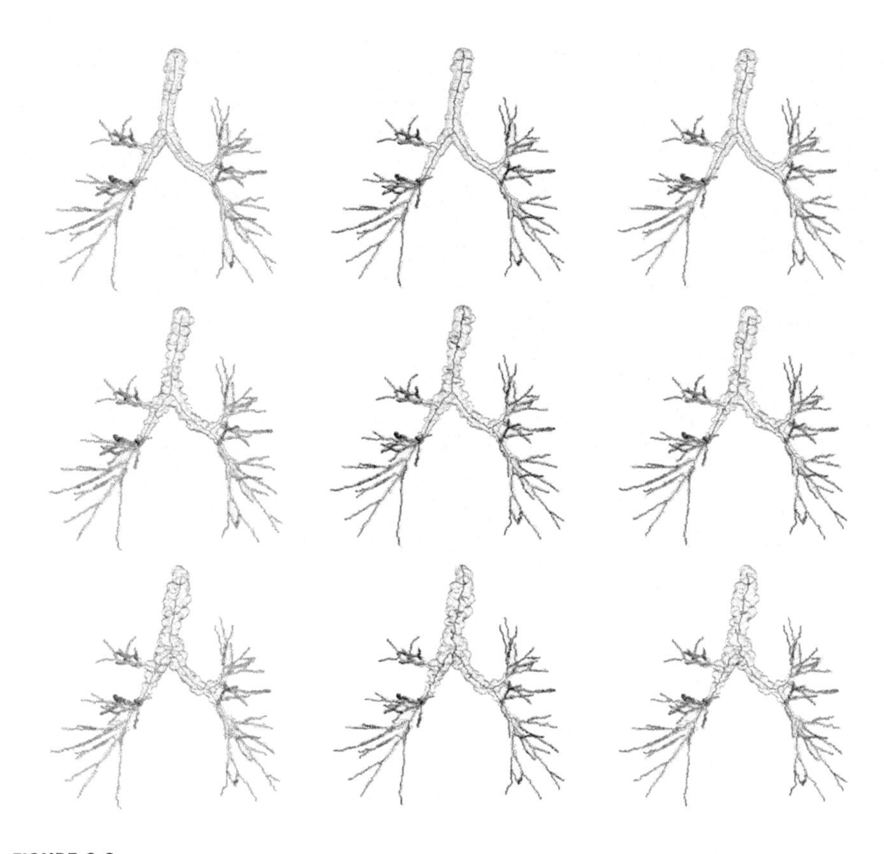

FIGURE 6.9

Results of curve skeletonization by different methods on 3-D airway phantoms at different levels of noisy granules simulating protrusions and dents. The three rows represent phantom objects at low, medium, and high noise. Results on the left column were generated using a recent minimum-cost path method [34], whereas those on middle and right columns were generated using two popular conventional methods [10] and [79], respectively. The minimum-cost path method produced no noisy branch even at the highest level of noise, whereas the two conventional methods produced noisy branches at all three levels of noise. All three methods captured every visually meaningful branch.

and the algorithm automatically generates the path of flight for camera viewing along the centerline connecting the two end points. See Fig. 6.10 for an example of bronchoscopy. An example of centerline generation for virtual colonoscopy is shown in the right bottom corner of Fig. 6.8. An advantage of the minimum-cost path approach as compared to convention methods of cure skeletonization is that the former produces a smooth centerline curve that passes through the central region of a target organ despite large and complex deformations on its boundary. This feature of a minimum-cost path approach benefits the planning of camera flight generating re-

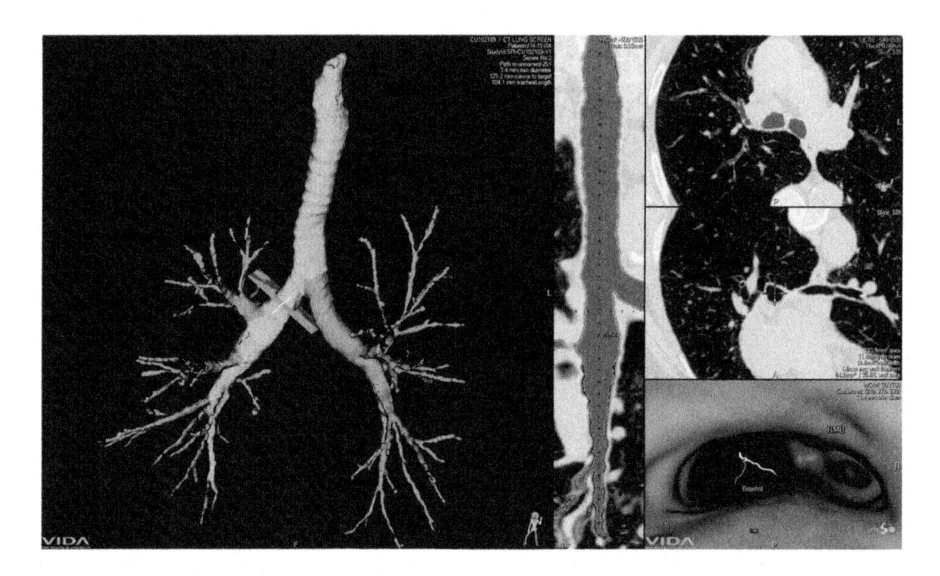

FIGURE 6.10

Illustration of the interactive graphical user interface under the virtual lung bronchoscopy planning application of a commercial software VIDAlvision® from the VIDA Diagnostics, Inc. Coralville, Iowa, USA.

alistic views of the inner wall. See [50,80] for further discussion and more examples on colonoscopy and bronchoscopy.

6.4.2 ANATOMICAL LABELING OF HUMAN AIRWAY TREES

Anatomic labeling of human airway is a common precursory step for developing an anatomic reference system for regional analysis of various physiologic and functional metrics in both cross-sectional and longitudinal studies. There is a consensus that topologic and geometric variation within the first five or six anatomic generations of the human airway tree is relatively small. A widely acknowledged anatomical nomenclature is available for 31 segments and 42 subsegments. These segmental and subsegmental branches are covered by first fifth or sixth generations of human airway branching. Beyond the subsegmental level, the human airway tree branching pattern and their structural features widely differ from one subject to another. These physiologic observations were described in an early lung anatomy book [92].

Recent advancements in computed tomography (CT) imaging allow *in vivo* imaging of human lung and airway tree, which is a common imaging modality in chronic obstructive pulmonary disease used for clinical and research diagnostic, planning, and monitoring purposes [93]. A labeled airway tree helps simplifying airway tree navigation in virtual bronchoscopy and identifying different lobar and sublobar structures, which can, for example, be used for surgical planning.

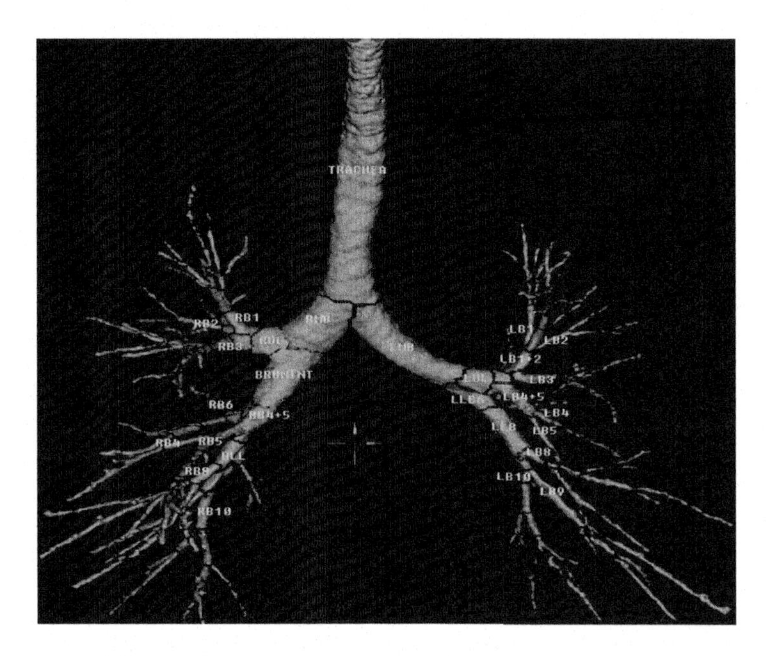

FIGURE 6.11

Anatomic labeling of segmental airway branches of a human airway tree using in vivo MDCT imaging generated by a commercial software VIDAlvision® from the VIDA Diagnostics, Inc. Coralville, Iowa, USA.

Several methods are available for anatomical labeling of human airway trees, which aim to locate and assign anatomic labels to the 31 segmental airway-tree branches illustrated in Fig. 6.11. A trivial solution to this problem is to use manual labeling through a graphical user interface. However, manual labeling is a tedious task and prone to subjective errors, especially, where nonexpert users are used in large studies. Several automatic or semiautomatic anatomic labeling algorithms are available [94–97]. In these methods, curve skeletonization is applied as the first step to generate a compact representation of the topologic and geometric features of an airway tree. In this context, it is necessary that the algorithm generates no false branches while capturing all meaningful branches. Otherwise, manual interaction in needed to remove the false branches in order for the labeling algorithm to work. A minimum-cost path-based curve skeletonization algorithm is a suitable method in this application because of its improved performance in terms of stopping noisy branches while capturing all meaningful ones.

6.4.3 MORPHOLOGIC ASSESSMENT OF PULMONARY ARTERIES

Pulmonary vascular dysfunction is gaining increased attention in regards to the role played in the development and progression of smoking associated parenchymal de-

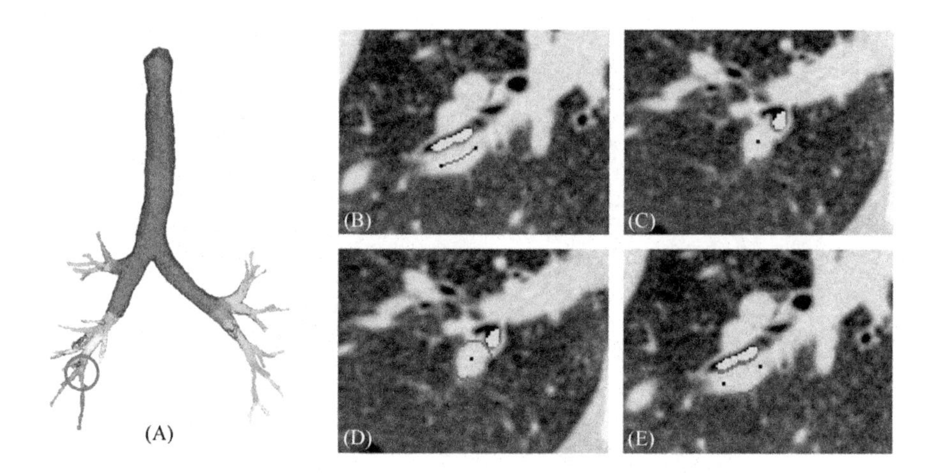

FIGURE 6.12

Results of intermediate steps for centerline tracing and quantitative morphometric analysis of an arterial segment linked to the anatomic airway branch RB10 shown by an arrow in (A). (B) Coronal plane of the user-specified arterial end-voxels (black) together with the joining centerline computed using a minimum-cost path approach. (C) Initial arterial boundary on the transverse plane computed with a half-max criterion for edge location. (D, E) Final arterial boundary after outlier analysis of the radial distance on the transverse (D) and coronal (E) plane. See [85] for detail.

struction (emphysema) [98–101], and there is an urgent need to quantify the arterial morphology at different anatomic lung regions.

Recently, a semiautomatic framework has been reported for segmenting the pulmonary arterial branch associated to a specific anatomic airway tree branch and computing morphometric properties of the arterial branch [85]. This algorithm uses a centered minimum-cost path approach to automatically compute the arterial centerline joining two user-specified endpoints for the target arterial segments. This computed centerline serves as the basis for computing orthogonal planes and other morphometric measures. Different steps in this algorithm are illustrated in Fig. 6.12.

6.5 CONCLUSIONS

Curve skeletonization reduces a 3-D volumetric object into its centerline representation that is highly compact and simple in terms of its topology and geometry. In several applications, where the target object is tubular or quasi-tubular, its curve skeleton provides an effective object representation that captures most useful features of the original object. Unlike Blum's skeletonization, curve skeletonization attempts to reduce the dimension of a 3-D object by two, which adds several instability in its theoretical formulation. For example, there are many 3-D objects whose curve skeletons do not exit, and there are many others whose curve skeletons fail to represent basic geometric features of original objects.

A large number of curve skeletonization methods are available in the literature. Sometimes the definition of a curve skeleton adopted in these algorithms are different. In this chapter, we focused on methods of curve skeletonization using minimum-cost path approach. An advantage of minimum-cost path-based curve skeletonization algorithms is that global information is naturally utilized, which makes skeletal branches less sensitive to perturbations and noise at object boundary; hence, it significantly reduces spurious branches in the computed skeleton requiring no post-pruning steps. Another benefit of a minimum-cost path approach is that it generates smooth skeletal branches, which are less affected by small changes at object boundary. A bottleneck with the minimum-cost path approach is its high computation time, where one skeletal branch is added in each iteration. A recent algorithm [34] has overcome this problem by using simultaneous skeletal branch addition from independent subtrees, which reduces the computation complexity from N to $\log_2 N$, where N is the number of terminal tree branches.

Several centerline extraction algorithms using minimum cost-path described in Section 6.2.3 directly work on raw intensity images requiring no prior object segmentation. Skeletonization method under this category are primarily for tracing and segmentation of tubular objects in 3-D images. Other minimum-cost path-based algorithms discussed in Section 6.3 require binary or fuzzy segmentation of the target objects. Skeletonization algorithms under this category are useful for local morphologic and global structural analysis and for object representation. Robustness of recent algorithms under noise (see Fig. 6.9) suggests that the computed skeletons should be resilient to locally confined segmentation errors. Obviously, the skeleton results will be affected if the segmentation error alter object topology or geometric features. A large number of medical and nonmedical imaging applications have used minimum-cost path approach to trace centerlines for elongated tubular and quasi-tubular objects.

REFERENCES

[1] L. Lam, S.-W. Lee, C.Y. Suen, Thinning methodologies – a comprehensive survey, IEEE Trans. Pattern Anal. Mach. Intell. 14 (1992) 869–885.

[2] K. Siddiqi, S.M. Pizer, Medial Representations: Mathematics, Algorithms and Applications, vol. 37, Springer, 2008.

[3] P.K. Saha, G. Borgefors, G. Sanniti di Baja, A survey on skeletonization algorithms and their applications, Pattern Recognit. Lett. 76 (2016) 3–12.

[4] N.D. Cornea, D. Silver, P. Min, Curve-skeleton properties, applications, and algorithms, IEEE Trans. Vis. Comput. Graph. 13 (2007) 530–548.

[5] H. Blum, A transformation for extracting new descriptors of shape, in: Models for the Perception of Speech and Visual Form, vol. 19, 1967, pp. 362–380.

[6] S.K. Pal, A. Rosenfeld, A fuzzy medial axis transformation based on fuzzy disks, Pattern Recognit. Lett. 12 (1991) 585–590.
[7] I. Bloch, Fuzzy skeleton by influence zones—application to interpolation between fuzzy sets, Fuzzy Sets Syst. 159 (2008) 1973–1990.
[8] D. Jin, P.K. Saha, A new fuzzy skeletonization algorithm and its applications to medical imaging, presented at the 17th International Conference on Image Analysis and Processing (ICIAP), Naples, Italy, 2013.
[9] Y. Tsao, K.S. Fu, A parallel thinning algorithm for 3-D pictures, Comput. Graph. Image Process. 17 (1981) 315–331.
[10] T.-C. Lee, R.L. Kashyap, C.-N. Chu, Building skeleton models via 3-D medial surface/axis thinning algorithm, CVGIP, Graph. Models Image Process. 56 (1994) 462–478.
[11] P.K. Saha, B.B. Chaudhuri, D.D. Majumder, A new shape preserving parallel thinning algorithm for 3D digital images, Pattern Recognit. 30 (1997) 1939–1955.
[12] K. Palágyi, A. Kuba, A parallel 3D 12-subiteration thinning algorithm, Graph. Models Image Process. 61 (1999) 199–221.
[13] G. Németh, P. Kardos, K. Palágyi, Thinning combined with iteration-by-iteration smoothing for 3D binary images, Graph. Models 73 (2011) 335–345.
[14] P.K. Saha, B.B. Chaudhuri, Detection of 3-D simple points for topology preserving transformations with application to thinning, IEEE Trans. Pattern Anal. Mach. Intell. 16 (1994) 1028–1032.
[15] P.K. Saha, B.B. Chaudhuri, B. Chanda, D.D. Majumder, Topology preservation in 3D digital space, Pattern Recognit. 27 (1994) 295–300.
[16] M. Sonka, M.D. Winniford, X. Zhang, S.M. Collins, Lumen centerline detection in complex coronary angiograms, IEEE Trans. Biomed. Eng. 41 (Jun. 1994) 520–528.
[17] H. Greenspan, M. Laifenfeld, S. Einav, O. Barnea, Evaluation of center-line extraction algorithms in quantitative coronary angiography, IEEE Trans. Med. Imaging 20 (2001) 928–941.
[18] O. Wink, W.J. Niessen, M.A. Viergever, Multiscale vessel tracking, IEEE Trans. Med. Imaging 23 (2004) 130–133.
[19] S. Wang, J. Wu, M. Wei, X. Ma, Robust curve skeleton extraction for vascular structures, Graph. Models 74 (2012) 109–120.
[20] G. Borgefors, Distance transform in arbitrary dimensions, Comput. Vis. Graph. Image Process. 27 (1984) 321–345.
[21] G. Borgefors, Distance transformations in digital images, Comput. Vis. Graph. Image Process. 34 (1986) 344–371.
[22] T.K. Dey, J. Sun, Defining and computing curve-skeletons with medial geodesic function, in: Symposium on Geometry Processing, 2006, pp. 143–152.
[23] D. Reniers, J.J. Van Wijk, A. Telea, Computing multiscale curve and surface skeletons of genus 0 shapes using a global importance measure, IEEE Trans. Vis. Comput. Graph. 14 (2008) 355–368.
[24] M.S. Hassouna, A.A. Farag, Variational curve skeletons using gradient vector flow, IEEE Trans. Pattern Anal. Mach. Intell. 31 (2009) 2257–2274.
[25] C. Arcelli, G. Sanniti di Baja, L. Serino, Distance-driven skeletonization in voxel images, IEEE Trans. Pattern Anal. Mach. Intell. 33 (Apr. 2011) 709–720.
[26] K. Palágyi, A. Kuba, A 3D 6-subiteration thinning algorithm for extracting medial lines, Pattern Recognit. Lett. 19 (1998) 613–627.
[27] C. Pudney, Distance-ordered homotopic thinning: a skeletonization algorithm for 3D digital images, Comput. Vis. Image Underst. 72 (1998) 404–413.

[28] G. Borgefors, I. Nyström, G. Sanniti di Baja, Computing skeletons in three dimensions, Pattern Recognit. 32 (1999) 1225–1236.

[29] J.-H. Chuang, C.-H. Tsai, M.-C. Ko, Skeletonization of three-dimensional object using generalized potential field, IEEE Trans. Pattern Anal. Mach. Intell. 22 (2000) 1241–1251.

[30] I. Bitter, A.E. Kaufman, M. Sato, Penalized-distance volumetric skeleton algorithm, IEEE Trans. Vis. Comput. Graph. 7 (2001) 195–206.

[31] N.D. Cornea, D. Silver, X. Yuan, R. Balasubramanian, Computing hierarchical curve-skeletons of 3D objects, Vis. Comput. 21 (2005) 945–955.

[32] M.S. Hassouna, A.A. Farag, Robust centerline extraction framework using level sets, in: Computer Vision and Pattern Recognition, 2005. IEEE Computer Society Conference on CVPR 2005, 2005, pp. 458–465.

[33] A.C. Jalba, A. Sobiecki, A.C. Telea, An unified multiscale framework for planar, surface, and curve skeletonization, IEEE Trans. Pattern Anal. Mach. Intell. 38 (2016) 30–45.

[34] D. Jin, K.S. Iyer, C. Chen, E.A. Hoffman, P.K. Saha, A robust and efficient curve skeletonization algorithm for tree-like objects using minimum cost paths, Pattern Recognit. Lett. 76 (2016) 32–40.

[35] A.D. Ward, G. Hamarneh, The groupwise medial axis transform for fuzzy skeletonization and pruning, IEEE Trans. Pattern Anal. Mach. Intell. 32 (2010) 1084–1096.

[36] G. Borgefors, I. Nyström, G. Sanniti di Baja, Skeletonizing volume objects part II: from surface to curve skeleton, in: Advances in Pattern Recognition, Springer, 1998, pp. 220–229.

[37] S. Svensson, I. Nyström, G. Sanniti di Baja, Curve skeletonization of surface-like objects in 3D images guided by voxel classification, Pattern Recognit. Lett. 23 (2002) 1419–1426.

[38] C. Arcelli, G. Sanniti di Baja, L. Serino, From 3D discrete surface skeletons to curve skeletons, in: Image Analysis and Recognition, Springer, 2008, pp. 507–516.

[39] P.K. Saha, Y. Xu, H. Duan, A. Heiner, G. Liang, Volumetric topological analysis: a novel approach for trabecular bone classification on the continuum between plates and rods, IEEE Trans. Med. Imaging 29 (Nov. 2010) 1821–1838.

[40] G. Malandain, G. Bertrand, N. Ayache, Topological segmentation of discrete surfaces, Int. J. Comput. Vis. 10 (1993) 183–197.

[41] P.K. Saha, B.B. Chaudhuri, 3D digital topology under binary transformation with applications, Comput. Vis. Image Underst. 63 (1996) 418–429.

[42] A. Telea, A. Vilanova, A robust level-set algorithm for centerline extraction, in: Proceedings of the Symposium on Data Visualisation 2003, 2003, pp. 185–194.

[43] S. Bouix, K. Siddiqi, A. Tannenbaum, Flux driven automatic centerline extraction, Med. Image Anal. 9 (2005) 209–221.

[44] D.W. Paglieroni, A unified distance transform algorithm and architecture, Mach. Vis. Appl. 5 (1992) 47–55.

[45] R.L. Ogniewicz, O. Kübler, Hierarchic Voronoi skeletons, Pattern Recognit. 28 (1995) 343–359.

[46] L. da F. Costa, R.M. Cesar Jr., Shape Analysis and Classification: Theory and Practice, CRC Press, Inc., 2000.

[47] K. Siddiqi, S. Bouix, A. Tannenbaum, S.W. Zucker, Hamilton–Jacobi skeletons, Int. J. Comput. Vis. 48 (2002) 215–231.

[48] I. Bitter, M. Sato, M. Bender, K.T. McDonnell, A. Kaufman, M. Wan, CEASAR: a smooth, accurate and robust centerline extraction algorithm, in: Proceedings of the Conference on Visualization'00, 2000, pp. 45–52.

[49] M. Sato, I. Bitter, M.A. Bender, A.E. Kaufman, M. Nakajima, TEASAR: tree-structure extraction algorithm for accurate and robust skeletons, in: The Eighth Pacific Conference on Computer Graphics and Applications, 2000. Proceedings, 2000, pp. 281–449.

[50] M. Wan, Z. Liang, Q. Ke, L. Hong, I. Bitter, A. Kaufman, Automatic centerline extraction for virtual colonoscopy, IEEE Trans. Med. Imaging 21 (2002) 1450–1460.

[51] H. Li, A. Yezzi, Vessels as 4-D curves: global minimal 4-D paths to extract 3-D tubular surfaces and centerlines, IEEE Trans. Med. Imaging 26 (2007) 1213–1223.

[52] D. Lesage, E.D. Angelini, I. Bloch, G. Funka-Lea, A review of 3D vessel lumen segmentation techniques: models, features and extraction schemes, Med. Image Anal. 13 (Dec. 2009) 819–845.

[53] S.D. Olabarriaga, M. Breeuwer, W.J. Niessen, Minimum cost path algorithm for coronary artery central axis tracking in CT images, in: International Conference on Medical Image Computing and Computer-Assisted Intervention, 2003, pp. 687–694.

[54] M. Schaap, C.T. Metz, T. van Walsum, A.G. van der Giessen, A.C. Weustink, N.R. Mollet, et al., Standardized evaluation methodology and reference database for evaluating coronary artery centerline extraction algorithms, Med. Image Anal. 13 (Oct. 2009) 701–714.

[55] S.R. Aylward, E. Bullitt, Initialization, noise, singularities, and scale in height ridge traversal for tubular object centerline extraction, IEEE Trans. Med. Imaging 21 (2002) 61–75.

[56] T. Deschamps, L.D. Cohen, Fast extraction of minimal paths in 3D images and applications to virtual endoscopy, Med. Image Anal. 5 (2001) 281–299.

[57] O. Wink, A.F. Frangi, B. Verdonck, M.A. Viergever, W.J. Niessen, 3D MRA coronary axis determination using a minimum cost path approach, Magn. Reson. Med. 47 (2002) 1169–1175.

[58] F. Benmansour, L.D. Cohen, Tubular structure segmentation based on minimal path method and anisotropic enhancement, Int. J. Comput. Vis. 92 (2011) 192–210.

[59] Y. Rouchdy, L.D. Cohen, Geodesic voting for the automatic extraction of tree structures. Methods and applications, Comput. Vis. Image Underst. 117 (2013) 1453–1467.

[60] E.R. Antúnez, L.J. Guibas, Robust extraction of 1D skeletons from grayscale 3D images, in: ICPR, 2008, pp. 1–4.

[61] G. Peyré, M. Péchaud, R. Keriven, L.D. Cohen, Geodesic methods in computer vision and graphics, Found. Trends Comput. Graph. Vis. 5 (2010) 197–397.

[62] D.S. Paik, C.F. Beaulieu, R.B. Jeffrey, G.D. Rubin, S. Napel, Automated flight path planning for virtual endoscopy, Med. Phys. 25 (1998) 629–637.

[63] D. Jin, K.S. Iyer, E.A. Hoffman, P.K. Saha, A new approach of arc skeletonization for tree-like objects using minimum cost path, in: 22nd International Conference on Pattern Recognition, Stockholm, Sweden, 2014, pp. 942–947.

[64] R. Kimmel, A. Amir, A.M. Bruckstein, Finding shortest paths on surfaces using level sets propagation, IEEE Trans. Pattern Anal. Mach. Intell. 17 (1995) 635–640.

[65] L.D. Cohen, R. Kimmel, Global minimum for active contour models: a minimal path approach, Int. J. Comput. Vis. 24 (1997) 57–78.

[66] L. Cohen, Minimal paths and fast marching methods for image analysis, in: N. Paragios, Y. Chen, O. Faugeras (Eds.), Handbook of Mathematical Models in Computer Vision, Springer, 2006, pp. 97–111.

[67] A. Dufour, O. Tankyevych, B. Naegel, H. Talbot, C. Ronse, J. Baruthio, et al., Filtering and segmentation of 3D angiographic data: advances based on mathematical morphology, Med. Image Anal. 17 (2013) 147–164.

[68] L.D. Cohen, Multiple contour finding and perceptual grouping using minimal paths, J. Math. Imaging Vis. 14 (2001) 225–236.

[69] P.K. Saha, R. Strand, G. Borgefors, Digital topology and geometry in medical imaging: a survey, IEEE Trans. Med. Imaging 34 (Sept. 2015) 1940–1964.

[70] M. Sonka, V. Hlavac, R. Boyle, Image Processing, Analysis, and Machine Vision, 3rd ed., Thomson Engineering, Toronto, Canada, 2007.

[71] P.K. Saha, F.W. Wehrli, B.R. Gomberg, Fuzzy distance transform: theory, algorithms, and applications, Comput. Vis. Image Underst. 86 (2002) 171–190.

[72] G. Sanniti di Baja, Well-shaped, stable, and reversible skeletons from the (3, 4)-distance transform, J. Vis. Commun. Image Represent. 5 (1994) 107–115.

[73] P.K. Saha, F.W. Wehrli, Fuzzy distance transform in general digital grids and its applications, presented at the 7th Joint Conference on Information Sciences, Research Triangular Park, NC, 2003.

[74] S. Svensson, Centres of maximal balls extracted from a fuzzy distance transform, in: 8th International Symposium on Mathematical Morphology, Rio de Janeiro, Brazil, 2007, pp. 19–20.

[75] J.A. Sethian, Level Set Methods and Fast Marching Methods, Cambridge University Press, 1996.

[76] E. Dijkstra, A note on two problems in connection with graphs, Numer. Math. 1 (1959) 269–271.

[77] D. Jin, C. Chen, P.K. Saha, Filtering non-significant quench points using collision impact in grassfire propagation, in: Proc. of 18th Intern. Conf. on Image Anal. and Process. (ICIAP), 2015, pp. 432–443.

[78] P. Flajolet, A. Odlyzko, The average height of binary trees and other simple trees, J. Comput. Syst. Sci. 25 (1982) 171–213.

[79] K. Palágyi, J. Tschirren, E.A. Hoffman, M. Sonka, Quantitative analysis of pulmonary airway tree structures, Comput. Biol. Med. 36 (2006) 974–996.

[80] A.P. Kiraly, J.P. Helferty, E.A. Hoffman, G. McLennan, W.E. Higgins, Three-dimensional path planning for virtual bronchoscopy, IEEE Trans. Med. Imaging 23 (Nov. 2004) 1365–1379.

[81] S.R. Aylward, J. Jomier, S. Weeks, E. Bullitt, Registration and analysis of vascular images, Int. J. Comput. Vis. 55 (2003) 123–138.

[82] S.M. Pizer, G. Gerig, S. Joshi, S.R. Aylward, Multiscale medial shape-based analysis of image objects, Proc. IEEE 91 (2003) 1670–1679.

[83] J. Tschirren, E.A. Hoffman, G. McLennan, M. Sonka, Segmentation and quantitative analysis of intrathoracic airway trees from computed tomography images, Proc. Am. Thorac. Soc. 2 (2005) 484–487, 503–504.

[84] D. Jin, K.S. Iyer, E.A. Hoffman, P.K. Saha, Automated assessment of pulmonary arterial morphology in multi-row detector CT imaging using correspondence with anatomic airway branches, in: Proc. of 11th Intern. Symp. on Visual Computing (ISVC), Las Vegas, NV, 2014, pp. 521–530.

[85] D. Jin, J. Guo, T.M. Dougherty, K.S. Iyer, E.A. Hoffman, P.K. Saha, A semi-automatic framework of measuring pulmonary arterial metrics at anatomic airway locations using CT imaging, in: SPIE Medical Imaging, 2016, 978816.

[86] J. Guo, C. Wang, K.-S. Chan, D. Jin, P.K. Saha, J.P. Sieren, et al., A controlled statistical study to assess measurement variability as a function of test object position and configuration for automated surveillance in a multicenter longitudinal COPD study (SPIROMICS), Med. Phys. 43 (2016) 2598–2610.

[87] N. Gagvani, D. Silver, Animating volumetric models, Graph. Models 63 (2001) 443–458.

[88] J. Bloomenthal, Medial-based vertex deformation, in: Proceedings of the 2002 ACM SIGGRAPH/Eurographics Symposium on Computer Animation, 2002, pp. 147–151.

[89] D. Reniers, A. Telea, Skeleton-based hierarchical shape segmentation, in: IEEE International Conference on Shape Modeling and Applications, 2007, SMI'07, 2007, pp. 179–188.

[90] L. Serino, C. Arcelli, G. Sanniti di Baja, From skeleton branches to object parts, Comput. Vis. Image Underst. 129 (2014) 42–51.

[91] H. Sundar, D. Silver, N. Gagvani, S. Dickinson, Skeleton based shape matching and retrieval, in: Shape Modeling International, 2003, 2003, pp. 130–139.

[92] E. Boyden, Segmental Anatomy of the Lungs, McGraw-Hill Book Co., New York, 1955.

[93] W.J. Kim, E.K. Silverman, E. Hoffman, G.J. Criner, Z. Mosenifar, F.C. Sciurba, et al., CT metrics of airway disease and emphysema in severe COPD, Chest 136 (2009) 396–404.

[94] K. Mori, J. Hasegawa, Y. Suenaga, J. Toriwaki, Automated anatomical labeling of the bronchial branch and its application to the virtual bronchoscopy system, IEEE Trans. Med. Imaging 19 (Feb. 2000) 103–114.

[95] H. Kitaoka, Y. Park, J. Tschirren, J. Reinhardt, M. Sonka, G. McLennan, et al., Automated nomenclature labeling of the bronchial tree in 3D-CT lung images, in: Medical Image Computing and Computer-Assisted Intervention—MICCAI 2002, Springer, 2002, pp. 1–11.

[96] J. Tschirren, G. McLennan, K. Palagyi, E.A. Hoffman, M. Sonka, Matching and anatomical labeling of human airway tree, IEEE Trans. Med. Imaging 24 (Dec. 2005) 1540–1547.

[97] A. Feragen, J. Petersen, M. Owen, P. Lo, L.H. Thomsen, M.M. Wille, et al., A hierarchical scheme for geodesic anatomical labeling of airway trees, in: Medical Image Computing and Computer-Assisted Intervention—MICCAI 2012, Springer, 2012, pp. 147–155.

[98] S.K. Alford, E.J. van Beek, G. McLennan, E.A. Hoffman, Heterogeneity of pulmonary perfusion as a mechanistic image-based phenotype in emphysema susceptible smokers, Proc. Natl. Acad. Sci. 107 (2010) 7485–7490.

[99] R.G. Barr, D.A. Bluemke, F.S. Ahmed, J.J. Carr, P.L. Enright, E.A. Hoffman, et al., Percent emphysema, airflow obstruction, and impaired left ventricular filling, N. Engl. J. Med. 362 (Jan. 21 2010) 217–227.

[100] J.M. Wells, G.R. Washko, M.K. Han, N. Abbas, H. Nath, A.J. Mamary, et al., Pulmonary arterial enlargement and acute exacerbations of COPD, N. Engl. J. Med. 367 (Sep. 6, 2012) 913–921.

[101] K.S. Iyer, J.D. Newell Jr., D. Jin, M.K. Fuld, P.K. Saha, S. Hansdottir, et al., Quantitative dual energy computed tomography supports a vascular etiology of smoking induced inflammatory lung disease, Am. J. Respir. Crit. Care Med. 193 (2016) 652–661.

CHAPTER

7 Parallel skeletonization algorithms in the cubic grid based on critical kernels

Gilles Bertrand, Michel Couprie

Université Paris-Est, LIGM, Équipe A3SI, ESIEE Paris, France

Contents

7.1 Introduction ... 181
7.2 Voxel Complexes and Simple Voxels ... 183
7.3 Critical Cliques ... 184
7.4 Decreasing Rank Strategy ... 186
7.5 Asymmetric Thinning ... 187
7.5.1 Generic Parallel Asymmetric Thinning Scheme ... 187
7.5.2 Isthmus-Based Asymmetric Thinning ... 189
7.5.3 Comparison with Other Parallel Curve Skeletonization Methods ... 191
7.6 Symmetric Thinning ... 193
7.6.1 Generic Parallel Symmetric Thinning Scheme ... 194
7.6.2 Isthmus-Based Symmetric Thinning ... 195
7.7 Characterization of Critical Cliques and k-Isthmuses ... 197
7.8 Isthmus Persistence and Skeleton Filtering ... 200
7.9 Hierarchies of Skeletons ... 202
7.10 Complexity ... 206
7.11 Conclusion ... 207
References ... 208

7.1 INTRODUCTION

Parallel thinning in discrete grids is a topic that has been studied since the pioneering years of digital image processing. The most essential requirement for a skeletonization method is topology preservation, hence the abundant literature devoted to the study of conditions under which a thinning method meets this requirement. Among these works, the notions of minimal nonsimple set [1], P-simple point [2], and critical kernel [3] constitute major contributions to a systematic study of topology preservation in parallel thinning. Critical kernels provide, to our knowledge, the most general framework for the design of topologically sound thinning algorithms. In particular,

Skeletonization. DOI: 10.1016/B978-0-08-101291-8.00008-0

we proved in [4] that the notions of minimal nonsimple sets and P-simple points can be characterized in terms of critical kernels.

An important distinction must be made, among parallel thinning algorithms, between symmetric and asymmetric algorithms. Symmetric algorithms [5–7], on one hand, are the ones that produce skeletons that are uniquely defined and invariant under 90-degree rotations. Such skeletons may contain parts that are not as thin as possible, like, for example, two-voxel-wide ribbons. On the other hand, asymmetric algorithms [8–17] produce thinner skeletons, but the price to pay for the improved thinness is the loss of 90-degree rotation invariance. Both kinds of thinning algorithms are useful in applications. Choosing between symmetric and asymmetric algorithms is a matter of context.

Let us outline briefly and informally the framework that we will detail in Section 7.3, derived from the notion of critical kernel. First of all, instead of considering single voxels, we consider cliques. A clique is a set of mutually adjacent voxels. In a voxel set X, certain cliques, called critical cliques, cannot be removed from X without altering its topological characteristics. The critical cliques of X can be identified using local conditions. The main theorem of the critical kernels framework [3,18] implies that any subset of the object X can be removed in parallel, provided that at least one voxel of every critical clique is preserved, and this guarantees topology preservation.

Since each clique can contain between one and eight voxels, there usually exists a huge number of combinations for the choice of the voxels that are kept during one step of thinning.

Consider, for example, the case of a critical clique made of two voxels a, b. A first possibility is to keep both of them, thus avoiding to make an arbitrary choice between a and b. Such a strategy applied to all the critical cliques leads to symmetric thinning algorithms.

On the other hand, if we prefer thinner skeletons, then we may use a systematic rule to decide which voxel among a and b will be removed. This leads to the so-called asymmetric thinning algorithm.

In addition, various constraints can easily be added to preserve some voxels having certain geometrical properties, for example, to obtain curve or surface skeletons. The notions of 1-isthmus and 2-isthmus (Section 7.5.2) will allow us to detect such voxels.

In this chapter, after providing the necessary background notions in Section 7.2 and recalling the notion of critical clique in Section 7.3, we discuss in Section 7.4 a general strategy to design parallel thinning algorithms in this framework. Then, we propose an asymmetric (Section 7.5) and a symmetric (Section 7.6) generic thinning scheme. These schemes can be used to obtain different kinds of skeletons, for example, ultimate curve or surface skeletons by choosing appropriate parameters.

It is well known that skeletons are highly sensitive to contour noise, that is, a small perturbation of the contour may provoke the appearing or the disappearing of an arbitrarily long skeleton branch. In Section 7.8, we propose a natural extension of our thinning methods that copes with the robustness to noise issue. This extension is based on a notion of "isthmus persistence" [19], which has been introduced in the framework of arbitrary 3D complexes (objects that are not necessarily made of voxels). Isthmus persistence is natural in our framework and could not be adapted to other approaches based, e.g., on end voxel detection.

We also propose in Section 7.9 a way to efficiently compute a hierarchy of nested filtered skeletons, corresponding to the different values of isthmus persistence.

We provide in Section 7.7 some local characterizations that allow for the effective implementation of our methods. All the algorithms proposed in this chapter can be implemented to run in linear time complexity (Section 7.10).

This chapter is essentially based on the contents of the papers [20,21,18]. However, it offers a new perspective on these works, thanks to a compact, synthetic and self-contained presentation.

7.2 VOXEL COMPLEXES AND SIMPLE VOXELS

Let us first give some basic definitions for voxel complexes; see also [22,23]. Let $\mathbb{Z}$ be the set of integers. We consider the families of sets $\mathbb{F}_0^1$ and $\mathbb{F}_1^1$ such that $\mathbb{F}_0^1 = \{\{a\} \mid a \in \mathbb{Z}\}$ and $\mathbb{F}_1^1 = \{\{a, a+1\} \mid a \in \mathbb{Z}\}$. A subset f of $\mathbb{Z}^n$, $n \geq 2$, that is the Cartesian product of exactly d elements of $\mathbb{F}_1^1$ and $(n-d)$ elements of $\mathbb{F}_0^1$ is called a *face* or a *d-face* of $\mathbb{Z}^n$, where d is the *dimension of f*, and we write $dim(f) = d$.

In the illustrations of this chapter, unless explicitly stated, a 3-face (resp. 2-face, 1-face, 0-face) is depicted by a cube (resp. square, segment, dot); see, e.g., Fig. 7.3.

A 3-face of $\mathbb{Z}^3$ is also called a *voxel*. A finite set that is composed solely of voxels is called a *(voxel) complex* (see Fig. 7.1). We denote by $\mathbb{V}^3$ the collection of all voxel complexes.

Notice that, by identifying each voxel with its center of mass, we get an equivalence between the data of a voxel complex and the one of a subset of $\mathbb{Z}^3$.

We say that two voxels x, y are *adjacent* if $x \cap y \neq \emptyset$. We write $\mathcal{N}(x)$ for the set of all voxels that are adjacent to a voxel x; $\mathcal{N}(x)$ is called the *neighborhood of x*. Note that, for each voxel x, we have $x \in \mathcal{N}(x)$. We set $\mathcal{N}^*(x) = \mathcal{N}(x) \setminus \{x\}$.

Let $d \in \{0, 1, 2\}$. We say that two voxels x, y are *d-neighbors* if $x \cap y$ is a d-face. Thus, two distinct voxels x and y are adjacent if and only if they are d-neighbors for some $d \in \{0, 1, 2\}$. For example, in Fig. 7.1, voxels b and c are 2-neighbors, voxels a and b are 1-neighbors but not 2-neighbors, and voxels b and d are 0-neighbors but neither 2-neighbors nor 1-neighbors.

Let $X \in \mathbb{V}^3$. We say that X is *connected* if, for any $x, y \in X$, there exists a sequence $\langle x_0, \ldots, x_k \rangle$ of voxels in X such that $x_0 = x$, $x_k = y$, and x_i is adjacent to x_{i-1}, $i = 1, \ldots, k$.

Intuitively, a voxel x of a complex X is called a simple voxel if its removal from X "does not change the topology of X." This notion may be formalized with the help of the following recursive definition introduced in [18]; see also [24,25] for other recursive approaches for the notion of simple point.

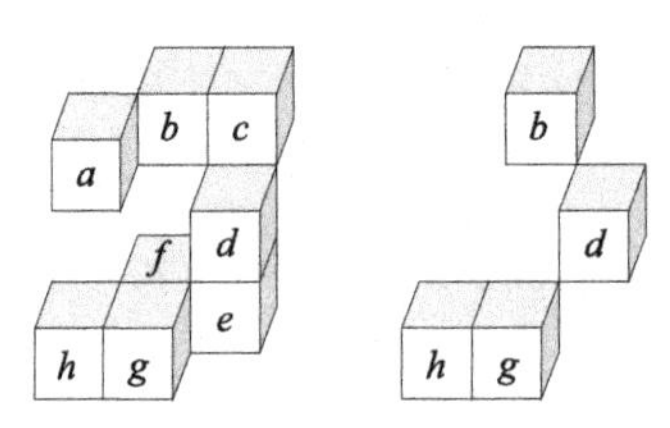

FIGURE 7.1

Left: A complex X made of 8 voxels. Right: A complex $Y \subseteq X$, which is a thinning of X.

Definition 1. Let $X \in \mathbb{V}^3$. We say that X is *reducible* if either

i) X is composed of a single voxel or
ii) there exists $x \in X$ such that $\mathcal{N}^*(x) \cap X$ is reducible and $X \setminus \{x\}$ is reducible.

Definition 2. Let $X \in \mathbb{V}^3$. A voxel $x \in X$ is *simple for* X if $\mathcal{N}^*(x) \cap X$ is reducible. If $x \in X$ is simple for X, then we say that $X \setminus \{x\}$ is an *elementary thinning of* X.

Thus, a complex $X \in \mathbb{V}^3$ is reducible if and only if it is possible to reduce X to a single voxel by iteratively removing simple voxels. Observe that a reducible complex is necessarily nonempty and connected.

In Fig. 7.1 (left), the voxel a is simple for X ($\mathcal{N}^*(a) \cap X = \{b\}$ is made of a single voxel), the voxel d is not simple for X ($\mathcal{N}^*(d) \cap X = \{c, e, f, g\}$ is not connected), the voxel h is simple for X ($\mathcal{N}^*(h) \cap X = \{f, g\}$ is made of two voxels that are 2-neighbors and is reducible).

In [18], it was shown that the above definition of a simple voxel is equivalent to classical characterizations based on connectivity properties of the voxel neighborhood [26–30]. An equivalence was also established with a definition based on the operation of collapse [31]; this operation is a discrete analogue of a continuous deformation (a homotopy); see [24,3,30].

The notion of a simple voxel allows us to define thinnings of a complex; see an illustration in Fig. 7.1 (right).

Let $X, Y \in \mathbb{V}^3$. We say that Y *is a thinning of* X or that X is *reducible to* Y if there exists a sequence $\langle X_0, \ldots, X_k \rangle$ such that $X_0 = X$, $X_k = Y$, and X_i is an elementary thinning of X_{i-1}, $i = 1, \ldots, k$. Thus, a complex X is reducible if and only if it is reducible to a set made of a single voxel.

7.3 CRITICAL CLIQUES

Let X be a voxel complex. It is well known that if we remove simultaneously (in parallel) simple voxels from X, then we may "change the topology" of the original object X. For example, the two voxels f and g are simple for the object X depicted in Fig. 7.1 (left). Nevertheless, $X \setminus \{f, g\}$ has two connected components, whereas X is connected.

In this section, we recall a framework for thinning in parallel discrete objects with the warranty that we do not alter the topology of these objects [3,32,18].

Let $d \in \{0, 1, 2, 3\}$, and let $C \in \mathbb{V}^3$. We say that C is a *d-clique* or a *clique* if $\cap\{x \in C\}$ is a d-face. If C is a d-clique, then we say that d is the *rank of* C.

For example, in Fig. 7.1, $\{d, e, f, g\}$ is a 0-clique, $\{f, g, e\}$ is a 1-clique, $\{b, c\}$ is a 2-clique, and $\{a\}$ is a 3-clique.

If C is made of solely two distinct voxels x and y, then we note that C is a d-clique if and only if x and y are d-neighbors with $d \in \{0, 1, 2\}$.

Let $X \in \mathbb{V}^3$, and let $C \subseteq X$ be a clique. We say that C is *essential for* X if we have $C = D$ whenever D is a clique such that:

i) $C \subseteq D \subseteq X$, and
ii) $\cap\{x \in C\} = \cap\{x \in D\}$.

In other words, C is essential for X if it is maximal with respect to the inclusion among all the cliques D in X such that ii) holds.

Observe that any complex C that is made of a single voxel is a clique (a 3-clique). Furthermore, any such clique in a complex X is essential for X.

In Fig. 7.1 (left), $\{f, g\}$ is a 2-clique that is essential for X, $\{b, d\}$ is a 0-clique that is not essential for X, $\{b, c, d\}$ is a 0-clique essential for X, and $\{e, f, g\}$ is a 1-clique essential for X.

Definition 3. Let $S \in \mathbb{V}^3$. The *$\mathcal{K}$-neighborhood of* S, written $\mathcal{K}(S)$, is the set made of all voxels that are adjacent to each voxel in S. We set $\mathcal{K}^*(S) = \mathcal{K}(S) \setminus S$.

We note that we have $\mathcal{K}(S) = \mathcal{N}(x)$ whenever S is made of a single voxel x. We also observe that we have $S \subseteq \mathcal{K}(S)$ whenever S is a clique.

Definition 4. Let $X \in \mathbb{V}^3$, and let C be a clique that is essential for X. We say that the clique C is *regular for* X if $\mathcal{K}^*(C) \cap X$ is reducible. We say that C is *critical for* X if C is not regular for X.

Notice that if $C = \{x\}$ is a 3-clique in X, then C is regular for X if and only if x is simple for X. We can thus say that the notion of regular clique generalizes the one of simple voxel.

In Fig. 7.1 (left), the cliques $C_1 = \{b, c, d\}$, $C_2 = \{f, g\}$, and $C_3 = \{g, h\}$ are essential for X. We have $\mathcal{K}^*(C_1) \cap X = \emptyset$, $\mathcal{K}^*(C_2) \cap X = \{d, e, h\}$, and $\mathcal{K}^*(C_3) \cap X = \{f\}$. Thus, C_1 and C_2 are critical for X, whereas C_3 is regular for X.

The following result is a consequence of theorem 16 of [18] and a general theorem that holds for complexes of arbitrary dimensions [3].

Theorem 5. *Let $X \in \mathbb{V}^3$, and let $Y \subseteq X$. The complex Y is a thinning of X if any clique that is critical for X contains at least one voxel of Y.*

See an illustration in Fig. 7.1, where the complexes X and Y satisfy the condition of Theorem 5. For example, the voxel d is a nonsimple voxel for X, thus $\{d\}$ is a critical 3-clique for X, and d belongs to Y. Also, Y contains voxels in the critical cliques $C_1 = \{b, c, d\}$, $C_2 = \{f, g\}$, and the other ones.

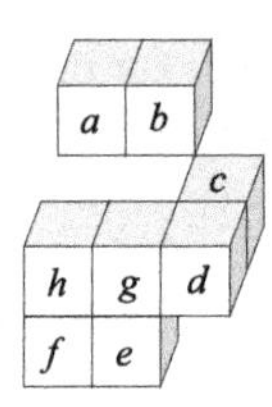

FIGURE 7.2

A complex X.

The notion of critical kernel has been defined in the framework of simplicial or cubical complexes [3,18]. However, it is possible to give a characterization of the critical kernel of a voxel complex X in terms of cliques. The *trace of a clique* C is the face $f = \cap\{x \in C\}$. The critical kernel of X is the set formed of the traces of the critical cliques of X (see an illustration in Fig. 7.3) and of all the faces included in these ones. Thus, Theorem 5 can be restated in the following terms: The complex Y is a thinning of X if each face of the critical kernel of X is in at least one voxel of Y.

Notice that definitions of simple voxels and critical cliques in this chapter take advantage of special properties of faces in $\mathbb{Z}^d$ and they cannot be transposed to other kinds of complexes (such as simplicial complexes).

7.4 DECREASING RANK STRATEGY

A parallel thinning algorithm consists of repeatedly applying thinning steps, each of which is aimed at identifying and removing a subset of the current object. Equivalently, we have to define the subset of the object that will be kept until the end of the current step. To ensure topology preservation, when designing such an algorithm, we have to define a subset Y of a voxel complex X that is guaranteed to include at least one voxel of each clique that is critical for X. By Theorem 5 this subset Y is a thinning of X.

There are many possible choices for defining such a subset Y. Among all the possible choices, we prefer the ones that provide sets Y having smaller sizes. Such a strategy will provide us with efficient thinning algorithms in the sense where few thinning steps will be needed to achieve the final result.

Suppose that we examine critical cliques by decreasing rank (see Fig. 7.2 for the illustrations):

- A critical 3-clique is a clique composed of a unique nonsimple voxel (like voxel b). This is the simplest case since no choice needs to be made: this voxel must be preserved.
- A critical 2-clique (like clique $\{a, b\}$) is necessarily composed of two voxels that intersect in a 2-face. If one of these two voxels forms a critical 3-clique (i.e., is

not simple like voxel b), then this voxel has been detected previously and must be preserved. It is not necessary to keep the other one.

- A critical 1-clique (like clique $\{e, f, g, h\}$) may include critical 3-cliques or critical 2-cliques (like here, clique $\{e, g\}$). If it is the case, then at least one voxel of the included critical clique(s) has been chosen to be preserved, and no other voxel needs to be kept.

Following this reasoning, which also applies to critical 0-cliques, we see that to reduce both the number of decisions and the number of kept voxels, we have to examine critical cliques by decreasing rank and to keep track of the decisions that have been made for the cliques of higher ranks. This strategy will guide us for the design of all the algorithms proposed in this chapter.

7.5 ASYMMETRIC THINNING

We first introduce our 3D parallel asymmetric thinning scheme (see also [32,4,18]). Then, in Section 7.5.2, we show how to use this scheme to compute curve or surface skeletons, based on the notion of isthmus that we will introduce. We show in Section 7.5.3 some experimental results for curve thinning, which allows us to compare our approach with respect to all previously published methods of the same kind.

7.5.1 GENERIC PARALLEL ASYMMETRIC THINNING SCHEME

The main features of the proposed scheme (algorithm `AsymThinningScheme`) are the following:

- Following the strategy presented in Section 7.4, critical cliques are considered according to their decreasing ranks (step 3). Thus, each iteration includes four subiterations (steps 3–6). Voxels previously selected are stored in a set Y (step 6). At a given subiteration, we consider only critical cliques included in $X \setminus Y$ (step 4).
- *Select* is a function from $\mathbb{V}^3$ to V^3, the set of all voxels. More precisely, *Select* associates with each set S of voxels a unique voxel x of S. We refer to such a function as a *selection function*. This function allows us to select a voxel in any given critical clique (step 5). A possible choice is to take for $Select(S)$, the first voxel of S in the lexicographic order of the voxels coordinates.
- In order to compute curve or surface skeletons, we have to keep other voxels than the ones that are necessary for the preservation of the topology of the object X. In the scheme, the set K corresponds to a set of features that we want to be preserved by a thinning algorithm (thus, we have $K \subseteq X$). This set K, called the *constraint set*, is updated dynamically at step 8. The parameter $Skel_X$ is a function from X on $\{True, False\}$, which allows us to detect some *skeletal voxels* of X, e.g., some voxels belonging to parts of X that are surfaces or curves. For example, if we want to obtain curve skeletons, then a popular solution is to set $Skel_X(x) = True$ whenever x is a so-called *end voxel* of X: an end voxel is a voxel that has exactly

one neighbor inside X. However, this solution is limited and does extend easily to the case of surface skeletons. Alternative choices for such a function will be introduced in Section 7.5.2.

By construction, at each iteration, the complex Y at step 7 satisfies the conditions of Theorem 5. Thus, the result of the scheme is a thinning of the original complex X, whatever the choices of parameters K and $Skel_X$. Observe also that, except step 3, each step of the scheme may be computed in parallel.

Algorithm 1 (AsymThinningScheme($X, K, Skel_X$)).

```
   Data: X ∈ V³, K ⊆ X, Skel_X is a function from X on {True, False}
   Result: X
1  repeat
2  |   Y := K;
3  |   for d ← 3 to 0 do
4  |   |   B := set of all d-cliques that are critical for X and included in X \ Y;
5  |   |   A := {Select(C) | C ∈ B};
6  |   |_  Y := Y ∪ A;
7  |   X := Y;
8  |   foreach voxel x ∈ X \ K such that Skel_X(x) = True do  K := K ∪ {x};
9  until stability;
```

Fig. 7.3 provides an illustration of AsymThinningScheme. Let us consider the complex X depicted in (A). We suppose in this example that we do not keep any skeletal voxel, i.e., the initial constraint set K is empty, and for any x in X, we set $Skel_X(x) = False$. The set of traces of the cliques that are critical for X is represented in (B). Thus, the set of the cliques that are critical for X is precisely composed of six 0-cliques, two 1-cliques, three 2-cliques, and one 3-clique. In (C), the different subiterations of the first iteration of the scheme (lines 3–5) are illustrated:

- when $d = 3$, only one clique is considered, the dark-gray voxel is selected whatever the selection function;
- when $d = 2$, all the three 2-cliques are considered since none of these cliques contains the above voxel. Voxels that could be selected by a selection function are depicted in medium gray;
- when $d = 1$, only one clique is considered, a voxel that could be selected is depicted in light gray;
- when $d = 0$, no clique is considered since each of the 0-cliques contains at least one voxel that has been previously selected.

After these subiterations, we obtain the complex depicted in (D). Figures (E) and (F) illustrate the second iteration, and at the end of this iteration, the complex is reduced to a single voxel.

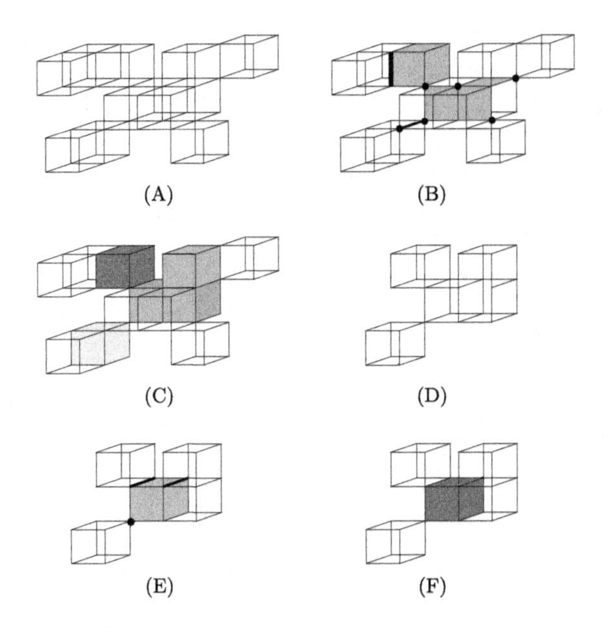

FIGURE 7.3

(A) A complex X made of precisely 12 voxels. (B) The set of traces of the cliques that are critical for X (i.e., the maximal faces of the critical kernel of X). (C) Voxels that have been selected by the algorithm. (D) The result Y of the first iteration. (E) The set of traces of the cliques that are critical for Y. (F) The result of the second iteration.

Of course, the result of the scheme may depend on the choice of the selection function. This is the price to be paid if we try to obtain thin skeletons. For example, some arbitrary choices have to be made for reducing a two-voxel-wide ribbon to a simple curve.

In the sequel of the chapter, we take for $Select(S)$, the first voxel of S in the lexicographic order of the voxels coordinates.

Fig. 7.4 shows another illustration, on bigger objects, of `AsymThinningScheme`. Here also, we give $\emptyset$ as parameter K, and for any x in X, we have $Skel_X(x) = False$ (no skeletal voxel). The result is called an ultimate asymmetric skeleton.

7.5.2 ISTHMUS-BASED ASYMMETRIC THINNING

In this section, we show how to use our generic scheme `AsymThinningScheme` to get a procedure that computes either curve or surface skeletons. This thinning procedure preserves a constraint set K that is made of "isthmuses."

Intuitively, a voxel x of an object X is said to be a 1-isthmus (resp. a 2-isthmus) if the neighborhood of x corresponds, up to a thinning, to the one of a point belonging to a curve (resp. a surface) [18].

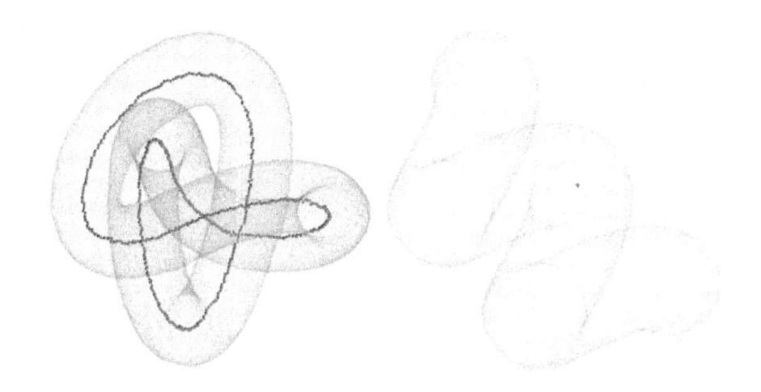

FIGURE 7.4

Ultimate asymmetric skeletons obtained by using `AsymThinningScheme`. On the left, the object is a solid cylinder bent to form a knot. Its ultimate skeleton S is a discrete curve, i.e., for any x in S, $\mathcal{N}^*(x) \cap S$ is made of exactly two voxels. On the right, the object is connected and without holes and cavities. Its ultimate skeleton is a single voxel.

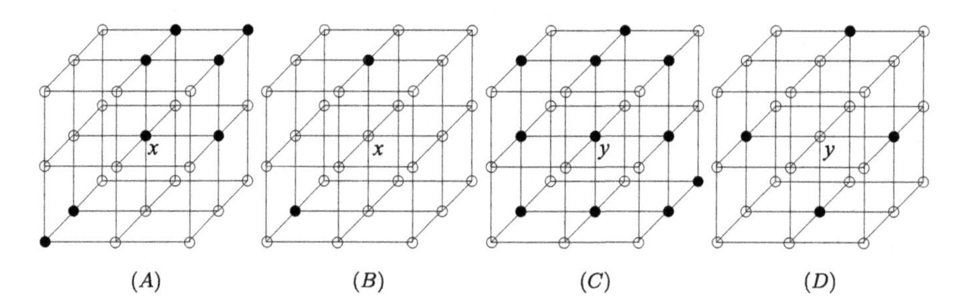

FIGURE 7.5

In this figure, a voxel is represented by its central point. (A) A voxel x and the set $\mathcal{N}(x) \cap X$ (black points). (B) A set S that is a 0-surface, $\mathcal{N}^*(x) \cap X$ is reducible to S, and thus x is a 1-isthmus for X. (C) A voxel y and the set $\mathcal{N}(y) \cap X$ (black points). (D) A set S that is a 1-surface, $\mathcal{N}^*(y) \cap X$ is reducible to S, and thus y is a 2-isthmus for X.

We say that $X \in \mathbb{V}^3$ is a 0-*surface* if X is precisely made of two voxels x and y such that $x \cap y = \emptyset$.

We say that $X \in \mathbb{V}^3$ is a 1-*surface* (or a *simple closed curve*) if:

i) X is connected, and
ii) for each $x \in X$, $\mathcal{N}^*(x) \cap X$ is a 0-surface.

Definition 6. Let $X \in \mathbb{V}^3$, and let $x \in X$. We say that:

the voxel x is a 1-*isthmus for* X if $\mathcal{N}^*(x) \cap X$ is reducible to a 0-surface,
the voxel x is a 2-*isthmus for* X if $\mathcal{N}^*(x) \cap X$ is reducible to a 1-surface, and
the voxel x is a 2^+-*isthmus for* X if x is a 1-isthmus or a 2-isthmus for X.

See Fig. 7.5 for an illustration of 1- and 2-isthmuses.

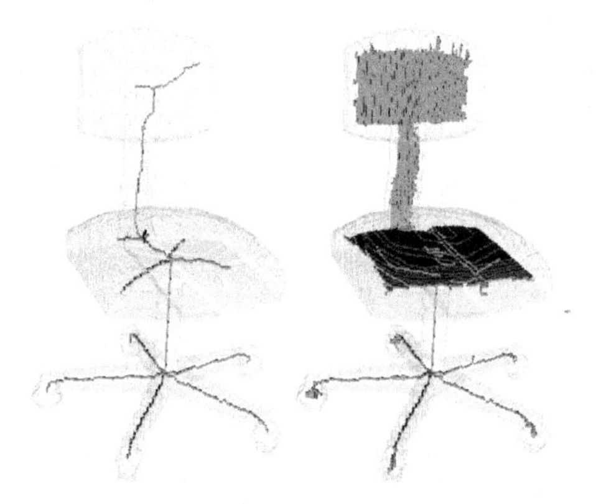

FIGURE 7.6

Asymmetric skeletons obtained by using `AsymThinningScheme`. Left: curve skeleton. The function $Skel_X$ is based on 1-isthmuses. Right: surface skeleton. The function $Skel_X$ is based on 2^+-isthmuses.

Notice that statement 1) of forthcoming Theorem 8 (and the fact that 0- and 1-surfaces are not reducible) will imply that every 1- or 2-isthmus voxel is nonsimple.

Our aim is to thin an object while preserving a constraint set K, initially empty, that is made of voxels that are detected as k-isthmuses during the thinning process. We obtain curve skeletons with $k = 1$ and surface skeletons with $k = 2^+$. These two kinds of skeletons may be obtained by using `AsymThinningScheme` with the function $Skel_X$ defined as follows:

$$Skel_X(x) = \begin{cases} True & \text{if } x \text{ is a } k\text{-isthmus,} \\ False & \text{otherwise.} \end{cases}$$

Observe that a voxel may belong to a k-isthmus at a given step of the algorithm, but not at further steps. This is why previously detected isthmuses are stored (see line 8 of `AsymThinningScheme`).

In Fig. 7.6, we show a curve skeleton and a surface skeleton obtained by our method from the same object.

7.5.3 COMPARISON WITH OTHER PARALLEL CURVE SKELETONIZATION METHODS

To validate our approach, we made some experiments to compare our isthmus-based curve thinning method using `AsymThinningScheme` with all other existing methods of the same kind. To make a fair comparison, we considered only parallel asymmetric thinning methods that produce curve skeletons of voxel objects and that have no parameter, namely, [8–17].

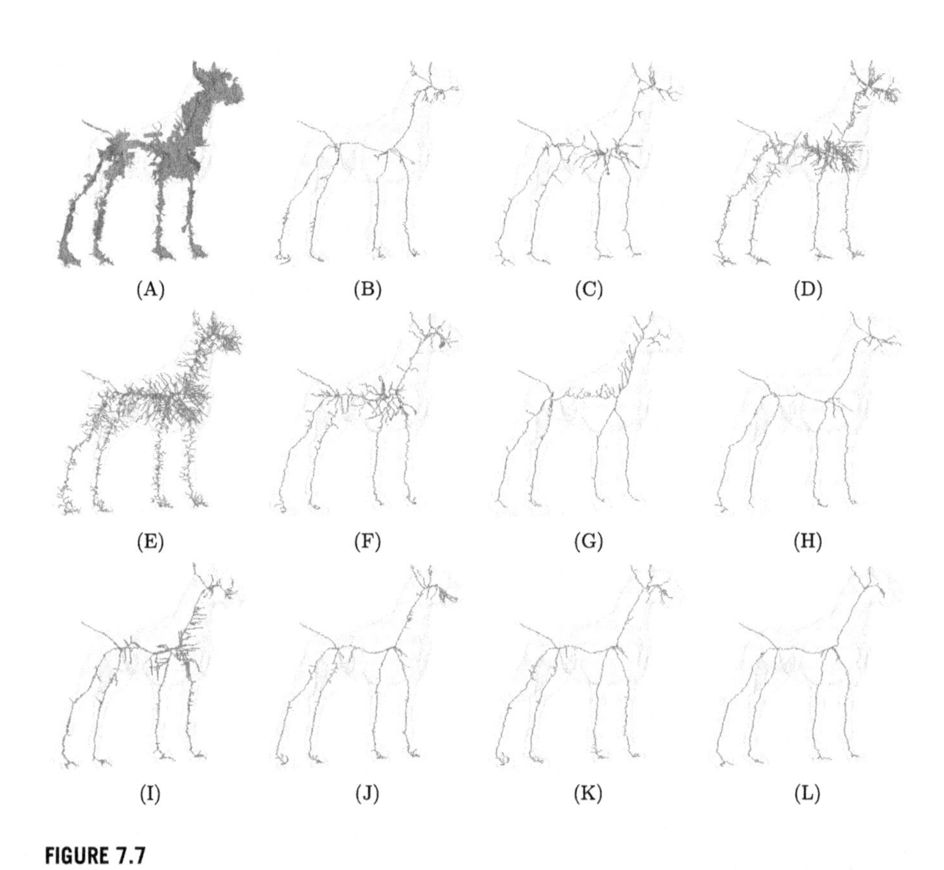

FIGURE 7.7

Curve skeletons of the same object obtained through different methods: (A) [8], (B) [9], (C) [10], (D) [11], (E) [12], (F) [13], (G) [14], (H) [15], (I) [16], (J) [17], (K) [17], (L) `AsymThinningScheme`.

In Figs. 7.7 and 7.8, we show the results of all these methods and ours for the same shape. We notice in particular that some methods, for example, [8] (A), are not sufficiently powerful to produce results that may be interpreted as curve skeletons. This illustrates the difficulty of designing a method that keeps enough voxels in order to preserve topology, and at the same time, deletes a sufficient number of voxels to produce thin curve skeletons. This difficulty is indeed high when these two opposite constraints are not made explicit in the algorithm. The strength of our approach lies in a complete separation of these constraints.

The example of Fig. 7.8 illustrates very well the sensitivity to contour noise of the methods. The original object is a discretized thickened spiral, and its curve skeleton should ideally be a simple curve. Any extra branch of the skeleton must undoubtedly be considered as spurious. To count the number of spurious branches for this example, we can simply count the number of end points and subtract 2 since the ideal skeleton

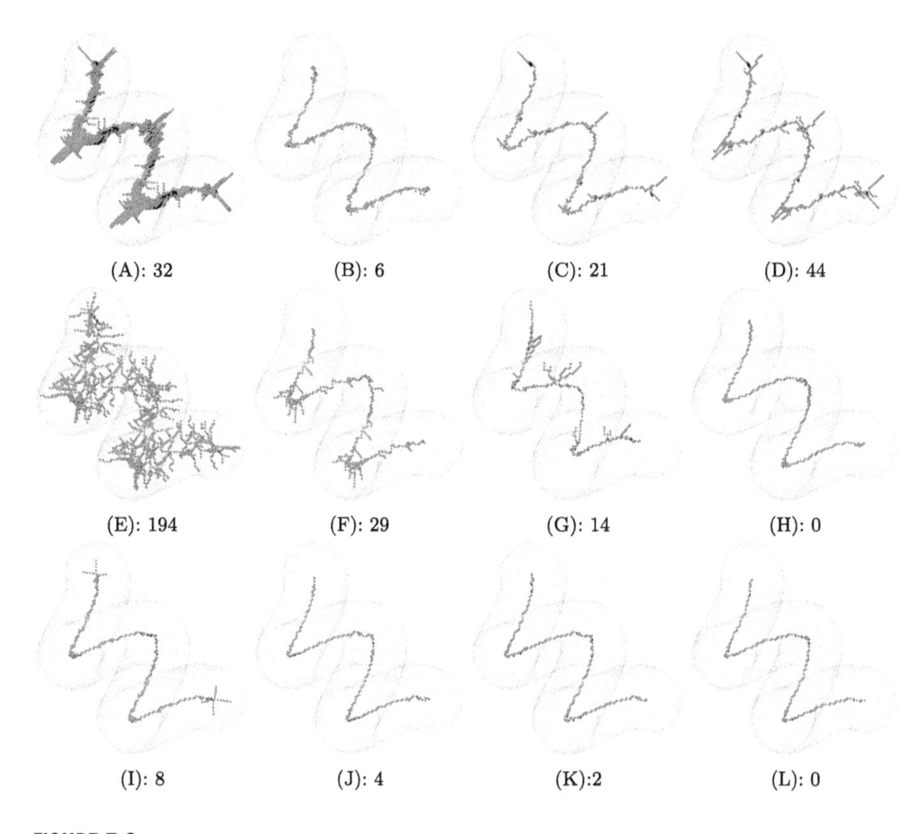

FIGURE 7.8

Idem Fig. 7.7, the number of spurious branches is given for each skeleton.

has exactly two extremities. In Fig. 7.8, we indicate for each skeleton the number of spurious branches. Notice that only [15] and `AsymThinningScheme` produce a skeleton free of spurious branches for this object.

Furthermore, in [21], we describe a quantitative study that shows that our method outperforms the other ones with respect to contour noise sensitivity.

7.6 SYMMETRIC THINNING

As in the previous section, our goal here is to define a subset of a voxel complex X that is guaranteed to include at least one voxel of each clique that is critical for X. By Theorem 5 this subset will be a thinning of X.

But now we want our method to be independent of arbitrary choices, in particular, of a choice of specific voxels in a given critical clique. This will ensure that the obtained skeletons are invariant with respect to 90-degree rotations.

To obtain this result, we do not consider individual voxels anymore, but only cliques. We denote by $\mathcal{C}(X)$ the set of all cliques included in X.

7.6.1 GENERIC PARALLEL SYMMETRIC THINNING SCHEME

As in Section 7.5.1, we first propose a generic algorithm (algorithm SymThinningScheme) that takes as input parameters a voxel complex X, a constraint set $K \subseteq X$, and a boolean function $Skel_X$, which serves to dynamically detect the cliques that must be preserved during the thinning.

Following the strategy presented in Section 7.4, critical cliques are considered according to their decreasing ranks (step 3).

Algorithm 2 (SymThinningScheme(X, K, $Skel_X$)).

Data: $X \in \mathbb{V}^3$, $K \subseteq X$, $Skel_X$ a function from $\mathcal{C}(X)$ on $\{True, False\}$
Result: X

```
1 repeat
2 |   Y := K;
3 |   for d ← 3 to 0 do
4 |   |   A := union of all d-cliques that are critical for X and included in X \ Y;
5 |   |   Y := Y ∪ A;
6 |   X := Y;
7 |   foreach clique C ∈ C(X) such that Skel_X(C) = True do K := K ∪ C;
8 until stability;
```

Let us illustrate this scheme with a commonly used kind of constraint set, namely, the medial axis of X.

The notion of medial axis has been introduced by Blum in the 1960s [33]. Let S be a subset of $\mathbb{R}^n$ or $\mathbb{Z}^n$. The medial axis of S consists of the centers of the n-dimensional balls that are included in S but that are not included in any other n-dimensional ball included in S. The medial axis of a voxel set X may be readily defined by identifying each voxel of X with its center.

In the continuous space $\mathbb{R}^n$, the medial axis of any set S is known to be homotopy-equivalent to S under certain conditions of "smoothness" [34]. However, this is not the case in discrete grids, whatever the distance that is chosen to define the notion of ball. This fact is illustrated in 2D in Fig. 7.9A.

To ensure topology preservation, we use SymThinningScheme to thin the original object X with the only constraint of preserving the medial axis of X; see Fig. 7.9B. Notice that, by its very definition, the medial axis is invariant to 90-degree rotations. Thus, using SymThinningScheme, we ensure that the skeleton finally obtained is invariant to 90-degree rotations.

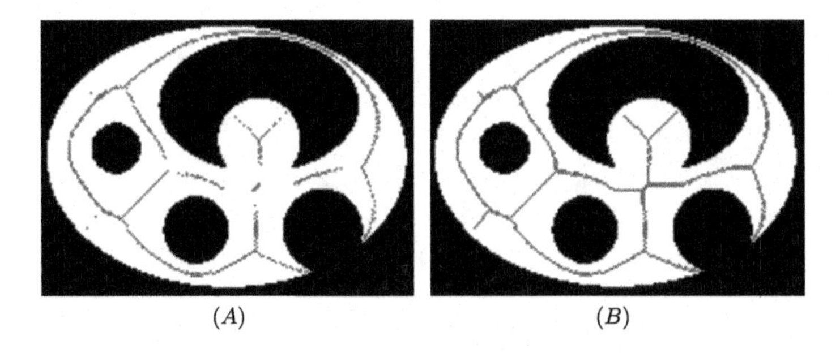

FIGURE 7.9

(A) A voxel set X and its medial axis MA(X), based on the so-called city block distance or 4-distance. (B) Result of SymThinningScheme(X, MA(X), $Skel_X$) with $Skel_X(C) = False$ for any $C \in \mathcal{C}(X)$.

7.6.2 ISTHMUS-BASED SYMMETRIC THINNING

In Section 7.5.2, we defined 1-isthmuses and 2-isthmuses as voxels satisfying a certain condition. To handle thick structures, as, for example, two-voxel-thick ribbons that may exist in symmetric curve skeletons, we replace the notion of a voxel by the one of a clique. Intuitively, and generalizing Definition 6, a critical clique C of an object X is said to be a 1-isthmus (resp. a 2-isthmus) if the neighborhood of C corresponds, up to a thinning, to the one of a point belonging to a curve (resp. a surface).

Definition 7. Let $X \in \mathbb{V}^3$, and let C be a clique that is essential for X. We say that:

the clique C is a 1-*isthmus for* X if $\mathcal{K}^*(C) \cap X$ is reducible to a 0-surface,
the clique C is a 2-*isthmus for* X if $\mathcal{K}^*(C) \cap X$ is reducible to a 1-surface, and
the clique C is a 2^+-*isthmus for* X if C is a 1-isthmus or a 2-isthmus for X.

Recall that a clique C is critical for X if C is essential for X and if $\mathcal{K}^*(C) \cap X$ is not reducible to a single voxel (Definition 4). It will follow from statement 1) of forthcoming Theorem 8 that a k-isthmus, $k \in \{1, 2, 2^+\}$, is necessarily critical. See Figs. 7.10 and 7.5, where several examples of 1- and 2-isthmuses are given.

Curve skeletons (with $k = 1$) and surface skeletons (with $k = 2^+$) may be obtained by using SymThinningScheme with the function $Skel_X$ defined for any C in $\mathcal{C}(X)$ as follows:

$$Skel_X(C) = \begin{cases} True & \text{if } C \text{ is a } k\text{-isthmus,} \\ False & \text{otherwise.} \end{cases}$$

In Figs. 7.11 and 7.12, we illustrate the above algorithm for computing curve and surface skeletons, respectively.

To our knowledge, the only other parallel and symmetric method for obtaining both types of skeletons was proposed in [6]. It is based on P-simple points, introduced by one of the authors [2]. The obtained skeletons are very close to the ones

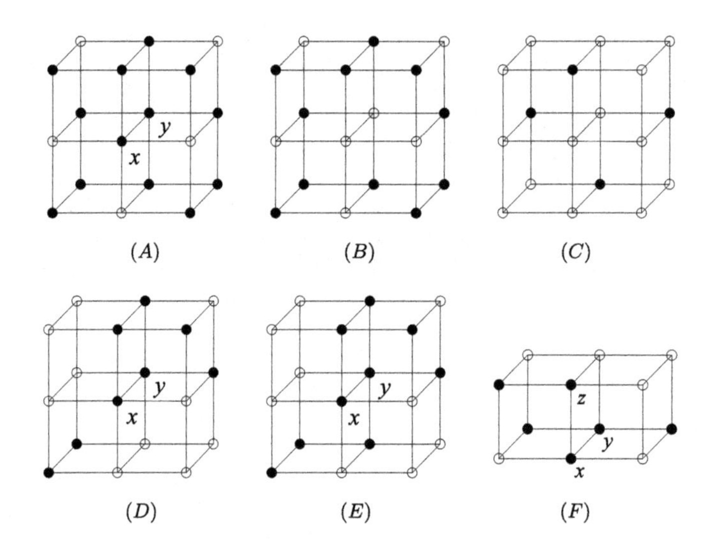

FIGURE 7.10

In this figure, a voxel is represented by its central point. (A) A 2-clique $C = \{x, y\}$ and the set $\mathcal{K}(C) \cap X$ (black points). It can be seen that C is essential for X. (B) The set $D = \mathcal{K}^*(C) \cap X$. (C) A set E, D is reducible to E. The set E is a 1-surface, and thus C is a 2-isthmus for X. (D) A 2-clique $C = \{x, y\}$ and the set $\mathcal{K}(C) \cap X$ (black points), C is a 1-isthmus for X. (E) A 2-clique $C = \{x, y\}$ and the set $\mathcal{K}(C) \cap X$, C is neither a 1-isthmus nor a 2-isthmus for X. In fact, C is regular for X. (F) A 1-clique $C = \{x, y, z\}$ and the set $\mathcal{K}(C) \cap X$, C is a 1-isthmus for X.

FIGURE 7.11

Illustrations of isthmus-based curve thinning using `SymThinningScheme`.

produced using the methods described in this chapter. However, in [6], the preservation of salient object features is achieved through the detection of end voxels. Using this strategy, it is not possible to define a notion of persistence (see Section 7.8) that allows one to filter skeletons. Filtering is mandatory in almost all applications dealing

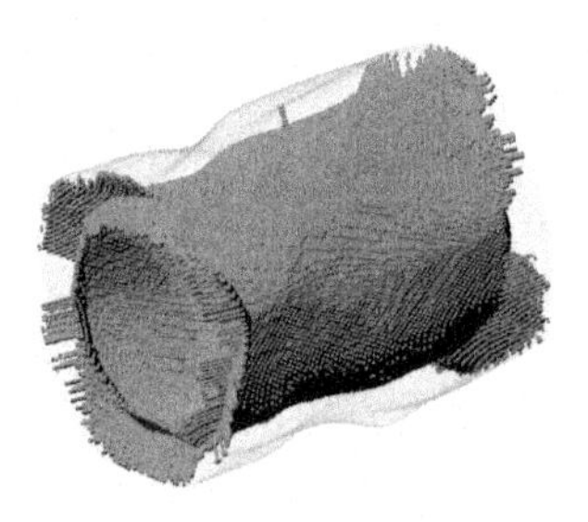

FIGURE 7.12

Illustration of isthmus-based surface thinning using `SymThinningScheme`.

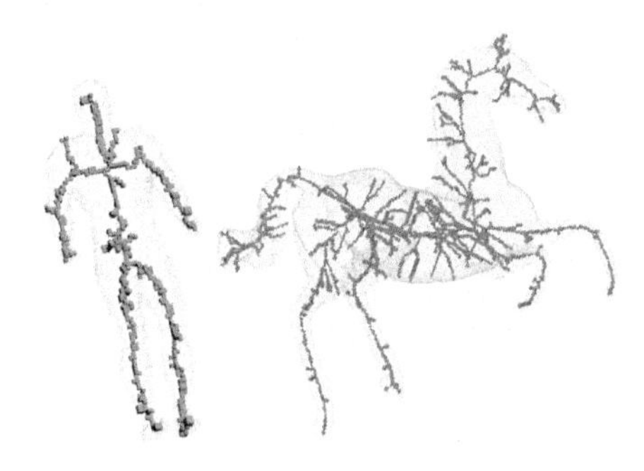

FIGURE 7.13

Illustrations of isthmus-based curve thinning using `SymThinningScheme` with noisy shapes (a small amount of random noise has been added to the contours of the shapes).

with skeletons since noise is usually affecting real data and skeletons are notoriously sensitive to noise (see Figs. 7.11, 7.12, 7.13). This problem will be studied, and a solution will be proposed in the following section (see Fig. 7.14).

7.7 CHARACTERIZATION OF CRITICAL CLIQUES AND k-ISTHMUSES

A key point in the implementation of the algorithms proposed in this chapter is the detection of critical cliques and k-isthmuses.

In this section, we will see that it is possible to detect these cliques with efficient algorithms operating on four different kinds of neighborhoods. See also [18], where a set of masks is proposed for critical cliques.

First of all, recall that a clique that is either critical or a k-isthmus for a complex X in $\mathbb{V}^3$ is necessarily essential for X.

FIGURE 7.14

Preview of the filtered skeletons obtained using the method of Section 7.8 on the same noisy shapes as in Fig. 7.13.

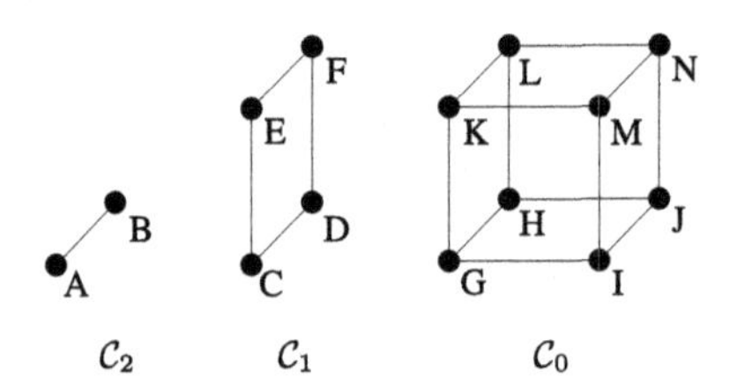

FIGURE 7.15

Masks for 2-cliques ($\mathcal{C}_2$), 1-cliques ($\mathcal{C}_1$), and 0-cliques ($\mathcal{C}_0$). Here, a voxel is represented by its central point.

Up to $\pi/2$ rotations, the three configurations $\mathcal{C}_2$, $\mathcal{C}_1$, and $\mathcal{C}_0$ given in Fig. 7.15 may be used for the detection of a clique that is essential for a complex $X \in \mathbb{V}^3$ (in this figure, a voxel is represented by a point). Let $C \subseteq X$. It may be seen that, up to $\pi/2$ rotations:

- C is a 2-clique that is essential for X if and only if $C = \mathcal{C}_2$.
- C is a 1-clique that is essential for X if and only if $C = \mathcal{C}_1 \cap X$ and there exist two voxels of C that are 1-neighbors.
- C is a 0-clique that is essential for X if and only if $C = \mathcal{C}_0 \cap X$ and there exist two voxels of C that are 0-neighbors or three voxels of C that are mutually 1-neighbors.

The $\mathcal{K}$-neighborhoods of the configurations $\mathcal{C}_2$, $\mathcal{C}_1$, and $\mathcal{C}_0$ are given Fig. 7.16. Observe that we have $\mathcal{K}_0 = \mathcal{C}_0$.

Recall that a set made of a single voxel x of an object X constitutes a clique (a 3-clique) that is essential for X. Therefore, there are precisely four different kinds of neighborhoods for an essential clique ($\mathcal{K}_0$, $\mathcal{K}_1$, $\mathcal{K}_2$, and $\mathcal{N}(x)$).

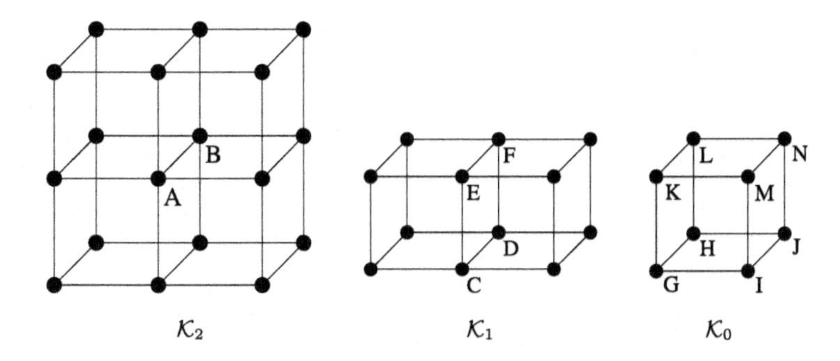

FIGURE 7.16

$\mathcal{K}$-neighborhoods for 2-cliques ($\mathcal{K}_2$), 1-cliques ($\mathcal{K}_1$), and 0-cliques ($\mathcal{K}_0$). A voxel is represented by its central point.

Thus, the neighborhoods involved in the definitions of critical cliques and k-isthmuses (Definitions 4 and 7) are fully specified.

Nevertheless, we observe that, in these two definitions, we have to check if a given complex S is reducible to a certain complex T determined by specific properties.

For example, for checking whether an essential clique C is a 2-isthmus for X, we have to verify whether $S = \mathcal{K}^*(C) \cap X$ is reducible to a simple closed curve (a 1-surface).

Let $S \in \mathbb{V}^3$ be an arbitrary complex that is reducible to a complex T. In fact, there is the possibility that S is reducible to a complex R with $T \subseteq R$ but that R is not reducible to T.

For example, such a situation occurs when S is a cuboid, T is made of a single voxel, and R is an object such as the so-called *dunce hat* [35] or *the house with two rooms* [36]. In [30], this kind of object is referred to as *a lump*.

As a consequence of this fact, for testing if an arbitrary complex S is reducible to T, it is a priori necessary to compute all the complexes R such that S is reducible to R and $T \subseteq R$. In other words, the "reducibility problem" is, in general, a complex problem from the algorithmic point of view. See, e.g., [37] for this issue.

In [30], it was shown that it is possible to find lumps in the neighborhood of a cubical element (a face) of $\mathbb{Z}^5$. The following result, which has been proved with the help of a computer program as explained in [20], shows that, in $\mathbb{V}^3$, there is not enough space for such objects to lie in the $\mathcal{K}$-neighborhood of a clique. Note that point 1) of Theorem 8 corresponds to Theorem 17 of [18].

Theorem 8 ([20]). *Let $C \in \mathbb{V}^3$ be a clique, let $S \subseteq \mathcal{K}^*(C)$, and let x be a voxel that is simple for S.*

1. *If S is reducible, then $S \setminus \{x\}$ is reducible.*
2. *If S is reducible to a 0-surface, then $S \setminus \{x\}$ is reducible to a 0-surface.*
3. *If S is reducible to a 1-surface, then $S \setminus \{x\}$ is reducible to a 1-surface.*

As a consequence of Theorem 8, we can use a greedy algorithm for testing if an essential clique C is critical, a 1-isthmus, a 2-isthmus, or not critical for X. We can arbitrarily select and remove a voxel that is simple for the set $S = \mathcal{K}^*(C) \cap X$. After repeating this operation until stability, we check if C is critical (S is not made of a single voxel), if C is a 1-isthmus (S is made of two voxels), or if C is a 2-isthmus (S is a simple closed curve).

Testing whether a voxel x is simple or not for an object $S \in \mathbb{V}^3$ can be done in constant time. This may be done by using previously proposed characterizations [30] or by using a precomputed look-up table. By Theorem 8, testing the status of a clique may be performed in linear time with respect to the size of its neighborhood (at most 26). Again, the results can be precomputed and stored in look-up tables to speed up the test. Each test (simple, critical, 1-isthmus, 2-isthmus) can thus be performed in constant time.

7.8 ISTHMUS PERSISTENCE AND SKELETON FILTERING

In this section and the following one, we focus on symmetric thinning. However, the persistence-based thinning method presented next is adaptable with only very minor changes to the asymmetric thinning algorithm presented in Section 7.5.2.

Even in the continuous framework, the skeleton suffers from its sensitivity to small contour perturbations; in other words, it lacks stability. This fact, among others, explains why it is often necessary to add a filtering step (or pruning step) to any method that aims at computing the skeleton. Hence, there is a rich literature devoted to skeleton pruning, in which different criteria were proposed in order to discard "spurious" skeleton points or branches; see, e.g., [20], Section 7.

A particularly appealing approach to prevent the appearance of spurious skeleton branches (or surfaces) is based on the notion of isthmus persistence. Its advantages are the following:

- it applies, both in 2D and 3D cases, to curve and surface skeletons,
- it is governed by a single parameter,
- it is effective (although measuring this criterion is difficult because the quality of a skeleton is always a tradeoff between fidelity and low complexity, see [38], and therefore its assessment depends very much on the application),
- it is quite simple to compute in the framework of a thinning procedure.

This notion first appeared, to our knowledge, in the context of cubical complexes, in the work of Liu et al. [19].

The idea is the following. When a voxel is detected as belonging to an isthmus for the first time during the thinning process, the number of thinning steps that have been performed at this moment is recorded and called the *birth date* of this voxel. Intuitively, it gives an information about the local thickness of the object around this voxel (see Fig. 7.17 for an illustration in 2D). Then, later on during the thinning

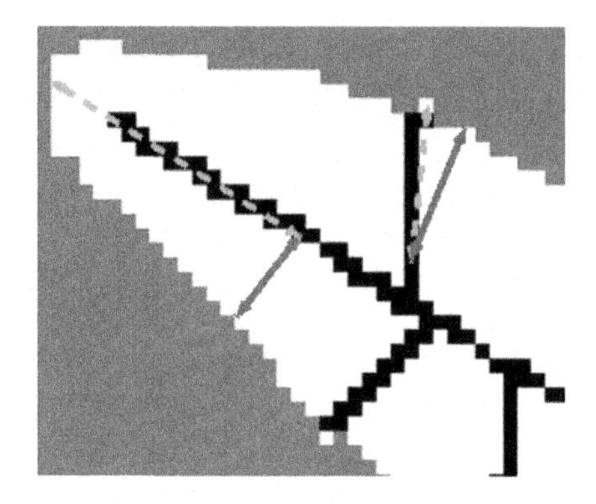

FIGURE 7.17

The intervals depicted with a solid line correspond to the birth dates, the dotted lines to the death dates.

process, the same voxel may become a candidate for deletion, that is, it is no longer in an isthmus. At this time, we record the number of thinning steps that were performed, and we call this number the *death date* of the voxel. The difference *death date* − *birth date* is called the *persistence* of the voxel (see also [39]). A voxel that has a short persistence is very likely to be part of a spurious skeleton branch, whereas voxels with long persistences indicate the presence of robust skeleton parts (Fig. 7.17). In this approach, only isthmuses having a persistence greater than a given threshold will be kept.

The works [19] and [39] present some asymmetric thinning procedures that take place in the framework of general cubical complexes, i.e., complexes that are not necessarily made of voxels. With our definitions of 1D and 2D isthmuses, either single-voxel or multiple-voxel cliques, it becomes easy to implement this strategy to detect robust skeleton elements in the context of symmetric or asymmetric parallel thinning of objects made of voxels.

In the following algorithm, k stands for the type of the considered isthmuses (1 or 2^+), and p is the parameter that sets the persistence threshold. The function b associates to certain voxels their birth date, and K is a constraint set that is dynamically updated by adding those voxels whose persistence is greater than the threshold p (lines 9–10).

In line 8, the birth date $b(x)$ of each new isthmus voxel x is recorded. Notice also that the voxels in P (line 9) are deletable. The test $b(x) > 0$ means that such a voxel x was an isthmus voxel formerly, and its death date is i. These voxels are kept in X and added to the constraint set K (line 10) because of their long persistence (at least p).

In Figs. 7.18 and 7.19, we show outcomes of algorithm `PersistenceSymThinning` with $k = 1$ and $k = 2^+$, respectively, for different values of parameter p.

Algorithm 3 (PersistenceSymThinning(X, k, p)).

Data: $X \in \mathbb{V}^3, k \in \{1, 2^+\}, p \in \mathbb{N} \cup \{+\infty\}$
Result: X

1 $i := 0$; $K := \emptyset$; **foreach** $x \in X$ **do** $b(x) := 0$;
2 **repeat**
3 $\quad i := i + 1$; $Y := K$; $Z := \emptyset$;
4 $\quad$ **for** $d \leftarrow 3$ **to** 0 **do**
5 $\quad\quad A :=$ union of all d-cliques that are critical for X and included in $X \setminus Y$;
6 $\quad\quad B :=$ union of all d-cliques that are k-isthmuses for X and included in $X \setminus Y$;
7 $\quad\quad Y := Y \cup A$; $Z := Z \cup B$;
8 $\quad$ **foreach** $x \in Z$ *such that* $b(x) = 0$ **do** $b(x) := i$;
9 $\quad P := \{x \in X \setminus Y \mid b(x) > 0 \text{ and } i - b(x) \geqslant p\}$;
10 $\quad K := K \cup P$; $X := Y \cup P$;
11 **until** *stability*;

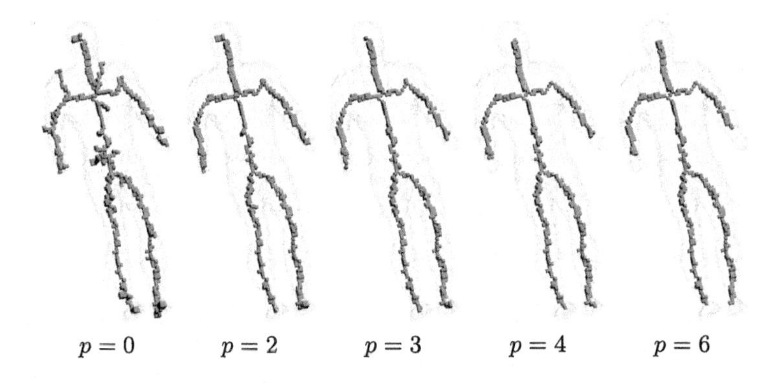

FIGURE 7.18

Outcomes of algorithm PersistenceSymThinning with $k = 1$ for different values of parameter p.

7.9 HIERARCHIES OF SKELETONS

With the preceding examples, we saw that, intuitively, the greater the persistence, the smaller the filtered skeleton. By varying the value of the persistence parameter we may compute a whole family of nested homotopic skeletons to the cost of applying the previous algorithm once for each parameter value ("naive" approach). Once computed, such a family can be efficiently stored as a function, i.e., a grayscale image, obtained by stacking all the filtered skeletons. Thresholding this function (named

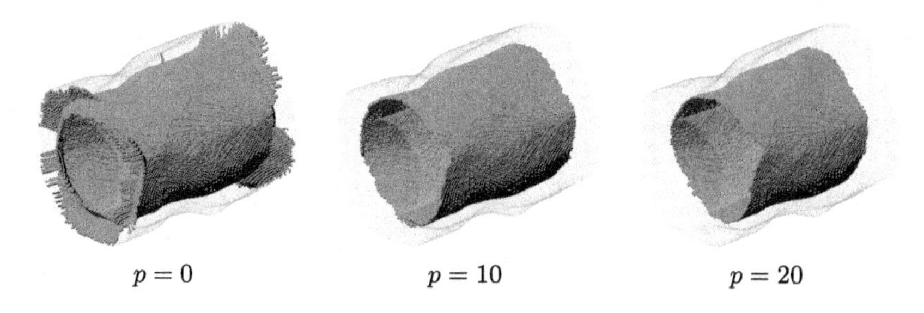

FIGURE 7.19

Outcome of algorithm `PersistenceSymThinning` with $k = 2^+$ for different values of parameter p.

skeleton stack function or *SSF* in the sequel) at a given value would give back the filtered skeleton corresponding to this value of the parameter.

More specifically, given a voxel complex X, we want to compute a function Ψ from X to $\mathbb{N} \cup \{+\infty\}$. Let Ψ_k denote the threshold of Ψ at level k, i.e., $\Psi_k = \{x \in X \mid \Psi(x) \geq k\}$. The function Ψ must meet the two following requirements:

i) Ψ_0 is a symmetric skeleton of X (unfiltered), and
ii) for any k and any ℓ in $\mathbb{N}$ such that $k > \ell$, Ψ_k is a symmetric thinning of Ψ_ℓ.

It may be interesting to compute an SSF instead of a single filtered skeleton in cases, which are frequent, where there is no trivial way to fix a priori a single parameter value for a set of images. In such cases, the precomputed SSF may allow a user to interactively visualize the effect of each parameter value on the filtering. Alternatively, a postprocessing algorithm may be applied in certain applications to automatically choose a threshold of the SSF that satisfies a given criterion (for example, a chosen number of skeleton branches).

Now, we are facing the problem of efficiently computing an SSF. We show in the rest of this section that the cost of this computation can be dramatically reduced with respect to the one of the naive method evoked at the beginning of this section.

A first idea consists of storing, during the thinning process, the function Π that associates its persistence to each voxel that is detected as an isthmus at any step. The algorithm `Persistence` (Algorithm 4), derived from algorithm `PersistenceSymThinning`, plays this role. This algorithm computes the birth and death dates of all nonpermanent isthmus voxels and stores the differences (lines 8–12). This corresponds to the case where the parameter p is set to infinity in algorithm `PersistenceSymThinning`.

What happens if we threshold the map Π at a given level π? We could expect to get a filtered skeleton as computed by algorithm `PersistenceSymThinning` with parameter $p = \pi$, but this is not always the case: see Fig. 7.20 for a counterexample. We also see with this example that the topological characteristics are not always preserved by this procedure.

Algorithm 4 (`Persistence`(X, k)).

```
   Data: X ∈ V^3, k ∈ {1, 2+}
   Result: Π : a function from X into ℕ ∪ {+∞}
1  i := 0; foreach x ∈ X do Π(x) := 0;
2  repeat
3  |  i := i + 1;
4  |  Y := ∅; Z := ∅; for d ← 3 to 0 do
5  |  |  A := union of all d-cliques that are critical for X and included in X \ Y;
6  |  |  B := union of all d-cliques that are k-isthmuses for X and included in
   |  |  X \ Y;
7  |  |  Y := Y ∪ A; Z := Z ∪ B;
8  |  foreach x ∈ Z such that Π(x) = 0 do
9  |  |  Π(x) := i; // birth
11 |  foreach x ∈ X \ Y such that Π(x) ≠ 0 do
12 |  |  Π(x) := i − Π(x); // death
14 |  X := Y;
15 until stability;
16 foreach x ∈ X do Π(x) := +∞;
```

To compute a skeleton stack based on the persistence map, we introduce algorithm `PersistenceSymThinning` (Algorithm 5). This algorithm first computes a persistence function Π using algorithm `Persistence` (line 1). Then, it computes the lowest function Ψ that is above Π and satisfies requirements i) and ii) of an SSF. To do this, the algorithm performs a symmetric parallel thinning of X guided by Π (that is, voxels are, loosely speaking, considered in increasing order of the value $\Pi(x)$; see lines 6 and 8–14. Notice however that when a voxel x is deleted at some iteration, each undeleted neighbor y of x will be considered for possible deletion at a future iteration even if $\Pi(y) < \Pi(x)$; see lines 15 and 17–18). For each deleted voxel x, the value of a variable ψ is stored as $\Psi(x)$ (line 16). The value of ψ increases monotonically during the thinning as slowly as possible, but at each iteration of the main loop, line 7 ensures that it will not be less than λ, which is the persistence value $\Pi(x)$ of the voxels x that are to be considered for deletion at that iteration.

We illustrate this algorithm in 2D since it is much easier to visualize a 2D map than a 3D one. Fig. 7.21 displays an object X, its SSF computed by the above algorithm, and two thresholds of this map. The first threshold, at the value 1, is indeed identical to the result of algorithm `IsthmusSymmetricThinning` with $k = 1$ (curve skeleton). The second threshold, at the value 3, provides a filtered skeleton that captures the main structure of the object.

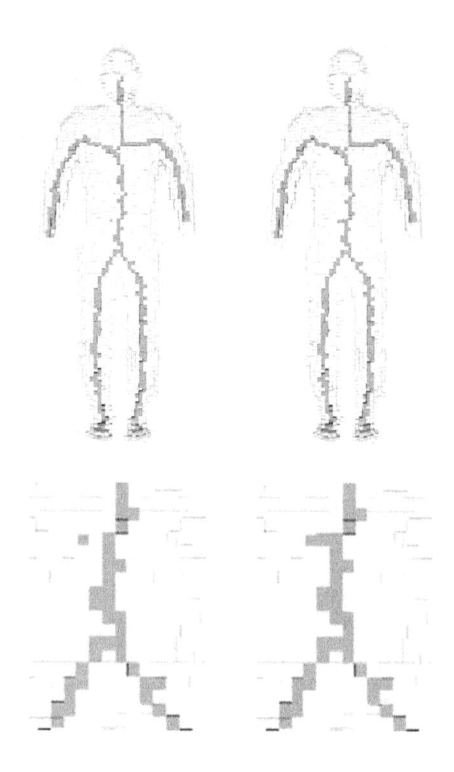

FIGURE 7.20

Left: a threshold of the map Π, computed by algorithm `Persistence` at value 3. Right: outcome of algorithm `PersistenceSymThinning` with $p = 3$. Bottom row: detail (zoomed).

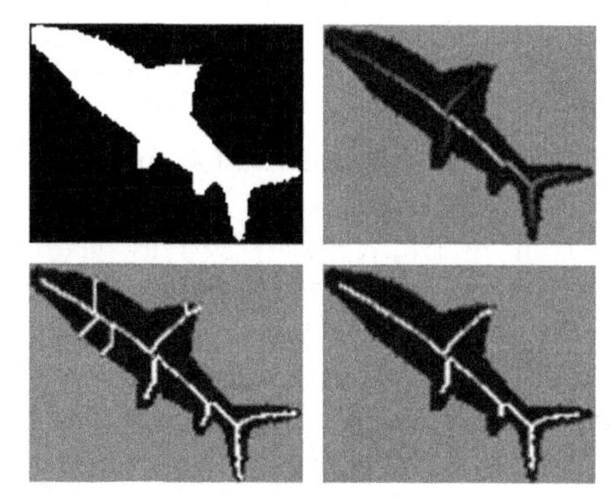

FIGURE 7.21

First row: an object X (left) and its SSF Ψ (right). The darker the pixel, the lowest the value. Notice that the brightest value corresponds to the value $+\infty$, which is given to pixels that belong to any filtered skeleton. Second row: thresholds of the map Ψ at values 1 (left) and 3 (right).

Algorithm 5 (`PersistenceSSF`(X, k)).

```
    Data: X ∈ 𝕍³, k ∈ {1, 2⁺}
    Result: Ψ : a function from X into ℕ ∪ {+∞}
 1  Π := Persistence(X, k);
 2  ψ := −∞;
 3  foreach x ∈ X do Ψ(x) := +∞;
 4  Q := {(x, Π(x)) | x ∈ X};
 5  repeat
 6      λ := min{π | (., π) ∈ Q};
 7      if λ > ψ then ψ := λ;
 8      P := {x ∈ X | (x, λ) ∈ Q};
 9      Q := Q \ {(x, λ) | x ∈ P};
10      Y := X \ P;
11      for d ← 3 to 0 do
12          A := set of all voxels belonging to any d-clique which is critical for X
            and included in X \ Y;
13          Y := Y ∪ A;
14      V := X \ Y; X := Y;
15      foreach x ∈ V do
16          Ψ(x) := ψ;
17          foreach y ∈ 𝒩*(x) ∩ X do
18              Q := Q ∪ {(y, Π(y))};
19  until Q = ∅;
```

7.10 COMPLEXITY

In this section, we discuss the time complexity of the five algorithms presented in the chapter.

Algorithms 1 (`AsymThinningScheme`), 2 (`SymThinningScheme`): These algorithms may be computed in $O(t.n)$ time on a sequential computer, where t is the number of steps to thin the object (which corresponds, intuitively, to its thickness), and in $O(t)$ time on an ideal parallel computer.

Alternatively, a breadth-first strategy may be employed in order to achieve a linear ($O(n)$) processing time for these algorithms with a sequential computer. It consists of maintaining a list of the voxels whose status changes during an iteration and using it to restrict the work to be done, in the next iteration, to their neighborhoods.

Algorithm 3 (`PersistenceSymThinning`): On an ideal parallel computer, the instructions in lines 5–7 may be executed in constant time. Thus, the overall complexity of the algorithm is in $O(t)$, where t represents the number of steps of the repeat loop needed to reach stability (intuitively, this corresponds to the thickness of the object X). On a sequential computer, a direct implementation leads to an algorithm in $O(t \times n)$, where n is the number of voxels in the boolean array representation of X. However, the use of a breadth-first strategy leads to a linear-time ($O(n)$) algorithm.

Algorithm 4 (`Persistence`):
The complexity analysis is the same as for Algorithm 3 (`PersistenceSymThinning`).

Algorithm 5 (`PersistenceSSF`):
To obtain the best complexity for this algorithm, a breadth-first strategy must be employed to compute the sets Y as in Algorithm 3 (`PersistenceSymThinning`). Also, an adapted priority queue data structure must be used to represent the set Q. Since the priorities are small integers (step indices), which may be sorted in linear time by counting sort, Q can be managed to perform insertion and min extraction operations in constant time (see [40]). Furthermore, observe that the sets V (lines 14–15) at different iterations of the repeat loop are disjoint and that, for each voxel x, there are at most 26 neighboring voxels y (line 17). Overall, the algorithm may be implemented to run in $O(n)$ complexity.

7.11 CONCLUSION

We presented two generic parallel thinning schemes acting in the cubic 3D grid. The first scheme is asymmetric, with the aim of producing "thin" skeletons. The second one is symmetric and produces skeletons that are uniquely defined and invariant to 90-degree rotations.

We showed how these schemes can be used to produce ultimate curve or surface skeletons of 3D objects made of voxels, based in particular on the notions of 1- and 2-isthmus. However, it should be noticed that any static constraint set and any dynamic skeletal voxel detection rule (for example, an application-dependent one) can be used. In any case, topology preservation is ensured by the properties of critical kernels.

We performed some experiments in order to compare our asymmetric curve skeletonization method with all methods of the same class found in the literature. The results show clearly that our method outperforms the other ones with respect to robustness to noise. All our algorithms can be implemented to run in linear time.

Furthermore, we showed that an effective filtering can be easily performed within our framework thanks to the notion of persistence. In this approach, the filtering is done dynamically with very little added cost and is governed by a unique parameter. Persistence is closely linked to the notion of isthmus, and we stress that this kind of filtering cannot be adapted to the other methods considered in our experiments.

REFERENCES

[1] C. Ronse, Minimal test patterns for connectivity preservation in parallel thinning algorithms for binary digital images, Discrete Appl. Math. 21 (1) (1988) 67–79.

[2] G. Bertrand, On P-simple points, C. R. Acad. Sci., Ser. 1 Math. I (321) (1995) 1077–1084.

[3] G. Bertrand, On critical kernels, C. R. Acad. Sci., Ser. 1 Math. I (345) (2007) 363–367.

[4] G. Bertrand, M. Couprie, On parallel thinning algorithms: Minimal non-simple sets, P-simple points and critical kernels, J. Math. Imaging Vis. 35 (1) (2009) 23–35.

[5] A. Manzanera, T. Bernard, F. Prêteux, B. Longuet, n-dimensional skeletonization: a unified mathematical framework, J. Electron. Imaging 11 (1) (2002) 25–37.

[6] C. Lohou, G. Bertrand, Two symmetrical thinning algorithms for 3D binary images, Pattern Recognit. 40 (2007) 2301–2314.

[7] K. Palágyi, A 3D fully parallel surface-thinning algorithm, Theor. Comput. Sci. 406 (1–2) (2008) 119–135.

[8] Y. Tsao, K. Fu, A parallel thinning algorithm for 3D pictures, Comput. Graph. Image Process. 17 (4) (1981) 315–331.

[9] Y. Tsao, K. Fu, A 3D parallel skeletonwise thinning algorithm pictures, in: Proceedings PRIP 82: IEEE Computer Society Conference on Pattern Recognition and Image Processing, IEEE, 1982, pp. 678–683.

[10] K. Palágyi, A. Kuba, A 3D 6-subiteration thinning algorithm for extracting medial lines, Pattern Recognit. Lett. 19 (7) (1998) 613–627.

[11] C.M. Ma, S.Y. Wan, Parallel thinning algorithms on 3D (18, 6) binary images, Comput. Vis. Image Underst. 80 (2000) 364–378.

[12] C.M. Ma, S.Y. Wan, J.D. Lee, Three-dimensional topology preserving reduction on the 4-subfields, IEEE Trans. Pattern Anal. Mach. Intell. 24 (12) (2002) 1594–1605.

[13] C. Ma, S. Wan, H. Chang, Extracting medial curves on 3D images, Pattern Recognit. Lett. 23 (8) (2002) 895–904.

[14] C. Lohou, G. Bertrand, A 3D 12-subiteration thinning algorithm based on P-simple points, Discrete Appl. Math. 139 (2004) 171–195.

[15] C. Lohou, G. Bertrand, A 3D 6-subiteration curve thinning algorithm based on P-simple points, Discrete Appl. Math. 151 (2005) 198–228.

[16] G. Németh, P. Kardos, K. Palágyi, Topology preserving 2-subfield 3D thinning algorithms, in: Proceedings Signal Processing, Pattern Recognition and Applications (SPPRA 2010), ACTA Press, 2010, pp. 311–316.

[17] G. Németh, P. Kardos, K. Palágyi, Topology preserving 3D thinning algorithms using four and eight subfields, in: A. Campilho, M. Kamel (Eds.), Image Analysis and Recognition, in: Lect. Notes Comput. Sci., vol. 6111, Springer, Berlin/Heidelberg, 2010, pp. 316–325.

[18] G. Bertrand, M. Couprie, Powerful parallel and symmetric 3D thinning schemes based on critical kernels, J. Math. Imaging Vis. 48 (1) (2014) 134–148.

[19] L. Liu, E.W. Chambers, D. Letscher, T. Ju, A simple and robust thinning algorithm on cell complexes, Comput. Graph. Forum 29 (7) (2010) 2253–2260.

[20] G. Bertrand, M. Couprie, Isthmus based parallel and symmetric 3D thinning algorithms, Graph. Models 80 (2015) 1–15.

[21] M. Couprie, G. Bertrand, Asymmetric parallel 3D thinning scheme and algorithms based on isthmuses, in: Special Issue on Skeletonization and Its Application, Pattern Recognit. Lett. 76 (2016) 22–31.

[22] V. Kovalevsky, Finite topology as applied to image analysis, Comput. Vis. Graph. Image Process. 46 (1989) 141–161.

[23] T.Y. Kong, A. Rosenfeld, Digital topology: introduction and survey, Comput. Vis. Graph. Image Process. 48 (1989) 357–393.

[24] T.Y. Kong, Topology-preserving deletion of 1's from 2-, 3- and 4-dimensional binary images, in: Proceedings Discrete Geometry for Computer Imagery, in: Lect. Notes Comput. Sci., vol. 1347, Springer, 1997, pp. 3–18.

[25] G. Bertrand, New notions for discrete topology, in: Proceedings Discrete Geometry for Computer Imagery, in: Lect. Notes Comput. Sci., vol. 1568, Springer, 1999, pp. 218–228.

[26] G. Bertrand, G. Malandain, A new characterization of three-dimensional simple points, Pattern Recognit. Lett. 15 (2) (1994) 169–175.

[27] G. Bertrand, Simple points, topological numbers and geodesic neighborhoods in cubic grids, Pattern Recognit. Lett. 15 (1994) 1003–1011.

[28] P. Saha, B. Chaudhuri, B. Chanda, D. Dutta Majumder, Topology preservation in 3D digital space, Pattern Recognit. 27 (1994) 295–300.

[29] T.Y. Kong, On topology preservation in 2-D and 3-D thinning, Int. J. Pattern Recognit. Artif. Intell. 9 (1995) 813–844.

[30] M. Couprie, G. Bertrand, New characterizations of simple points in 2D, 3D and 4D discrete spaces, IEEE Trans. Pattern Anal. Mach. Intell. 31 (4) (2009) 637–648.

[31] J. Whitehead, Simplicial spaces, nuclei and m-groups, Proc. Lond. Math. Soc. 45 (2) (1939) 243–327.

[32] G. Bertrand, M. Couprie, Two-dimensional thinning algorithms based on critical kernels, J. Math. Imaging Vis. 31 (1) (2008) 35–56.

[33] H. Blum, A transformation for extracting new descriptors of shape, in: W. Wathen-Dunn (Ed.), Models for the Perception of Speech and Visual Form, MIT Press, 1967, pp. 362–380.

[34] A. Lieutier, Any open bounded subset of R^n has the same homotopy type as its medial axis, Comput. Aided Des. 36 (11) (2004) 1029–1046.

[35] E. Zeeman, On the dunce hat, Topology 2 (1964) 341–358.

[36] R. Bing, Some aspects of the topology of 3-manifolds related to the Poincaré conjecture, in: Lectures Modern Math. II, 1964, pp. 93–128.

[37] R. Malgouyres, A. Francés, Deciding whether a simplicial 3-complex collapses to a 1-complex is NP-complete, in: Proceedings Discrete Geometry for Computer Imagery, in: Lect. Notes Comput. Sci., vol. 4992, Springer, 2008, pp. 177–188.

[38] D. Shaked, A.M. Bruckstein, Pruning medial axes, Comput. Vis. Image Underst. 69 (2) (1998) 156–169.

[39] J. Chaussard, Topological Tools for Discrete Shape Analysis, Ph.D. dissertation, Université Paris-Est, 2010.

[40] M. Thorup, Equivalence between priority queues and sorting, in: Proceedings 43rd Symposium on Foundations of Computer Science. FOCS '02, IEEE Computer Society, Washington, DC, USA, 2002, pp. 125–134.

CHAPTER

Critical kernels, minimal nonsimple sets, and hereditarily simple sets in binary images on n-dimensional polytopal complexes

T.Y. Kong
Department of Computer Science, Queens College, CUNY, Flushing, NY, USA

Contents

8.1 Introduction 212
8.1.1 Background and Prior Work 214
8.1.2 Contributions of This Chapter 217
8.2 Preliminaries: Complexes, Polyhedra, and Binary Images 218
8.2.1 Convex Polytopes 218
8.2.2 Complexes and Pure Complexes 219
8.2.3 Polyhedra 221
8.2.4 Binary Images on Pure Polytopal Complexes 222
8.3 Nonempty Topologically Simple (NETS) Polyhedra 222
8.3.1 Properties of NETS Polyhedra 223
8.3.2 The Axioms That Specify NETS Polyhedra 224
8.3.3 In a Strongly Normal Collection of Nonempty Convex Polytopes, the Union of Any Element with Its Neighbors Is NETS 225
8.3.4 Possible Definitions of NETS Polyhedra Using Homology Groups ... 226
8.4 Consequences of the NETS Axioms 228
8.4.1 A 0- or 1-Dimensional Complex Is NETS Just If It Is a Tree 229
8.4.2 Collapsible Polyhedra Are NETS, and NETS Polyhedra Never Collapse to Non-NETS Polyhedra 231
8.4.3 A Polyhedron in the Plane or in the Boundary of a 3-Dimensional Convex Polytope Is NETS If and Only If It Is Collapsible 232
8.5 Simple, Hereditarily Simple, and Minimal Nonsimple Sets 233
8.5.1 Simple 1s, Attachment Complexes, and Attachment Sets 233

Skeletonization. DOI: 10.1016/B978-0-08-101291-8.00009-2

8.5.2 Definitions and Basic Properties of Simple, Hereditarily Simple, and Minimal Nonsimple Sets.... 235
8.6 Cliques, Cores, and the Critical Kernel.... 236
8.6.1 $\mathbb{I}$-Induced Cliques and $\mathbb{I}$-Essential Cells.... 237
8.6.2 $\mathbb{I}$-Cores.... 238
8.6.3 $\mathbb{I}$-Regular and $\mathbb{I}$-Critical Cells; the Critical Kernel of $\mathbb{I}$.... 239
8.7 Characterizations of P-Simple 1s, Hereditarily Simple Sets, and Minimal Nonsimple Sets.... 240
8.8 A Proof of Theorem 8.5.5 and a Second Proof of the "If" Part of Theorem 8.7.1 Based on NETS-Collapsing.... 245
8.8.1 NETS-Collapsing and a Proof of Theorem 8.5.5.... 245
8.8.2 $\mathbb{I}$-Essential Complexes and Another Proof of the "If" Part of Theorem 8.7.1.... 251
8.9 Concluding Remarks.... 253
References.... 254

8.1 INTRODUCTION

Given a binary image $\mathbb{I} : \mathsf{V} \to \{0, 1\}$, we call an element v of V a *1* of $\mathbb{I}$ if $\mathbb{I}(\mathsf{v}) = 1$. If D is any set of 1s of $\mathbb{I}$, then we can *delete* D (from the 1s of $\mathbb{I}$) by changing the value of $\mathbb{I}$ at each element of D from 1 to 0. For example, each iteration of a parallel thinning algorithm will delete a certain set of 1s of the binary image that results from previous iterations.

A *simple* set of 1s has the property that key topological attributes of the binary image are preserved when that entire set of 1s is deleted. (A mathematically precise definition of a simple set will be given in Subsection 8.5.2.) The most fundamental example of a simple set is a simple singleton set. If v is a 1 such that the singleton set {v} is a simple set of the binary image, then we call v a simple 1 of the image. We note, however, that a set of 1s may be nonsimple even if each of its elements is a simple 1.

Thinning algorithms are expected to satisfy the condition that the set of 1s they delete from an input binary image is always a simple set of that image. This is to ensure that the algorithm is topologically sound—i.e., that it always preserves key topological properties of the input image.

Unfortunately, it can be quite hard to verify that a proposed parallel thinning algorithm is indeed topologically sound in this sense. There is a systematic and fairly general method of doing this, a method of Ronse [41] based on the concept of a *minimal nonsimple set*. Although this method is mathematically correct (i.e., it always gives correct results when used with sufficient care), in practice it can be tedious and quite error-prone when applied to parallel thinning algorithms for three-dimensional images. Ronse's method would be even more difficult to apply to higher-dimensional thinning algorithms.

However, there is a newer approach to designing parallel thinning algorithms that entirely avoids the need to verify the topological soundness of the developed algorithms. This approach is due to Bertrand and Couprie [7,9,10,15], and is based on Bertrand's concept of the *critical kernel* of a binary image [3]. (The critical kernel of a binary image consists of all those 1s that are *not* simple, together with certain lower-dimensional cells.) The idea of this approach is to design each iteration of the parallel thinning algorithm so that, for each cell e of the critical kernel of that iteration's input binary image, not all of the 1s of the image that contain e are deleted (i.e., at least one 1 that contains e is preserved at that iteration). If every iteration of a parallel thinning algorithm satisfies this condition, then the set of 1s deleted at each iteration is guaranteed to be simple, and so the algorithm is guaranteed to be topologically sound. For binary images on 2-, 3-, and 4-dimensional Cartesian arrays, Bertrand and Couprie's paper [8] established key facts about the critical kernel and the relationship between its elements and the minimal nonsimple sets of the image.

The above-mentioned work raises the question of how the concept of critical kernel can best be generalized to images of arbitrary dimension. We note here that Bertrand's definition of the critical kernel and the above-mentioned approach to developing parallel thinning algorithms that are guaranteed to be topologically sound is mathematically valid in any number of dimensions. However, although Bertrand's critical kernel can be efficiently found in 2D and 3D images, and in 4D images on a Cartesian array [14], for images of dimension greater than 4 the author is not aware of any algorithm that has been shown to find Bertrand's critical kernel with reasonable worst-case efficiency.

Moreover, the mathematical arguments used by Bertrand and Couprie [8] to establish fundamental properties of Bertrand's critical kernels of images on 2-, 3-, and 4-dimensional Cartesian arrays do not generalize to images of arbitrary dimension. This is because their arguments depend on "confluence" properties[1] of collapsing that hold for all subcomplexes of the face complexes of 2-, 3-, and 4-cubes [14], but which do not hold for all subcomplexes of face complexes of cubes of arbitrary dimension.

For these reasons, we will use new definitions of critical kernel, which we refer to as "NETS definitions" because they are based on our concept of a "NETS polyhedron" that will be axiomatically defined in Subsection 8.3.2.

In 2D and 3D images, our NETS definitions of the critical kernel are mathematically equivalent to Bertrand's definition of the critical kernel, and the author is convinced that this is also the case in images on 4D Cartesian arrays. However, our critical kernels may be proper subsets of Bertrand's critical kernels in images of high dimension, as a result of which thinning algorithms based on our critical kernels may not always preserve all homotopy properties of a high-dimensional image. On the other hand, using a homology-based NETS definition of critical kernel (see Subsection 8.3.4) instead of Bertrand's definition has the computational advantage that

[1]These confluence properties are stated as Theorems 5 and 6 in [8].

critical kernels of images of arbitrarily high dimension can be found using homology group computation algorithms such as those in [25, Ch. 3].

In this chapter, we show how the main results established in [8] for Bertrand's critical kernels and minimal nonsimple sets in images on Cartesian arrays of dimension ≤ 4 can be generalized to images of any dimension when we use a NETS definition of critical kernel.

This work has potential applications to the design of topologically sound parallel thinning algorithms for 5- and higher-dimensional images, because of the above-mentioned computational advantage of using a homology-based NETS definition of critical kernel instead of Bertrand's definition. In addition, the results established in this chapter advance our understanding of critical kernels in images on non-Cartesian complexes (such as 3D images on face-centered cubic or body-centered cubic grids, when we identify each grid point with its Voronoi neighborhood in $\mathbb{R}^3$) since images on non-Cartesian complexes were not considered in [8].

Although almost all reported applications of image thinning have used 2D or 3D images on Cartesian image arrays, over the years there has been some work relating to thinning of images of high dimension (see, e.g., [23,24,27,31,34,38,47]) and images on 3D face-centered and body-centered cubic grids (see, e.g., [13,18,52] and [53, Sec. 6.5]). In particular, [24] describes how higher-dimensional image thinning may possibly be applied to robot path planning. Here the number of dimensions is the number of parameters needed to specify the dispositions (positions, orientations, etc.) of a collection of moving robots.

8.1.1 BACKGROUND AND PRIOR WORK

The concept of a simple 1 has a fairly long history. It was introduced in the late 1960s by Rosenfeld [42] for images on a 2-dimensional Cartesian array. (Simple 1s are called "deletable" in [42] and some other works.) The concept was extended in the early 1980s to 3-dimensional images [37,54], and later to 4-dimensional images [27,29]. Although simple 1s have been characterized in different ways, there now seems to be no disagreement as to just which 1s of a binary image on a 2-, 3-, or 4-dimensional Cartesian array are simple 1s.

A *hereditarily simple* set of 1s is a simple set C of 1s with the property that *every* proper subset of C is also a simple set [27,28]. Thus, for every subset D of a hereditarily simple set C, key topological properties of the binary image are preserved when we delete all the 1s in D. It is evident that any subset of a hereditarily simple set is itself a hereditarily simple set.

To see why hereditarily simple sets are of interest, consider an iteration of a parallel thinning algorithm that first identifies a set C of 1s as candidates for deletion and then selects and deletes just those 1s of C that satisfy some (possibly empty) set of additional constraints. Regardless of what the additional constraints (if any) are, key topological properties of the image will be preserved when the selected 1s are deleted if C is hereditarily simple.

The concept of a hereditarily simple set is closely related to Bertrand's concept of a *P-simple* point [2]. Using terminology that is more consistent with [8], a *P-simple* element for a set C of 1s in a binary image $\mathbb{I}$ is an element c of C such that, for *every* subset D of C \ {c}, p is a simple 1 of the binary image we get from $\mathbb{I}$ by deleting the set D.

For *any* set C of 1s of $\mathbb{I}$, the subset of C consisting of all those elements of C that are P-simple for C is a hereditarily simple set of $\mathbb{I}$. (This is an easy consequence of the precise definition of a simple set.) So, assuming that we are able to efficiently identify the 1s that are P-simple for a set C, we can efficiently identify a hereditarily simple subset of any given set C of 1s.

A *minimal nonsimple* set is a set M of 1s that is not a simple set but has the property that every proper subset of M is a simple set. In this chapter, binary images are assumed to have finite domains, which implies that the set of all 1s is finite and hence that every nonsimple set contains at least one minimal nonsimple set. It follows that a set of 1s is hereditarily simple just if it contains no minimal nonsimple set.

The concept of a minimal nonsimple set was introduced by Ronse [41] in the late 1980s for binary images on a 2-dimensional Cartesian array. For such images, his work determined the forms of all possible minimal nonsimple sets. This provided the basis for a systematic method of verifying conclusively that the set of 1s deleted at each iteration of any given parallel thinning algorithm is always a hereditarily simple set, so that the thinning algorithm is topologically sound: The method consists of verifying, for each of the possible forms of a minimal nonsimple set, that it is impossible for a minimal nonsimple set of that form to be deleted at some iteration of the thinning algorithm. We will refer to this as the *Ronse method* of verifying the topological soundness of a given parallel thinning algorithm.

There is now a significant literature on the above-mentioned topics. For binary images on a 3-dimensional Cartesian array, Bertrand's 1994 paper [2] gave convenient local characterizations of the 1s that are P-simple for any set C of 1s. (As we have seen, this enables us to efficiently identify a hereditarily simple subset of C.) We mentioned above that Ronse's paper [41] (which made use of results from [40]) determined the forms of all possible minimal nonsimple sets in binary images on 2-dimensional Cartesian arrays. In the 1990s and early 2000s, Hall, Ma, Gau, and the author determined the forms of all possible minimal nonsimple sets in binary images on a 2-dimensional hexagonal array [20] (implicitly), a 3-dimensional Cartesian array [28,35], a 3-dimensional face-centered cubic array [18], and a 4-dimensional Cartesian array [19,32]. The author later solved the problem more generally [30,31]: Specifically, for $n = 2, 3$, or 4, he determined the forms of all possible minimal nonsimple sets in binary images $\mathbb{I} : \mathsf{V} \to \{0, 1\}$ where V is the set of n-cells of any n-dimensional pure polytopal complex that satisfies certain mild conditions [31]. For such 2-, 3-, and 4-dimensional binary images, these mathematical results enable us to verify the topological soundness of parallel thinning algorithms using Ronse's method.

But, for images of dimension greater than 2, Ronse's method is likely to require consideration of many cases, and it is rather easy to overlook cases in which a minimal nonsimple set is deleted by the algorithm. When the latter happens, we may (mistakenly) conclude that the thinning algorithm is topologically sound even though it actually is not.

As mentioned before, Bertrand and Couprie have pioneered a new approach to designing parallel thinning algorithms that is based on Bertrand's concept of *critical kernel*. This approach avoids the difficulties referred to in the previous paragraph because it develops parallel thinning algorithms that are guaranteed to be topologically sound.

Critical kernels are defined for cell complexes, but the concept is easily applied to binary images. This is because the domain of a binary image on an n-dimensional Cartesian array can be thought of as the set of n-cells of an n-dimensional cubical complex, and then the 1s of the image (each of which is now an n-cell of the complex) induce a subcomplex of this cubical complex. Let us call the subcomplex induced by the 1s of a binary image $\mathbb{I}$ the *foreground complex* of $\mathbb{I}$, so that the 1s are exactly the n-cells of the foreground complex. Then the critical kernel of $\mathbb{I}$'s foreground complex can be thought of as the critical kernel of the binary image $\mathbb{I}$. These remarks also apply to binary images on non-Cartesian arrays such as the 2-dimensional hexagonal and the 3-dimensional face-centered cubic and body-centered cubic arrays, except that the cells of the complexes we use have different shapes for different types of arrays.

The definition of Bertrand's critical kernel is based on a clever generalization of the well-established concept of a nonsimple 1 (in this context, a nonsimple n-cell of the n-dimensional foreground complex) to the concept of a *critical essential cell*. Here an *essential cell* is a cell of the foreground complex that either is a 1 of the image or is the intersection of two or more 1s. (Any nonempty intersection of two or more distinct 1s will be a k-cell of the foreground complex for some $k < n$.)

Just as a nonsimple 1 is a 1 whose detachment[2] from the foreground complex fails to preserve key topological properties, a critical essential cell is an essential cell e such that, if we write $\mathbf{F}(e)$ to denote the complex induced by e itself and all the 1s that do *not* contain e, then detachment of e from $\mathbf{F}(e)$ fails to preserve key topological properties. (Note that if e is a 1 of the binary image, then e is critical just if e is a nonsimple 1 because $\mathbf{F}(e)$ is just the foreground complex itself in that case.)

The *critical kernel* is the complex induced by the critical essential cells; it contains, for example, all the nonsimple 1s and all the essential 0-cells, as these cells are always critical. It does not contain any simple 1.

For binary images on 2-, 3-, and 4-dimensional Cartesian arrays, Bertrand and Couprie's 2009 paper [8] presents elegant and useful characterizations of hereditarily

[2]Detachment of a 1 from the foreground complex is the operation that removes the 1 and also removes each of its proper faces that is not a face of another 1. Similarly, detachment of the essential cell e from the complex $\mathbf{F}(e)$ defined in this paragraph is the operation that removes the cell e and also removes each of e's proper faces that is not a face of a 1 in $\mathbf{F}(e) \setminus \{e\}$. These are special cases of Bertrand's detachment operation [3].

simple sets, P-simple 1s, and minimal nonsimple sets in terms of critical kernels. This paper represents the culmination of much earlier work on "topology-preserving" parallel deletion of 1s from binary images.

It follows from one of Bertrand and Couprie's results [8, Prop. 22] that, in binary images on 2D, 3D, and 4D Cartesian arrays, a set C of 1s is hereditarily simple if and only if (i) every 1 in C is simple and (ii) each (inclusion-)maximal cell of the critical kernel is contained in at least one 1 that is not in C. (Since any nonsimple 1 is a maximal cell of the critical kernel, (ii) actually implies (i).) Thus an iteration of a parallel thinning algorithm is guaranteed to be topologically sound if, for each maximal cell e of the critical kernel of the current image, at least one 1 that contains e is preserved by that iteration.

In this subsection, we have given an overview of prior work that is particularly relevant to the rest of this chapter. Much other work relating to "topology preserving" deletion of 1s from binary images is discussed by Saha, Borgefors, and Sanniti Di Baja in Sections 3 and 4 of their recent survey article [44].

8.1.2 CONTRIBUTIONS OF THIS CHAPTER

This chapter generalizes the main theorems of Bertrand and Couprie's paper [8] to binary images on pure polytopal complexes of any dimension. However, for reasons given above, we will not define critical kernels in exactly the same way as Bertrand: Whereas Bertrand defines a critical essential cell to be an essential cell that does not collapse to its core, in this chapter a critical essential cell is defined to be an essential cell e for which the union of e's core is not a *NETS polyhedron*.

NETS polyhedra are not uniquely defined but are specified by four axioms (the axioms **NETS-1–NETS-4** in Subsection 8.3.2) that NETS polyhedra must satisfy: These four NETS axioms are the only properties of NETS polyhedra that are assumed in the proofs of our theorems. Any class of polyhedra that satisfies the NETS axioms will be a valid definition of NETS polyhedra, and any valid definition of NETS polyhedra will give a definition of critical kernel for which the arguments and results of this chapter are valid. Thus our results are proved for a *class* of definitions of critical kernel; any definition in this class will be called a *NETS definition* of critical kernel.

For binary images on 2-, 3-, and 4-dimensional Cartesian array complexes, Bertrand's definition of critical kernel is equivalent to a NETS definition, as we will see in Subsection 8.3.4. Moreover, in the 2- and 3-dimensional cases, it will follow from a result established in Subsection 8.4.3 that there is essentially no other NETS definition of critical kernel (i.e., there is no NETS definition that is inequivalent to Bertrand's definition). The author is convinced that this is also the case for binary images on a 4-dimensional Cartesian array complex, though he does not have a complete proof.

8.2 PRELIMINARIES: COMPLEXES, POLYHEDRA, AND BINARY IMAGES

8.2.1 CONVEX POLYTOPES

A *convex polytope* is a set that is the convex hull of a finite set of points in Euclidean n-space $\mathbb{R}^n$ (for some positive integer n). It can be shown [55] that the intersection of any nonempty finite collection of convex polytopes is a convex polytope. Although the empty set is a convex polytope, we will frequently exclude the empty set by referring to nonempty convex polytopes. For example, the definition of a complex in Subsection 8.2.2 will require that each element of a complex be a nonempty convex polytope.

For any convex polytope P that contains more than one point, we write $\dim(P)$ to denote the least integer n such that $\mathbb{R}^n$ contains a convex polytope that is congruent to P. For a convex polytope P that contains just one point, we define $\dim(P) = 0$. We define $\dim(\emptyset) = -1$. A convex polytope P such that $\dim(P) = k$ is said to be *k-dimensional*. Thus a 0-dimensional convex polytope consists of just a single point, a 1-dimensional convex polytope is a closed straight line segment of positive length, and a 2-dimensional convex polytope is a convex polygon with three or more vertices.

Any convex polytope P such that $\dim(P) \geq 1$ has *proper faces* (which will be defined in the sequel). We will see that each proper face of P is a *nonempty* convex polytope that is strictly contained in P. In fact, P has k-dimensional proper faces for every integer k such that $0 \leq k \leq \dim(P) - 1$. Before giving a precise definition of a proper face, we give some examples to illustrate this concept:

Example 8.2.1. A cube is a 3-dimensional convex polytope that has 26 proper faces: It has 6 different 2-dimensional proper faces (each of which is a square), 12 different 1-dimensional proper faces (each of which is an edge of the cube), and 8 different 0-dimensional proper faces (each consisting of a vertex of the cube).

If P is any convex polytope in $\mathbb{R}^n$ (so that $\dim(P) \leq n$), then the *proper faces* of P are the nonempty convex polytopes F such that $F = P \cap H \neq P$ for some $(n-1)$-dimensional supporting hyperplane[3] of P. If P is a nonempty convex polytope and either $Q = P$ or Q is a proper face of P, then we say that Q is a *face* of P. In this chapter, we do *not* consider the empty set to be a face of a convex polytope.

We write $\mathbf{Faces}(P)$ to denote the set of all the faces of a convex polytope P. Thus $\mathbf{Faces}(P)$ is a set of nonempty convex polytopes each of which is a subset of P, $P \in \mathbf{Faces}(P)$, $\emptyset \notin \mathbf{Faces}(P)$, and the set of all the proper faces of P is $\mathbf{Faces}(P) \setminus \{P\}$.

If F is a face of a convex polytope P such that $\dim(F) = k$, then we say F is a *k-face* of P; it can be shown [55] that if $0 < k < \dim(P)$, then each $(k-1)$-face of P is a face of more than one k-face of P.

[3]Let H be an $(n-1)$-dimensional hyperplane of $\mathbb{R}^n$, and let H^+ and H^- be the two closed half-spaces of $\mathbb{R}^n$ that are bounded by H. Then H is said to be a *supporting hyperplane* of P if $H \cap P \neq \emptyset$ and either $P \subset H^+$ or $P \subset H^-$.

The union of all the proper faces of a convex polytope P is called the *boundary* of P; this set is also the union of all the $(\dim(P) - 1)$-faces of P. By the *interior* of a convex polytope P we mean the set of all those points of P that do not lie on the boundary of P. It can be shown that the boundary of a nonempty convex polytope P is a proper subset of P, and that it is empty if (and only if) P is 0-dimensional. Equivalently, the interior of a nonempty convex polytope P is nonempty, and it is P itself if (and only if) P is 0-dimensional.

8.2.2 COMPLEXES AND PURE COMPLEXES

For any positive integer n, a *complex* (also called a *polytopal complex*) is a possibly empty finite collection $\mathbf{K}$ of nonempty convex polytopes in $\mathbb{R}^n$ that satisfies the following two conditions:

P1: If $P \in \mathbf{K}$, then $\mathbf{Faces}(P) \subseteq \mathbf{K}$.
P2: If $P_1, P_2 \in \mathbf{K}$ and $P_1 \cap P_2 \neq \emptyset$, then $P_1 \cap P_2 \in \mathbf{Faces}(P_1) \cap \mathbf{Faces}(P_2)$.

It follows from **P1** and **P2** that any nonempty finite intersection of members of $\mathbf{K}$ is a member of $\mathbf{K}$. Moreover, the following is an easy consequence of **P2** and the definition of a face of a convex polytope:

P2a: If $P_1, P_2 \in \mathbf{K}$, then $P_1 \subseteq P_2$ if and only if $P_1 \in \mathbf{Faces}(P_2)$.

Now if P_1 and P_2 are any convex polytopes such that $P_1 \cap P_2 \neq \emptyset$, then each member of $\mathbf{Faces}(P_1 \cap P_2)$ is a subset both of P_1 and of P_2, and each member of $\mathbf{Faces}(P_1) \cap \mathbf{Faces}(P_2)$ is a subset of $P_1 \cap P_2$. So we can deduce the following from **P2a**, **P1**, and **P2**:

P2b: If $P_1, P_2 \in \mathbf{K}$ and $P_1 \cap P_2 \neq \emptyset$, then $\mathbf{Faces}(P_1 \cap P_2) = \mathbf{Faces}(P_1) \cap \mathbf{Faces}(P_2)$.

A *subcomplex* of a complex $\mathbf{K}$ is a complex $\mathbf{L}$ such that $\mathbf{L} \subseteq \mathbf{K}$. A *proper subcomplex* of a complex $\mathbf{K}$ is a complex $\mathbf{L}$ such that $\mathbf{L} \subsetneq \mathbf{K}$. Readily, if $\mathbf{A}$ and $\mathbf{B}$ are any two complexes, then $\mathbf{A} \cap \mathbf{B}$ is a subcomplex of each of $\mathbf{A}$ and $\mathbf{B}$. It is also easy to verify that if each of $\mathbf{A}$ and $\mathbf{B}$ is a subcomplex of a complex $\mathbf{C}$, then $\mathbf{A} \cup \mathbf{B}$ is also a subcomplex of $\mathbf{C}$.

Since any face of a face of a convex polytope P is a subset of P, we see from **P2a** and **P1** that $\mathbf{Faces}(P)$ is a subcomplex of every complex that contains a convex polytope P. It can in fact be shown that, for any convex polytope P, $\mathbf{Faces}(P)$ is a complex [55].

Each element of a complex $\mathbf{K}$ will be called a *cell* of $\mathbf{K}$; an m-dimensional cell of $\mathbf{K}$ will be called an *m-cell* of $\mathbf{K}$. If $\mathbf{K} \neq \emptyset$, then $\dim(\mathbf{K})$ will denote the integer $\max_{P \in \mathbf{K}} \dim(P)$; we define $\dim(\emptyset) = -1$. We say that $\mathbf{K}$ is *k-dimensional* if $\dim(\mathbf{K}) = k$. So, for any integer $k \geq 0$, if a complex $\mathbf{K}$ is k-dimensional, then $\mathbf{K}$ contains at least one k-cell but does not contain a $(k+1)$-cell.

Let C be any set of cells of a complex $\mathbf{K}$. Then the set $\bigcup\{\mathbf{Faces}(P) \mid P \in C\}$, which consists of the cells in C and all their proper faces, is a subcomplex of $\mathbf{K}$. This complex is called the complex *induced by* the set $\mathcal{C}$ of cells. It is evidently the smallest complex that contains all the cells in $\mathcal{C}$.

The *Euler characteristic* of an n-dimensional complex $\mathbf{K}$ is the integer $\sum_{i=0}^{n}(-1)^n c(i, \mathbf{K})$, where $c(i, \mathbf{K})$ denotes the number of i-cells in $\mathbf{K}$. For example, if P is a cube and $\mathbf{K}$ is the 2-dimensional complex $\mathbf{Faces}(P) \setminus \{P\}$, then (as we saw in Example 8.2.1) $c(0, \mathbf{K}) = 8$, $c(1, \mathbf{K}) = 12$, and $c(2, \mathbf{K}) = 6$, and so the Euler characteristic of $\mathbf{K}$ is $8 - 12 + 6 = 2$. If, instead, $\mathbf{K}$ is the 3-dimensional complex $\mathbf{Faces}(P)$, then $c(0, \mathbf{K}) = 8$, $c(1, \mathbf{K}) = 12$, $c(2, \mathbf{K}) = 6$, and $c(3, \mathbf{K}) = 1$, and so the Euler characteristic of $\mathbf{K}$ is $8 - 12 + 6 - 1 = 1$.

A *facet* of a complex $\mathbf{K}$ is an inclusion-maximal cell of $\mathbf{K}$; in other words, a facet of $\mathbf{K}$ is a cell of $\mathbf{K}$ that is not a face of any other cell of $\mathbf{K}$. Evidently, every n-cell of an n-dimensional complex $\mathbf{K}$ is a facet of $\mathbf{K}$, but $\mathbf{K}$ might have other facets too. We write Facets($\mathbf{K}$) to denote the set of all the facets of a complex $\mathbf{K}$. A complex $\mathbf{K}$ is uniquely determined by Facets($\mathbf{K}$); indeed, $\mathbf{K}$ is the complex induced by Facets($\mathbf{K}$).

An n-dimensional complex $\mathbf{K}$ is said to be *pure* if every facet of $\mathbf{K}$ is an n-cell or, equivalently, if every cell of $\mathbf{K}$ is a face of an n-cell of $\mathbf{K}$. Example 8.2.2 defines a simple and important class of pure complexes.

Example 8.2.2 (Cartesian Array Complexes). For any positive integers $m_1, \ldots, m_n$, there is a pure n-dimensional complex whose facets are the $m_1 \times \cdots \times m_n$ closed n-cubes $[k_1, k_1 + 1] \times \cdots \times [k_n, k_n + 1]$ in which each k_i $(1 \leq i \leq n)$ is one of the integers $0, \ldots, m_i - 1$. In this chapter, such a complex will be called a *Cartesian array complex*.

A facet $[k_1, k_1 + 1] \times \cdots \times [k_n, k_n + 1]$ of the complex specified in the previous paragraph will be called a *border facet* of the complex if there is some i in $\{1, \ldots, n\}$ for which $k_i = 0$ or $k_i + 1 = m_i$.

Many of the results in this chapter can be understood as generalizations to arbitrary pure complexes of results (about minimal nonsimple sets, P-simple 1s, and critical kernels) that were established by Bertrand and Couprie [8] for 2-, 3-, or 4-dimensional Cartesian array complexes.

Now let $\mathbf{K}$ be any pure n-dimensional complex. Then two facets of $\mathbf{K}$ are said to be *weakly adjacent*, and each is said to be a *weak neighbor* of the other, if the two facets are distinct but share a proper face.[4] A set T of facets of $\mathbf{K}$ is said to be *weakly connected* if, for all distinct elements t and t' of T, there exists a sequence $\mathsf{t}_1, \ldots, \mathsf{t}_k$ of elements of T such that $\mathsf{t}_1 = \mathsf{t}$, $\mathsf{t}_k = \mathsf{t}'$, and t_i is weakly adjacent to t_{i+1} for $1 \leq i \leq k - 1$. Such a sequence $\mathsf{t}_1, \ldots, \mathsf{t}_k$ is called a *weak path* in T from t to t'. A *weak component* of a nonempty set of facets of $\mathbf{K}$ is a maximal weakly connected subset of that set.

[4]Another natural adjacency relation on Facets($\mathbf{K}$) is the *strong adjacency* relation of, e.g., [31]: Two facets of a pure n-dimensional complex $\mathbf{K}$ are strongly adjacent just if they are distinct but share an $(n - 1)$-dimensional face. However, this adjacency relation will not be used in the present chapter.

Example 8.2.3. If **K** is a 2-dimensional Cartesian array complex, then the weak neighbors of a facet of **K** are just the 8-neighbors of that facet. So if T is a set of facets of **K**, then the weak components of T are the 8-components of T.

Example 8.2.4. If **K** is a 3-dimensional Cartesian array complex, then the weak neighbors of a facet of **K** are just the 26-neighbors of that facet. So if T is a set of facets of **K**, then the weak components of T are the 26-components of T.

Example 8.2.5. Let **K** be an n-dimensional Cartesian array complex for any positive integer n. Then two facets of **K** are weakly adjacent just if they are 0-adjacent in the terminology of, e.g., Klette and Rosenfeld's text [26] (though [26] defines this terminology only in the cases $n = 2$ and $n = 3$). In the terminology of Herman's text [22], two facets of **K** are weakly adjacent just if their centers are α_n-adjacent.

8.2.3 POLYHEDRA[5]

The *polyhedron* of a complex **K**, also called the *underlying set* of **K**, is the set $\bigcup \mathbf{K} = \bigcup \text{Facets}(\mathbf{K})$. This is the set of all points that lie in at least one cell of **K** (and therefore lie in at least one facet of **K**). We say that a complex **K** is *connected* if its polyhedron is connected. Readily, each point in the polyhedron of a complex **K** lies in the interior of exactly one cell of **K**, and that cell is a proper face of any other cell of **K** that contains the point in question.

We use the term *polyhedron* to mean a set that is the polyhedron of some complex. Note that the empty set is a polyhedron. It can be shown that a subset of $\mathbb{R}^n$ is a polyhedron if and only if it is the union of a finite set of convex polytopes. It follows that if each of S and T is a polyhedron, then $S \cap T$ and $S \cup T$ are also polyhedra.

A polyhedron is said to be k*-dimensional* if it is the polyhedron of a k-dimensional complex. (If S is a polyhedron that is not a finite set of isolated points, then there will be different complexes **K** such that S is the polyhedron $\bigcup \mathbf{K}$ of **K**, but dim(**K**) will have the same value for all such complexes **K**.)

The Euler characteristic of a polyhedron P is defined to be the Euler characteristic of any complex whose polyhedron in P. (It follows from the Euler–Poincaré formula [17, p. 85] that two complexes that have the same polyhedron will also have the same Euler characteristic.)

We end this subsection with a list of facts about the polyhedra of complexes; these facts will be used in our mathematical arguments. Let **A** and **B** be any two subcomplexes of some complex **C**, let f be any cell of **C**, and let p be any point in the interior of the cell f. Bearing in mind that each point in the polyhedron $\bigcup \mathbf{K}$ of a complex **K** lies in the interior of exactly one cell of **K**, it is readily confirmed that the four polyhedra $\bigcup \mathbf{A}$, $\bigcup \mathbf{B}$, $\bigcup(\mathbf{A} \cap \mathbf{B})$, and $\bigcup(\mathbf{A} \cup \mathbf{B})$ have the following properties:

$$f \in \mathbf{A} \iff \mathbf{Faces}(f) \subseteq \mathbf{A} \iff f \subseteq \bigcup \mathbf{A} \iff p \in \bigcup \mathbf{A}, \tag{8.1}$$

[5] The term *polyhedron* has more than one meaning in the literature. In this chapter, it is used in the sense of, e.g., [17], and not in the sense of, e.g., [55].

$$\mathbf{A} \subseteq \mathbf{B} \iff \bigcup \mathbf{A} \subseteq \bigcup \mathbf{B}, \tag{8.2}$$

$$\mathbf{A} = \mathbf{B} \iff \bigcup \mathbf{A} = \bigcup \mathbf{B}, \tag{8.3}$$

$$(\bigcup \mathbf{A}) \cap (\bigcup \mathbf{B}) = \bigcup (\mathbf{A} \cap \mathbf{B}), \tag{8.4}$$

$$(\bigcup \mathbf{A}) \cup (\bigcup \mathbf{B}) = \bigcup (\mathbf{A} \cup \mathbf{B}). \tag{8.5}$$

Whereas (8.1)–(8.4) need not be true if **A** and **B** are not complexes or if they are not subcomplexes of the same complex, it is easy to see that (8.5) is true for arbitrary collections **A** and **B** of sets.

The reader may have noticed that in (8.1) a convex polytope is denoted by a lowercase letter (namely f). We will use lowercase letters to denote convex polytopes that are being thought of as cells of complexes. Other polyhedra will usually be denoted by uppercase letters such as S.

8.2.4 BINARY IMAGES ON PURE POLYTOPAL COMPLEXES

Let **K** be a pure complex. Then any function $\mathbb{I}$: Facets(**K**) $\rightarrow$ $\{0, 1\}$ will be called a *binary image* on **K**. We say that such a binary image $\mathbb{I}$ is n-dimensional if **K** is n-dimensional; correspondingly, we define $\dim(\mathbb{I}) = \dim(\mathbf{K})$. (Although it would be possible to apply these definitions to complexes **K** that are not pure, we will not do so in this chapter.)

Let $\mathbb{I}$ be any binary image on the pure complex **K**. Then a facet v of **K** will be called a 1 of $\mathbb{I}$ if $\mathbb{I}(\mathsf{v}) = 1$. Thus $\mathbb{I}^{-1}[1]$ is the set of all 1s of $\mathbb{I}$. If $\mathsf{D} \subseteq \mathbb{I}^{-1}[1]$ (i.e., if D is any set of 1s of $\mathbb{I}$), then we write $\mathbb{I} - \mathsf{D}$ to denote the binary image on **K** whose set of 1s is $\mathbb{I}^{-1}[1] \setminus \mathsf{D}$. So, for all facets v of **K**, we have that

$$(\mathbb{I} - \mathsf{D})(\mathsf{v}) = \begin{cases} 1 & \text{if } \mathbb{I}(\mathsf{v}) = 1 \text{ and } \mathsf{v} \notin \mathsf{D}, \\ 0 & \text{if } \mathbb{I}(\mathsf{v}) = 0 \text{ or } \mathsf{v} \in \mathsf{D}. \end{cases}$$

If d is any 1 of $\mathbb{I}$, then $\mathbb{I} - \mathsf{d}$ will mean $\mathbb{I} - \{\mathsf{d}\}$. The subcomplex of **K** that is induced by $\mathbb{I}^{-1}[1]$ (i.e., the complex $\bigcup\{\mathbf{Faces}(\mathsf{d}) \mid \mathsf{d} \in \mathbb{I}^{-1}[1]\}$) will be called the *foreground complex* of the binary image $\mathbb{I}$. The polyhedron $\bigcup \mathbb{I}^{-1}[1]$ of $\mathbb{I}$'s foreground complex will be called the *foreground polyhedron* of $\mathbb{I}$.

8.3 NONEMPTY TOPOLOGICALLY SIMPLE (NETS) POLYHEDRA

In this section, we introduce the concept of a *nonempty topologically simple (NETS)* polyhedron, which is of great importance in this chapter because the definitions of a simple 1 and the critical kernel of a binary image will be based on this concept. For example, if v is a 1 of a binary image $\mathbb{I}$, then we will see in Subsection 8.5.1 that v is called a *simple* 1 of $\mathbb{I}$ just if v's $\mathbb{I}$-attachment set (which is the intersection of the convex polytope v with the union of all the other 1s of $\mathbb{I}$) is a NETS polyhedron.

8.3.1 PROPERTIES OF NETS POLYHEDRA

We will not adopt any single definition of NETS polyhedra. Instead, we will state four axioms, the axioms **NETS-1–NETS-4** in Subsection 8.3.2, that NETS polyhedra must satisfy. These axioms are the *only* properties of NETS polyhedra that will be used in the proofs of our theorems. Thus, our results will be valid for all definitions of NETS polyhedra that satisfy our axioms.

While there are different (and inequivalent) definitions of NETS polyhedra that satisfy our axioms, it turns out that all possible definitions of NETS polyhedra are in fact equivalent when we restrict our attention to polyhedra in the plane and polyhedra in the boundary of a 2- or 3-dimensional convex polytope: For any such polyhedron S, we will see in Subsection 8.4.3 that S is NETS if and only if S is *collapsible*. (The meaning of "collapsible" will be explained in Subsection 8.4.2.)

It can be deduced from this fact that a polyhedron S in the plane $\mathbb{R}^2$ is NETS if and only if S satisfies the following three conditions:

1. S is nonempty.
2. S is connected.
3. The complement of S in the plane is connected (i.e., S has no holes).

Similarly, a polyhedron in the boundary of any 2- or 3-dimensional convex polytope P is NETS if and only if S satisfies the following conditions:

1′. S is nonempty and is not the entire boundary of the 2- or 3-dimensional convex polytope P.
2′. S is connected.
3′. The complement of S in the boundary of P is connected.

When the convex polytope P is 2-dimensional (i.e., P is a convex polygon), we only need *one* of conditions $2'$ and $3'$ because it is easy to see that each of these two conditions implies the other if S satisfies $1'$. When P is 3-dimensional, we need all three conditions; it is easy to see that S may satisfy any two of the conditions without satisfying the remaining condition.

Example 8.3.1. Let P be a square, so that P has eight proper faces: four 0-faces (one for each vertex of P) and four 1-faces (the four edges of P). Let S be the union of a set of proper faces of P. Then S is NETS in just thc following cases:

- S is a 0-face of P.
- S is a 1-face of P.
- S is the union of two 1-faces of P that share a 0-face (i.e., the union of two adjacent edges of P).
- S is the union of three 1-faces of P.

In all other cases, it is readily confirmed that S is not NETS either because S is empty, or because S is disconnected, or because S is the whole boundary of P.

Example 8.3.2. Let P be a cube, so that P has six 2-faces, twelve 1-faces, and eight 0-faces. Here are a few examples of unions of proper faces of P that are NETS, and unions of proper faces of P that are not NETS:

- If S is the union of two disjoint proper faces of P, then S is not NETS (because S is disconnected).
- If S is the union of two disjoint 2-faces of P with one of the four 1-faces of P that intersect both of those 2-faces, then S is NETS.
- If S is the union of two disjoint 2-faces of P with two, three, or all four of the four 1-faces of P that intersect both of those 2-faces, then S is not NETS (because the complement of S in the boundary of P is disconnected).
- If S is the union of a set of proper faces of P that all contain a certain vertex of P, then S is NETS.

8.3.2 THE AXIOMS THAT SPECIFY NETS POLYHEDRA

Nonempty topologically simple (NETS) polyhedra are required to satisfy the following axioms:

NETS-1 Every NETS polyhedron is nonempty and connected.
NETS-2 Every nonempty convex polytope is NETS.
NETS-3 If A and B are polyhedra such that $A \cap B$ is NETS, then $A \cup B$ is NETS if and only if each of A and B is NETS.
NETS-4 If each of A and B is a NETS polyhedron but $A \cap B$ is not NETS, then $A \cup B$ is not NETS.

We say that a complex **K** is NETS if its polyhedron $\bigcup \mathbf{K}$ is NETS.

Note that any set consisting of just a single point in $\mathbb{R}^n$ is a 0-dimensional convex polytope and is therefore a NETS polyhedron (by **NETS-2**). However, any other finite set of points in $\mathbb{R}^n$ is an example of a polyhedron that is not NETS (by **NETS-1**).

It is readily confirmed that the axioms **NETS-3** and **NETS-4** are together equivalent to the following lemma.

Lemma 8.3.3. *Let A and B be polyhedra that satisfy two of the following three conditions:*

1. *Each of A and B is NETS.*
2. *$A \cap B$ is NETS.*
3. *$A \cup B$ is NETS.*

Then A and B satisfy all three of these conditions. □

It can be seen from this lemma that our axioms **NETS-3** and **NETS-4** may be reformulated as *completion* properties in the sense of Bertrand [4–6].[6] Moreover, it

[6]Specifically, **NETS-3** and **NETS-4** are together equivalent to the assertion that if $\mathcal{K}$ is the set of all NETS polyhedra, then $\mathcal{K}$ satisfies the completion properties $\langle D1 \rangle$ and $\langle D2 \rangle$ of [5,6] and the following

follows from Theorem 2 of [4] and the Mayer–Vietoris exact sequence for reduced homology [21, Sect. 2.2] that the *smallest* collection of polyhedra that satisfies the NETS axioms is the collection of nonempty connected polyhedra whose integral homology groups are trivial in all positive dimensions: This collection of polyhedra satisfies the NETS axioms and is contained in any collection of polyhedra that satisfies those axioms.

The next lemma states consequences of Lemma 8.3.3 that will be much used in this chapter. Note that each of conditions 1 and 2 of Lemma 8.3.4 implies that every member of the collection $\mathcal{S}$ is NETS since we may let $\mathcal{T}$ consist of just one member of $\mathcal{S}$.

Lemma 8.3.4. *Let $\mathcal{S}$ be a finite collection of polyhedra. Then the following are equivalent:*

1. *$\bigcap \mathcal{T}$ is NETS whenever $\emptyset \neq \mathcal{T} \subseteq \mathcal{S}$.*
2. *$\bigcup \mathcal{T}$ is NETS whenever $\emptyset \neq \mathcal{T} \subseteq \mathcal{S}$.*
3. *Every member of $\mathcal{S}$ is NETS, and every set that is obtainable from members of $\mathcal{S}$ by applying one or more union and/or intersection operations is NETS.*

Proof. This result can be proved in the same way as statements 1(a), 1(b), and 1(e) of Lemma 6.2 in [31] are shown to be equivalent in that paper, except that we substitute "NETS" for "contractible" throughout the argument and appeal to Lemma 8.3.3 where that proof appeals to Fact 6.1 of [31]. We can begin by using induction on the cardinality of $\mathcal{S}$ to show that statement 1 implies $\bigcup \mathcal{S}$ is NETS. An analogous inductive argument is given in detail in the proof of Lemma 1 of [30, Sect. 5]. ☐

8.3.3 IN A STRONGLY NORMAL COLLECTION OF NONEMPTY CONVEX POLYTOPES, THE UNION OF ANY ELEMENT WITH ITS NEIGHBORS IS NETS

Saha, Rosenfeld, and others (including the author) have studied collections of sets that have a property called *strong normality* [12,34,47–51], which will be defined below. It can be shown that an n-dimensional Cartesian array complex is an example of a strongly normal collection.[7]

Let $\mathcal{T}$ be any collection of sets, and let $T \in \mathcal{T}$. Then any element of $\mathcal{T} \setminus \{T\}$ that intersects T will be called a *$\mathcal{T}$-neighbor* of T. For example, if $\mathcal{T}$ is any set of facets of a pure complex and $T \in \mathcal{T}$, then the $\mathcal{T}$-neighbors of T are just the elements of $\mathcal{T}$ that are weak neighbors of T in the complex.

---

completion property: If $S, T \in \mathcal{K}$ and $S \subseteq T$, then $A, B \in \mathcal{K}$ whenever A and B are polyhedra such that $A \cap B = S$ and $A \cup B = T$.

[7] This can be deduced from the fact that if $c, c_1, \ldots, c_k$ are any cells of an n-dimensional Cartesian array complex such that $\bigcap_{i=1}^{k} c_i \neq \emptyset$ but $c \cap \bigcap_{i=1}^{k} c_i = \emptyset$, then there must exist an $(n-1)$-dimensional hyperplane in $\mathbb{R}^n$ that is orthogonal to a coordinate axis and strictly separates c from at least one of the cells $c_1, \ldots, c_k$. Any $(n-1)$-dimensional hyperplane in $\mathbb{R}^n$ that is orthogonal to a coordinate axis and strictly separates c from the cell $\bigcap_{i=1}^{k} c_i$ will have this property.

The collection $\mathcal{T}$ is said to be *strongly normal* if each $T \in \mathcal{T}$ has only finitely many $\mathcal{T}$-neighbors, and every nonempty intersection of those $\mathcal{T}$-neighbors intersects T. It is not hard to show [34, Cor. 2.3] that $\mathcal{T}$ is strongly normal just if each $T \in \mathcal{T}$ has only finitely many $\mathcal{T}$-neighbors and every pairwise intersecting subcollection of $\mathcal{T}$ has nonempty intersection. In combinatorics, a collection $\mathcal{T}$ with the latter property is called a "Helly family of order 2."

Statement 1 of the following proposition justifies the heading of this subsection. Statement 2, which is a consequence of statement 1 and Lemma 8.3.3, will be used in Subsections 8.5.1 and 8.6.3: When $\mathbb{I}$ is a binary image on a strongly normal pure complex, statement 2 will allow us to characterize[8] the simple 1s of $\mathbb{I}$ as those 1s whose weak neighbors have a NETS union, and will also allow us to give an analogous characterization of $\mathbb{I}$-regular proper faces of 1s.

Proposition 8.3.5. *Let $\mathcal{C}$ be any strongly normal finite collection of nonempty convex polytopes, let C be any element of $\mathcal{C}$, and let $N_1, \ldots, N_k$ be any $\mathcal{C}$-neighbors of C. Then:*

1. $C \cup \bigcup_{i=1}^{k} N_i$ *is NETS.*
2. $C \cap \bigcup_{i=1}^{k} N_i$ *is NETS if and only if* $\bigcup_{i=1}^{k} N_i$ *is NETS.*

Proof. Let $\mathcal{C}' = \{C \cup N_i \mid 1 \leq i \leq k\}$. Then we claim that $\bigcap \mathcal{C}''$ is NETS for all nonempty subcollections $\mathcal{C}''$ of $\mathcal{C}'$. Statement 1 will follow from this claim and Lemma 8.3.4 since $\bigcup \mathcal{C}' = C \cup \bigcup_{i=1}^{k} N_i$.

To justify our claim, let $\mathcal{C}'' = \{C \cup N_{i_r} \mid 1 \leq r \leq m\}$ be any nonempty subcollection of $\mathcal{C}'$. Then $\bigcap \mathcal{C}'' = C \cup \bigcap_{r=1}^{m} N_{i_r}$, and we just need to show that this set is NETS. Now if $\bigcap_{r=1}^{m} N_{i_r} = \emptyset$, then all is well (as $\bigcap \mathcal{C}''$ is then a nonempty convex polytope C and is therefore NETS by **NETS-2**), so let us assume that $\bigcap_{r=1}^{m} N_{i_r} \neq \emptyset$. Then, since $\mathcal{C}$ is strongly normal, we have that $C \cap \bigcap_{r=1}^{m} N_{i_r} \neq \emptyset$. Thus each of the sets C, $\bigcap_{r=1}^{m} N_{i_r}$, and $C \cap \bigcap_{r=1}^{m} N_{i_r}$ is a nonempty convex polytope and is therefore NETS. It follows from this and **NETS-3** that $C \cup \bigcap_{r=1}^{m} N_{i_r} = \bigcap \mathcal{C}''$ is NETS. This justifies our claim.

As mentioned before, statement 1 follows from the claim and Lemma 8.3.4. Since C is a nonempty convex polytope and is therefore NETS, statement 2 follows from statement 1 and Lemma 8.3.3. □

8.3.4 POSSIBLE DEFINITIONS OF NETS POLYHEDRA USING HOMOLOGY GROUPS

For arbitrary polyhedra (i.e., polyhedra that are not necessarily in the plane or the boundary of a 3-dimensional convex polytope), it can be shown[9] that, for any nontrivial Abelian group G, our NETS axioms (**NETS-1–NETS-4**) are all satisfied if we

[8] This characterization is closely related to [34, Prop. 4.1], which generalizes a result of Saha and Rosenfeld [50].

[9] This fact is a consequence of the Mayer–Vietoris exact sequence for reduced homology [21, Sect. 2.2].

define a NETS polyhedron to be a nonempty connected polyhedron whose homology groups with coefficients in G are trivial in all positive dimensions.

Assuming that NETS polyhedra are defined in this way, if S is a polyhedron in 3-space or in the boundary of a 2-, 3-, or 4-dimensional convex polytope, then it can be shown that, regardless of our choice of coefficient group G, the polyhedron S is NETS just if S is *contractible*[10] (i.e., just if S can be continuously deformed over itself to a set that consists of a single point).

As mentioned before, we are interested in NETS polyhedra partly because of their role in our definition of a simple 1 of a binary image $\mathbb{I}$: If v is a 1 of $\mathbb{I}$, then the question of whether v is simple or nonsimple is equivalent to the question of whether the $\mathbb{I}$-attachment set of the convex polytope v is NETS or is not NETS.

Under the assumption that NETS polyhedra are defined as described above, it is computationally easy to determine whether or not v is simple when the dimension of the binary image $\mathbb{I}$ is no greater than 4. Indeed, since the $\mathbb{I}$-attachment set of v is a polyhedron in the boundary of the convex polytope v, when $\dim(\mathsf{v}) = \dim(\mathbb{I}) \leq 4$ any polyhedron S in the boundary of v will be NETS (under our assumption) if and only if[11] S has all three of the following properties:

1. S is connected.
2. The complement of S in the boundary of v is connected.
3. The Euler characteristic of S is 1.

When $\dim(\mathsf{v}) = \dim(\mathbb{I}) > 4$, it is less easy to determine whether v is a simple 1 of $\mathbb{I}$, but the question can be answered (under the assumption that NETS polyhedra are defined as described above) in no more than cubic time with respect to the number of faces of the convex polytope v. This is because the homology groups of a complex can be computed in no more than cubic time with respect to the number of cells in the complex, and the $\mathbb{I}$-attachment set of v is the polyhedron of a subcomplex of $\mathbf{Faces}(\mathsf{v}) \setminus \{\mathsf{v}\}$.

We also mention that if we adopt the above definition of NETS polyhedra with $G = \mathbb{Z}$, then v is a simple 1 of $\mathbb{I}$ in the sense of this chapter just if v is homology-

[10] Any contractible polyhedron S must be nonempty and connected, and its homology groups must be trivial in all positive dimensions. The converse is false for an arbitrary polyhedron S but is true under our hypotheses, which imply that S is a polyhedron that can be embedded in $\mathbb{R}^3$: For any such polyhedron, it follows from [21, Cor. 3.45] that the polyhedron's integral homology groups are free Abelian groups. This fact and the Universal Coefficient Theorem for homology (e.g., [21, Thm. 3A.3]) imply that, if the polyhedron's homology groups with coefficients in G are trivial in all positive dimensions for some nontrivial Abelian group G, then the same is true for all other Abelian groups G. In this case the polyhedron must also be simply connected (see, e.g., [33]) and hence contractible by [36, Thm. 8.3.10].

[11] Condition 3 implies that S is nonempty and that S is not the entire boundary of v. Bearing this in mind, we can deduce from the Euler–Poincaré formula and the Alexander Duality Theorem for mod 2 homology [17] that conditions 1–3 all hold just if S is nonempty and connected, and the homology groups of S with coefficients in $\mathbb{Z}_2$ are trivial in all positive dimensions. For any nontrivial Abelian group G, the homology groups of S with coefficients in G are trivial in all positive dimensions just if the homology groups of S with coefficients in $\mathbb{Z}_2$ are trivial in all positive dimensions, as explained in footnote 10.

simple in the sense of Niethammer et al. [38] or, equivalently, $(3^n - 1, 2n)$-simple in the sense of [27, Def. 13].

We now address the relationship between the definitions of simple 1 and critical kernel used by Bertrand and Couprie [8] for binary images on 2-, 3- and 4-dimensional Cartesian array complexes and the definitions used in this chapter. The following paragraphs assume the reader is acquainted with the notion of a *collapsible* complex; this will be defined in Subsection 8.4.2.

It follows from [14, Thm. 12] (on putting $A = \mathbf{Faces}(\mathsf{v})$) that the definitions of Bertrand and Couprie will be equivalent to ours for binary images on an n-dimensional Cartesian array complex (where $n = 2,\ 3,\ \text{or}\ 4$) just if the following condition is satisfied.

Condition 8.3.6. *If* v *is an* n*-cube and* $\mathbf{K}$ *is any subcomplex of* $\mathbf{Faces}(\mathsf{v})$*, then the polyhedron* $\bigcup \mathbf{K}$ *is NETS if and only if* $\mathbf{K}$ *is collapsible.*

It can be deduced from the NETS-axioms—*without* assuming that NETS polyhedra are defined in the manner described above—that Condition 8.3.6 holds in the cases $n = 2$ and $n = 3$ and that the "if" part of the condition holds regardless of the value of n: The former fact will be established in Subsection 8.4.3, and the latter fact will follow from Proposition 8.4.2.

In the case $n = 4$, the "only if" part of Condition 8.3.6 holds (as well as the "if" part) if NETS polyhedra are defined in the way described above. This is because a polyhedron S in the boundary of a 4-cube v will then be NETS just if S has properties 1–3 above, and if $S = \bigcup \mathbf{K}$ for a subcomplex $\mathbf{K}$ of $\mathbf{Faces}(\mathsf{v})$, then properties 1–3 are sufficient [29, Sect. 7.3] for $\mathbf{K}$ to be collapsible.

For binary images on 2-, 3-, and 4-dimensional Cartesian array complexes, we see from the above that the definitions of simple 1 and critical kernel used by Bertrand and Couprie belong to the class of definitions of simple 1 and critical kernel for which the arguments and results of this chapter are valid. In the 2- and 3-dimensional cases, we also see that their definitions are the only definitions in our class.

In fact, the author is convinced (though he does not have a complete proof) that Condition 8.3.6 can be shown to hold in the case $n = 4$ without assuming anything about NETS polyhedra other than the NETS axioms, so that Bertrand and Couprie's definitions of simple 1 and critical kernel are again the only definitions in our class for binary images on a 4-dimensional Cartesian array complex.

We end this section by reiterating that our mathematical arguments in the sequel will not assume any properties of NETS polyhedra other than our axioms **NETS-1–NETS-4**. In particular, our arguments will not assume that NETS polyhedra are defined in the manner described at the beginning of this subsection.

8.4 CONSEQUENCES OF THE NETS AXIOMS

In this section, we deduce some properties of NETS polyhedra from the axioms **NETS-1–NETS-4**. Important goals are to establish the facts stated in the headings of Subsections 8.4.2 and 8.4.3. As mentioned in Subsection 8.3.4, it follows from the latter result and [14, Thm. 12] that, for binary images on 2- and 3-dimensional Cartesian array complexes, our NETS definitions of simple 1 and critical kernel are equivalent to Bertrand and Couprie's definitions of those concepts.

Here we do not assume that NETS polyhedra are defined in the way that was described in Subsection 8.3.4: Indeed, if we assume that NETS polyhedra are defined in that way, then the properties of NETS polyhedra we establish in this section will be restatements or easy consequences of well-known topological facts.

8.4.1 A 0- OR 1-DIMENSIONAL COMPLEX IS NETS JUST IF IT IS A TREE

A 0- or 1-dimensional complex $\mathbf{G}$ can be regarded as a graph whose vertices are the 0-cells of $\mathbf{G}$ and whose edges are the 1-cells of $\mathbf{G}$, in which a vertex v is an endvertex of an edge e just if the 0-cell v is a face of the 1-cell e. This allows us to use the familiar terminology of graph theory when discussing 0- and 1-dimensional complexes. In the present subsection, we will take advantage of this.

A graph $\mathbf{G}$ is called a *tree* if $\mathbf{G}$ is connected and acyclic. The Euler characteristic of a graph is just the number of its vertices minus the number of its edges. It is well known that a graph $\mathbf{G}$ is a tree if and only if $\mathbf{G}$ is connected and $\mathbf{G}$'s Euler characteristic is 1.

Theorem 8.4.1. *Let* $\mathbf{G}$ *be a* 0- *or* 1-*dimensional complex. Then* $\bigcup \mathbf{G}$ *is NETS if and only if* $\mathbf{G}$ *is a tree (or, equivalently, if and only if* $\mathbf{G}$ *is connected and* $\mathbf{G}$*'s Euler characteristic is* 1*).*

Proof. Suppose first of all that $\mathbf{G}$ has no edge, i.e., $\mathbf{G}$ is 0-dimensional. If $\mathbf{G}$ has just one vertex, then $\mathbf{G}$ is a tree, and $\bigcup \mathbf{G}$ (which consists of just one point) is NETS by **NETS-2**. If $\mathbf{G}$ has two or more vertices, then $\mathbf{G}$ is not a tree, and $\bigcup \mathbf{G}$ is not NETS (by **NETS-1**). So the theorem is true when $\mathbf{G}$ has no edge.

Now suppose there is a counterexample to the theorem, and suppose $\mathbf{G}$ is a counterexample that *has as few edges as possible*. Then $\mathbf{G}$ has at least one edge. Since $\mathbf{G}$ is a counterexample, either $\mathbf{G}$ is a tree but $\bigcup \mathbf{G}$ is not NETS, or $\mathbf{G}$ is not a tree but $\bigcup \mathbf{G}$ is NETS.

Let e be any edge of $\mathbf{G}$, let v and w be the endvertices of e, and consider the graph $\mathbf{G} \setminus \{e\}$. Now either there is a path in $\mathbf{G} \setminus \{e\}$ from v to w, or there is no such path. In both cases, we will show that $\mathbf{G}$ is not in fact a counterexample to the theorem, and this contradiction will prove the theorem.

Case 1: There is a path in $\mathbf{G} \setminus \{e\}$ *from* v *to* w.

In this case, $\mathbf{G}$ is not a tree because it is not acyclic: Indeed, let $\mathbf{C}$ be the subgraph of $\mathbf{G}$ whose vertices are the vertices on a shortest path in $\mathbf{G} \setminus \{e\}$ from v to w and whose edges are the edges on that path together with e itself. Then $\mathbf{C}$ is a cycle.

Now we will show $\bigcup \mathbf{G}$ is not NETS, contrary to our hypothesis that $\mathbf{G}$ is a counterexample to the theorem. Since $\mathbf{C}$ is a cycle, the graph $\mathbf{C} \setminus \{e\}$ is a tree. Moreover, this tree $\mathbf{C} \setminus \{e\}$ has fewer edges than $\mathbf{G}$. So (since $\mathbf{G}$ is a counterexample to the theorem with as few edges as possible) we deduce that

$$\bigcup(\mathbf{C} \setminus \{e\}) \text{ is NETS.} \tag{8.6}$$

But the line segment e is NETS (by **NETS-2**), whereas **NETS-1** implies that $e \cap \bigcup(\mathbf{C} \setminus \{e\})$ is not NETS because this intersection is just the union of the two endvertices v and w of e.

Hence it follows from **NETS-4** that $e \cup \bigcup(\mathbf{C} \setminus \{e\}) = \bigcup \mathbf{C}$ is not NETS. On the other hand, $(\bigcup \mathbf{C}) \cap \bigcup(\mathbf{G} \setminus \{e\}) = \bigcup(\mathbf{C} \setminus \{e\})$ is NETS, by (8.6). Therefore $\bigcup \mathbf{G} = (\bigcup \mathbf{C}) \cup \bigcup(\mathbf{G} \setminus \{e\})$ is not NETS (by **NETS-3**), so that $\mathbf{G}$ is not a counterexample to the theorem, contrary to our hypothesis.

Case 2: There is no path in $\mathbf{G} \setminus \{e\}$ *from* v *to* w.

Note that $\mathbf{G}$ must be a connected graph; otherwise, $\bigcup \mathbf{G}$ is not NETS (by **NETS-1**) and $\mathbf{G}$ is not a tree, contrary to our hypothesis that $\mathbf{G}$ is a counterexample to the theorem.

If neither v nor w is incident with an edge of $\mathbf{G} \setminus \{e\}$, then (since $\mathbf{G}$ is connected) $\mathbf{G}$ consists just of e and its endvertices v and w, so $\mathbf{G}$ is a tree and $\bigcup \mathbf{G}$ is NETS (by **NETS-2**). But then $\mathbf{G}$ is not a counterexample to the theorem, contrary to our hypothesis. Hence we may assume without loss of generality that v is incident with an edge e' of $\mathbf{G} \setminus \{e\}$.

Now let $\mathbf{G}_v$ be the connected component of the graph $\mathbf{G} \setminus \{e\}$ that contains v, and $\mathbf{G}_w$ the connected component of $\mathbf{G} \setminus \{e\}$ that contains w. Since $\mathbf{G}$ is connected, we readily have that $\mathbf{G}_v \cup \mathbf{G}_w = \mathbf{G} \setminus \{e\}$.

Moreover, $\mathbf{G}_v \cap \mathbf{G}_w = \emptyset$; otherwise, $\mathbf{G}_v$ and $\mathbf{G}_w$ would share a vertex x and so we could catenate a path from v to x in $\mathbf{G}_v$ with a path from x to w in $\mathbf{G}_w$ to produce a path in $\mathbf{G} \setminus \{e\}$ from v to w, contrary to our assumption that no such path exists. Hence $e' \notin \mathbf{G}_w$ (as $e' \in \mathbf{G}_v$).

Let $\mathbf{H} = \mathbf{G}_w \cup \{e, v\}$, so $\mathbf{H} \cup \mathbf{G}_v = \mathbf{G}$ and $\mathbf{H} \cap \mathbf{G}_v = \{v\}$. Then $(\bigcup \mathbf{H}) \cap (\bigcup \mathbf{G}_v) = v$ is NETS and $(\bigcup \mathbf{H}) \cup (\bigcup \mathbf{G}_v) = \bigcup \mathbf{G}$ by (8.4), **NETS-2**, and (8.5). From this and **NETS-3** it follows that

$$\bigcup \mathbf{G} = (\bigcup \mathbf{H}) \cup (\bigcup \mathbf{G}_v) \text{ is NETS if and only if each of } \bigcup \mathbf{H} \text{ and } \bigcup \mathbf{G}_v \text{ is NETS.} \tag{8.7}$$

But, since each of $\mathbf{H}$ and $\mathbf{G}_v$ has fewer edges than $\mathbf{G}$ (as $e' \notin \mathbf{H}$ and $e \notin \mathbf{G}_v$), neither $\mathbf{H}$ nor $\mathbf{G}_v$ is a counterexample to the theorem. So each of $\bigcup \mathbf{H}$ and $\bigcup \mathbf{G}_v$ is NETS just if each of $\mathbf{H}$ and $\mathbf{G}_v$ is a tree. Hence, by (8.7), $\bigcup \mathbf{G}$ is NETS just if each of $\mathbf{H}$ and $\mathbf{G}_v$ is a tree. Moreover, since $\mathbf{H}$ and $\mathbf{G}_v$ share just one vertex, it is also the case that $\mathbf{G} = \mathbf{H} \cup \mathbf{G}_v$ is a tree just if each of $\mathbf{H}$ and $\mathbf{G}_v$ is a tree. Thus $\bigcup \mathbf{G}$ is NETS just if $\mathbf{G}$ is a tree, contrary to our hypothesis that $\mathbf{G}$ is a counterexample to the theorem. □

8.4.2 COLLAPSIBLE POLYHEDRA ARE NETS, AND NETS POLYHEDRA NEVER COLLAPSE TO NON-NETS POLYHEDRA

In this subsection and the next, we present some properties of NETS polyhedra that relate to collapsing. We first define this concept for those readers who are unfamiliar with it.

Let **K** be a polytopal complex. Then a *free pair* of cells of **K** is an ordered pair of cells (f, g) that satisfies the following conditions:

1. f is a facet of **K**.
2. g is proper face of f and $\dim(g) = \dim(f) - 1$.
3. f is the only cell of **K** that has g as a proper face.

Condition 1 and the second part of condition 2 are in fact redundant: If g is a proper face of a cell f of **K** and condition 3 holds, then f must be a facet of **K** such that $\dim(g) = \dim(f) - 1$.

If (f, g) is a free pair of cells of a complex **K**, then **K** is said to *collapse elementarily to* the complex $\mathbf{K} \setminus \{f, g\}$, and the transformation of **K** to $\mathbf{K} \setminus \{f, g\}$ is called an *elementary collapse* operation. A complex **K** is said to *collapse to* a complex **L** if $\mathbf{L} = \mathbf{K}$ or if there is a sequence of complexes $\mathbf{C}_0, \mathbf{C}_1, \ldots, \mathbf{C}_n$ (where $n \geq 1$) such that $\mathbf{C}_0 = \mathbf{K}$, $\mathbf{C}_n = \mathbf{L}$, and $\mathbf{C}_i$ collapses elementarily to $\mathbf{C}_{i+1}$ for $0 \leq i \leq n - 1$.

If a complex **K** collapses to a complex **L**, then it follows readily from the above definitions that:

C1: If **L** collapses to a complex **M**, then **K** also collapses to **M**.

C2: Each connected component of $\bigcup \mathbf{K}$ contains exactly one connected component of $\bigcup \mathbf{L}$. In particular, $\bigcup \mathbf{L}$ is connected if and only if $\bigcup \mathbf{K}$ is connected.

C3: **K** and **L** have the same Euler characteristic.

C4: If $\bigcup \mathbf{K} \subsetneq B$, where B is $\mathbb{R}^n$ or the boundary of a convex polytope, then each connected component of $B \setminus \bigcup \mathbf{L}$ contains exactly one connected component of $B \setminus \bigcup \mathbf{K}$. In particular, $B \setminus \bigcup \mathbf{L}$ is connected if and only if $B \setminus \bigcup \mathbf{K}$ is connected.

A complex **K** is said to be *collapsible* if **K** collapses to a complex that consists of just a single 0-cell. A polyhedron S is said to be *collapsible* if there exists a collapsible complex **K** such that $\bigcup \mathbf{K} = S$.

The following proposition justifies the heading of this subsection.

Proposition 8.4.2. *Let* **K** *be any complex. Then:*

1. *If* **K** *collapses to a complex* **L**, *then* $\bigcup \mathbf{K}$ *is NETS if and only if* $\bigcup \mathbf{L}$ *is NETS.*
2. *If* **K** *is collapsible, then* $\bigcup \mathbf{K}$ *is NETS.*

Proof. Since statement 2 follows from the case of statement 1 in which **L** consists of a single 0-cell (since $\bigcup \mathbf{L}$ is NETS in that case by **NETS-2**), it suffices to prove statement 1.

Suppose that statement 1 is false and **K** is a minimal complex that does not satisfy statement 1. Then **K** collapses to a complex $\mathbf{L} \subsetneq \mathbf{K}$ such that just one of $\bigcup \mathbf{K}$

and $\bigcup\mathbf{L}$ is NETS. Since $\mathbf{K}$ collapses to $\mathbf{L}$, there must exist a sequence of complexes $\mathbf{C}_0, \mathbf{C}_1, \ldots, \mathbf{C}_n$ (where $n \geq 1$) such that $\mathbf{C}_0 = \mathbf{K}$, $\mathbf{C}_n = \mathbf{L}$, and $\mathbf{C}_i$ collapses elementarily to $\mathbf{C}_{i+1}$ for $0 \leq i \leq n-1$.

Now there must be a free pair (f, g) of $\mathbf{K} = \mathbf{C}_0$ for which $\mathbf{C}_1 = \mathbf{K} \setminus \{f, g\}$, so that:

- $f \cup \bigcup\mathbf{C}_1 = \bigcup\mathbf{K}$.
- $f \cap \bigcup\mathbf{C}_1 = \bigcup(\mathbf{Faces}(f) \cap \mathbf{C}_1) = \bigcup(\mathbf{Faces}(f) \setminus \{f, g\})$.

Since (f, g) is a free pair of $\mathbf{K}$, g is a $(\dim(f) - 1)$-dimensional face of the convex polytope f, and it can be shown that this implies the complex $\mathbf{Faces}(f) \setminus \{f, g\}$ is collapsible.[12] Since $\mathbf{K}$ is a minimal complex for which statement 1 is false and $\mathbf{Faces}(f) \setminus \{f, g\}$ is a collapsible proper subcomplex of $\mathbf{K}$, $\mathbf{Faces}(f) \setminus \{f, g\}$ is a collapsible complex that satisfies statement 1 (and therefore also satisfies statement 2). Hence $\bigcup(\mathbf{Faces}(f) \setminus \{f, g\})$ is NETS. Equivalently, $f \cap \bigcup\mathbf{C}_1$ is NETS. Moreover, f is NETS (by **NETS-2**). Therefore $\bigcup\mathbf{K} = f \cup \bigcup\mathbf{C}_1$ is NETS if and only if $\bigcup\mathbf{C}_1$ is NETS (by **NETS-3**). But, since $\mathbf{C}_1$ is a proper subcomplex of $\mathbf{K}$ that collapses to $\mathbf{C}_n = \mathbf{L}$ and $\mathbf{K}$ is a minimal complex for which statement 1 is false, $\bigcup\mathbf{C}_1$ is NETS if and only if $\bigcup\mathbf{L}$ is NETS. It follows that $\bigcup\mathbf{K}$ is NETS if and only if $\bigcup\mathbf{L}$ is NETS. Thus it is not the case that just one of $\bigcup\mathbf{K}$ and $\bigcup\mathbf{L}$ is NETS. This contradiction shows that statement 1 is in fact true. □

8.4.3 A POLYHEDRON IN THE PLANE OR IN THE BOUNDARY OF A 3-DIMENSIONAL CONVEX POLYTOPE IS NETS IF AND ONLY IF IT IS COLLAPSIBLE

Let $\mathbf{S}$ be any complex whose cells all lie in the boundary of a 3-dimensional convex polytope P. We will now show that the polyhedron $\bigcup\mathbf{S}$ is NETS just if $\mathbf{S}$ is collapsible.[13] This implies that, if $\mathbf{T}$ is any complex whose cells all lie in the plane $\mathbb{R}^2$, then $\bigcup\mathbf{T}$ is NETS just if $\mathbf{T}$ is collapsible (because, for any such $\mathbf{T}$, there is a 3-dimensional convex polytope whose boundary contains all the cells of $\mathbf{T}$).

Suppose first of all that $\bigcup\mathbf{S}$ is not the entire boundary of P. If $\mathbf{S} = \emptyset$, then the stated result is true since $\emptyset$ is neither NETS nor collapsible, so let us assume that $\mathbf{S} \neq \emptyset$. If $\mathbf{S}$ has a 2-cell, then it is not hard to show that some 2-cell in $\mathbf{S}$ is a member of a free pair of cells of $\mathbf{S}$. We can deduce from this, by induction on the number of 2-cells in $\mathbf{S}$, that $\mathbf{S}$ must collapse to some 0- or 1-dimensional complex $\mathbf{S}'$.

---

[12] The fact that $\mathbf{Faces}(f) \setminus \{f, g\}$ is collapsible follows, e.g., from [55, Cor. 8.13] and [16, Lemma 17].

[13] This raises the following question: Can we give a reasonably short proof, which does not use homology theory, that our axioms **NETS-1–NETS-4** hold when "NETS" is replaced by "collapsible" if we confine our attention to polyhedra in the boundary of a 3-dimensional convex polytope? Note that such a proof would also be a proof, which does not involve homology theory, that our NETS axioms are logically consistent if we confine our attention to polyhedra in the boundary of a 3-dimensional convex polytope. The answer is yes. For example, a proof could be based on a version of Theorem 2 in Passat et al.'s paper [39] that allows 2-cells to be arbitrary convex polygons; this version of their theorem can be proved in essentially the same way as their version.

By Proposition 8.4.2, $\bigcup \mathbf{S}$ is NETS just if $\bigcup \mathbf{S}'$ is NETS. By Theorem 8.4.1, $\bigcup \mathbf{S}'$ is NETS just if $\mathbf{S}'$ is a tree or, equivalently, just if $\mathbf{S}'$ is a connected graph whose Euler characteristic is 1.

Thus if $\bigcup \mathbf{S}$ is NETS, then $\mathbf{S}'$ is a tree, which readily implies that $\mathbf{S}'$ is collapsible; we leave the details to the reader. But if $\mathbf{S}'$ is collapsible, then $\mathbf{S}$ is also collapsible (by **C1**). So if $\bigcup \mathbf{S}$ is NETS, then $\mathbf{S}$ is collapsible.

Now suppose $\bigcup \mathbf{S}$ is not NETS. Then $\bigcup \mathbf{S}'$ is not NETS. So either $\mathbf{S}'$ is disconnected or the Euler characteristic of $\mathbf{S}'$ is not 1, and the same must be true of $\mathbf{S}$ (by **C2** and **C3**). This implies that $\mathbf{S}$ is not collapsible (by **C2** and **C3** again) since a complex that consists of just a single 0-cell is connected and has an Euler characteristic of 1. So if $\bigcup \mathbf{S}$ is not NETS, then $\mathbf{S}$ is not collapsible.

We have now established that the stated result is true if $\bigcup \mathbf{S}$ is not the entire boundary of the convex polytope P.

Finally, we consider the case where $\bigcup \mathbf{S}$ *is* the entire boundary of P. In this case, $\mathbf{S}$ is not collapsible (because it has no free pair). We will now show that $\bigcup \mathbf{S}$ is also not NETS.

Let f be a 2-cell in $\mathbf{S}$. Since $\bigcup \mathbf{S}$ is the entire boundary of the convex polytope P, we readily see that $\bigcup(\mathbf{S} \setminus \{f\})$ is connected. The complement of $\bigcup(\mathbf{S} \setminus \{f\})$ in the boundary of P is also connected (as it is just the interior of the 2-cell f). Moreover, $\bigcup(\mathbf{S} \setminus \{f\})$ is not the entire boundary of P, and so (as explained in the second paragraph of this subsection) $\mathbf{S} \setminus \{f\}$ must collapse to a 0- or 1-dimensional complex $\mathbf{S}''$.

Now $\bigcup \mathbf{S}''$ and its complement in the boundary of P are both connected (by **C2** and **C4**). It can be deduced from this and the Jordan Curve Theorem that $\mathbf{S}''$ is a connected acyclic graph, so that $\bigcup \mathbf{S}''$ is NETS (by Theorem 8.4.1). This and Proposition 8.4.2 imply that $\bigcup(\mathbf{S} \setminus \{f\})$ is NETS. On the other hand, $f \cap \bigcup(\mathbf{S} \setminus \{f\})$ is the boundary of the 2-cell f, and we see from Theorem 8.4.1 that this boundary is not NETS. Since f is NETS (by **NETS-2**) and $\bigcup(\mathbf{S} \setminus \{f\})$ is NETS but $f \cap \bigcup(\mathbf{S} \setminus \{f\})$ is not NETS, we see from **NETS-4** that $f \cup \bigcup(\mathbf{S} \setminus \{f\}) = \bigcup \mathbf{S}$ is not NETS.

We have now justified the heading of this subsection. It should be noted, however, that a NETS polyhedron in 3-space need not be collapsible. This follows from the fact that there exist collapsible complexes $\mathbf{K}$ in 3-space that collapse to a 2-dimensional complex $\mathbf{L}$ that has no free pair. (An example is shown in [8, Fig. 4].) For any such complexes $\mathbf{K}$ and $\mathbf{L}$, we see from Proposition 8.4.2 that the polyhedron $\bigcup \mathbf{L}$ is NETS, even though $\bigcup \mathbf{L}$ is not collapsible.

8.5 SIMPLE, HEREDITARILY SIMPLE, AND MINIMAL NONSIMPLE SETS

8.5.1 SIMPLE 1S, ATTACHMENT COMPLEXES, AND ATTACHMENT SETS

Let v be any 1 of a binary image $\mathbb{I}$ on a pure complex. Then the $\mathbb{I}$-*attachment complex* of v, denoted by **Attachment**$(\mathsf{v}, \mathbb{I})$, is the set of faces of v that lie in the foreground complex of the binary image $\mathbb{I} - \mathsf{v}$. In other words, the complex **Attachment**$(\mathsf{v}, \mathbb{I})$ contains each face of v that is a face of at least one other 1 of $\mathbb{I}$ and contains no other cell. In view of property **P2b** of complexes, we have that

$$\begin{aligned}\mathbf{Attachment}(\mathsf{v}, \mathbb{I}) &= \mathbf{Faces}(\mathsf{v}) \cap \bigcup\{\mathbf{Faces}(\mathsf{w}) \mid \mathsf{w} \in \mathbb{I}^{-1}[1] \setminus \{\mathsf{v}\}\} \\ &= \bigcup\{\mathbf{Faces}(\mathsf{v} \cap \mathsf{w}) \mid \mathsf{w} \in \mathbb{I}^{-1}[1] \setminus \{\mathsf{v}\}\}.\end{aligned}$$

From this we can deduce

$$\text{Facets}(\mathbf{Attachment}(\mathsf{v}, \mathbb{I})) \subseteq \{\mathsf{v} \cap \mathsf{w} \mid \mathsf{w} \in \mathbb{I}^{-1}[1] \setminus \{\mathsf{v}\}\} \subseteq \mathbf{Attachment}(\mathsf{v}, \mathbb{I}). \quad (8.8)$$

The polyhedron $\bigcup \mathbf{Attachment}(\mathsf{v}, \mathbb{I})$ is called the $\mathbb{I}$-*attachment set* of v. Since $\bigcup \mathbf{Attachment}(\mathsf{v}, \mathbb{I}) = \bigcup \text{Facets}(\mathbf{Attachment}(\mathsf{v}, \mathbb{I}))$, it follows from (8.8) that

$$\bigcup \mathbf{Attachment}(\mathsf{v}, \mathbb{I}) = \bigcup\{\mathsf{v} \cap \mathsf{w} \mid \mathsf{w} \in \mathbb{I}^{-1}[1] \setminus \{\mathsf{v}\}\} = \mathsf{v} \cap \bigcup(\mathbb{I}^{-1}[1] \setminus \{\mathsf{v}\}), \quad (8.9)$$

and so the $\mathbb{I}$-attachment set of v is exactly the intersection of the convex polytope v with the foreground polyhedron of the binary image $\mathbb{I} - \mathsf{v}$.

We say that v is a *simple* 1 of $\mathbb{I}$ if $\bigcup \mathbf{Attachment}(\mathsf{v}, \mathbb{I})$ is NETS and say that v is a *nonsimple* 1 of $\mathbb{I}$ if $\bigcup \mathbf{Attachment}(\mathsf{v}, \mathbb{I})$ is not NETS. Examples 8.5.1–8.5.4 below explain how this concept is equivalent to previous concepts of simple elements when $\mathbb{I}$ is a binary image on a Cartesian array complex.

We also mention that, if v is a 1 of a binary image $\mathbb{I}$ on a Cartesian array complex or any other strongly normal pure complex, then we have that:

- v is a simple 1 of $\mathbb{I}$ if and only if the union of all the 1s of $\mathbb{I}$ that are weakly adjacent to v is NETS.

This can be deduced from (8.9) and statement 2 of Proposition 8.3.5 when we put $C = \mathsf{v}$ and let $N_1, \ldots, N_k$ be the 1s of $\mathbb{I}$ that are weakly adjacent to v.

Example 8.5.1. Let **K** be a 2-dimensional Cartesian array complex, and let p be a 1 of a binary image $\mathbb{I}$ on **K** such that p is not a border facet of **K**. Then p is a simple 1 of $\mathbb{I}$ if and only if p is 8-simple (in the usual sense, as defined, for example, in [43, pp. 232–233]). This is easily deduced from Example 8.3.1.

Example 8.5.2. Let **K** be a 3-dimensional Cartesian array complex, and let p be a 1 of a binary image $\mathbb{I}$ on **K** such that p is not a border facet of **K**. Then p is a simple 1 of $\mathbb{I}$ if and only if p is 26-simple in the usual sense. This follows from the fact that a polyhedron in the boundary of a 3-dimensional convex polytope is NETS if and only if it satisfies conditions $1'$, $2'$, and $3'$ in Subsection 8.3.1, and characterizations of 26-simple voxels by Bertrand and Malandain [1,11] and Saha et al. [45,46].

Example 8.5.3. Let **K** be a 2-, 3-, or 4-dimensional Cartesian array complex, let p be a 1 of a binary image $\mathbb{I}$ on **K**, and suppose we define NETS polyhedra in the way that is described in Subsection 8.3.4. Then p is a simple 1 of $\mathbb{I}$ if and only if **Faces**(p) is simple for the foreground complex of $\mathbb{I}$ in the sense of Bertrand and Couprie [3,8,14]. This can be deduced from [14, Thm. 16]; in the 4-dimensional case, one part of the proof of this theorem relied on exhaustive case-checking by computer.

Example 8.5.4. Let **K** be an n-dimensional Cartesian array complex (for any positive integer n), let p be a 1 of a binary image $\mathbb{I}$ on **K** such that p is not a border facet of **K**, and suppose we define NETS polyhedra in the way described in Subsection 8.3.4 with $G = \mathbb{Z}$. Then, as mentioned in Subsection 8.3.4, p is a simple 1 of $\mathbb{I}$ if and only if p is homology-simple in the sense of Niethammer et al. [38] or, equivalently, if and only if p is $(3^n - 1, 2n)$-simple in the sense of [27, Def. 13].

8.5.2 DEFINITIONS AND BASIC PROPERTIES OF SIMPLE, HEREDITARILY SIMPLE, AND MINIMAL NONSIMPLE SETS

As before, let $\mathbb{I}$ be a binary image on a pure complex. A *simple sequence* of $\mathbb{I}$ is a nonempty sequence $\mathsf{q}_0, \ldots, \mathsf{q}_n$ of distinct 1s of $\mathbb{I}$ such that q_0 is a simple 1 of $\mathbb{I}$ and, for $1 \leq i \leq n$, q_i is a simple 1 of $\mathbb{I} - \{\mathsf{q}_j \mid 0 \leq j \leq i - 1\}$. A *simple set* of $\mathbb{I}$ is a set D of 1s of $\mathbb{I}$ such that either $\mathsf{D} = \emptyset$ or there exists a simple sequence $\mathsf{d}_0, \ldots, \mathsf{d}_n$ such that $\mathsf{D} = \{\mathsf{d}_0, \ldots, \mathsf{d}_n\}$. Here n may be 0: A simple sequence or simple set may consist of just a single simple 1 of $\mathbb{I}$. (In this chapter, complexes are required to be finite and so our binary images have only finitely many 1s. Our definition of a simple set reflects this assumption.)

A *hereditarily simple set* of $\mathbb{I}$ is a simple set D of $\mathbb{I}$ such that every proper subset of D is also a simple set of $\mathbb{I}$. A *minimal nonsimple set* of $\mathbb{I}$ is a set D of 1s of $\mathbb{I}$ such that D is not a simple set of $\mathbb{I}$ but every proper subset of D is a simple set of $\mathbb{I}$. Here are some immediate consequences of the above definitions:

1. $\emptyset$ is a hereditarily simple set of $\mathbb{I}$.
2. Every element of a hereditarily simple set of $\mathbb{I}$ is a simple 1 of $\mathbb{I}$.
3. If d is any 1 of $\mathbb{I}$, then the singleton set $\{\mathsf{d}\}$ is a hereditarily simple set of $\mathbb{I}$ or a minimal nonsimple set of $\mathbb{I}$ according to whether d is a simple 1 or a nonsimple 1 of $\mathbb{I}$.
4. A set Q of 1s of $\mathbb{I}$ is a minimal nonsimple set of $\mathbb{I}$ if and only if Q is not a hereditarily simple set of $\mathbb{I}$ but every proper subset of Q is a hereditarily simple set of $\mathbb{I}$.
5. Any set of 1s of $\mathbb{I}$ that is not a hereditarily simple set of $\mathbb{I}$ must contain at least one minimal nonsimple set of $\mathbb{I}$.

The next theorem characterizes the concept of a hereditarily simple set of $\mathbb{I}$ in terms of Bertrand's concept of *P-simpleness*. An element d of a set D of 1s of $\mathbb{I}$ is said to be *P-simple* for D in $\mathbb{I}$ if d is a simple 1 of $\mathbb{I} - \mathsf{D}'$ for all $\mathsf{D}' \subseteq \mathsf{D} \setminus \{\mathsf{d}\}$.

Theorem 8.5.5. *Let $\mathbb{I}$ be a binary image on a pure complex, and let Q be a set of 1s of $\mathbb{I}$. Then the following are equivalent:*

1. *Q is a hereditarily simple set of $\mathbb{I}$.*
2. *Every q in Q is P-simple for Q in $\mathbb{I}$.*
3. *Every nonempty sequence of distinct elements of Q is a simple sequence of $\mathbb{I}$.*

Our proof of this theorem will be based on Corollary 8.8.10 in Subsection 8.8.1 and will be given at the end of that subsection. It is actually quite easy to see from the definitions of a simple sequence and a simple set that assertion 3 implies assertion 1. It is also a straightforward matter to deduce from the definitions of a P-simple 1 and a hereditarily simple set that assertion 2 implies assertion 3. However, it is not as easy to prove that assertion 1 implies assertion 2.

Corollary 8.5.6 (NETS-based generalization of part of [31, Thm. 5.3]). *Let $\mathbb{I}$ be a binary image on a pure complex, and let Q be a nonempty set of 1s of $\mathbb{I}$. Then Q is a minimal nonsimple set of $\mathbb{I}$ if and only if both of the following are true for every $\mathsf{q} \in \mathsf{Q}$:*

1. *q is a nonsimple 1 of $\mathbb{I} - (\mathsf{Q} \setminus \{\mathsf{q}\})$.*
2. *q is a simple 1 of $\mathbb{I} - \mathsf{Q}'$ for every proper subset Q' of $\mathsf{Q} \setminus \{\mathsf{q}\}$.*

Proof. It is readily confirmed that the corollary is true if $|\mathsf{Q}| = 1$, so let us assume that $|\mathsf{Q}| \geq 2$.

To see that the "only if" part of the corollary is true, suppose that Q is a minimal nonsimple set of $\mathbb{I}$ and q is any element of Q. Then $\mathsf{Q} \setminus \{\mathsf{q}\}$ is a nonempty simple set of $\mathbb{I}$, so there is a simple sequence $\mathsf{q}_0, \ldots, \mathsf{q}_k$ such that $\{\mathsf{q}_0, \ldots, \mathsf{q}_k\} = \mathsf{Q} \setminus \{\mathsf{q}\}$. Hence condition 1 must hold, for otherwise $\mathsf{q}_0, \ldots, \mathsf{q}_k, \mathsf{q}$ would be a simple sequence of $\mathbb{I}$ such that $\{\mathsf{q}_0, \ldots, \mathsf{q}_k, \mathsf{q}\} = \mathsf{Q}$, and Q would be a simple set of $\mathbb{I}$. Moreover, since every proper subset of Q is a hereditarily simple set of $\mathbb{I}$ (as Q is a minimal nonsimple set of $\mathbb{I}$), if Q' is any proper subset of $\mathsf{Q} \setminus \{\mathsf{q}\}$, then $\mathsf{Q}' \cup \{\mathsf{q}\}$ is a hereditarily simple set of $\mathbb{I}$, whence q is P-simple for $\mathsf{Q}' \cup \{\mathsf{q}\}$ in $\mathbb{I}$ (by Theorem 8.5.5), and so q is a simple 1 of $\mathbb{I} - \mathsf{Q}'$. Thus condition 2 also holds.

To see that the "if" part of the corollary is true, suppose conditions 1 and 2 hold for all $\mathsf{q} \in \mathsf{Q}$. Since condition 2 holds for all $\mathsf{q} \in \mathsf{Q}$, q is P-simple for Q'' whenever $\mathsf{q} \in \mathsf{Q}'' \subsetneq \mathsf{Q}$, which (by Theorem 8.5.5) implies that Q'' is a hereditarily simple set of $\mathbb{I}$ whenever $\mathsf{Q}'' \subsetneq \mathsf{Q}$. But Q cannot be a simple set of $\mathbb{I}$, for otherwise there would exist a simple sequence $\mathsf{q}_0, \ldots, \mathsf{q}_n$ of $\mathbb{I}$ such that $\{\mathsf{q}_0, \ldots, \mathsf{q}_n\} = \mathsf{Q}$, whence q_n would be a simple 1 of $\mathbb{I} - \{\mathsf{q}_0, \ldots, \mathsf{q}_{n-1}\} = \mathbb{I} - (\mathsf{Q} \setminus \{\mathsf{q}_n\})$ and condition 1 would not hold for $\mathsf{q} = \mathsf{q}_n$. Since every proper subset of Q is a simple set of $\mathbb{I}$ but Q is not, Q is a minimal nonsimple set of $\mathbb{I}$. □

8.6 CLIQUES, CORES, AND THE CRITICAL KERNEL

This section defines some key concepts that were introduced by Bertrand and Couprie [3,7,8]. The only really significant difference between our definitions and theirs is that whereas they define a *regular* (*critical*) essential cell as an essential cell whose face complex collapses (does not collapse) to its core, we will define a regular (critical) cell to be an essential cell whose core is NETS (is not NETS).

Throughout this section, $\mathbb{I}$ denotes an arbitrary binary image on a pure complex **K**.

8.6.1 $\mathbb{I}$-INDUCED CLIQUES AND $\mathbb{I}$-ESSENTIAL CELLS

Let c be any cell of $\mathbb{I}$'s foreground complex $\bigcup\{\mathbf{Faces}(\mathsf{d}) \mid \mathsf{d} \in \mathbb{I}^{-1}[1]\}$. Then we define

$$\mathsf{Clique}(c, \mathbb{I}) = \{\mathsf{v} \in \mathbb{I}^{-1}[1] \mid c \subseteq \mathsf{v}\}.$$

This set of one or more 1s of $\mathbb{I}$ will be called the $\mathbb{I}$-*induced clique* of the cell c. (In the terminology of [9,10,15], these $\mathbb{I}$-induced cliques are the cliques that are *essential* for $\mathbb{I}^{-1}[1]$.) Note that:

- $c \subseteq \bigcap \mathsf{Clique}(c, \mathbb{I})$.
- $\mathsf{Clique}(\mathsf{v}, \mathbb{I}) = \{\mathsf{v}\}$ if v is any 1 of $\mathbb{I}$.

We say that a cell d of **K** is $\mathbb{I}$-*essential* if d is a cell of $\mathbb{I}$'s foreground complex and $d = \bigcap \mathsf{Clique}(d, \mathbb{I})$. We observe that a cell d of **K** is $\mathbb{I}$-essential if and only if *either* d is a 1 of $\mathbb{I}$ *or* d is equal to the intersection of two or more 1s of $\mathbb{I}$. (The "only if" part of this observation is true because $\mathsf{Clique}(d, \mathbb{I})$ is a set of one or more 1s of $\mathbb{I}$. The "if" part of the observation is true because if $d = \bigcap \mathsf{Q}$ for some set Q of 1s of $\mathbb{I}$, then each element of Q contains d, and so $\mathsf{Q} \subseteq \mathsf{Clique}(d, \mathbb{I})$, whence $\bigcap \mathsf{Clique}(d, \mathbb{I})$ is no larger than $\bigcap \mathsf{Q} = d$.) Two immediate consequences of this observation are:

- The intersection of any two $\mathbb{I}$-essential cells is an $\mathbb{I}$-essential cell.
- If D is any set of 1s of $\mathbb{I}$, then every $(\mathbb{I} - \mathsf{D})$-essential cell is also an $\mathbb{I}$-essential cell.

However, the converse of the second fact is false: An $\mathbb{I}$-essential cell need not be $(\mathbb{I} - \mathsf{D})$-essential.

The "only if" parts of assertions 1 and 2 of the following lemma state easy (but important) consequences of the definition of an $\mathbb{I}$-essential cell. The "if" parts would be true even if e and e' were cells of $\mathbb{I}$'s foreground complex that were not $\mathbb{I}$-essential.

Lemma 8.6.1. *Let $\mathbb{I}$ be a binary image on a pure complex, and let each of e and e' be an $\mathbb{I}$-essential cell. Then:*

1. $\mathsf{Clique}(e', \mathbb{I}) = \mathsf{Clique}(e, \mathbb{I})$ *if and only if* $e = e'$.
2. $\mathsf{Clique}(e', \mathbb{I}) \subseteq \mathsf{Clique}(e, \mathbb{I})$ *if and only if* $e' \supseteq e$.

Proof. Since two sets are equal just if each set contains the other, assertion 2 implies assertion 1, and so it suffices to verify assertion 2. The "if" part of assertion 2 follows immediately from the definition of $\mathsf{Clique}(c, \mathbb{I})$. The "only if" part of assertion 2 is true because if $\mathsf{Clique}(e', \mathbb{I}) \subseteq \mathsf{Clique}(e, \mathbb{I})$, then $e' = \bigcap \mathsf{Clique}(e', \mathbb{I}) \supseteq \bigcap \mathsf{Clique}(e, \mathbb{I}) = e$ since e and e' are $\mathbb{I}$-essential. □

We see from this lemma that there is a natural (inclusion-reversing) bijective correspondence between the $\mathbb{I}$-essential cells and their $\mathbb{I}$-induced cliques. Thus any property of an $\mathbb{I}$-essential cell can also be thought of as a property of its $\mathbb{I}$-induced clique, and vice versa.

Moreover, if d is any cell of $\mathbb{I}$'s foreground complex, then there is a unique $\mathbb{I}$-essential cell e such that $\mathsf{Clique}(e, \mathbb{I}) = \mathsf{Clique}(d, \mathbb{I})$: It is readily confirmed that $e = \bigcap \mathsf{Clique}(d, \mathbb{I})$ is the $\mathbb{I}$-essential cell that has this property. Thus every cell's $\mathbb{I}$-induced clique is the $\mathbb{I}$-induced clique of an $\mathbb{I}$-essential cell.

8.6.2 $\mathbb{I}$-CORES

The $\mathbb{I}$-*core* of an $\mathbb{I}$-essential cell e is the complex induced by the $\mathbb{I}$-essential proper faces of e: It is the subcomplex of $\mathbf{Faces}(e) \setminus \{e\}$ that consists of the $\mathbb{I}$-essential proper faces of e and all of their proper faces. This complex will be denoted by $\mathbf{Core}(e, \mathbb{I})$. The $\mathbb{I}$-core of an $\mathbb{I}$-essential cell is a generalization of the $\mathbb{I}$-attachment complex of a 1 of $\mathbb{I}$:

- If d is any 1 of $\mathbb{I}$, then $\mathbf{Core}(\mathsf{d}, \mathbb{I}) = \mathbf{Attachment}(\mathsf{d}, \mathbb{I})$.

This will be seen from the following proposition, statements 1, 2, and 4 of which express $\mathbf{Core}(e, \mathbb{I})$ and its polyhedron in terms of e and $\mathbb{I}^{-1}[1] \setminus \mathsf{Clique}(e, \mathbb{I})$. (Note that $\mathbb{I}^{-1}[1] \setminus \mathsf{Clique}(e, \mathbb{I})$ is the set of all the 1s of $\mathbb{I}$ that do *not* contain e.)

Proposition 8.6.2. *Let $\mathbb{I}$ be any binary image on a pure complex, and e any $\mathbb{I}$-essential cell. Then:*

1. $\mathbf{Core}(e, \mathbb{I}) = \bigcup\{\mathbf{Faces}(e \cap \mathsf{w}) \mid \mathsf{w} \in \mathbb{I}^{-1}[1] \setminus \mathsf{Clique}(e, \mathbb{I})\}$.
2. $\mathbf{Core}(e, \mathbb{I}) = \mathbf{Faces}(e) \cap \bigcup\{\mathbf{Faces}(\mathsf{w}) \mid \mathsf{w} \in \mathbb{I}^{-1}[1] \setminus \mathsf{Clique}(e, \mathbb{I})\}$.
3. $\text{Facets}(\mathbf{Core}(e, \mathbb{I})) \subseteq \{e \cap \mathsf{w} \mid \mathsf{w} \in \mathbb{I}^{-1}[1] \setminus \mathsf{Clique}(e, \mathbb{I})\} \subseteq \mathbf{Core}(e, \mathbb{I})$.
4. $\bigcup \mathbf{Core}(e, \mathbb{I}) = \bigcup\{e \cap \mathsf{w} \mid \mathsf{w} \in \mathbb{I}^{-1}[1] \setminus \mathsf{Clique}(e, \mathbb{I})\} = e \cap \bigcup(\mathbb{I}^{-1}[1] \setminus \mathsf{Clique}(e, \mathbb{I}))$.

Proof. To establish assertion 1, we just need to show:

(i) If w is any element of $\mathbb{I}^{-1}[1] \setminus \mathsf{Clique}(e, \mathbb{I})$, then $\mathbf{Faces}(e \cap \mathsf{w}) \subseteq \mathbf{Core}(e, \mathbb{I})$.
(ii) If x is any cell of $\mathbf{Core}(e, \mathbb{I})$, then there is some w in $\mathbb{I}^{-1}[1] \setminus \mathsf{Clique}(e, \mathbb{I})$ such that $x \in \mathbf{Faces}(e \cap \mathsf{w})$.

To verify (i), let w be any element of $\mathbb{I}^{-1}[1] \setminus \mathsf{Clique}(e, \mathbb{I})$. Then, since e is $\mathbb{I}$-essential and w is a 1 of $\mathbb{I}$, the cell $e \cap \mathsf{w}$ is $\mathbb{I}$-essential. Moreover, $e \not\subseteq \mathsf{w}$ (as $\mathsf{w} \notin \mathsf{Clique}(e, \mathbb{I})$), and so $e \cap \mathsf{w} \subsetneq e$. Hence $e \cap \mathsf{w} \in \mathbf{Core}(e, \mathbb{I})$, and so $\mathbf{Faces}(e \cap \mathsf{w}) \subseteq \mathbf{Core}(e, \mathbb{I})$.

To verify (ii), let x be any cell of $\mathbf{Core}(e, \mathbb{I})$. Then x is a face of some $\mathbb{I}$-essential cell y such that $y \subsetneq e$. Since y and e are $\mathbb{I}$-essential cells such that $y \subsetneq e$, we have that $\mathsf{Clique}(y, \mathbb{I}) \supsetneq \mathsf{Clique}(e, \mathbb{I})$ (by Lemma 8.6.1), and so there exists an element w of $\mathsf{Clique}(y, \mathbb{I}) \setminus \mathsf{Clique}(e, \mathbb{I})$. Now $\mathsf{w} \in \mathbb{I}^{-1}[1] \setminus \mathsf{Clique}(e, \mathbb{I})$ and $y \subseteq \mathsf{w}$. Since

$y \subsetneq e$ and $y \subseteq \mathsf{w}$, we have that $y \subseteq e \cap \mathsf{w}$ and hence that $x \subseteq e \cap \mathsf{w}$ (since x is a face of y). Therefore $x \in \mathbf{Faces}(e \cap \mathsf{w})$ by property **P2a** of complexes.

This completes the proof of assertion 1. Assertion 2 follows from assertion 1 and property **P2b** of complexes. It is also easy to see that assertion 1 implies assertion 3. Since $\bigcup \text{Facets}(\mathbf{Core}(e, \mathbb{I})) = \bigcup \mathbf{Core}(e, \mathbb{I})$, assertion 3 implies assertion 4. □

8.6.3 $\mathbb{I}$-REGULAR AND $\mathbb{I}$-CRITICAL CELLS; THE CRITICAL KERNEL OF $\mathbb{I}$

An $\mathbb{I}$-*regular* cell is an $\mathbb{I}$-essential cell whose $\mathbb{I}$-core is NETS. An $\mathbb{I}$-*critical* cell is an $\mathbb{I}$-essential cell that is not $\mathbb{I}$-regular (i.e., an $\mathbb{I}$-essential cell whose $\mathbb{I}$-core is not NETS). Since the $\mathbb{I}$-core of a 1 of $\mathbb{I}$ is just the $\mathbb{I}$-attachment complex of that 1 of $\mathbb{I}$, the concept of an $\mathbb{I}$-regular cell is a generalization of the concept of an $\mathbb{I}$-simple 1:

- If d is any 1 of $\mathbb{I}$, then d is $\mathbb{I}$-regular just if d is a simple 1 of $\mathbb{I}$.

It should also be noted that every $\mathbb{I}$-essential cell whose $\mathbb{I}$-core is empty is $\mathbb{I}$-critical (because the empty set is not NETS). Equivalently, every minimal $\mathbb{I}$-essential cell (i.e., every $\mathbb{I}$-essential cell that has no $\mathbb{I}$-essential proper face) is $\mathbb{I}$-critical. It follows that every $\mathbb{I}$-essential 0-cell is $\mathbb{I}$-critical and that every $\mathbb{I}$-regular cell has an $\mathbb{I}$-critical proper face.

We mentioned earlier that if v is a 1 of a binary image $\mathbb{I}$ on a Cartesian array complex or any other strongly normal pure complex, then v is simple 1 of $\mathbb{I}$ if and only if the union of all the 1s of $\mathbb{I}$ that are weakly adjacent to v is NETS. More generally, if e is any $\mathbb{I}$-essential cell of a binary image $\mathbb{I}$ on a Cartesian array complex or any other strongly normal pure complex, then we have that:

- e is $\mathbb{I}$-regular if and only if the union of all the 1s of $\mathbb{I}$ that intersect but do not contain e is NETS.

This fact can be deduced from statement 4 of Proposition 8.6.2 and statement 2 of Proposition 8.3.5 when we put $C = e$ and let $N_1, \ldots, N_k$ be the 1s of $\mathbb{I}$ that intersect but do not contain e (i.e., the elements of $\mathbb{I}^{-1}[1] \setminus \mathsf{Clique}(e, \mathbb{I})$ that intersect e). The strong normality hypothesis also implies that if e is an $\mathbb{I}$-essential cell, then a 1 of $\mathbb{I}$ intersects $e = \bigcap \mathsf{Clique}(e, \mathbb{I})$ if (and, trivially, only if) that 1 intersects every element of $\mathsf{Clique}(e, \mathbb{I})$.[14]

The *critical kernel* of $\mathbb{I}$ is the complex whose elements are the $\mathbb{I}$-critical cells and their proper faces. Thus the critical kernel of $\mathbb{I}$ is the subcomplex of $\mathbb{I}$'s foreground complex $\bigcup\{\mathbf{Faces}(\mathsf{v}) \mid \mathsf{v} \in \mathbb{I}^{-1}[1]\}$ that is induced by the nonsimple 1s of $\mathbb{I}$ and the lower-dimensional $\mathbb{I}$-critical cells.

[14] In view of this, [9, Thm. 16], [14, Thm. 12], and the result of Subsection 8.4.3, when $\mathbb{I}$ is a binary image on a 3-dimensional Cartesian array complex we have that the set of all the 1s of $\mathbb{I}$ that intersect but do not contain an $\mathbb{I}$-essential cell e is a set of 3-cells that is *reducible* in the sense of Bertrand and Couprie [9,10,15] if and only if the union of those 1s is NETS.

8.7 CHARACTERIZATIONS OF P-SIMPLE 1S, HEREDITARILY SIMPLE SETS, AND MINIMAL NONSIMPLE SETS

In this section, we achieve our goal of generalizing the main theorems of Bertrand and Couprie's paper [8] to binary images on arbitrary pure polytopal complexes. (However, some of our arguments will depend on Theorem 8.5.5, which will be proved in the next section.)

Theorem 8.7.1 (NETS-based generalization of [8, Thm. 21]). *Let $\mathbb{I}$ be a binary image on a pure complex, let* D *be a set of* 1*s of* $\mathbb{I}$*, and let* d *be any element of* D*. Then* d *is P-simple for* D *in* $\mathbb{I}$ *if and only if there is no* $\mathbb{I}$*-critical cell* f *such that* $\mathsf{d} \in \mathsf{Clique}(f, \mathbb{I}) \subseteq \mathsf{D}$.

The "only if" part of this theorem will be deduced from the following proposition.

Proposition 8.7.2. *Let* $\mathbb{I}$ *be a binary image on a pure complex, and let* f *be an* $\mathbb{I}$*-critical cell. Then no element of* $\mathsf{Clique}(f, \mathbb{I})$ *is P-simple for* $\mathsf{Clique}(f, \mathbb{I})$ *in* $\mathbb{I}$.

Proof. Let $\mathsf{d}_0, \ldots, \mathsf{d}_k$ be any enumeration of $\mathsf{Clique}(f, \mathbb{I})$, so that $\mathsf{d}_0 \cap \cdots \cap \mathsf{d}_k = f$ (since f is $\mathbb{I}$-essential). For $1 \leq i \leq k$, let $a_i = \mathsf{d}_i \cap \mathsf{d}_0$. Let $R = \bigcup(\mathbb{I}^{-1}[1] \setminus \mathsf{Clique}(f, \mathbb{I}))$, and let $A = R \cap \mathsf{d}_0$. Then we claim:

- **A.** $A \cap a_1 \cap \cdots \cap a_k = R \cap \mathsf{d}_0 \cap \cdots \cap \mathsf{d}_k = R \cap f = \bigcup \mathbf{Core}(f, \mathbb{I})$.
- **B.** $A \cap a_1 \cap \cdots \cap a_k$ is not NETS.
- **C.** There is a nonempty subcollection of $\{A, a_1, \ldots, a_k\}$ whose union is not NETS.
- **D.** The intersection of any nonempty subcollection of $\{a_1, \ldots, a_k\}$ is NETS.
- **E.** The union of any nonempty subcollection of $\{a_1, \ldots, a_k\}$ is NETS.
- **F.** d_0 is not P-simple for $\mathsf{Clique}(f, \mathbb{I})$ in $\mathbb{I}$.

Now we justify these assertions. Assertion A follows from the assertion 4 of Proposition 8.6.2. Since f is $\mathbb{I}$-critical, assertion B follows from assertion A. Assertion C follows from assertion B and Lemma 8.3.4. Assertion D is true because the intersection of any nonempty subcollection of $\{a_1, ..., a_k\}$ is nonempty (since it contains f) and is therefore a nonempty convex polytope, which is NETS. Assertion E follows from assertion D and Lemma 8.3.4. Assertions C and E imply that there exist $a_{i_1}, \ldots, a_{i_m} \in \{a_1, \ldots, a_k\}$ such that $A \cup a_{i_1} \cup \cdots \cup a_{i_m}$ is not NETS.

But $A \cup a_{i_1} \cup \cdots \cup a_{i_m} = \bigcup \mathbf{Attachment}(\mathsf{d}_0, \mathbb{I} - (\mathsf{Clique}(f, \mathbb{I}) \setminus \{\mathsf{d}_0, \mathsf{d}_{i_1}, \ldots, \mathsf{d}_{i_m}\}))$ by (8.9). So, since $A \cup a_{i_1} \cup \cdots \cup a_{i_m}$ is not NETS, we have that d_0 is not a simple 1 of $\mathbb{I} - (\mathsf{Clique}(f, \mathbb{I}) \setminus \{\mathsf{d}_0, \mathsf{d}_{i_1}, \ldots, \mathsf{d}_{i_m}\})$, which implies assertion F. Since $\mathsf{d}_0, \ldots, \mathsf{d}_k$ is an arbitrary enumeration of $\mathsf{Clique}(f, \mathbb{I})$, assertion F implies the proposition. □

Proof of Theorem 8.7.1. The "only if" part of the theorem follows from Proposition 8.7.2: Indeed, if there is an $\mathbb{I}$-critical cell f such that $\mathsf{d} \in \mathsf{Clique}(f, \mathbb{I}) \subseteq \mathsf{D}$, then we see from Proposition 8.7.2 that d is not P-simple for $\mathsf{Clique}(f, \mathbb{I})$ in $\mathbb{I}$, whence d is not P-simple for D in $\mathbb{I}$.

To prove the "if" part, suppose that there is no $\mathbb{I}$-critical cell f such that $\mathsf{d} \in \mathsf{Clique}(f, \mathbb{I}) \subseteq \mathsf{D}$, and let $\mathsf{d}_1, \ldots, \mathsf{d}_k$ be any enumeration of the elements of $\mathsf{D} \setminus \{\mathsf{d}\}$.

Then we just need to show that d is a simple 1 of $\mathbb{I} - \{d_{i_1}, \ldots, d_{i_m}\}$ for every subsequence $i_1, \ldots, i_m$ of $1, \ldots, k$ (including the empty sequence).

Since there is no $\mathbb{I}$-critical cell f such that $d \in \mathsf{Clique}(f, \mathbb{I}) \subseteq D$ (and since $d \in \mathsf{Clique}(d, \mathbb{I}) = \{d\} \subseteq D$), d is an $\mathbb{I}$-regular cell (i.e., d is a simple 1 of $\mathbb{I}$) and so $\bigcup \mathbf{Core}(d, \mathbb{I})$ is NETS.

Now let $R = \bigcup(\mathbb{I}^{-1}[1] \setminus D)$. Note that if a cell f of $\mathbb{I}$'s foreground complex does not lie in $R = \bigcup(\mathbb{I}^{-1}[1] \setminus D)$, then $\mathsf{Clique}(f, \mathbb{I}) \subseteq D$. So, for any permutation $i_1, \ldots, i_k$ of $1, \ldots, k$ and $1 \leq m \leq k$, if $d \cap d_{i_1} \cap \cdots \cap d_{i_m} \not\subseteq R$, then we have that $d \in \mathsf{Clique}(d \cap d_{i_1} \cap \cdots \cap d_{i_m}, \mathbb{I}) \subseteq D$ and hence that $d \cap d_{i_1} \cap \cdots \cap d_{i_m}$ is an $\mathbb{I}$-regular cell, whence $\bigcup \mathbf{Core}(d \cap d_{i_1} \cap \cdots \cap d_{i_m}, \mathbb{I})$ is NETS.

We see from the above two paragraphs that:

(i) $\bigcup \mathbf{Core}(d, \mathbb{I})$ is NETS.

(ii) For any permutation $i_1, \ldots, i_k$ of $1, \ldots, k$ and $1 \leq m \leq k$, either $d \cap d_{i_1} \cap \cdots \cap d_{i_m} \subseteq R$ or $\bigcup \mathbf{Core}(d \cap d_{i_1} \cap \cdots \cap d_{i_m}, \mathbb{I})$ is NETS.

Let $A = R \cap d$, and, for $i = 1, \ldots, k$, let $a_i = d_i \cap d$. Then by Proposition 8.6.2 we have that $\bigcup \mathbf{Core}(d, \mathbb{I}) = (R \cup d_1 \cup \cdots \cup d_k) \cap d = A \cup a_1 \cup \cdots \cup a_k$. This and fact (i) imply:

A. $A \cup a_1 \cup \cdots \cup a_k$ is NETS.

Now let $i_1, \ldots, i_k$ be any permutation of $1, \ldots, k$, and let $1 \leq m \leq k$.

Whenever $d \cap d_{i_1} \cap \cdots \cap d_{i_m} \not\subseteq R \cup d_{i_{m+1}} \cup \cdots \cup d_{i_k}$, we see from Proposition 8.6.2 that the set $\bigcup \mathbf{Core}(d \cap d_{i_1} \cap \cdots \cap d_{i_m}, \mathbb{I})$ is just $(R \cup d_{i_{m+1}} \cup \cdots \cup d_{i_k}) \cap (d \cap d_{i_1} \cap \cdots \cap d_{i_m}) = (A \cup a_{i_{m+1}} \cup \cdots \cup a_{i_k}) \cap (a_{i_1} \cap \cdots \cap a_{i_m})$, and (since $d \cap d_{i_1} \cap \cdots \cap d_{i_m} \not\subseteq R$) we see from fact (ii) that this set is NETS. Note also that $d \cap d_{i_1} \cap \cdots \cap d_{i_m} \subset R \cup d_{i_{m+1}} \cup \cdots \cup d_{i_k}$ just if $d \cap d_{i_1} \cap \cdots \cap d_{i_m} \subseteq d \cap (R \cup d_{i_{m+1}} \cup \cdots \cup d_{i_k})$ or, equivalently, just if $a_{i_1} \cap \cdots \cap a_{i_m} \subseteq A \cup a_{i_{m+1}} \cup \cdots \cup a_{i_k}$. So we conclude that:

B. Either $(A \cup a_{i_{m+1}} \cup \cdots \cup a_{i_k}) \cap (a_{i_1} \cap \cdots \cap a_{i_m})$ is NETS or $a_{i_1} \cap \cdots \cap a_{i_m} \subseteq A \cup a_{i_{m+1}} \cup \cdots \cup a_{i_k}$.

Now we claim that the following are also true for all permutations $i_1, \ldots, i_k$ of $1, \ldots, k$ and $1 \leq m \leq k$:

C. $(A \cup a_{i_{m+1}} \cup \cdots \cup a_{i_k}) \cup (a_{i_1} \cap \cdots \cap a_{i_m})$ is NETS.
D. $(A \cup a_{i_{m+1}} \cup \cdots \cup a_{i_k})$ is NETS.

Note that C implies D in view of B and Lemma 8.3.3. We also see that D is true if $m = 0$ (since D is equivalent to A when $m = 0$). To establish C and D in the cases $1 \leq m \leq k$, we now assume as an induction hypothesis that D holds for *all* permutations $i_1, \ldots, i_k$ of $1, \ldots, k$ and $0 \leq m < r$ (where $r \in \{1, \ldots, k\}$) and deduce that C holds for any permutation $i_1, \ldots, i_k$ of $1, \ldots, k$ and $m = r$.

For any permutation $i_1, \ldots, i_k$ of $1, \ldots, k$ and $m = r$, $(A \cup a_{i_{m+1}} \cup \cdots \cup a_{i_k}) \cup (a_{i_1} \cap \cdots \cap a_{i_m}) = \bigcap_{j=1}^{r} P_j$, where $P_j = A \cup a_{i_j} \cup a_{i_{r+1}} \cup \cdots \cup a_{i_k}$ for $1 \leq j \leq r$. To show that C holds for this permutation, we just need to show that $\bigcap_{j=1}^{r} P_j$ is NETS. But the induction hypothesis implies that the collection of sets $\mathcal{P} = \{P_j \mid 1 \leq j \leq r\}$

has the property that the union of any nonempty subcollection of $\mathcal{P}$ is NETS. Hence it follows from Lemma 8.3.4 that $\bigcap_{j=1}^{r} P_j$ is NETS, as required. This justifies our claim.

Since D is true for all permutations $i_1, \ldots, i_k$ of $1, \ldots, k$ and $0 \leq m \leq k$, and $A \cup a_{i_{m+1}} \cup \cdots \cup a_{i_k}$ is just the set $\bigcup$ **Attachment**$(\mathsf{d}, \mathbb{I} - \{\mathsf{d}_{i_1}, \ldots, \mathsf{d}_{i_m}\})$, we have that d is a simple 1 of $\mathbb{I} - \{\mathsf{d}_{i_1}, \ldots, \mathsf{d}_{i_m}\}$ for every subsequence $i_1, \ldots, i_m$ of $1, \ldots, k$, and so d is P-simple for D in $\mathbb{I}$. □

Corollary 8.7.3. *Let $\mathbb{I}$ be a binary image on a pure complex, and let* D *be a set of* 1s *of $\mathbb{I}$. Then:*

1. D *is a hereditarily simple set of $\mathbb{I}$ if and only if no $\mathbb{I}$-critical cell f satisfies* $\mathsf{Clique}(f, \mathbb{I}) \subseteq \mathsf{D}$.
2. D *is a hereditarily simple set of $\mathbb{I}$ if and only if there is no cell f of the critical kernel of $\mathbb{I}$ such that* $\mathsf{Clique}(f, \mathbb{I}) \subseteq \mathsf{D}$.
3. D *is a hereditarily simple set of $\mathbb{I}$ if and only if* $\bigcup(\mathbb{I}^{-1}[1] \setminus \mathsf{D})$ *contains all cells of $\mathbb{I}$'s critical kernel.*

Proof. Assertion 1 follows from Theorems 8.5.5 and 8.7.1. Assertion 2 is equivalent to assertion 1 since there is an $\mathbb{I}$-critical cell f such that $\mathsf{Clique}(f, \mathbb{I}) \subseteq \mathsf{D}$ if and only if there is a cell f of the critical kernel of $\mathbb{I}$ such that $\mathsf{Clique}(f, \mathbb{I}) \subseteq \mathsf{D}$: Here the "only if" part is true since every $\mathbb{I}$-critical cell is a cell of the critical kernel of $\mathbb{I}$, while the "if" part is true because if f is a cell of the critical kernel of $\mathbb{I}$ such that $\mathsf{Clique}(f, \mathbb{I}) \subseteq \mathsf{D}$, then f is a face of some $\mathbb{I}$-critical cell f', and any such f' satisfies $\mathsf{Clique}(f', \mathbb{I}) \subseteq \mathsf{Clique}(f, \mathbb{I}) \subseteq \mathsf{D}$.

Assertion 3 is equivalent to assertion 2: Indeed, a cell f of $\mathbb{I}$'s foreground complex does not lie or does lie in $\bigcup(\mathbb{I}^{-1}[1] \setminus \mathsf{D})$ according to whether there does not exist or does exist an element of $\mathbb{I}^{-1}[1] \setminus \mathsf{D}$ that has f as a face, or, equivalently, according to whether $\mathsf{Clique}(f, \mathbb{I}) \subseteq \mathsf{D}$ or $\mathsf{Clique}(f, \mathbb{I}) \not\subseteq \mathsf{D}$. □

The next theorem characterizes the minimal nonsimple sets of any binary image $\mathbb{I}$ on a pure complex as the $\mathbb{I}$-induced cliques of (inclusion-)maximal $\mathbb{I}$-critical cells. Note that the maximal $\mathbb{I}$-critical cells are just the facets of the critical kernel of $\mathbb{I}$. (In the terminology of Bertrand and Couprie [7,8], the $\mathbb{I}$-induced cliques of maximal $\mathbb{I}$-critical cells are the *crucial cliques* for $\mathbb{I}$'s foreground complex.)

Theorem 8.7.4 (NETS-based generalization of [8, Thm. 27])**.** *Let $\mathbb{I}$ be a binary image on a pure complex, and let* D *be any set of* 1s *of $\mathbb{I}$. Then* D *is a minimal nonsimple set of $\mathbb{I}$ if and only if there is a maximal $\mathbb{I}$-critical cell e such that* $\mathsf{Clique}(e, \mathbb{I}) = \mathsf{D}$.

Proof. We will deduce this theorem from the following claims:

(a) If e is any $\mathbb{I}$-critical cell, then $\mathsf{Clique}(e, \mathbb{I})$ is not a hereditarily simple set of $\mathbb{I}$.
(b) If D is a minimal nonsimple set of $\mathbb{I}$, then there is some $\mathbb{I}$-critical cell e such that $\mathsf{Clique}(e, \mathbb{I}) \subseteq \mathsf{D}$.
(c) If D is a minimal nonsimple set of $\mathbb{I}$, then there is no $\mathbb{I}$-critical cell e such that $\mathsf{Clique}(e, \mathbb{I}) \subsetneq \mathsf{D}$.

(d) If e is any nonmaximal $\mathbb{I}$-critical cell, then $\mathsf{Clique}(e, \mathbb{I})$ is not a minimal nonsimple set of $\mathbb{I}$.

Claims (a) and (b) follow from assertion 1 of Corollary 8.7.3: Claim (a) follows from the "only if" part of that assertion, while claim (b) follows from the "if" part (since a minimal nonsimple set of $\mathbb{I}$ is not a hereditarily simple set of $\mathbb{I}$). Claim (c) follows from claim (a) and the fact that, if D is a minimal nonsimple set of $\mathbb{I}$, then every proper subset of D is a hereditarily simple set of $\mathbb{I}$. Claim (d) follows from claim (c) because if e' is an $\mathbb{I}$-critical cell that strictly contains another $\mathbb{I}$-essential cell e, then $\mathsf{Clique}(e', \mathbb{I}) \subsetneq \mathsf{Clique}(e, \mathbb{I})$ (by Lemma 8.6.1).

The "only if" part of the theorem follows from (b), (c), and (d): Indeed, if D is a minimal nonsimple set of $\mathbb{I}$, then (b) and (c) imply that there is some $\mathbb{I}$-critical cell e such that $\mathsf{Clique}(e, \mathbb{I}) = \mathsf{D}$, while (d) implies that such a cell e must be a maximal $\mathbb{I}$-critical cell.

To see that the "if" part of the theorem is also true, let e be any maximal $\mathbb{I}$-critical cell. Then we see from Lemma 8.6.1 that there is no $\mathbb{I}$-critical cell e' such that $\mathsf{Clique}(e', \mathbb{I}) \subsetneq \mathsf{Clique}(e, \mathbb{I})$. From this and from the "only if" part of the theorem we deduce that no proper subset of $\mathsf{Clique}(e, \mathbb{I})$ is a minimal nonsimple set of $\mathbb{I}$. This implies every proper subset of $\mathsf{Clique}(e, \mathbb{I})$ is a hereditarily simple set of $\mathbb{I}$. But $\mathsf{Clique}(e, \mathbb{I})$ itself is not a hereditarily simple set of $\mathbb{I}$, by (a). Hence $\mathsf{Clique}(e, \mathbb{I})$ is a minimal nonsimple set of $\mathbb{I}$. □

For any binary image $\mathbb{I}$ on a pure complex, we now give a characterization of those $\mathbb{I}$-essential cells x for which $\mathsf{Clique}(x, \mathbb{I})$ is a weak component of the 1s of $\mathbb{I}$.

Theorem 8.7.5 (NETS-based generalization of [8, Thm. 20]). *Let $\mathbb{I}$ be a binary image on a pure complex, and let x be an $\mathbb{I}$-essential cell. Then the following are equivalent:*

1. **Core**$(x, \mathbb{I}) = \emptyset$*, and x is a maximal $\mathbb{I}$-critical cell.*
2. $\mathsf{Clique}(x, \mathbb{I})$ *is a weak component of the 1s of $\mathbb{I}$.*

Proof. Suppose condition 2 holds. Then $(\bigcup \mathsf{Clique}(x, \mathbb{I})) \cap \bigcup(\mathbb{I}^{-1}[1] \setminus \mathsf{Clique}(x, \mathbb{I})) = \emptyset$, from which it follows that $x \cap \bigcup(\mathbb{I}^{-1}[1] \setminus \mathsf{Clique}(x, \mathbb{I})) = \emptyset$ and hence (by Proposition 8.6.2) that $\mathbf{Core}(x, \mathbb{I}) = \emptyset$ and x is $\mathbb{I}$-critical. To complete the proof that condition 2 implies condition 1, we now show that x is a maximal $\mathbb{I}$-critical cell. Suppose there is an $\mathbb{I}$-essential cell e that strictly contains x. Then we just need to show that e must be an $\mathbb{I}$-regular cell. By Lemma 8.6.1, $\mathsf{Clique}(e, \mathbb{I}) \subsetneq \mathsf{Clique}(x, \mathbb{I})$ and—since $e \subseteq \bigcup \mathsf{Clique}(e, \mathbb{I}) \subsetneq \bigcup \mathsf{Clique}(x, \mathbb{I})$ and $(\bigcup \mathsf{Clique}(x, \mathbb{I})) \cap \bigcup(\mathbb{I}^{-1}[1] \setminus \mathsf{Clique}(x, \mathbb{I})) = \emptyset$, which imply $e \cap \mathsf{w} = \emptyset$ for all $\mathsf{w} \in \mathbb{I}^{-1}[1] \setminus \mathsf{Clique}(x, \mathbb{I})$—we see from Proposition 8.6.2 that $\bigcup \mathbf{Core}(e, \mathbb{I}) = \bigcup \mathcal{P}$, where $\mathcal{P} = \{e \cap \mathsf{w} \mid \mathsf{w} \in \mathsf{Clique}(x, \mathbb{I}) \setminus \mathsf{Clique}(e, \mathbb{I})\}$. Since $\mathcal{P}$ is a collection of convex polytopes that contain x, the intersection of any nonempty subcollection of $\mathcal{P}$ is a nonempty convex polytope, and so is NETS, by **NETS-2**. Moreover, $\mathcal{P} \neq \emptyset$ since $\mathsf{Clique}(e, \mathbb{I}) \subsetneq \mathsf{Clique}(x, \mathbb{I})$. Hence $\bigcup \mathbf{Core}(e, \mathbb{I}) = \bigcup \mathcal{P}$ is NETS, by Lemma 8.3.4, and so e is an $\mathbb{I}$-regular cell, as required.

To prove that condition 1 implies condition 2, we now suppose that $\mathbf{Core}(x, \mathbb{I}) = \emptyset$ but $\mathsf{Clique}(x, \mathbb{I})$ is a *not* a weak component of the 1s of $\mathbb{I}$, and deduce that x is not a maximal $\mathbb{I}$-critical cell.

Since $\mathsf{Clique}(x, \mathbb{I})$ is not a weak component of the 1s of $\mathbb{I}$, some element of $\mathsf{Clique}(x, \mathbb{I})$ intersects the set $\bigcup(\mathbb{I}^{-1}[1] \setminus \mathsf{Clique}(x, \mathbb{I}))$. Let M be a *maximal* subset of $\mathsf{Clique}(x, \mathbb{I})$ such that the $\mathbb{I}$-essential cell $\bigcap \mathsf{M}$ intersects the set $\bigcup(\mathbb{I}^{-1}[1] \setminus \mathsf{Clique}(x, \mathbb{I}))$. Now $x \subseteq \bigcap \mathsf{M}$ (as $\mathsf{M} \subseteq \mathsf{Clique}(x, \mathbb{I})$) and $\mathsf{Clique}(\bigcap \mathsf{M}, \mathbb{I}) = \mathsf{M}$ (because if $\mathsf{Clique}(\bigcap \mathsf{M}, \mathbb{I}) \supsetneq \mathsf{M}$, then M would not be a *maximal* subset of $\mathsf{Clique}(x, \mathbb{I})$ with the stated property). Since $\mathbf{Core}(x, \mathbb{I}) = \emptyset$, we see from Proposition 8.6.2 that x does *not* intersect $\bigcup(\mathbb{I}^{-1}[1] \setminus \mathsf{Clique}(x, \mathbb{I}))$, and so $x \neq \bigcap \mathsf{M}$. We now complete the proof by showing that the $\mathbb{I}$-essential cell $\bigcap \mathsf{M}$, which strictly contains x, is $\mathbb{I}$-critical, so that x is *not* a maximal $\mathbb{I}$-critical cell. In view of **NETS-1**, we can do this by showing that $\bigcup \mathbf{Core}(\bigcap \mathsf{M}, \mathbb{I})$ is not connected.

Since $x \subsetneq \bigcap \mathsf{M}$, Lemma 8.6.1 implies $\mathsf{Clique}(x, \mathbb{I}) \supsetneq \mathsf{Clique}(\bigcap \mathsf{M}, \mathbb{I}) = \mathsf{M}$, so $\mathsf{Clique}(x, \mathbb{I}) \setminus \mathsf{M} \neq \emptyset$. Consider the sets $A = \{\mathsf{w} \cap \bigcap \mathsf{M} \mid \mathsf{w} \in \mathsf{Clique}(x, \mathbb{I}) \setminus \mathsf{M}\}$ and $B = \{\mathsf{w} \cap \bigcap \mathsf{M} \mid \mathsf{w} \in \mathbb{I}^{-1}[1] \setminus \mathsf{Clique}(x, \mathbb{I})\}$. Since $(\mathsf{Clique}(x, \mathbb{I}) \setminus \mathsf{M}) \cup (\mathbb{I}^{-1}[1] \setminus \mathsf{Clique}(x, \mathbb{I})) = \mathbb{I}^{-1}[1] \setminus \mathsf{M}$ and $\mathsf{M} = \mathsf{Clique}(\bigcap \mathsf{M}, \mathbb{I})$, Proposition 8.6.2 implies that $\mathrm{Facets}(\mathbf{Core}(\bigcap \mathsf{M}, \mathbb{I})) \subseteq A \cup B \subseteq \mathbf{Core}(\bigcap \mathsf{M}, \mathbb{I})$. Hence:

- $\bigcup \mathbf{Core}(\bigcap \mathsf{M}, \mathbb{I}) = (\bigcup A) \cup (\bigcup B)$.

Now $\bigcup A \neq \emptyset$ because $\mathsf{Clique}(x, \mathbb{I}) \setminus \mathsf{M} \neq \emptyset$, and, for any $\mathsf{w} \in \mathsf{Clique}(x, \mathbb{I})$, the cell x lies in $\mathsf{w} \cap \bigcap \mathsf{M}$. Moreover, $\bigcup B \neq \emptyset$ because the very definition of M implies that there is some $\mathsf{w} \in \mathbb{I}^{-1}[1] \setminus \mathsf{Clique}(x, \mathbb{I})$ such that $\mathsf{w} \cap \bigcap \mathsf{M} \neq \emptyset$. However, we claim that $(\bigcup A) \cap (\bigcup B) = \emptyset$.

Indeed, suppose not. Then there exist $a \in A$ and $b \in B$ such that $a \cap b \neq \emptyset$. Since $a \in A$ and $b \in B$, there exist $\mathsf{w}_a \in \mathsf{Clique}(x, \mathbb{I}) \setminus \mathsf{M}$ and $\mathsf{w}_b \in \mathbb{I}^{-1}[1] \setminus \mathsf{Clique}(x, \mathbb{I})$ such that $a = \mathsf{w}_a \cap \bigcap \mathsf{M}$, $b = \mathsf{w}_b \cap \bigcap \mathsf{M}$, and $\mathsf{w}_a \cap \mathsf{w}_b \cap \bigcap \mathsf{M} = a \cap b \neq \emptyset$. But we also have that $\mathsf{w}_a \cap \mathsf{w}_b \cap \bigcap \mathsf{M} = \mathsf{w}_b \cap \bigcap(\{\mathsf{w}_a\} \cup \mathsf{M}) = \emptyset$ because $\mathsf{w}_b \subseteq \bigcup(\mathbb{I}^{-1}[1] \setminus \mathsf{Clique}(x, \mathbb{I}))$, $\mathsf{w}_a \in \mathsf{Clique}(x, \mathbb{I}) \setminus \mathsf{M}$, and M is a *maximal* subset of $\mathsf{Clique}(x, \mathbb{I})$ such that $\bigcup(\mathbb{I}^{-1}[1] \setminus \mathsf{Clique}(x, \mathbb{I}))$ intersects $\bigcap \mathsf{M}$. This contradiction justifies our claim.

Since $\bigcup A \neq \emptyset$, $\bigcup B \neq \emptyset$, and $(\bigcup A) \cap (\bigcup B) = \emptyset$, and since $\bigcup A$ and $\bigcup B$ are closed sets, we deduce that $\bigcup \mathbf{Core}(\bigcap \mathsf{M}, \mathbb{I}) = (\bigcup A) \cup (\bigcup B)$ is not connected and therefore not NETS. Hence $\bigcap \mathsf{M}$ is $\mathbb{I}$-critical, and so the proper subset x of $\bigcap \mathsf{M}$ is not a maximal $\mathbb{I}$-critical cell. □

Corollary 8.7.6. *Let $\mathbb{I}$ be a binary image on a pure complex, and let D be a weak component of the 1s of $\mathbb{I}$ such that $\bigcap \mathsf{D} \neq \emptyset$. Then D is a minimal nonsimple set of $\mathbb{I}$.*

Proof. We first observe that $\mathsf{D} = \mathsf{Clique}(\bigcap \mathsf{D}, \mathbb{I})$: Indeed, if $\mathsf{D} \neq \mathsf{Clique}(\bigcap \mathsf{D}, \mathbb{I})$, then (since $\mathsf{D} \subseteq \mathsf{Clique}(\bigcap \mathsf{D}, \mathbb{I})$) there would be some $\mathsf{v} \in \mathsf{Clique}(\bigcap \mathsf{D}, \mathbb{I}) \setminus \mathsf{D}$; any such v is not in D but intersects $\bigcup \mathsf{D}$, which is impossible because D is a weak component of the 1s of $\mathbb{I}$. As $\mathsf{D} = \mathsf{Clique}(\bigcap \mathsf{D}, \mathbb{I})$ is a weak component of the 1s of $\mathbb{I}$, it follows from Theorems 8.7.4 and 8.7.5 that D is a minimal nonsimple set of $\mathbb{I}$. □

Corollary 8.7.7. *Let $\mathbb{I}$ be a binary image on a pure complex* **K**, *and let* D *be a set of* 1*s of* $\mathbb{I}$ *such that* $\bigcap$ D *is a* 0*-cell of* **K**. *Then* D *is a minimal nonsimple set of* $\mathbb{I}$ *if and only if* D *is a weak component of the* 1*s of* $\mathbb{I}$.

Proof. If D is a weak component of the 1s of $\mathbb{I}$, then it follows from Corollary 8.7.6 that D is a minimal nonsimple set of $\mathbb{I}$. Conversely, suppose D is a minimal nonsimple set of $\mathbb{I}$. Then, by Theorem 8.7.4, D $=$ Clique$(e, \mathbb{I})$ for some maximal $\mathbb{I}$-critical cell e. Since e is $\mathbb{I}$-essential, we have that $e = \bigcap$ Clique$(e, \mathbb{I}) = \bigcap$ D, so that e is a 0-cell and therefore **Core**$(e, \mathbb{I}) = \emptyset$. Hence we see from Theorem 8.7.5 that D $=$ Clique$(e, \mathbb{I})$ is a weak component of the 1s of $\mathbb{I}$. □

8.8 A PROOF OF THEOREM 8.5.5 AND A SECOND PROOF OF THE "IF" PART OF THEOREM 8.7.1 BASED ON NETS-COLLAPSING

In the previous section, we made essential use of Theorem 8.5.5 to establish Corollary 8.7.3, which was then used to prove a number of other results. (One of those results was Theorem 8.7.4, which generalized Bertrand and Couprie's characterization of the minimal nonsimple sets of a binary image $\mathbb{I}$ on a 2-, 3-, or 4-dimensional Cartesian array complex—as the $\mathbb{I}$-induced cliques of maximal $\mathbb{I}$-critical cells—to binary images $\mathbb{I}$ on arbitrary pure polytopal complexes.) However, Theorem 8.5.5 has not itself been proved. Accordingly, the most important objective of this section is to prove that theorem.

To do this, we will define a generalized version of collapsing, which we call *NETS-collapsing*, in such a way that if $\mathbb{I}$ is a binary image on any pure complex and $\mathbb{I}'$ is a binary image obtained from $\mathbb{I}$ by sequential deletion of simple 1s, then the foreground complex of $\mathbb{I}$ NETS-collapses to the foreground complex of $\mathbb{I}'$. (The previous sentence would *not* be true for arbitrary pure complexes if we replaced "NETS-collapses" with "collapses": In fact, this is a notable difference between our NETS definition of simple 1s and the definition used by Bertrand and Couprie.) NETS-collapsing has another nice property, stated in Theorem 8.8.9, which will be just what we need to establish Theorem 8.5.5.

After proving Theorem 8.5.5, we will go on to establish an analog, for NETS-collapsing, of an important result of Bertrand [3, Thm. 4.2] regarding collapsing and the critical kernel. This will be used to give a second proof of the "if" part of Theorem 8.7.1.

8.8.1 NETS-COLLAPSING AND A PROOF OF THEOREM 8.5.5

Recall that a cell c of a complex **C** is called a *facet* of **C** if c is not a proper face of another cell of **C**. In a pure n-dimensional complex, the facets are just the n-cells. A complex that is not pure will also have other facets.

Let f be a facet of a complex $\mathbf{C}$ (which need not be pure), and let $\mathbf{C}'$ be any subcomplex of $\mathbf{C}$ that satisfies the following conditions:

1. $f \notin \mathbf{C}'$.
2. $f \cup \bigcup \mathbf{C}' = \bigcup \mathbf{C}$.
3. $f \cap \bigcup \mathbf{C}'$ is NETS.

Then we call the transformation of $\mathbf{C}$ to $\mathbf{C}'$ an *elementary NETS-collapse* operation on $\mathbf{C}$ that removes the facet f, and we say that $\mathbf{C}$ *NETS-collapses elementarily to* $\mathbf{C}'$ *through removal of* f. We see from (8.3), (8.5), and the fact $f = \bigcup \mathbf{Faces}(f)$ that condition 2 is equivalent to:

2′. $\mathbf{Faces}(f) \cup \mathbf{C}' = \mathbf{C}$.

Thus the original complex $\mathbf{C}$ is uniquely determined by the resultant complex $\mathbf{C}'$ and the removed facet f of $\mathbf{C}$. We also see from 2′ that every facet of $\mathbf{C}$ other than f lies in $\mathbf{C}'$, and so the removed facet is uniquely determined by the original complex and the resultant complex.

However, we emphasize that the resultant complex $\mathbf{C}'$ is *not* uniquely determined by the original complex $\mathbf{C}$ and the removed facet f: For a given complex $\mathbf{C}$ and facet f of $\mathbf{C}$, there may be more than one elementary NETS-collapse operation on $\mathbf{C}$ that removes f.

The definition of an elementary NETS-collapse operation is motivated by the following examples:

Example 8.8.1. Let $\mathbb{I}$ be a binary image on a pure complex, and let v be a 1 of $\mathbb{I}$. Then it follows from (8.9) that v is a simple 1 of $\mathbb{I}$ if and only if the complex induced by $\mathbb{I}^{-1}[1]$ NETS-collapses elementarily to the complex induced by $\mathbb{I}^{-1}[1] \setminus \{\mathsf{v}\}$ through removal of the facet v.

Example 8.8.2. Any elementary collapse operation on a complex that removes a free pair of cells (f, g) is an elementary NETS-collapse operation that removes the facet f. This is because when a convex polytope g is a $(\dim(f)-1)$-dimensional face of a convex polytope f, the complex $\mathbf{Faces}(f) \setminus \{f, g\}$ is collapsible (as mentioned in footnote 12) and is therefore NETS (by Proposition 8.4.2).

Proposition 8.8.3. *Let* $\mathbf{C} \supsetneq \mathbf{C}'$ *be complexes, and* f *a facet of* $\mathbf{C}$ *such that* $f \notin \mathbf{C}'$ *and* $f \cup \bigcup \mathbf{C}' = \bigcup \mathbf{C}$. *Suppose further that at least two of the following three conditions are satisfied:*

1. $\mathbf{C}$ *NETS-collapses elementarily to* $\mathbf{C}'$ *through removal of* f*; equivalently,* $f \cap \bigcup \mathbf{C}'$ *is NETS.*
2. $\bigcup \mathbf{C}$ *is NETS.*
3. $\bigcup \mathbf{C}'$ *is NETS.*

Then all three of these conditions are satisfied.

Proof. Let $A = f$, and let $B = \bigcup \mathbf{C}'$. Then:

- Condition 1 is equivalent to condition 2 of Lemma 8.3.3.
- Since $\bigcup \mathbf{C} = f \cup \bigcup \mathbf{C}'$, condition 2 is equivalent to condition 3 of Lemma 8.3.3.
- Since $A = f$ is a cell of the complex $\mathbf{C}$ and is therefore NETS, condition 3 is equivalent to condition 1 of Lemma 8.3.3.

So the proposition follows from Lemma 8.3.3. □

A sequence of complexes $\mathbf{C}_0, \mathbf{C}_1, \ldots, \mathbf{C}_n$ (where $n \geq 0$) will be called a *NETS-collapsing sequence* (of $\mathbf{C}_0$ to $\mathbf{C}_n$) if, for $0 \leq i \leq n-1$, *either* $\mathbf{C}_{i+1} = \mathbf{C}_i$, *or* $\mathbf{C}_i$ NETS-collapses elementarily to $\mathbf{C}_{i+1}$.

We say that a complex $\mathbf{D}$ *NETS-collapses to* a subcomplex $\mathbf{D}'$ of $\mathbf{C}$ if there exists a NETS-collapsing sequence of $\mathbf{D}$ to $\mathbf{D}'$. (Note that every complex NETS-collapses to itself.) The following proposition is the most important reason why we are interested in this concept.

Proposition 8.8.4. *Let $\mathbb{I}$ be a binary image on a pure complex, and let D be a simple set of 1s of $\mathbb{I}$. Then the foreground complex of $\mathbb{I}$ NETS-collapses to the foreground complex of $\mathbb{I} - \mathsf{D}$.*

Proof. Since D is a simple set of 1s of $\mathbb{I}$, $\mathbb{I}$ has a simple sequence $\mathsf{d}_1, \ldots, \mathsf{d}_n$ such that $\mathsf{D} = \{\mathsf{d}_i \mid 1 \leq i \leq n\}$. For k in $\{0, \ldots, n\}$, let $\mathbf{C}_k$ be the foreground complex of $\mathbb{I} - \{\mathsf{d}_i \mid 1 \leq i \leq k\}$. Since $\mathsf{d}_1, \ldots, \mathsf{d}_n$ is a simple sequence of $\mathbb{I}$, for k in $\{1, \ldots, n\}$ we have that d_k is a simple 1 of $\mathbb{I} - \{\mathsf{d}_i \mid 1 \leq i \leq k-1\}$ or, equivalently (as we noted in Example 8.8.1), that $\mathbf{C}_{k-1}$ NETS-collapses elementarily to $\mathbf{C}_k$ through removal of the facet d_k. Hence $\mathbf{C}_0, \ldots, \mathbf{C}_n$ is a NETS-collapsing sequence of $\mathbf{C}_0$ (which is the foreground complex of $\mathbb{I}$) to $\mathbf{C}_n$ (which is the foreground complex of $\mathbb{I} - \mathsf{D}$). □

Proposition 8.8.5. *Let $\mathbf{C}$ and $\mathbf{C}'$ be complexes such that $\mathbf{C}$ NETS-collapses to $\mathbf{C}'$. Then $\bigcup \mathbf{C}$ is NETS if and only if $\bigcup \mathbf{C}'$ is NETS.*

Proof. Let $\mathbf{C}_0, \mathbf{C}_1, \ldots, \mathbf{C}_n$ be a NETS-collapsing sequence of $\mathbf{C}$ to $\mathbf{C}'$. Then, for all i in $\{0, \ldots, n-1\}$, it follows from Proposition 8.8.3 that $\bigcup \mathbf{C}_{i+1}$ is NETS if and only if $\bigcup \mathbf{C}_i$ is NETS. Hence $\bigcup \mathbf{C}_n = \bigcup \mathbf{C}'$ is NETS if and only if $\bigcup \mathbf{C}_0 = \bigcup \mathbf{C}$ is NETS. □

Lemma 8.8.6. *Let $\mathbf{X}$ and $\mathbf{Y}$ be two subcomplexes of some complex, and let $\mathbf{C}_0, \mathbf{C}_1, \ldots, \mathbf{C}_n$ be a NETS-collapsing sequence in which $\mathbf{X} \cup \mathbf{Y} \supseteq \mathbf{C}_i \supseteq \mathbf{X} \cap \mathbf{Y}$ for $0 \leq i \leq n$. Then:*

1. *For $i \in \{0, \ldots, n-1\}$ such that $\mathbf{C}_i$ NETS-collapses elementarily to $\mathbf{C}_{i+1}$ through removal of a facet f of $\mathbf{C}_i$ in $\mathbf{X}$, we have that:*
   (i) *$\mathbf{C}_i \cap \mathbf{X}$ NETS-collapses elementarily to $\mathbf{C}_{i+1} \cap \mathbf{X}$ through removal of the same facet f, and $f \cap \bigcup(\mathbf{C}_{i+1} \cap \mathbf{X}) = f \cap \bigcup \mathbf{C}_{i+1}$.*
   (ii) *$\mathbf{C}_{i+1} \cup \mathbf{X} = \mathbf{C}_i \cup \mathbf{X}$.*
2. *For $i \in \{0, \ldots, n-1\}$ such that $\mathbf{C}_i$ NETS-collapses elementarily to $\mathbf{C}_{i+1}$ through removal of a facet f of $\mathbf{C}_i$ in $\mathbf{Y}$, we have that:*

(i) $\mathbf{C}_i \cup \mathbf{X}$ *NETS-collapses elementarily to* $\mathbf{C}_{i+1} \cup \mathbf{X}$ *through removal of the same facet* f, *and* $f \cap \bigcup(\mathbf{C}_{i+1} \cup \mathbf{X}) = f \cap \bigcup \mathbf{C}_{i+1}$.

(ii) $\mathbf{C}_{i+1} \cap \mathbf{X} = \mathbf{C}_i \cap \mathbf{X}$.

3. $\mathbf{C}_0 \cap \mathbf{X}, \mathbf{C}_1 \cap \mathbf{X}, \ldots, \mathbf{C}_n \cap \mathbf{X}$ *is a NETS-collapsing sequence.*
4. $\mathbf{C}_0 \cup \mathbf{X}, \mathbf{C}_1 \cup \mathbf{X}, \ldots, \mathbf{C}_n \cup \mathbf{X}$ *is a NETS-collapsing sequence.*

Proof. It suffices to prove assertions 1 and 2, because assertion 3 follows from the fact that either 1(i) or 2(ii) holds for each $i \in \{0, \ldots, n-1\}$ such that $\mathbf{C}_{i+1} \neq \mathbf{C}_i$, and assertion 4 follows from the fact that either 1(ii) or 2(i) holds for each $i \in \{0, \ldots, n-1\}$ such that $\mathbf{C}_{i+1} \neq \mathbf{C}_i$.

To prove assertions 1 and 2, let $i \in \{0, \ldots, n-1\}$ and suppose $\mathbf{C}_i$ NETS-collapses elementarily to $\mathbf{C}_{i+1}$ through removal of a facet f of $\mathbf{C}_i$, so that $f \notin \mathbf{C}_{i+1}$, $f \cup \bigcup \mathbf{C}_{i+1} = \bigcup \mathbf{C}_i$, $\mathbf{Faces}(f) \cup \mathbf{C}_{i+1} = \mathbf{C}_i$, and $f \cap \bigcup \mathbf{C}_{i+1}$ is NETS.

Proof of Assertion 1. To prove assertion 1, we assume that $f \in \mathbf{X}$. Then $\mathbf{Faces}(f) \subseteq \mathbf{X}$, and so $\mathbf{C}_i \cup \mathbf{X} = (\mathbf{Faces}(f) \cup \mathbf{C}_{i+1}) \cup \mathbf{X} = \mathbf{C}_{i+1} \cup \mathbf{X}$. Thus 1(ii) holds.

We now claim that: (a) f is a facet of $\mathbf{C}_i \cap \mathbf{X}$; (b) $f \notin \mathbf{C}_{i+1} \cap \mathbf{X}$; (c) $f \cup \bigcup(\mathbf{C}_{i+1} \cap \mathbf{X}) = \bigcup(\mathbf{C}_i \cap \mathbf{X})$; and (d) $f \cap \bigcup(\mathbf{C}_{i+1} \cap \mathbf{X}) = f \cap \bigcup \mathbf{C}_{i+1}$. Since $f \cap \bigcup \mathbf{C}_{i+1}$ is NETS, 1(i) holds if these four claims are valid.

Claims (a) and (b) are valid since f is a facet of $\mathbf{C}_i$, $f \in \mathbf{X}$, and $f \notin \mathbf{C}_{i+1}$. Since $f \in \mathbf{X}$, we have that $f \subseteq \bigcup \mathbf{X}$. So, in view of (8.4), $f \cup \bigcup(\mathbf{C}_{i+1} \cap \mathbf{X}) = f \cup ((\bigcup \mathbf{C}_{i+1}) \cap \bigcup \mathbf{X})) = (f \cup \bigcup \mathbf{C}_{i+1}) \cap \bigcup \mathbf{X} = (\bigcup \mathbf{C}_i) \cap \bigcup \mathbf{X} = \bigcup(\mathbf{C}_i \cap \mathbf{X})$. So claim (c) is valid. Moreover, $f \cap \bigcup(\mathbf{C}_{i+1} \cap \mathbf{X}) = f \cap (\bigcup \mathbf{C}_{i+1}) \cap \bigcup \mathbf{X} = f \cap \bigcup \mathbf{C}_{i+1}$, where the first equality follows from (8.4) and the second from the fact that $f \subseteq \bigcup \mathbf{X}$. So claim (d) is valid, and 1(i) holds.

Proof of Assertion 2. To prove assertion 2, we assume that $f \in \mathbf{Y}$. Then $\mathbf{Faces}(f) \subseteq \mathbf{Y}$, and so we have that $\mathbf{Faces}(f) \cap \mathbf{X} \subseteq \mathbf{Y} \cap \mathbf{X} \subseteq \mathbf{C}_{i+1} \cap \mathbf{X}$. We can deduce 2(ii) from this:

$$\mathbf{C}_i \cap \mathbf{X} = (\mathbf{Faces}(f) \cup \mathbf{C}_{i+1}) \cap \mathbf{X} = (\mathbf{Faces}(f) \cap \mathbf{X}) \cup (\mathbf{C}_{i+1} \cap \mathbf{X}) = \mathbf{C}_{i+1} \cap \mathbf{X}.$$

To verify 2(i), we claim that: (a) f is a facet of $\mathbf{C}_i \cup \mathbf{X}$; (b) $f \notin \mathbf{C}_{i+1} \cup \mathbf{X}$; (c) $f \cup \bigcup(\mathbf{C}_{i+1} \cup \mathbf{X}) = \bigcup(\mathbf{C}_i \cup \mathbf{X})$; and (d) $f \cap \bigcup(\mathbf{C}_{i+1} \cup \mathbf{X}) = f \cap \bigcup \mathbf{C}_{i+1}$. Since $f \cap \bigcup \mathbf{C}_{i+1}$ is NETS, 2(i) holds if these four claims are valid.

Note that $f \notin \mathbf{X}$ because $f \in \mathbf{Y}$ but $f \notin \mathbf{C}_{i+1} \supseteq \mathbf{X} \cap \mathbf{Y}$. Hence claim (b) is valid. Since $f \notin \mathbf{X}$, f is not a face of any cell in $\mathbf{X}$. So, since f is a facet of $\mathbf{C}_i$, f is also a facet of $\mathbf{C}_i \cup \mathbf{X}$. Hence claim (a) is valid. Claim (c) is valid because $f \cup \bigcup(\mathbf{C}_{i+1} \cup \mathbf{X}) = f \cup (\bigcup \mathbf{C}_{i+1}) \cup \bigcup \mathbf{X}$, $f \cup \bigcup \mathbf{C}_{i+1} = \bigcup \mathbf{C}_i$, and $(\bigcup \mathbf{C}_i) \cup \bigcup \mathbf{X} = \bigcup(\mathbf{C}_i \cup \mathbf{X})$. In view of (8.4), we also have that $f \cap \bigcup(\mathbf{C}_{i+1} \cup \mathbf{X}) = f \cap ((\bigcup \mathbf{C}_{i+1}) \cup \bigcup \mathbf{X}) = (f \cap \bigcup \mathbf{C}_{i+1}) \cup (f \cap \bigcup \mathbf{X}) = (f \cap \bigcup \mathbf{C}_{i+1}) \cup (f \cap (\bigcup \mathbf{Y}) \cap \bigcup \mathbf{X}) = (f \cap \bigcup \mathbf{C}_{i+1}) \cup (f \cap \bigcup(\mathbf{Y} \cap \mathbf{X})) = f \cap \bigcup \mathbf{C}_{i+1}$, where the third-last equality follows from the fact that $f \subseteq \bigcup \mathbf{Y}$, and the last equality from the fact that $\mathbf{Y} \cap \mathbf{X} \subseteq \mathbf{C}_{i+1}$. So claim (d) is valid, and 2(i) holds. □

Corollary 8.8.7. *Let* $\mathbf{X}$ *and* $\mathbf{Y}$ *be two subcomplexes of some complex. Let* $\mathbf{C}$ *and* $\mathbf{C}'$ *be complexes such that* $\mathbf{X} \cup \mathbf{Y} \supseteq \mathbf{C} \supseteq \mathbf{C}' \supseteq \mathbf{X} \cap \mathbf{Y}$ *and such that* $\mathbf{C}$ *NETS-collapses to* $\mathbf{C}'$*. Then* $\mathbf{C} \cap \mathbf{Y}$ *NETS-collapses to* $\mathbf{C}' \cap \mathbf{Y}$*, and* $\mathbf{C} \cup \mathbf{Y}$ *NETS-collapses to* $\mathbf{C}' \cup \mathbf{Y}$*.*

Proof. Let $\mathbf{C}_0, \mathbf{C}_1, \ldots, \mathbf{C}_n$ be a NETS-collapsing sequence of $\mathbf{C}$ to $\mathbf{C}'$. Then it follows from assertions 3 and 4 of Lemma 8.8.6 (when we switch the roles of $\mathbf{X}$ and $\mathbf{Y}$) that $\mathbf{C}_0 \cap \mathbf{Y}, \mathbf{C}_1 \cap \mathbf{Y}, \ldots, \mathbf{C}_n \cap \mathbf{Y}$ and $\mathbf{C}_0 \cup \mathbf{Y}, \mathbf{C}_1 \cup \mathbf{Y}, \ldots, \mathbf{C}_n \cup \mathbf{Y}$ are NETS-collapsing sequences of $\mathbf{C} \cap \mathbf{Y}$ to $\mathbf{C}' \cap \mathbf{Y}$ and of $\mathbf{C} \cup \mathbf{Y}$ to $\mathbf{C}' \cup \mathbf{Y}$, respectively. □

Lemma 8.8.8. *For any two complexes* $\mathbf{X}$ *and* $\mathbf{X}'$ *such that* $\mathbf{X}' \subseteq \mathbf{X}$*, there exists a collapsible complex* $\mathbf{Y}$ *such that* $\dim(\mathbf{Y}) = \dim(\mathbf{X}') + 1$*,* $\mathbf{X} \cup \mathbf{Y}$ *is a complex, and* $\mathbf{X} \cap \mathbf{Y} = \mathbf{X}'$*.*

Proof. We may assume without loss of generality that $\mathbf{X}$ is a complex in the k-dimensional hyperplane $\mathbb{R}^k \times \{1\}$ of $\mathbb{R}^{k+1}$ for some positive integer k.

Let o be the 0-cell in $\mathbb{R}^{k+1}$ whose only point is the origin $(0, \ldots, 0)$ of $\mathbb{R}^{k+1}$. For any complex $\mathbf{C}$ in $\mathbb{R}^k \times \{1\}$, let $\mathbf{Cone}(\mathbf{C})$ denote the complex $\mathbf{C} \cup \{o\} \cup \{\mathrm{conv}(c \cup o) \mid c \in \mathbf{C}\}$, where $\mathrm{conv}(c \cup o)$ denotes the convex hull of the set $c \cup o$. Then it is not hard to verify by induction on the number of facets in $\mathbf{C}$ that $\mathbf{Cone}(\mathbf{C})$ is a collapsible complex for any complex $\mathbf{C}$ in $\mathbb{R}^k \times \{1\}$. (This follows from the observations that $\mathbf{Cone}(\emptyset) = \{o\}$ and, for any facet f of any complex $\mathbf{C}$ in $\mathbb{R}^k \times \{1\}$, $(\mathrm{conv}(f \cup o), f)$ is a free pair of cells of $\mathbf{Cone}(\mathbf{C})$, so that $\mathbf{Cone}(\mathbf{C})$ collapses elementarily to the complex $\mathbf{Cone}(\mathbf{C}) \setminus \{\mathrm{conv}(f \cup o), f\} = \mathbf{Cone}(\mathbf{C} \setminus \{f\})$.) Now define $\mathbf{Y} = \mathbf{Cone}(\mathbf{X}')$. Then it is readily confirmed that $\mathbf{Y}$ is a complex with the stated properties. □

Theorem 8.8.9. *Let* $\mathbf{X}$*,* $\mathbf{B}$*, and* $\mathbf{X}'$ *be complexes such that* $\mathbf{X} \supsetneq \mathbf{B} \supsetneq \mathbf{X}'$ *and there is a facet* f *of* $\mathbf{B}$ *for which* $f \cup \bigcup \mathbf{X}' = \bigcup \mathbf{B}$*. Suppose further that* $\mathbf{X}$ *NETS-collapses to* $\mathbf{B}$ *and that* $\mathbf{X}$ *NETS-collapses to* $\mathbf{X}'$*. Then* $\mathbf{B}$ *NETS-collapses elementarily to* $\mathbf{X}'$ *through removal of the facet* f*.*

Proof. Since $\bigcup \mathbf{B} \supsetneq \bigcup \mathbf{X}'$ (because $\mathbf{B} \supsetneq \mathbf{X}'$) and the facet f of $\mathbf{B}$ satisfies $f \cup \bigcup \mathbf{X}' = \bigcup \mathbf{B}$, we have that $f \notin \mathbf{X}'$. So we see from the definition of an elementary NETS-collapse operation that the theorem will be proved if we can establish the following:

$$f \cap \bigcup \mathbf{X}' \text{ is NETS.} \tag{8.10}$$

By Lemma 8.8.8 there is a collapsible complex $\mathbf{Y}$ such that $\mathbf{X} \cup \mathbf{Y}$ is a complex and $\mathbf{X} \cap \mathbf{Y} = \mathbf{X}'$ (which implies $\mathbf{X}' \cup \mathbf{Y} = \mathbf{Y}$). Since $f \subseteq \bigcup \mathbf{X}$ (as $f \in \mathbf{B} \subsetneq \mathbf{X}$), it follows from (8.4) that $f \cap \bigcup \mathbf{Y} = f \cap (\bigcup \mathbf{X}) \cap \bigcup \mathbf{Y} = f \cap \bigcup (\mathbf{X} \cap \mathbf{Y}) = f \cap \bigcup \mathbf{X}'$. Hence (8.10) is true just if the following is true:

$$f \cap \bigcup \mathbf{Y} \text{ is NETS.} \tag{8.11}$$

Since $f \in \mathbf{X}$ but $f \notin \mathbf{X}' = \mathbf{X} \cap \mathbf{Y}$, and since $f \cup \bigcup \mathbf{X}' = \bigcup \mathbf{B}$, the facet f of $\mathbf{B}$ has the following properties:

(i) $f \notin \mathbf{Y}$.
(ii) f is a facet of $\mathbf{B} \cup \mathbf{Y}$.
(iii) $f \cup \bigcup \mathbf{Y} = f \cup \bigcup(\mathbf{X}' \cup \mathbf{Y}) = f \cup (\bigcup \mathbf{X}') \cup \bigcup \mathbf{Y} = (\bigcup \mathbf{B}) \cup (\bigcup \mathbf{Y}) = \bigcup(\mathbf{B} \cup \mathbf{Y})$.

Since $\mathbf{X}$ NETS-collapses to $\mathbf{B}$ and $\mathbf{X}$ NETS-collapses to $\mathbf{X}'$, and since $\mathbf{X} \cup \mathbf{Y} \supseteq \mathbf{X} \supsetneq \mathbf{B} \supsetneq \mathbf{X}' = \mathbf{X} \cap \mathbf{Y}$, it follows from Corollary 8.8.7 that:

1. $\mathbf{X} \cup \mathbf{Y}$ NETS-collapses to $\mathbf{B} \cup \mathbf{Y}$.
2. $\mathbf{X} \cup \mathbf{Y}$ NETS-collapses to $\mathbf{X}' \cup \mathbf{Y} = \mathbf{Y}$.

Since $\bigcup \mathbf{Y}$ is NETS (by Proposition 8.4.2, as $\mathbf{Y}$ is collapsible), we see from Proposition 8.8.5 and property 2 that $\bigcup(\mathbf{X} \cup \mathbf{Y})$ is NETS. We deduce from this, property 1, and Proposition 8.8.5 that $\bigcup(\mathbf{B} \cup \mathbf{Y})$ is NETS. Since $\bigcup \mathbf{Y}$ is NETS and $\bigcup(\mathbf{B} \cup \mathbf{Y})$ is NETS, we see from (i)–(iii) and Proposition 8.8.3 that (8.11) holds. □

Corollary 8.8.10. *Let $\mathbb{I}$ be a binary image on a pure complex, let D be a simple set of 1s of $\mathbb{I}$, and let $\mathsf{d} \in \mathsf{D}$ be such that $\mathsf{D} \setminus \{\mathsf{d}\}$ is a simple set of 1s of $\mathbb{I}$. Then d must be a simple 1 of $\mathbb{I} - (\mathsf{D} \setminus \{\mathsf{d}\})$.*

Proof. Let $\mathbf{X}$, $\mathbf{X}'$, and $\mathbf{B}$ be respectively the foreground complexes of $\mathbb{I}$, $\mathbb{I} - \mathsf{D}$, and $\mathbb{I} - (\mathsf{D} \setminus \{\mathsf{d}\})$. Then, by Proposition 8.8.4, we have that (i) $\mathbf{X}$ NETS-collapses to $\mathbf{X}'$ and (ii) $\mathbf{X}$ NETS-collapses to $\mathbf{B}$. On putting $f = \mathsf{d}$ in Theorem 8.8.9, we see from (i) and (ii) that $\mathbf{B}$ NETS-collapses elementarily to $\mathbf{X}'$ through removal of the facet d. Equivalently (as we observed in Example 8.8.1), d is a simple 1 of $\mathbb{I} - (\mathsf{D} \setminus \{\mathsf{d}\})$. □

We are now in a position to achieve the main objective of this section by proving Theorem 8.5.5, which we restate here.

Theorem 8.5.5. *Let $\mathbb{I}$ be a binary image on a pure complex, and let Q be a set of 1s of $\mathbb{I}$. Then the following are equivalent:*

1. *Q is a hereditarily simple set of $\mathbb{I}$.*
2. *Every q in Q is P-simple for Q in $\mathbb{I}$.*
3. *Every nonempty sequence of distinct elements of Q is a simple sequence of $\mathbb{I}$.*

Proof. If assertion 3 holds, then we see from the very definition of a simple set of $\mathbb{I}$ that every subset of Q is a simple set of $\mathbb{I}$. Hence assertion 3 implies assertion 1.

To see that assertion 2 implies assertion 3, suppose that assertion 2 holds and $\mathsf{q}_0, \ldots, \mathsf{q}_k$ is any nonempty sequence of distinct elements of Q. For every $i \in \{0, \ldots, k\}$, assertion 2 implies that q_i is P-simple for Q in $\mathbb{I}$ and hence that q_i is a simple 1 of $\mathbb{I} - \{\mathsf{q}_j \mid 0 \leq j \leq i - 1\}$. So $\mathsf{q}_0, \ldots, \mathsf{q}_k$ is a simple sequence of $\mathbb{I}$.

Finally, we prove that assertion 1 implies assertion 2. Suppose that Q is a hereditarily simple set of $\mathbb{I}$ (so every subset of Q is a simple set of $\mathbb{I}$), q is any element of Q, and Q' is any subset of $\mathsf{Q} \setminus \{\mathsf{q}\}$. Then each of Q' and $\mathsf{Q}' \cup \{\mathsf{q}\}$ is a subset of Q and is therefore a simple set of $\mathbb{I}$, whence it follows from Corollary 8.8.10 (on putting $\mathsf{D} = \mathsf{Q}' \cup \{\mathsf{q}\}$ and $\mathsf{d} = \mathsf{q}$) that q is a simple 1 of $\mathbb{I} - \mathsf{Q}'$. Since q is an arbitrary element of Q and Q' an arbitrary subset of $\mathsf{Q} \setminus \{\mathsf{q}\}$, we can conclude that every q in Q is P-simple for Q in $\mathbb{I}$. □

8.8.2 $\mathbb{I}$-ESSENTIAL COMPLEXES AND ANOTHER PROOF OF THE "IF" PART OF THEOREM 8.7.1

Let $\mathbb{I}$ be a binary image on a pure complex. Then a complex $\mathbf{L}$ is said to be $\mathbb{I}$-*essential* if $\mathbf{L}$ is the complex induced by some set of $\mathbb{I}$-essential cells. Readily, $\mathbf{L}$ is an $\mathbb{I}$-essential complex if and only if every facet of $\mathbf{L}$ is an $\mathbb{I}$-essential cell. This is another concept introduced by Bertrand [3].

Note that if D is any set of 1s of $\mathbb{I}$, then every $(\mathbb{I} - \mathsf{D})$-essential complex is also an $\mathbb{I}$-essential complex because every $(\mathbb{I} - \mathsf{D})$-essential cell is also an $\mathbb{I}$-essential cell.

$\mathbb{I}$-essential complexes have an interesting property that will be stated as Corollary 8.8.14 (which can be regarded as a version of [3, Thm. 4.2] for NETS-collapsing instead of collapsing). This property will be used to give a second proof of the "if" part of Theorem 8.7.1.

Example 8.8.11. For any set Q of 1s of $\mathbb{I}$, the complex $\bigcup\{\mathbf{Faces}(\mathsf{q}) \mid \mathsf{q} \in \mathsf{Q}\}$ (i.e., the complex induced by the set Q) is $\mathbb{I}$-essential.

Example 8.8.12. For any set D of 1s of $\mathbb{I}$ and any $(\mathbb{I}-\mathsf{D})$-essential cell f, the complex $\mathbf{Core}(f, \mathbb{I} - \mathsf{D})$ is $(\mathbb{I} - \mathsf{D})$-essential and is therefore also $\mathbb{I}$-essential.

Given any facet f of an $\mathbb{I}$-essential complex $\mathbf{L}$, there is a (unique) greatest $\mathbb{I}$-essential subcomplex of $\mathbf{L}$ that does not contain f, namely the complex induced by all the $\mathbb{I}$-essential cells of $\mathbf{L}$ other than f. We will see from Theorem 8.8.13 that the set $(\mathbf{L} \setminus \mathbf{Faces}(f)) \cup \mathbf{Core}(f, \mathbb{I})$ is exactly this subcomplex of $\mathbf{L}$. This set will be denoted by $\mathbf{L} \ominus_{\mathbb{I}} f$.

$\mathbf{L} \ominus_{\mathbb{I}} f$ has another fundamental property that is expressed by statement 3 of Theorem 8.8.13. Note that statement 3 does *not* imply $\mathbf{L}$ collapses to $\mathbf{L} \ominus_{\mathbb{I}} f$ if the facet f is $\mathbb{I}$-regular: That this is not always true is a key difference between our NETS definition of an $\mathbb{I}$-regular cell and the definition used by Bertrand and Couprie and is an important reason why we introduced the concept of NETS-collapsing.

Theorem 8.8.13. *Let $\mathbb{I}$ be a binary image on a pure complex, let $\mathbf{L}$ be any $\mathbb{I}$-essential complex, and let f be any facet of $\mathbf{L}$. Then the set $\mathbf{L} \ominus_{\mathbb{I}} f = (\mathbf{L} \setminus \mathbf{Faces}(f)) \cup \mathbf{Core}(f, \mathbb{I})$ has the following properties:*

1. *$\mathbf{L} \ominus_{\mathbb{I}} f$ is the complex induced by the set of all $\mathbb{I}$-essential cells of $\mathbf{L}$ other than the facet f.*
2. *$\mathbf{L} \ominus_{\mathbb{I}} f$ is the greatest $\mathbb{I}$-essential subcomplex of $\mathbf{L}$ that does not contain the facet f. In other words, $\mathbf{L} \ominus_{\mathbb{I}} f \subseteq \mathbf{L} \setminus \{f\}$, and if $\mathbf{M}$ is any $\mathbb{I}$-essential complex such that $\mathbf{M} \subseteq \mathbf{L} \setminus \{f\}$, then $\mathbf{M} \subseteq \mathbf{L} \ominus_{\mathbb{I}} f$.*
3. *If the facet f is $\mathbb{I}$-regular, then $\mathbf{L}$ NETS-collapses elementarily to $\mathbf{L} \ominus_{\mathbb{I}} f$ through removal of the facet f, and $\bigcup(\mathbf{L} \ominus_{\mathbb{I}} f)$ is NETS if and only if $\bigcup \mathbf{L}$ is NETS.*

Proof of statements 1 and 2 of Theorem 8.8.13. Since statement 2 is a straightforward consequence of statement 1, we need only verify statement 1—i.e., $x \in (\mathbf{L} \setminus \mathbf{Faces}(f)) \cup \mathbf{Core}(f, \mathbb{I})$ if and only if there is some $\mathbb{I}$-essential cell y in $\mathbf{L} \setminus \{f\}$ such that $x \in \mathbf{Faces}(y)$.

To verify the "if" part, let y be any $\mathbb{I}$-essential cell in $\mathbf{L} \setminus \{f\}$, and suppose that $x \in \mathbf{Faces}(y)$ but $x \notin (\mathbf{L} \setminus \mathbf{Faces}(f))$. Then we just need to check that $x \in \mathbf{Core}(f, \mathbb{I})$. Now $x \in \mathbf{Faces}(f) \cap \mathbf{Faces}(y) = \mathbf{Faces}(f \cap y)$ (by property **P2b** of Subsection 8.2.2). Since y is $\mathbb{I}$-essential and f is also $\mathbb{I}$-essential (because $\mathbf{L}$ is $\mathbb{I}$-essential), we have that $f \cap y$ is $\mathbb{I}$-essential. Moreover, $f \cap y \subsetneq f$ (because $f \not\subseteq y$ since f is a facet of $\mathbf{L}$ and $y \in \mathbf{L} \setminus \{f\}$). Hence $f \cap y \in \mathbf{Core}(f, \mathbb{I})$, whence $x \in \mathbf{Core}(f, \mathbb{I})$ as required.

To verify the "only if" part, let x be any element of $(\mathbf{L} \setminus \mathbf{Faces}(f)) \cup \mathbf{Core}(f, \mathbb{I})$. Then we just need to show that there is some $\mathbb{I}$-essential cell y in $\mathbf{L} \setminus \{f\}$ such that $x \in \mathbf{Faces}(y)$. This is certainly true if $x \in \mathbf{Core}(f, \mathbb{I})$ (by the very definition of $\mathbf{Core}(f, \mathbb{I})$), so let us suppose that $x \in \mathbf{L} \setminus \mathbf{Faces}(f)$. Let y be a facet of $\mathbf{L}$ such that $x \in \mathbf{Faces}(y)$. Then y has the required properties: $y \neq f$ because $x \notin \mathbf{Faces}(f)$ but $x \in \mathbf{Faces}(y)$, and the facet y is $\mathbb{I}$-essential because $\mathbf{L}$ is $\mathbb{I}$-essential. □

Proof of statement 3 of Theorem 8.8.13. Since $\mathbf{L} \ominus_{\mathbb{I}} f$ and $\mathbf{Faces}(f)$ are subcomplexes of $\mathbf{L}$ such that $\mathbf{Faces}(f) \cap (\mathbf{L} \ominus_{\mathbb{I}} f) = \mathbf{Core}(f, \mathbb{I})$ and $\mathbf{Faces}(f) \cup (\mathbf{L} \ominus_{\mathbb{I}} f) = \mathbf{L}$, it follows from (8.4) and (8.5) that:

(i) $f \cap \bigcup(\mathbf{L} \ominus_{\mathbb{I}} f) = (\bigcup \mathbf{Faces}(f)) \cap (\bigcup(\mathbf{L} \ominus_{\mathbb{I}} f)) = \bigcup(\mathbf{Faces}(f) \cap (\mathbf{L} \ominus_{\mathbb{I}} f)) = \bigcup \mathbf{Core}(f, \mathbb{I})$.

(ii) $f \cup \bigcup(\mathbf{L} \ominus_{\mathbb{I}} f) = (\bigcup \mathbf{Faces}(f)) \cup (\bigcup(\mathbf{L} \ominus_{\mathbb{I}} f)) = \bigcup(\mathbf{Faces}(f) \cup (\mathbf{L} \ominus_{\mathbb{I}} f)) = \bigcup \mathbf{L}$.

Now suppose f is $\mathbb{I}$-regular. Then $\bigcup \mathbf{Core}(f, \mathbb{I})$ is NETS, and so it follows from (i) and (ii) that $\mathbf{L}$ NETS-collapses elementarily to $\mathbf{L} \ominus_{\mathbb{I}} f$ through removal of the facet f. It follows from this and Proposition 8.8.5 that $\bigcup(\mathbf{L} \ominus_{\mathbb{I}} f)$ is NETS if and only if $\bigcup \mathbf{L}$ is NETS. □

Corollary 8.8.14 (NETS-based analog of [3, Thm. 4.2] and [8, Thm. 15]). *Let $\mathbb{I}$ be a binary image on a pure complex, let $\mathbf{L}$ be an $\mathbb{I}$-essential complex, and let $\mathbf{M}$ be an $\mathbb{I}$-essential subcomplex of $\mathbf{L}$ that contains every $\mathbb{I}$-critical cell in $\mathbf{L}$. Then $\mathbf{L}$ NETS-collapses to $\mathbf{M}$, and $\bigcup \mathbf{M}$ is NETS just if $\bigcup \mathbf{L}$ is NETS.*

Proof. In view of Proposition 8.8.5, it suffices to verify that $\mathbf{L}$ NETS-collapses to $\mathbf{M}$.

It follows from Theorem 8.8.13 that we can construct a sequence of $\mathbb{I}$-essential subcomplexes of $\mathbf{L}$ as follows: We define $\mathbf{L}_0 = \mathbf{L}$. For $i = 0, 1, 2, \ldots$ until we have that $\mathbf{L}_i = \mathbf{M}$, we define $\mathbf{L}_{i+1} = \mathbf{L}_i \ominus_{\mathbb{I}} f_i$ for some facet f_i of $\mathbf{L}_i$ such that $f_i \notin \mathbf{M}$; we stop when $\mathbf{L}_i = \mathbf{M}$. Since $\mathbf{M}$ contains all $\mathbb{I}$-critical cells of $\mathbf{L}$, when $\mathbf{L}_i \neq \mathbf{M}$, the facet f_i of $\mathbf{L}_i$ that is not in $\mathbf{L}_{i+1}$ must be $\mathbb{I}$-regular. So it follows from statement 3 of Theorem 8.8.13 that the constructed sequence is a NETS-collapsing sequence of $\mathbf{L}$ to $\mathbf{M}$. □

As promised above, we now use this result to give an alternative and perhaps more intuitive proof of the "if" part of the following earlier theorem:

Theorem 8.7.1 (NETS-based generalization of [8, Thm. 21]). *Let $\mathbb{I}$ be a binary image on a pure complex $\mathbf{K}$, let D be a set of 1s of $\mathbb{I}$, and let d be any element of D. Then*

d *is P-simple for* D *in* $\mathbb{I}$ *if and only if there is no* $\mathbb{I}$*-critical cell* f *of* $\mathbf{K}$ *such that* $\mathsf{d} \in \mathsf{Clique}(f, \mathbb{I}) \subseteq \mathsf{D}$.

Second proof of Theorem 8.7.1. As mentioned in the first proof of this theorem, the "only if" part of the theorem follows from Proposition 8.7.2. We now deduce the "if" part of the theorem from Corollary 8.8.14.

Suppose there is no $\mathbb{I}$-critical cell f such that $\mathsf{d} \in \mathsf{Clique}(f, \mathbb{I}) \subseteq \mathsf{D}$. Let D' be any subset of $\mathsf{D} \setminus \{\mathsf{d}\}$. Then what we need to show is that d is a simple 1 of $\mathbb{I} - \mathsf{D}'$. Equivalently, we just need to show that $\bigcup \mathbf{Attachment}(\mathsf{d}, \mathbb{I} - \mathsf{D}')$ is NETS.

Now $\mathbf{Attachment}(\mathsf{d}, \mathbb{I} - \mathsf{D}') = \mathbf{Core}(\mathsf{d}, \mathbb{I} - \mathsf{D}')$ is an $(\mathbb{I} - \mathsf{D}')$-essential subcomplex (and hence an $\mathbb{I}$-essential subcomplex) of the $\mathbb{I}$-essential complex $\mathbf{Faces}(\mathsf{d})$. Since no $\mathbb{I}$-critical cell f is such that $\mathsf{d} \in \mathsf{Clique}(f, \mathbb{I}) \subseteq \mathsf{D}$, every $\mathbb{I}$-critical cell in $\mathbf{Faces}(\mathsf{d})$ is also a face of a 1 of $\mathbb{I} - \mathsf{D}$ (which will also be a 1 of $\mathbb{I} - \mathsf{D}'$). So every $\mathbb{I}$-critical cell in $\mathbf{Faces}(\mathsf{d})$ is a cell of $\mathbf{Attachment}(\mathsf{d}, \mathbb{I} - \mathsf{D}')$, and the hypotheses of Corollary 8.8.14 are satisfied when $\mathbf{L} = \mathbf{Faces}(\mathsf{d})$ and $\mathbf{M} = \mathbf{Attachment}(\mathsf{d}, \mathbb{I} - \mathsf{D}')$. Since $\bigcup \mathbf{Faces}(\mathsf{d}) = \mathsf{d}$ is NETS (by **NETS-2**), we see from Corollary 8.8.14 that $\bigcup \mathbf{Attachment}(\mathsf{d}, \mathbb{I} - \mathsf{D}')$ is NETS, as required. □

8.9 CONCLUDING REMARKS

An approach of Bertrand and Couprie to designing parallel thinning algorithms for 2D, 3D, and 4D binary images, based on Bertrand's *critical kernels* [3], has the advantage of eliminating the often difficult problem of verifying the topological soundness of the developed algorithms. For binary images on 2-, 3-, and 4-dimensional Cartesian array complexes, the paper [8] greatly clarified the relationships between Bertrand's critical kernel and concepts studied in earlier work on topological soundness of parallel thinning algorithms—specifically, the concepts of hereditarily simple sets, P-simple 1s, and minimal nonsimple sets. In this chapter, we have generalized the main theorems of [8] to binary images on arbitrary pure polytopal complexes of any dimension.

For reasons given in the Introduction, our definition of critical kernel in this chapter is not exactly the same as Bertrand's: We define a critical essential cell to be an essential cell e for which the union of the cells of e's core does not have our NETS property. In fact, this gives a *class of definitions* of critical kernel because the NETS property is specified by our axioms **NETS-1–NETS-4**: Any class of polyhedra that satisfies those axioms will be a valid definition of NETS polyhedra, and any valid definition of NETS polyhedra will in turn give a definition of critical kernel. The definitions of NETS polyhedra in Subsection 8.3.4 give definitions of critical kernel for which the critical kernels of images of arbitrary dimension can be found using homology group computation algorithms.

For binary images on 2D, 3D, and 4D Cartesian array complexes, we saw in Subsection 8.3.4 that Bertrand's definition of critical kernel falls into the class of definitions of critical kernel for which our arguments are valid. For binary images on

2D and 3D Cartesian array complexes, there is no other definition (i.e., no definition that is inequivalent to Bertrand's) in our class. Indeed, it is not hard to deduce from the result established in Subsection 8.4.3 that this is true for binary images on any 2D or 3D pure polytopal complex.

For binary images on 4D Cartesian array complexes, the author is convinced that Bertrand's definition is again the only definition of critical kernel in our class. Equivalently, the author is convinced that the axioms **NETS-1–NETS-4** imply Condition 8.3.6 in the case $n = 4$. However, he does not have a complete proof and leaves this as a matter for the interested reader to investigate.

REFERENCES

[1] G. Bertrand, Simple points, topological numbers, and geodesic neighborhoods in cubic grids, Pattern Recognit. Lett. 15 (1994) 1003–1011.
[2] G. Bertrand, On P-simple points, C. R. Acad. Sci., Ser. 1 Math. 321 (1995) 1077–1084.
[3] G. Bertrand, On critical kernels, C. R. Math. 345 (2007) 363–367.
[4] G. Bertrand, Completions and Simplicial Complexes, HAL-00761162, 2012.
[5] G. Bertrand, New structures based on completions, in: R. Gonzalez-Diaz, M.-J. Jimenez, B. Medrano (Eds.), Discrete Geometry for Computer Imagery: 17th IAPR International Conference, DGCI 2013, Seville, Spain, March 2013, Proceedings, Springer, 2013, pp. 83–94.
[6] G. Bertrand, Completions and simple homotopy, in: E. Barcucci, A. Frosini, S. Rinaldi (Eds.), Discrete Geometry for Computer Imagery: 18th IAPR International Conference, DGCI 2014, Siena, Italy, September 2014, Proceedings, Springer, 2014, pp. 63–74.
[7] G. Bertrand, M. Couprie, Two-dimensional parallel thinning algorithms based on critical kernels, J. Math. Imaging Vis. 31 (2008) 35–56.
[8] G. Bertrand, M. Couprie, On parallel thinning algorithms: minimal non-simple sets, P-simple points and critical kernels, J. Math. Imaging Vis. 35 (2009) 23–35.
[9] G. Bertrand, M. Couprie, Powerful parallel and symmetric 3D thinning schemes based on critical kernels, J. Math. Imaging Vis. 48 (2014) 134–148.
[10] G. Bertrand, M. Couprie, Isthmus based parallel and symmetric 3D thinning algorithms, Graph. Models 80 (2015) 1–15.
[11] G. Bertrand, G. Malandain, A new characterization of three-dimensional simple points, Pattern Recognit. Lett. 15 (1994) 169–175.
[12] P. Brass, On strongly normal tessellations, Pattern Recognit. Lett. 20 (1999) 957–960.
[13] D. Brunner, G. Brunnett, R. Strand, A High-Performance Parallel Thinning Approach Using a Non-Cubic Grid Structure, Chemnitzer Informatik-Berichte CSR-06-08, Faculty of Computer Science, Chemnitz University of Technology, 2006.
[14] M. Couprie, G. Bertrand, New characterizations of simple points in 2D, 3D, and 4D discrete spaces, IEEE Trans. Pattern Anal. Mach. Intell. 31 (2009) 637–648.
[15] M. Couprie, G. Bertrand, Asymmetric parallel 3D thinning scheme and algorithms based on isthmuses, Pattern Recognit. Lett. 76 (2016) 22–31.
[16] X. Dong, Alexander duality for projections of polytopes, Topology 41 (2002) 1109–1121.
[17] H. Edelsbrunner, J. Harer, Computational Topology: An Introduction, American Mathematical Society, 2010.

[18] C.J. Gau, T.Y. Kong, Minimal nonsimple sets of voxels in binary images on a face-centered cubic grid, Int. J. Pattern Recognit. Artif. Intell. 13 (1999) 485–502.

[19] C.J. Gau, T.Y. Kong, Minimal nonsimple sets in 4D binary images, Graph. Models 65 (2003) 112–130.

[20] R.W. Hall, Tests for connectivity preservation for parallel reduction operators, Topol. Appl. 46 (1992) 199–217.

[21] A. Hatcher, Algebraic Topology, Cambridge University Press, 2001.

[22] G.T. Herman, Geometry of Digital Spaces, Birkhäuser, 1998.

[23] P.P. Jonker, Discrete topology on N-dimensional square tessellated grids, Image Vis. Comput. 23 (2005) 213–225.

[24] P.P. Jonker, O. Vermeij, On skeletonization in 4D images, in: P. Perner, P. Wang, A. Rosenfeld (Eds.), Advances in Structural and Syntactical Pattern Recognition: 6th International Workshop, SSPR 1996, Leipzig, Germany, August 1996, Proceedings, Springer, 1996, pp. 79–89.

[25] T. Kaczynski, K. Mischaikow, M. Mrozek, Computational Homology, Springer, 2003.

[26] R. Klette, A. Rosenfeld, Digital Geometry, Morgan Kaufman, 2004.

[27] T.Y. Kong, On the problem of determining whether a parallel reduction operator for n-dimensional binary images always preserves topology, in: R.A. Melter, A.Y. Wu (Eds.), Vision Geometry II, Boston, USA, September 1993, Proceedings, in: Proc. SPIE, vol. 2060, 1993, pp. 69–77.

[28] T.Y. Kong, On topology preservation in 2D and 3D thinning, Int. J. Pattern Recognit. Artif. Intell. 9 (1995) 813–844.

[29] T.Y. Kong, Topology-preserving deletion of 1's from 2-, 3-, and 4-dimensional binary images, in: E. Ahronovitz, C. Fiorio (Eds.), Discrete Geometry for Computer Imagery: 7th International Workshop, DGCI 1997, Montpellier, France, December 1997, Proceedings, Springer, 1997, pp. 3–18.

[30] T.Y. Kong, Minimal non-simple and minimal non-cosimple sets in binary images on cell complexes, in: A. Kuba, L.G. Nyúl, K. Palágyi (Eds.), Discrete Geometry for Computer Imagery: 13th International Conference, DGCI 2006, Szeged, Hungary, October 2006, Proceedings, Springer, 2006, pp. 169–188.

[31] T.Y. Kong, Minimal non-deletable sets and minimal non-codeletable sets in binary images, Theor. Comput. Sci. 406 (2008) 97–118.

[32] T.Y. Kong, C.J. Gau, Minimal nonsimple sets in 4-dimensional binary images with $(8, 80)$-adjacency, in: R. Klette, J. Žunić (Eds.), Combinatorial Image Analysis: 10th International Workshop, IWCIA 2004, Auckland, New Zealand, December 2004, Proceedings, Springer, 2004, pp. 318–333.

[33] T.Y. Kong, A.W. Roscoe, Characterizations of simply-connected finite polyhedra in 3-space, Bull. Lond. Math. Soc. 17 (1985) 575–578.

[34] T.Y. Kong, P.K. Saha, A. Rosenfeld, Strongly normal sets of contractible tiles in N dimensions, Pattern Recognit. 40 (2007) 530–543.

[35] C.M. Ma, On topology preservation in 3D thinning, CVGIP, Image Underst. 59 (1994) 328–339.

[36] C.R.F. Maunder, Algebraic Topology, Dover Publications, 1996.

[37] D.G. Morgenthaler, Three-Dimensional Simple Points: Serial Erosion, Parallel Thinning and Skeletonization, TR-1005, Computer Vision Laboratory, University of Maryland, 1981.

[38] M. Niethammer, W.D. Kalies, K. Mischaikow, A. Tannenbaum, On the detection of simple points in higher dimensions using cubical homology, IEEE Trans. Image Process. 15 (2006) 2462–2469.
[39] N. Passat, M. Couprie, L. Mazo, G. Bertrand, Topological properties of thinning in 2-D pseudomanifolds, J. Math. Imaging Vis. 37 (2010) 27–39.
[40] C. Ronse, A topological characterization of thinning, Theor. Comput. Sci. 43 (1986) 31–41.
[41] C. Ronse, Minimal test patterns for connectivity preservation in parallel thinning algorithms for binary digital images, Discrete Appl. Math. 21 (1988) 67–79.
[42] A. Rosenfeld, Connectivity in digital pictures, J. Assoc. Comput. Mach. 17 (1970) 146–160.
[43] A. Rosenfeld, A.C. Kak, Digital Picture Processing, vol. 2, 2nd ed., Academic Press, 1982.
[44] P.K. Saha, G. Borgefors, G. Sanniti Di Baja, A survey on skeletonization algorithms and their applications, Pattern Recognit. Lett. 76 (2016) 3–12.
[45] P.K. Saha, B.B. Chaudhuri, Detection of 3D simple points for topology preserving transformation with applications to thinning, IEEE Trans. Pattern Anal. Mach. Intell. 16 (1994) 1028–1032.
[46] P.K. Saha, B.B. Chaudhuri, B. Chanda, D. Dutta Majumder, Topology preservation in 3D digital space, Pattern Recognit. 27 (1994) 295–300.
[47] P.K. Saha, T.Y. Kong, A. Rosenfeld, Strongly normal sets of tiles in N dimensions, Electron. Notes Theor. Comput. Sci. 46 (2001) 309–320.
[48] P.K. Saha, D. Dutta Majumder, A. Rosenfeld, Local topological parameters in a tetrahedral representation, CVGIP, Graph. Models Image Process. 60 (1998) 423–436.
[49] P.K. Saha, A. Rosenfeld, Determining simplicity and computing topological change in strongly normal partial tilings of R^2 or R^3, Pattern Recognit. 33 (2000) 105–118.
[50] P.K. Saha, A. Rosenfeld, The digital topology of sets of convex voxels, Graph. Models 62 (2000) 343–352.
[51] P.K. Saha, A. Rosenfeld, Local and global topology preservation in locally finite sets of tiles, Inf. Sci. 137 (2001) 303–311.
[52] R. Strand, Surface skeletons in grids with non-cubic voxels, in: Proceedings of the 17th International Conference on Pattern Recognition, ICPR 2004, Cambridge, UK, August 2004, vol. 1, IEEE Computer Society, 2004, pp. 548–551.
[53] R. Strand, Distance Functions and Image Processing on Point-Lattices With Focus on the 3D Face- and Body-centered Cubic Grids, Uppsala Dissertations from the Faculty of Science and Technology 79, Uppsala Universitet, 2008.
[54] Y.F. Tsao, K.S. Fu, A 3D parallel skeletonwise thinning algorithm, in: IEEE Computer Society Conference on Pattern Recognition and Image Processing, Las Vegas, June 1982, Proceedings, IEEE Computer Society, 1982, pp. 678–683.
[55] G.M. Ziegler, Lectures on Polytopes, Springer, 1995.

PART 2
Applications

CHAPTER

Skeletonization in natural images and its application to object recognition

Wei Shen*, Kai Zhao*, Jiang Yuan*, Yan Wang†, Zhijiang Zhang*, Xiang Bai‡

Key Laboratory of Specialty Fiber Optics and Optical Access Networks, Shanghai University, Shanghai, China Rapid-Rich Object Search Lab, Nanyang Technological University, Singapore† School of Electronic Information and Communications, Huazhong University of Science and Technology, Wuhan, China‡*

Contents

9.1 Introduction 260
9.2 Related Works 262
9.3 Methodology 264
9.3.1 Network Architecture 264
9.3.2 Skeleton Extraction by Fusing Scale-Associated Side Output 265
9.3.2.1 Training Phase 265
9.3.2.2 Testing Phase 269
9.3.3 Understanding of the Proposed Method 269
9.4 Experimental Results 269
9.4.1 Implementation Details 270
9.4.2 Performance Comparison 271
9.4.2.1 Evaluation Protocol 271
9.4.2.2 SK506 271
9.4.2.3 WH-SYMMAX 273
9.4.2.4 SYMMAX300 273
9.4.2.5 Cross Dataset Generalization 274
9.4.2.6 Symmetric Part Segmentation 275
9.4.2.7 Object Proposal Detection 277
9.4.2.8 Road Detection 277
9.4.2.9 Text Line Proposal Generation 277
9.5 Conclusion 283
Acknowledgments 283
References 284

Skeletonization. DOI: 10.1016/B978-0-08-101291-8.00011-0

FIGURE 9.1

Object skeleton extraction in natural images. The skeletons are in yellow.

9.1 INTRODUCTION

Object skeletonization (also referred to as "skeleton extraction") from natural images is a challenging problem in computer vision (Fig. 9.1). Here, the term "object" indicates a standalone thing with a well-defined boundary and center [1], including people, animals, or planes. Things like the sky, mountains, or even grass are not considered objects in this situation since they are amorphous. Since the skeleton (sometimes referred to as the symmetry axis) operates as a structure-based object descriptor, its extraction from natural images has significant application in fields as diverse as blood vessel detection [2], object recognition [3,4], text identification [5], and the detection of roads [2].

Skeleton extraction from presegmented images [6] had long been studied, which has been successfully applied to object matching and recognition based on shapes [7–9]. However, when being applied to natural images, such methods have severe limitations. Segmentation from natural images is still an unsolved problem.

Skeleton extraction from natural images is a much more difficult problem. The main challenges can be summarized in four points: (1) The diversity of object appearance: Objects in natural images may exhibit entirely different colors, textures, shapes, and sizes. (2) The complexity of natural scenes: Natural scenes can be very cluttered and contain distracted objects, such as fences, bricks, and even the shadows of objects, exhibit somewhat self-symmetry. (3) The variability of skeletons: local skeleton segments present a variety of patterns and shapes, such as straight lines, T-junctions, and Y-junctions. (4) The unknown-scale problem: A local skeleton segment naturally associates with a certain scale determined by the width of its corresponding object part. However, it is unknown in natural images. We term this problem as the unknown-scale problem in skeleton extraction.

Many researchers have attempted to address the challenges related to skeleton extraction. These researchers have largely utilized traditional image processing methods [10–13] (those develop a gradient intensity map according to geometric constraints between the skeletons and the edges) or learning-based methods [14–16, 2,17] (those learn a per-pixel classification or segment-linking model to generate the skeleton). These approaches both have problems. Traditional image processing methods cannot deal with intricate images because the prior of objects is not used; learning-based methods cannot extract complex objects with cluttered interior tex-

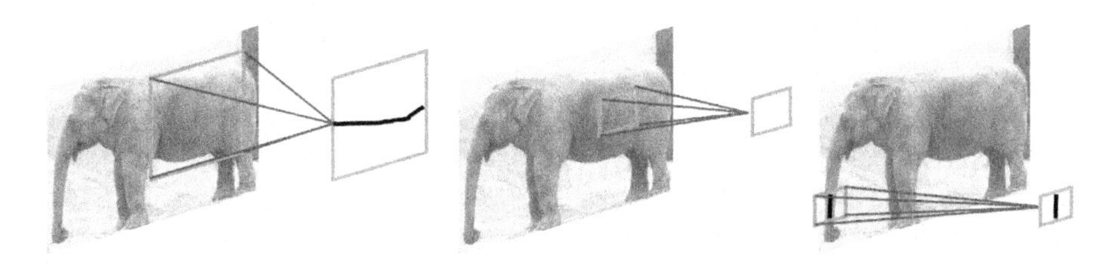

FIGURE 9.2

Using filters (the green squares on images) of multiple sizes for skeleton extraction. Only when the size of the filter is larger than the scale of current skeleton part, the filter can capture enough context feature to detect it.

tures. Besides, per-pixel approaches make predictions extremely time-consuming. The gap between the skeleton extraction methods and actual human perception is still large. Skeleton extraction has its unique aspect by looking into both local and global image context, which requires the use of integrated models that include multiscale feature learning and classifier learning, since the visual complexity increases exponentially with the size of the context field.

A holistically nested network with multiple scale-associated side outputs for skeleton extraction is developed to solve the aforementioned obstacles. The holistically nested network [18] is known as a deep fully convolutional network (FCN) [19], enabling prediction for per-pixel tasks and holistic image training. Here, to address the unknown-scale problem in skeleton extraction, a scale-associated side output is connected to each convolutional layer in the holistically nested network.

Let us suppose that we have a number of filters of varying sizes, such as the convolutional kernels in convolutional networks (see Fig. 9.2). To use these filters to detect skeleton pixels of a specific scale, the only responses that will appear are the ones for filters larger than that scale, whereas the other filters will not have response. The sequential convolutional layers in a holistically nested network are much like filters of increasing size, so each convolutional layer can capture the skeleton pixels with scales less than its receptive field size. The receptive field sizes of the sequential layers on the original image increase from shallow to deep, and this sequential increase creates the foundational principle to quantize the skeleton scale space. Based on these observations, we propose to impose additional supervision for each side output. Each side output is optimized toward a scale-associated groundtruth skeleton map, in which the skeleton pixels are the only ones that have scales smaller than the receptive field size of the side output. Thus, each side output is related to a scale. Each side output is therefore able to provide a certain number of scale-specific skeleton score maps (the score map for one specified quantized scale value).

The final predicted skeleton map can be obtained by fusing these scale-associated side outputs. A straightforward fusion method is to average them. However, a skeleton pixel with larger scale probably has a stronger response on a deeper side output, and a weaker response on a shallower side output; a skeleton pixel with smaller scale may have strong responses on both of the two side outputs. By considering this phenomenon, for each quantized scale value, we propose to use a scale-specific weight layer to fuse the corresponding scale-specific skeleton score map provided by each side output.

In summary, the core contribution of this chapter is the proposal of the scale-associated side output layer that enables both target learning and fusion in a scale-associated way. Therefore, our holistically nested network is able to localize skeleton pixels with multiple scales.

We construct a dataset, called SK506, to assess skeleton extraction methods. The SK506 dataset includes 506 natural images taken from the MS COCO dataset [20]. To obtain both the groundtruth skeleton maps and the groundtruth scale maps, a skeletonization method [21] is applied on the human-annotated foreground segmentation maps to generate them.

This chapter extends our preliminary work [22]. The main contribution is that we conduct much more experiments to verify the usefulness of the extracted skeletons for a variety of object recognition applications, including symmetric part segmentation, road detection, and text line generation.

9.2 RELATED WORKS

Recently, object skeleton extraction has garnered significant attention from researchers. However, many of the early works used presegmented images for skeleton extraction. These studies are not applicable to our work here since they required the use of object silhouettes.

Some works have been done to extract skeletons from the gradient intensity maps on natural images. Generally speaking, the gradient intensity map can be obtained by applying directional derivative operators to a gray-scale image smoothed by a Gaussian kernel. Lindeberg [12] determined the best size of the Gaussian kernel used for gradient computation by providing an automatic mechanism. Then he proposed to detect skeletons as the pixels for which the gradient intensity assumes a local maximum (minimum) in the direction of the main principal curvature. By iteratively minimizing an object function defined on the gradient intensity map Jang and Hong [11] extracted the skeleton from the pseudo-distance map. Yu and Bajaj [10] proposed to trace the ridges of the skeleton intensity map calculated from the diffused vector field of the gradient intensity map, which can remove undesirable biased skeletons. Because these methods lack object prior, they are only able to handle the images with simple scenes.

Recent learning-based skeleton extraction method is a more proper way to deal with the scene complexity problem in natural images, which can be categorized into two types. One is to formulate skeleton extraction to be a per-pixel classification problem. Tsogkas and Kokkinos [14] first computed the hand-designed features of multiscale and multiorientation at each pixel, and then they employed the multiple instance learning framework to determine whether it is symmetric.[1] Shen et al. [23] then improved their method by training MIL models on automatically learned scale- and orientation-related subspaces. To improve the skeleton localization accuracy, Sironi et al. [2] transformed the per-pixel classification problem to a regression one, which learns the distance to the closest skeleton segment in scale-space. The second type of learning-based method aims to learn the similarity between local skeleton segments (represented by superpixel [15,16] or spine model [17]) and links them by hierarchical clustering [15], dynamic programming [16], or particle filter [17]. Due to the limited power of the hand-designed features and traditional learning models, these methods are intractable to detect the skeleton pixels with large scales since much more context information is needed to be handled.

Our method is inspired by a holistically nested network for edge detection (HED) [18]. However, in edge detection, there is no unknown-scale problem. There will always have responses when applying a local filter to detect an edge pixel, either stronger or weaker, regardless of its size. So summing up the multiscale detection responses, which is adopted in the fusion layer in HED, is able to improve the performance of edge detection [24–26] while bringing noises across the scales for skeleton extraction. Our method differs from HED in two aspects: 1. We supervise the side outputs of the network with different scale-associated groundtruths, whereas HED uses the groundtruths with the same scale. 2. We use different scale-specific weight layers to fuse the corresponding scale-specific skeleton score maps provided by the side outputs, whereas HED fuses side outputs by a single weight layer. Such two changes utilize multiple stages in a network to detect the unknown scale explicitly, which HED is unable to deal with. With the added extra supervision to each layer, our method can provide a more informative result, that is, the predicted scale for each skeleton pixel. The obtained scale information can be useful for other potential applications, such as part segmentation and object proposal detection (we will show this in Sections 9.4.2.6 and 9.4.2.7), whereas the result of HED cannot to be applied to such applications.

We consider two other datasets for evaluation, the SYMMAX300 dataset [14] and the WH-SYMMAX dataset [23]. The SYMMAX300 dataset is used for the detection of the local reflection symmetry. This dataset is converted from the Berkeley Segmentation Benchmark (BSDS300) [27]. Local reflection symmetry is not a high-level image feature. Fig. 9.3A shows some examples from this dataset, and it is worth pointing out that many of the symmetries are outside of objects. Generally, the object skeletons are a subset of the local reflection symmetry. The WH-SYMMAX dataset is converted from the Weizmann Horse dataset [28] and is useful for evaluating object skeleton extraction methods. The problem with this dataset, as shown in Fig. 9.3B, is that it only contains the object category for one object, namely, the horse. The SK506 dataset, however, encompasses a variety of objects, including living things

[1] Although symmetry detection is not the same problem as skeleton extraction, we also compare the methods for it with ours since a skeleton can be considered a subset of symmetry.

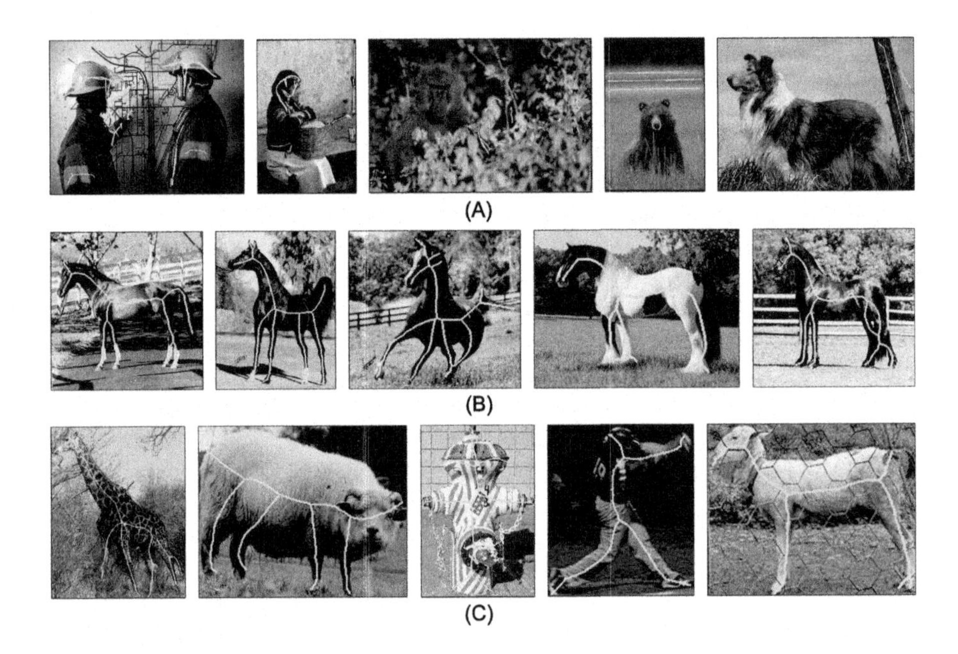

FIGURE 9.3

Some samples from three datasets. (A) The SYMMAX300 dataset [14]. (B) The WH-SYMMAX dataset [23]. (C) Our new dataset, the SK506 dataset. The groundtruths for skeleton or local reflection symmetry are in yellow.

like people and animals, and inanimate objects like hydrant (Fig. 9.3C). We use the WH-SYMMAX and SK506 datasets to assess various skeleton extraction and symmetry detection methods. Our results indicate that our method is superior to the others assessed in this chapter.

9.3 METHODOLOGY

In this section, we describe our methods for object skeleton extraction. First, the architecture of our holistically nested network is introduced. Then we will discuss how to optimize and fuse the multiple scale-associated side outputs in the network for skeleton extraction.

9.3.1 NETWORK ARCHITECTURE

The recent work [29] has demonstrated that to obtain a good performance on a new task, it is an efficient way to fine-tune well-pretrained deep neural networks. Therefore, we adopt the network architecture used in [18]. It should be noted that our network is converted from VGG 16-layer net [30] but significantly different from the net. We add additional side output layers and replace fully connected layers by convolutional layers with 1×1 kernel size. The outputs of all the stages are all the same size. This can be achieved because each convolutional layer is linked to an upsampling layer. Here, we make some modifications for our skeleton extraction tasks: a) All of the proposed scale-associated side output layers are linked to the final convolutional layer in each stage (except for the first one, which are conv2_2, conv3_3, conv4_3, conv5_3). The receptive field sizes of the sequential stages are 14, 40, 92, and 196, respectively. We leave out the first stage because of the small size of its receptive field. The receptive field size is only 5 pixels, and there are few skeleton pixels with scales less than such a small receptive field size. b) Each scale-associated side output layer is connected to a slice layer to obtain the skeleton score map for each scale. We use all the side output layers and a scale-specific weight layer to fuse the skeleton score maps for this scale. We are able to obtain the scale-specific weight layer via a convolutional layer with a 1×1 kernel size. The weight layers are used to fuse the skeleton score maps for different scales, and these fused maps are strung together to create the final version of the skeleton map. Figs. 9.4A and 9.4B show these two modifications. Our holistically nested network architecture has four stages, along with some scale-associated side output layers with strides of 2, 4, 8, and 16, respectively, and with different receptive field sizes. Five additional weight layers fuse the side outputs.

9.3.2 SKELETON EXTRACTION BY FUSING SCALE-ASSOCIATED SIDE OUTPUT

We can formulate skeleton extraction as a per-pixel classification problem. Given a raw input image $X = \{x_j, j = 1, \ldots, |X|\}$, our purpose is to predict its skeleton map $\hat{Y} = \{\hat{y}_j, j = 1, \ldots, |X|\}$, where $\hat{y}_j \in \{0, 1\}$ indicates the predicted label for each pixel x_j, that is, $\hat{y}_j = 1$ if x_j is predicted as a skeleton pixel and $\hat{y}_j = 0$ otherwise. Next, we will describe how we learn and fuse the scale-associated side outputs in the training phase and how we utilize the learned network in the testing phase, respectively.

9.3.2.1 *Training Phase*

Given a training dataset denoted by $\mathcal{S} = \{(X^{(n)}, Y^{(n)}), n = 1, \ldots, N\}$, where $X^{(n)} = \{x_j^{(n)}, j = 1, \ldots, |X^{(n)}|\}$ denotes a raw input image, and $Y^{(n)} = \{y_j^{(n)}, j = 1, \ldots, |X^{(n)}|\}$ ($y_j^{(n)} \in \{0, 1\}$) is its corresponding groundtruth skeleton map. First, we describe how to compute a quantized skeleton scale map for each training image, and this will be used for guiding the network training.

Skeleton scale quantization. According to the definition of skeleton [31], the scale of each skeleton pixel is defined as the diameter of the maximal disk centered at it, which can be obtained when computing the groundtruth skeleton map from the presegmented image. By setting the scale of each nonskeleton pixel to be zero we

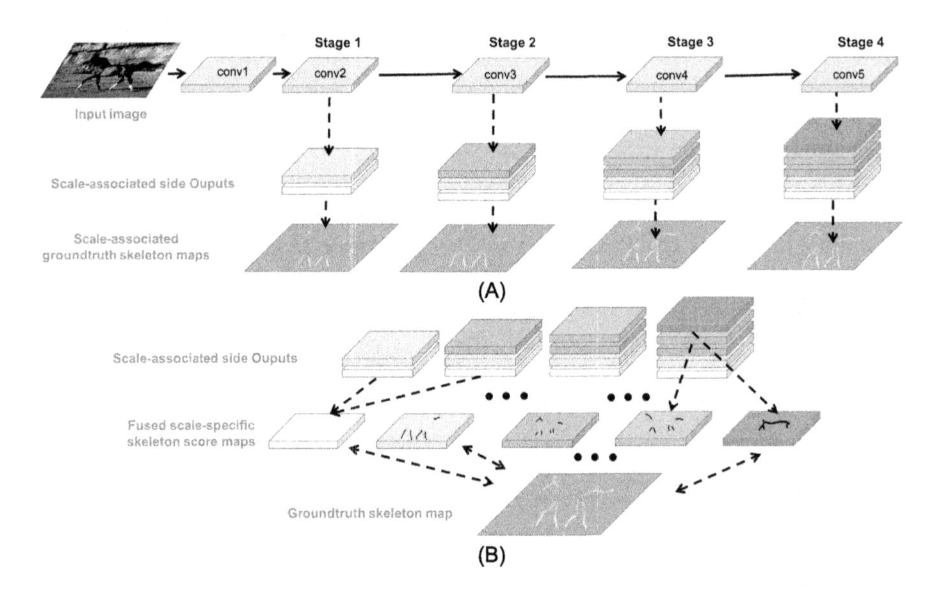

FIGURE 9.4

The proposed network architecture for skeleton extraction, which is converted from VGG 16-layer net [30]. (A) Scale-associated side outputs learning. Our network has four stages with scale-associated side output layers connected to the convolutional layers. The scale-associated side output in each stage is guided by a scale-associated groundtruth skeleton map (the skeleton pixels with different quantized scales are in different colors. Each block in a scale-associated side output is the activation map for one quantized scale, marked by the corresponding color). (B) Scale-specific fusion. Each scale-associated side output provides a certain number of scale-specific skeleton score maps (identified by stage number-quantized scale value pairs). The score maps of the same scales from different stages will be sliced and concatenated. Five scale-specific weighted-fusion layers are added to automatically fuse outputs from multiple stages.

build a scale map $S^{(n)} = \{s_j^{(n)}, j = 1, \ldots, |X^{(n)}|\}$ for each $Y^{(n)}$, and we have $y_j^{(n)} = \mathbf{1}(s_j^{(n)} > 0)$, where $\mathbf{1}(\cdot)$ is the indicator function. As we consider each image holistically, we drop the superscript n in our notation.

Our focus in this chapter is how to learn a holistically nested network with multiple stages of a convolutional layer connected to a scale-associated side output layer. Suppose that M such stages exist in this network. Here, the receptive field sizes of the convolutional layers increase sequentially, and $(r_i; i = 1, \ldots, M)$ is the sequence of the receptive field sizes. The convolutional layer can only capture the features of a skeleton pixel when its receptive field size is larger than the scale of that pixel. Thus, the scale of a skeleton pixel can be quantized into a discrete value and indicate which stages in the network are able to detect this skeleton pixel (we assume that r_M is sufficiently large for capturing the features of the skeleton pixels with the maximum scale). The quantized value z of a scale s is computed by

$$z = \begin{cases} \underset{i=1,\ldots,M}{\arg\min}\, i \text{ s.t. } r_i > \lambda s & \text{if } s > 0, \\ 0 & \text{if } s = 0, \end{cases} \tag{9.1}$$

where $\lambda > 1$ is the factor that ensures that the receptive field sizes are sufficiently large for computing feature. (We set $\lambda = 1.2$ in our experiments.) Now, for an image X, we can build a quantized scale value map $Z = \{z_j, j = 1, \ldots, |X|\}$ ($z_j \in \{0, 1, \ldots, M\}$).

Scale-associated side output learning. Whereas the groundtruth skeleton map Y can be trivially converted from Z: $Y = \mathbf{1}(Z > 0)$, this operation does not work in reverse. So, instead of Y, we would like to guide the network training by Z, as more supervision can be included. This actually converts a binary classification problem to a multiclass classification problem, where each class corresponds to a quantized scale. Each side output layer is then related to a softmax regression classifier. In our network, a stage can only detect the skeleton pixels with scales less than the corresponding receptive field size, which indicates that the side output is scale-associated. For the ith side output, we construct a scale-associated groundtruth skeleton map to guide it: $Z^{(i)} = Z \circ \mathbf{1}(Z \le i)$, where $\circ$ is an element-wise product operator. Let $K^{(i)}$ indicate the maximum value in $Z^{(i)}$, that is, $K^{(i)} = i$. Then we have $Z^{(i)} = \{z_j^{(i)}, j = 1, \ldots, |X|\}$, $z_j^{(i)} \in \{0, 1, \ldots, K^{(i)}\}$. Let $\ell_s^{(i)}(\mathbf{W}, \boldsymbol{\Phi}^{(i)})$ indicate the loss function for this scale-associated side output, where $\mathbf{W}$ and $\boldsymbol{\Phi}^{(i)}$ denote the layer parameters of the network and the parameters of the classifier of this stage. Because our network enables holistic image training, we compute the loss function over all pixels in the training image X and the scale-associated groundtruth skeleton map $Z^{(i)}$. Generally in an image, the distribution of skeleton pixels with different scales and nonskeleton pixels is biased. Therefore, a weighted softmax loss function to balance the loss between these multiple classes is defined:

$$\ell_s^{(i)}(\mathbf{W}, \boldsymbol{\Phi}^{(i)}) = \\ -\frac{1}{|X|} \sum_{j=1}^{|X|} \sum_{k=0}^{K^{(i)}} \beta_k^{(i)} \mathbf{1}(z_j^{(i)} = k) \log \Pr(z_j^{(i)} = k | X; \mathbf{W}, \boldsymbol{\Phi}^{(i)}), \tag{9.2}$$

where $\beta_k^{(i)}$ denotes the loss weight for the kth class, and $\Pr(z_j^{(i)} = k | X; \mathbf{W}, \boldsymbol{\Phi}^{(i)}) \in [0, 1]$ denotes the predicted score given by the classifier for how likely the quantized scale of x_j is k. $\mathcal{N}(\cdot)$ is the number of nonzero elements in a set; then β_k can be computed by

$$\beta_k^{(i)} = \frac{\frac{1}{\mathcal{N}(\mathbf{1}(Z_i == k))}}{\sum_{k=0}^{K^{(i)}} \frac{1}{\mathcal{N}(\mathbf{1}(Z_i == k))}}. \tag{9.3}$$

Let $a_{jk}^{(i)}$ be the activation of the ith side output associated with the quantized scale k for the input x_j. Then the softmax function [32] $\sigma(\cdot)$ is used to compute

$$\Pr(z_j^{(i)} = k|X; \mathbf{W}, \mathbf{\Phi}^{(i)}) = \sigma(a_{jk}^{(i)}) = \frac{\exp(a_{jk}^{(i)})}{\sum_{k=0}^{K^{(i)}} \exp(a_{jk}^{(i)})}. \tag{9.4}$$

We can obtain the partial derivation of $\ell_s^{(i)}(\mathbf{W}, \mathbf{\Phi}^{(i)})$ w.r.t. $a_{jl}^{(i)}$ ($l \in \{0, 1, \ldots, K^{(i)}\}$) by

$$\begin{aligned} \frac{\partial \ell_s^{(i)}(\mathbf{W}, \mathbf{\Phi}^{(i)})}{\partial a_{jl}^{(i)}} &= -\frac{1}{|X|}\Bigg(\beta_l^{(i)} \mathbf{1}(z_j^{(i)} = l) \\ &- \sum_{k=0}^{K^{(i)}} \beta_k^{(i)} \mathbf{1}(z_j^{(i)} = k) \Pr(z_j^{(i)} = l|X; \mathbf{W}, \mathbf{\Phi}^{(i)})\Bigg), \end{aligned} \tag{9.5}$$

where $\mathbf{\Phi} = (\mathbf{\Phi}^{(i)}; i = 1, \ldots, M)$ are the parameters of the classifiers in all the stages. Then the loss function for all the side outputs can be simply obtained by

$$\mathcal{L}_s(\mathbf{W}, \mathbf{\Phi}) = \sum_{i=1}^{M} \ell_s^{(i)}(\mathbf{W}, \mathbf{\Phi}^{(i)}). \tag{9.6}$$

Multiple scale-associated side outputs fusion. For an input pixel x_j, each scale-associated side output provides a predicted score $\Pr(z_j^{(i)} = k|X; \mathbf{W}, \mathbf{\Phi}^{(i)})$ (if $k \le K^{(i)}$), which represents how likely its quantized scale is k. A fused score f_{jk} can be obtained by simply summing them with weights $\mathbf{a}_k = (a_k^{(i)}; i = \max(k, 1), \ldots, M)$:

$$\begin{aligned} f_{jk} &= \sum_{i=\max(k,1)}^{M} a_k^{(i)} \Pr(z_j^{(i)} = k|X; \mathbf{W}, \mathbf{\Phi}^{(i)}) \\ &\text{s.t.} \sum_{i=\max(k,1)}^{M} a_k^{(i)} = 1. \end{aligned} \tag{9.7}$$

The above fusion process can be understood as follows: each scale-associated side output provides a certain number of scale-specific predicted skeleton score maps and $M + 1$ scale-specific weight layers; $\mathbf{A} = (\mathbf{a}_k; k = 0, \ldots, M)$ are utilized to fuse them. Similarly, we can define a fusion loss by

$$\begin{aligned} &\mathcal{L}_f(\mathbf{W}, \mathbf{\Phi}, \mathbf{A}) = \\ &-\frac{1}{|X|} \sum_{j=1}^{|X|} \sum_{k=0}^{M} \beta_k \mathbf{1}(z_j = k) \log \Pr(z_j = k|X; \mathbf{W}, \mathbf{\Phi}, \mathbf{a}_k), \end{aligned} \tag{9.8}$$

where we define β_k in the same way as in Eq. (9.3) and $\Pr(z_j = k|X; \mathbf{W}, \mathbf{\Phi}, \mathbf{w}_k) = \sigma(f_{jk})$.

Finally, the optimal parameters are obtained by

$$(\mathbf{W}, \mathbf{\Phi}, \mathbf{A})* = \arg\min(\mathcal{L}_s(\mathbf{W}, \mathbf{\Phi}) + \mathcal{L}_f(\mathbf{W}, \mathbf{\Phi}, \mathbf{A})). \quad (9.9)$$

9.3.2.2 Testing Phase

Given a testing image $X = \{x_j, j = 1, \ldots, |X|\}$ and the learned network $(\mathbf{W}, \mathbf{\Phi}, \mathbf{A})*$, its predicted skeleton map $\hat{Y} = \{\hat{y}_j, j = 1, \ldots, |X|\}$ is calculated by

$$\hat{y}_j = 1 - \Pr(z_j = 0|X; \mathbf{W}*, \mathbf{\Phi}*, \mathbf{a}_0*). \quad (9.10)$$

Recall that $z_j = 0$ and $z_j > 0$ denote that x_j is a nonskeleton/skeleton pixel, respectively. Our method is referred to as FSDS for fusing scale-associated deep side outputs.

9.3.3 UNDERSTANDING OF THE PROPOSED METHOD

Fig. 9.5 highlights the key aspects of our approach, and we also compare the intermediate results with those of HED to help understand our method more deeply. We used Eq. (9.10) to compute the response from each of the scale-associated side outputs. The response of each scale-associated side output is compared to the corresponding one in HED. (The side output 1 in HED is connected to conv1_2, whereas ours start from conv2_2.) With the additional scale-associated supervision, the responses of our side outputs are indeed related to scale. For example, the first one fires on the structure with small scales, such as the object boundaries and the interior textures; in the second one, the skeleton parts of the legs are clear, and meanwhile the noises on small scale structure are suppressed; in the third one, the skeleton parts of the torso reveal while the noises on small scale structure are further suppressed. In addition, scale-specific fusion is performed, by which each fused scale-specific skeleton score map indeed corresponds to one scale. This is particularly evident in the second and third response maps, which correspond to legs and torso, respectively. The side outputs of HED do not exhibit such performance since they are not able to accurately distinguish the skeleton pixels with various scales. Consequently, the first two respond on the whole body, which bring noises to the final fusion one.

9.4 EXPERIMENTAL RESULTS

In this section, we first discuss the implementation details and then compare the performance of our skeleton extraction methods with competitors.

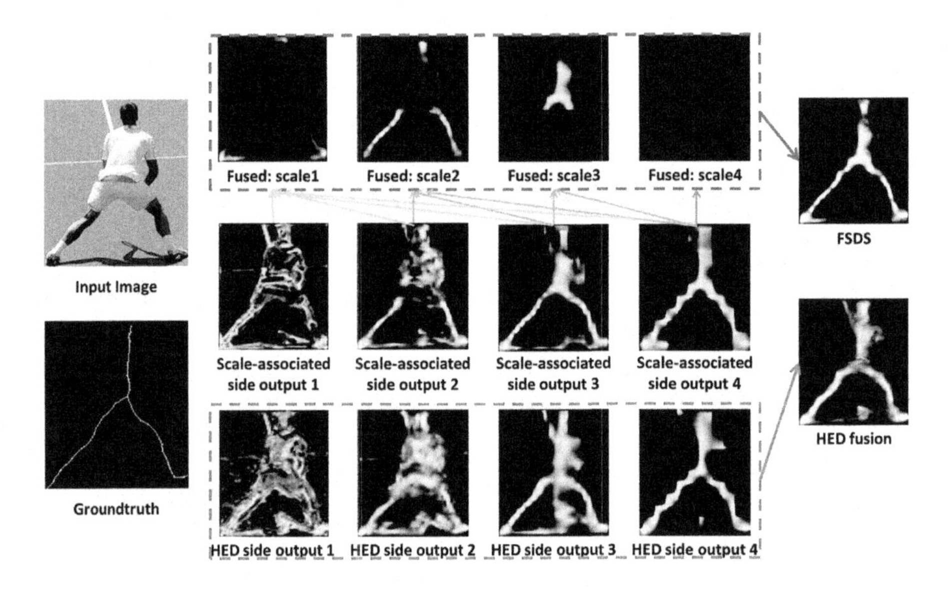

FIGURE 9.5

The comparison between the intermediate results of FSDS and HED. We can observe that the former are able to differentiate skeleton pixels with different scales, whereas the latter cannot.

9.4.1 IMPLEMENTATION DETAILS

The architecture of our network is built on the public available implementation of FCN [19] and HED [18]. We fine-tune the whole network from an initialization with the pretrained VGG 16-layer net [30].

Model parameters. The hyper-parameters of our network include: the base learning rate (1×10^{-6}), the mini-batch size (1), the momentum (0.9), the loss weight for each side-output (1), initialization of the scale-specific weighted fusion layer ($1/n$, where n is the number of sliced scale-specific map), initialization of the nested filters (0), the learning rate of the scale-specific weighted fusion layer (5×10^{-6}), the weight decay (2×10^{-4}), and the maximum number of training iterations (20,000).

Data augmentation. Data augmentation is a principal way to generate sufficient training data for learning a "good" deep network. We rotate the images to four different angles (0°, 90°, 180°, 270°) and flip with different axes (left-right, up-down, no flip) and then resize images to three different scales (0.8, 1.0, 1.2), totally leading to an augmentation factor of 36. It should be noted that when resizing a groundtruth skeleton map, the scales of the skeleton pixels in it should be multiplied by a resize factor accordingly.

9.4.2 PERFORMANCE COMPARISON

We conduct experiments by comparing our method FSDS with many others, including a traditional image processing method (Lindeberg's method [12]), three learning-based segment linking methods (Lee's method [16], Levinshtein's method [15], and particle filter [17]), three per-pixel classification/regression methods (MIL [14], distance regression [2], and MISL [23]) and a deep-learning-based method (HED [18]). For all theses methods, we use the source code provided by the authors under the default setting. For HED and FSDS, we perform sufficient numbers iterations to obtain optimal models, 15,000 and 18,000 iterations for FSDS and HED, respectively. A standard nonmaximal suppression algorithm [25] is applied to the response maps of HED and ours to obtain the thinned skeletons for performance evaluation.

9.4.2.1 *Evaluation Protocol*

We follow the evaluation protocol used in [14], where the performances of skeleton extraction methods are measured by their maximum F-measure ($\frac{2 \cdot \text{Precision} \cdot \text{Recall}}{\text{Precision}+\text{Recall}}$) and precision-recall curves with respect to the groundtruth skeleton map. To obtain the precision-recall curves, the detected symmetry response is first thresholded into a binary map, which is then matched with the groundtruth skeleton map. The matching allows small localization errors between detected positives and groundtruths. If a detected positive is matched with at least one groundtruth skeleton pixel, then it is classified as true positive. In contrast, pixels that do not correspond to any groundtruth skeleton pixel are false positives. By assigning different thresholds to the detected skeleton response we can obtain a sequence of precision and recall pair, which is used to plot the precision-recall curve.

9.4.2.2 *SK506*

We first conduct our experiments on the newly built SK506 Dataset. Object skeletons in this dataset have large variances in both scales and structures. This dataset is split into 300 training and 206 testing images. We report the F-measure and the average runtime per image of each method on this dataset in Table 9.1. Observe that both traditional image processing and per-pixel/segment learning methods do not perform well, indicating the difficulty of this task. In addition, the segment linking methods are extremely time consuming. Our proposed FSDS outperforms others significantly, even compared with the deep-learning-based method HED. Besides, thanks to the powerful convolution computation ability of GPU, our method is able to process images in real time, about 20 images per second. The precision/recall curves shown in Fig. 9.6 evidence again that FSDS is better than the competitors since FSDS shows both improved recall and precision at most of the precision-recall regimes. For qualitative comparison, we illustrate the skeleton extraction results obtained by several methods in Fig. 9.7. The qualitative examples show that our method hits on more groundtruth skeleton points and meanwhile suppresses the false positives. The false

Table 9.1 Performance comparison between different methods on the SK506 dataset

Method	F-measure	Avg runtime (s)
Levinshtein [15]	0.218	144.77
Particle Filter [17]	0.226	322.25†
Lindeberg [12]	0.227	4.03
Lee [16]	0.252	606.30
MIL [14]	0.392	42.38
HED [18]	0.542	0.05†
FSDS (ours)	**0.623**	**0.05†**

† *GPU time.*

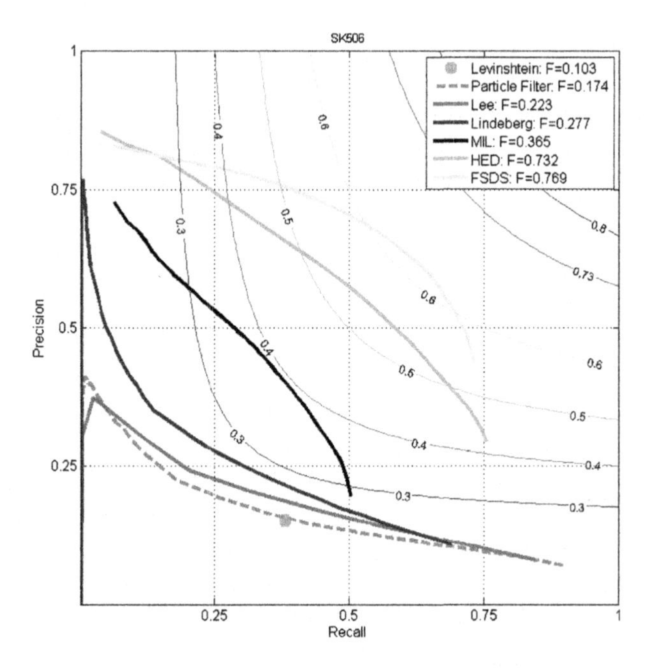

FIGURE 9.6

Evaluation of skeleton extractors on the SK506 dataset. Leading skeleton extraction methods are in an ascending order according to their best F-measure with respect to groundtruth skeletons. Our method FSDS achieves the top result and shows both improved recall and precision at most of the precision-recall regime. See Table 9.1 for more details about the other two quantities and method citations.

positives in the results of HED are probably introduced across response of different scales. Benefited from scale-associated learning and scale-specific fusion, our method is able to suppress such false positives.

FIGURE 9.7

Illustration of skeleton extraction results on the SK506 dataset for several selected images. The groundtruth skeletons are in yellow, and the thresholded extraction results are in red. Thresholds were optimized over the whole dataset.

9.4.2.3 *WH-SYMMAX*

The WH-SYMMAX dataset [23] contains a total of 328 images, among which the first 228 are used for training, and the rest are used for testing. The precision/recall curves of skeleton extraction methods are shown in Fig. 9.9, and summary statistics are presented in Table 9.2. Qualitative comparisons are illustrated in Fig. 9.10. Both quantitative and qualitative results demonstrate that our method is clearly better than others.

9.4.2.4 *SYMMAX300*

As we discussed in Section 9.1, a large number of groundtruths are labeled on nonobject parts in SYMMAX300 [14], which do not have organized structures as object skeletons. Our aim is to suppress those on nonobject parts, so that the obtained skeletons can be used for other potential applications. In addition, the groundtruths for scale are not provided by SYMMAX300. Therefore, we do not evaluate our method quantitatively on SYMMAX300. Even so, Fig. 9.8 shows that our method can obtain good skeletons of some objects in SYMMAX300. We also observe that the results obtained by our method have significantly less noises on background.

Table 9.2 Performance comparison between different methods on the WH-SYMMAX dataset [23]

Method	F-measure	Avg runtime (s)
Distance Regression [2]	0.103	5.78
Levinshtein [15]	0.174	105.51
Lee [16]	0.223	716.18
Lindeberg [12]	0.277	5.75
Particle Filter [17]	0.334	13.9†
MIL [14]	0.365	51.19
MISL [23]	0.402	78.41
HED [18]	0.732	0.06†
FSDS (ours)	**0.769**	0.07†

† *GPU time.*

FIGURE 9.8

Illustration of skeleton extraction results on the SYMMAX300 dataset [14] for several selected images. The groundtruth skeletons are in yellow, and the thresholded extraction results are in red. Thresholds were optimized over the whole dataset.

9.4.2.5 *Cross Dataset Generalization*

We may concern that the scale-associated side outputs learned from one dataset might lead to higher generalization error when applied to another dataset. To explore whether this is the case, we test the model on cross dataset, that is, learn the model from one dataset and test it on another one. For comparison, we list the cross dataset generalization results of MIL [14], HED [18], and our method in Table 9.3.

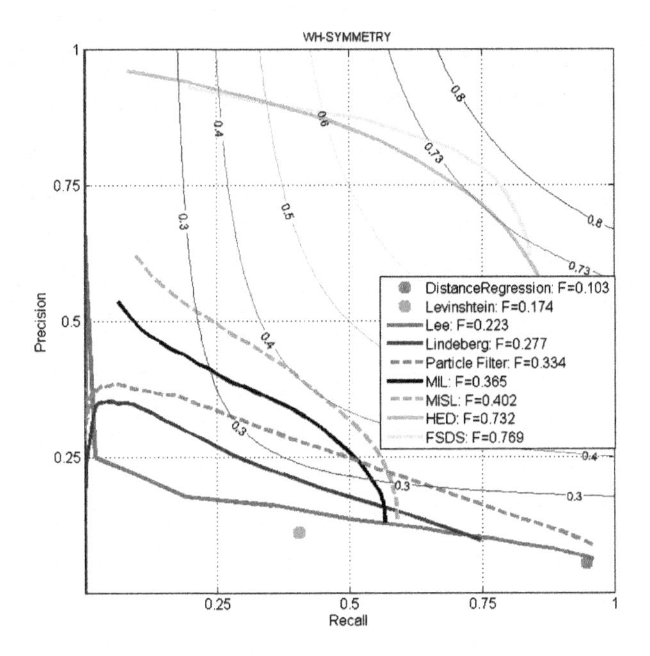

FIGURE 9.9

Evaluation of skeleton extractors on the WH-SYMMAX dataset [23]. Leading skeleton extraction methods are in an ascending order according to their best F-measure with respect to groundtruth skeletons. Our method FSDS achieves the top result and shows both improved recall and precision at most of the precision-recall regime. See Table 9.2 for more details about the other two quantities and method citations.

Table 9.3 Cross-dataset generalization results. TRAIN/TEST indicates the training/testing dataset used

Method	Train/Test	F-measure
MIL [14]	WH-SYMMAX/SK506	0.387
HED [18]	WH-SYMMAX/SK506	0.492
FSDS (ours)	WH-SYMMAX/SK506	**0.529**
MIL [14]	SK506/WH-SYMMAX	0.363
HED [18]	SK506/WH-SYMMAX	0.508
FSDS (ours)	SK506/WH-SYMMAX	**0.632**

Our method achieves a better cross dataset generalization result than both the "non-deep" method (MIL) and the "deep" method (HED).

9.4.2.6 Symmetric Part Segmentation

Part-based models are widely used for object detection and recognition in natural images [33,34]. To verify the usefulness of the extracted skeletons, we follow

FIGURE 9.10

Illustration of skeleton extraction results on the WH-SYYMAX dataset [23] for several selected images. The groundtruth skeletons are in yellow, and the thresholded extraction results are in red. Thresholds were optimized over the whole dataset.

the criteria in [15] for symmetric part segmentation. We evaluate the ability of our skeleton to find segmentation masks corresponding to object parts in a cluttered scene. Our network provides a predicted scale for each skeleton pixel (the fused skeleton score map for which scale has maximal response). With it we can recover object parts from skeletons. For each skeleton pixel x_j, we can predict its scale by $\hat{s}_j = \sum_{i=1}^{M} \Pr(z_j = i|X; \mathbf{\Theta}*, \mathbf{\Phi}*, \mathbf{a}_0*) r_i$. Then, for a skeleton segment $\{x_j, j = 1, \ldots, N\}$, where N is the number of the skeleton pixels in this segment, we can obtain a segmented object part mask by $\mathcal{M} = \bigcup_{j=1}^{N} D_j$, where D_j is the disk of center x_j and diameter $\hat{s}_j$. A confidence score is also assigned to each object part mask for quantitative evaluation: $P_{\mathcal{M}} = \frac{1}{N} \sum_{j=1}^{N} (1 - \Pr(z_j = 0|X; \mathbf{\Theta}*, \mathbf{\Phi}*, \mathbf{a}_0*))$. We compare our segmented part masks with Lee's method [16] and Levinshtein's method [15] on their BSDS-Parts dataset [16], which contains 36 images annotated with groundtruth masks corresponding to the symmetric parts of prominent objects. The segmentation results are evaluated by the protocol used in [16]: A segmentation mask $\mathcal{M}_{seg}$ is counted as a hit if its overlap with the groundtruth mask $\mathcal{M}_{gt}$ is greater than 0.4, where overlap is measured by intersection-over-union (IoU). A precision/recall curve is obtained by varying a threshold over the confidence scores of segmented masks. The quantitative evaluation results are summarized in Fig. 9.11, which indicate a significant improvement over the other two methods. Some qualitative results

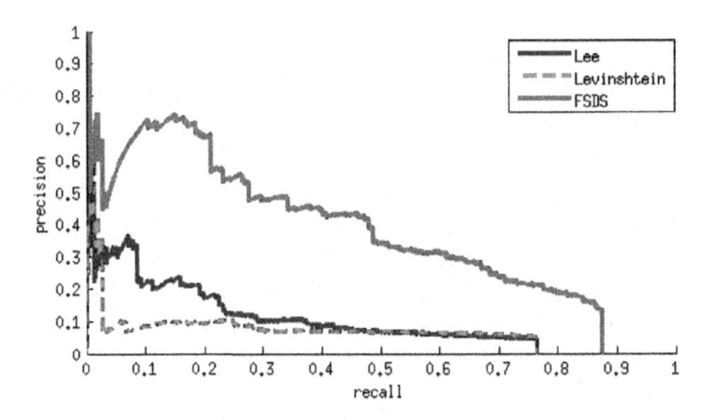

FIGURE 9.11

Symmetric part segmentation results on BSDS-Parts dataset [16].

on the BSDS-Parts dataset [16] and the Weizmann Horse dataset [28] are shown in Figs. 9.12 and 9.13.

9.4.2.7 Object Proposal Detection

To demonstrate the potential of the extracted skeletons in object detection, we do an experiment on object proposal detection. Let h_B^E be the objectness score of a bounding box B obtained by Edgebox [35]. We define our objectness score by $h_B = \frac{\sum_{\forall \mathcal{M} \cap B \neq \emptyset} \mathcal{M} \cap B}{\sum_{\forall \mathcal{M} \cap B \neq \emptyset} \mathcal{M} + \epsilon} \cdot h_B^E$, where ϵ is a very small number to ensure the denominator to be nonzero, and $\mathcal{M}$ is a part mask reconstructed by a detected skeleton segment. As shown in Figs. 9.14 and 9.15, the new scoring method achieves better object proposal detection results than Edge Boxes.

9.4.2.8 Road Detection

Road detection is useful in aerial imagery for developing georeferenced mosaics, route planning, and emergency management systems [37]. We test our method on a dataset of aerial images of road networks [2] to detect road centerlines. We follow the training/testing splits setting used in [2]. The qualitative and quantitative results are shown in Figs. 9.16 and 9.17, respectively. The performance of road detection is evaluated by the F-measure (Section 9.4.2.1) of the detected road centerlines. Our method clearly outperforms the distance regression method proposed in [2], and we can observe that our results have stronger responses on road centerlines and less noises on backgrounds.

9.4.2.9 Text Line Proposal Generation

As pointed out in [5], text lines in natural images always bear distinctive symmetry and self-similarity properties, which come from both themselves and their local backgrounds, as shown in Fig. 9.18. Therefore, we can perform our skeleton extraction method to detect the centerline of each text line and generate text line proposals

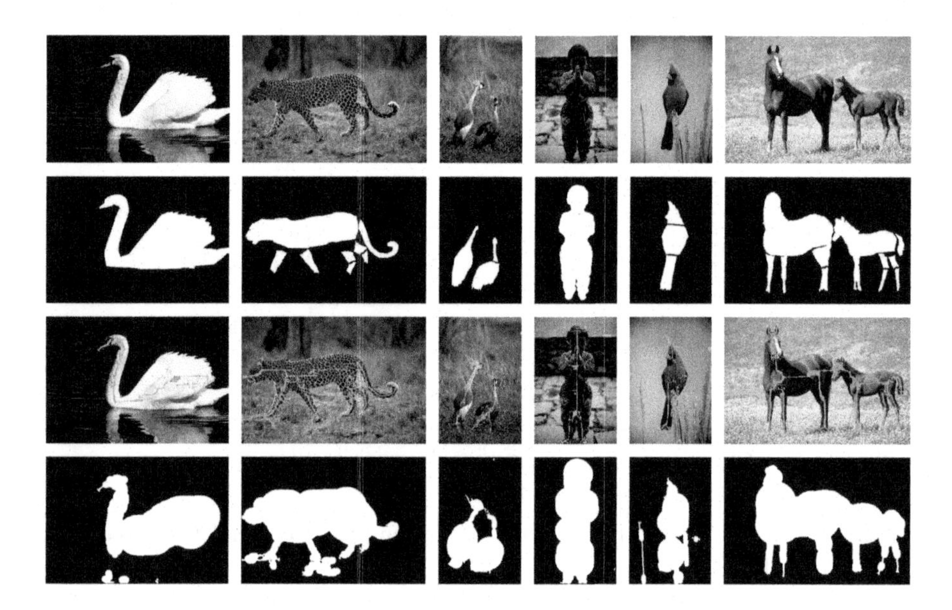

FIGURE 9.12

Illustration of symmetric part segmentation results on the BSDS-Parts dataset [16] for several selected images. In each column, we show the original image, the segmentation groundtruth, the thresholded extracted skeleton (in green), and the segmented masks recovered by the skeleton. Thresholds were optimized over the whole dataset.

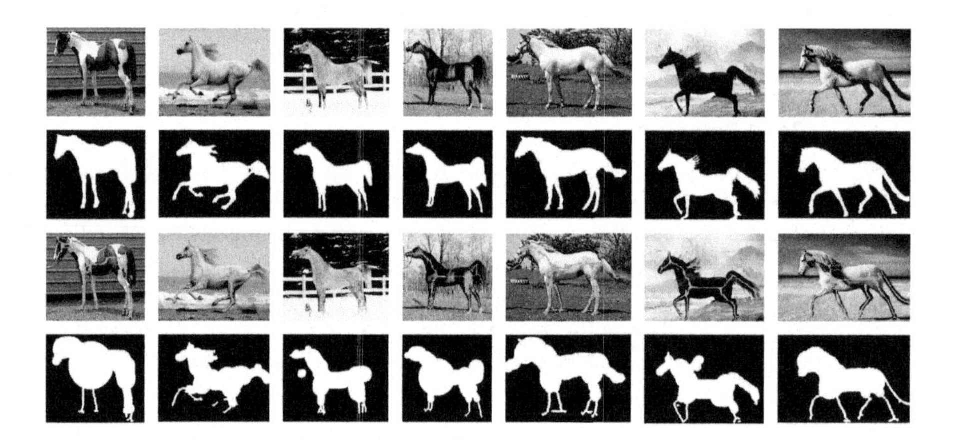

FIGURE 9.13

Illustration of symmetric part segmentation results on the Weizmann Horse dataset [28] for several selected images. In each column, we show the original image, the segmentation groundtruth, the thresholded extracted skeleton (in green), the segmented masks recovered by the skeleton. Thresholds were optimized over the whole dataset.

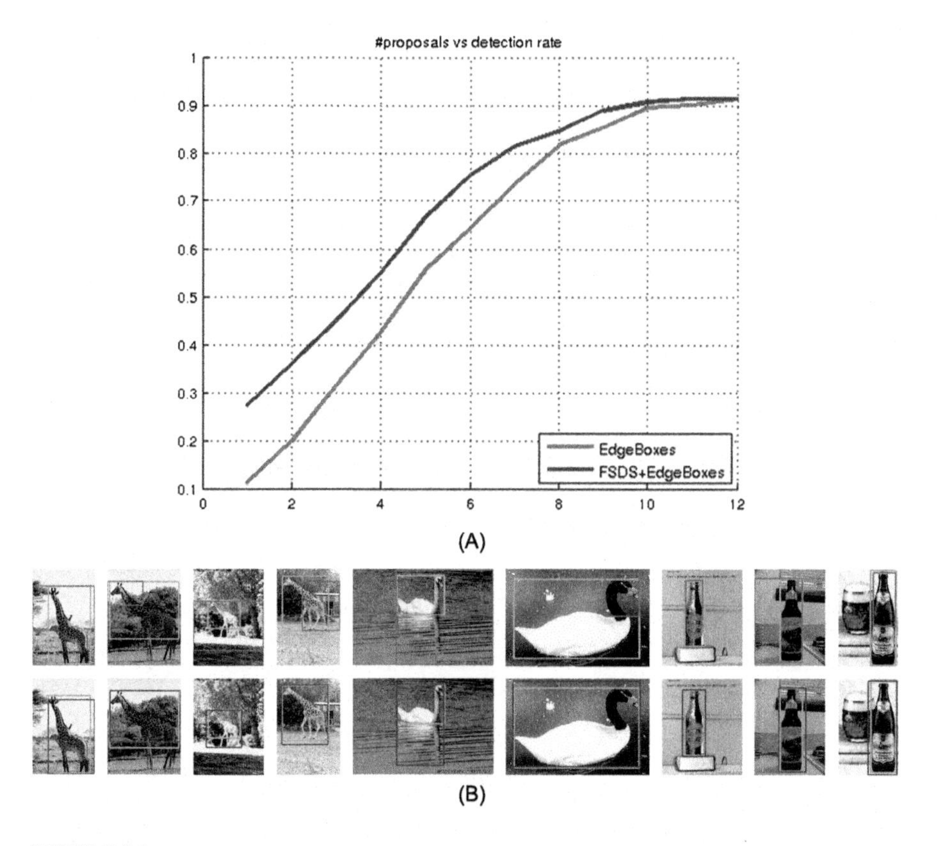

FIGURE 9.14

Object proposal results on ETHZ Shape Classes [36]. (A) The curve (IoU = 0.7). (B) Examples. Groundtruth (green), the closest proposal to groundtruth of Edgebox (red), and ours (blue).

for text detection. Toward this end, we replace the symmetry detection method used in [5] by FSDS. In the training stage, we use the centerline of the bounding box of each text line as its groundtruth symmetry axis.

In the testing stage, we resize a testing image X into images with multiple sizes $\{X^\gamma\}_{\gamma=1}^{\Gamma}$ and apply FSDS to them to detect the centerlines of text lines. In our experiments, we resize an image to $\Gamma = 3$ sizes, which is achieved by setting image height to 400, 600, and 800, respectively, and fixing the ratio between height and width. For a centerline pixel x_j^γ detected on X^γ, its predicted scale $\hat{s}_j^\gamma$ is computed by the same way described in Section 9.4.2.6. Then, for a centerline $C = \{x_j^\gamma, j = 1, \ldots; N\}$, where N is the number of the pixels in this centerline, the height and the width of the text line bounding box located on it are given by $\mathrm{h}_{bbx(C)} = \frac{2}{N}\sum_{j=1}^{N}\hat{s}_j^\gamma$ and $\mathrm{w}_{bbx(C)} = \max_{p,q\in\{1,\ldots,N\}} \mathrm{Px}(\|x_p^\gamma - x_q^\gamma\|)$, respectively, where $\mathrm{Px}(\cdot)$ is the projec-

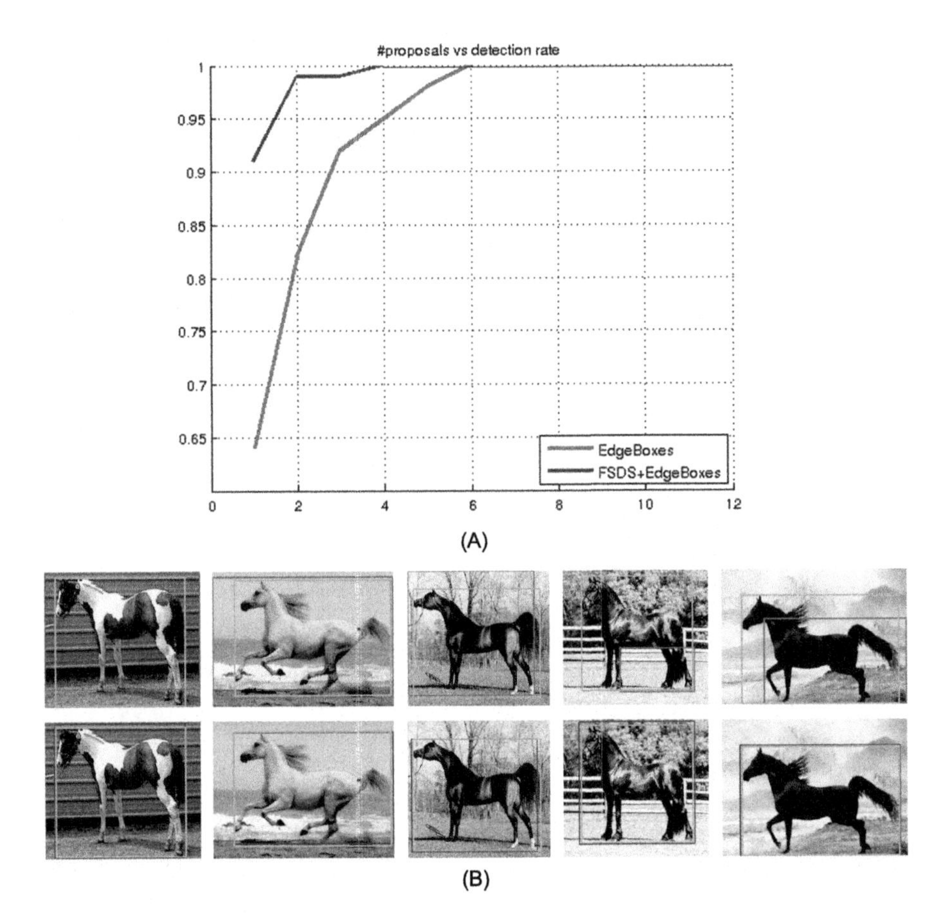

FIGURE 9.15

Object proposal results on the Weizmann Horse dataset [28]. (A) The curve (IoU = 0.7). (B) Examples. Groundtruth (green), the closest proposal to groundtruth of Edgebox (red), and ours (blue).

tion along the x-axis. Text line proposals detected from different sizes are merged, and nonmaximum suppression is applied to remove redundant detections. We test this text line proposal generate method on ICDAR2011 [38], which includes 299 training images and 255 testing images. Examples of generated text line proposals (colored bounding boxes) are demonstrated in Fig. 9.19.

To evaluate the ability of our method w.r.t. text line proposal generation, following [5], we adopt the following character detection rate measure:

$$R = \frac{\sum_{i=1}^{N} \sum_{j=1}^{|G_i|} \max_{k=1}^{|D_i|} M_s(G_i^{(j)}, D_i^{(k)})}{\sum_{i=1}^{N} |G_i|}, \tag{9.11}$$

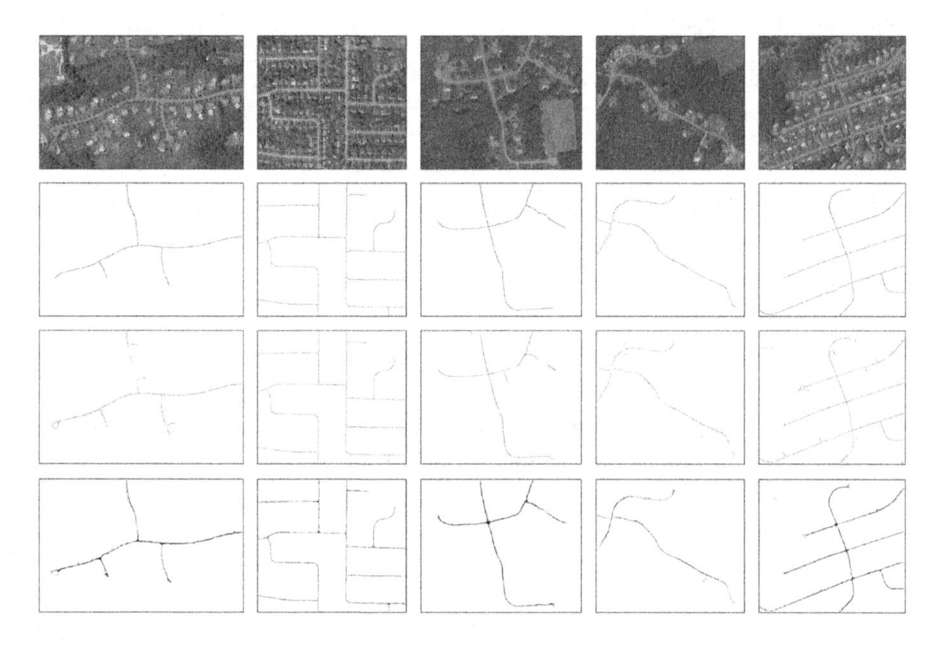

FIGURE 9.16

Road detection in Aerial images. In each column, we show from top to bottom the original image, the groundtruth road centerlines, the result of [2] and our result.

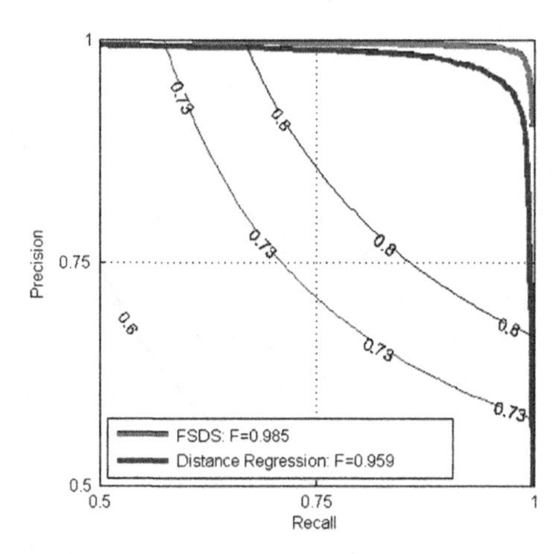

FIGURE 9.17

The precision-recall curves of road detection.

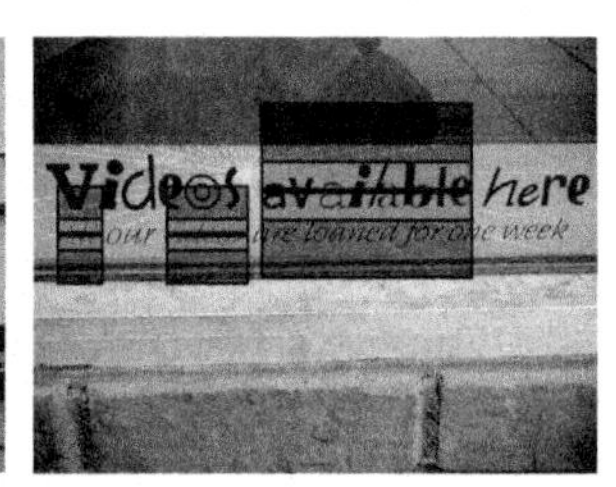

FIGURE 9.18

The symmetry properties of text lines in natural images.

FIGURE 9.19

Text line proposal generation on ICDAR2011 [38]. Proposals are represented by colored bounding boxes.

where N is the total number of images in a dataset, $G_i^{(j)}$ and $D_i^{(k)}$ are respectively the groundtruth set and the detection rectangles in ith image, and $M_s(G_i^{(j)}, D_i^{(k)})$ is the match score between the jth ground truth rectangle and the kth detection rectangle $D_i^{(k)}$. The match score is defined as

$$M_s(G_i^{(j)}, D_i^{(k)}) = \begin{cases} 1 & \dfrac{|G_i^{(j)} \cap D_i^{(k)}|}{|G_i^{(j)}|} \geq 0.8 \text{ and} \\ & \dfrac{\max(h(G_i^{(j)}), h(D_i^{(k)}))}{\min(h(G_i^{(j)}), h(D_i^{(k)}))} \leq 2.5, \\ 0 & \text{otherwise,} \end{cases} \tag{9.12}$$

where $h(\cdot)$ is the height of a rectangle. We compute the character detection rates w.r.t. different number of text line proposals. We compare our method with [5], a symmetry-based text detection method. By the comparison shown in Fig. 9.20 the text extraction ability of the proposed method is stronger than [5], evidencing the usefulness of the extracted skeletons in text detection.

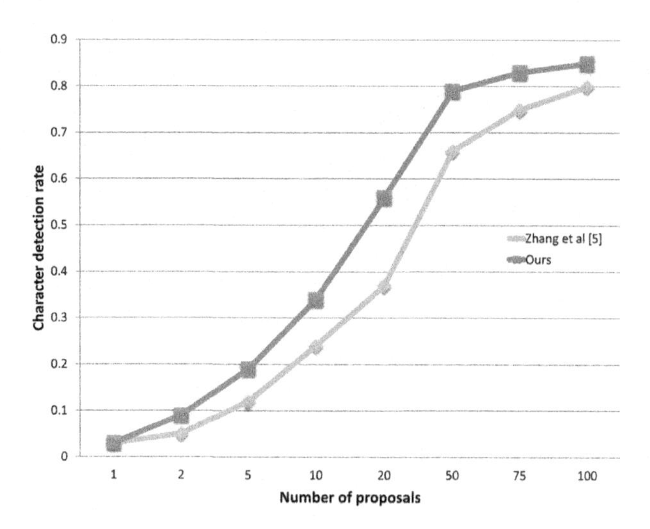

FIGURE 9.20

Character detection rate curves w.r.t. the number of text line proposals.

9.5 CONCLUSION

We have presented a fully convolutional network with multiple scale-associated side outputs to extract skeletons from natural images. We have elucidated the connection between the ability to detect skeletons of varying scales and the receptive field sizes of the sequential stages in the network. The proposed scale-associated side outputs play a important role in the multiscale feature learning and combining the scale-specific responses from various stages. The experimental results verify the usefulness of our method. Our method achieves significant improvements over all other competitors. The additional experiments on applications, such as symmetric part segmentation, object proposal detection, road detection, and text line proposal generation, verified the usefulness of the extracted skeletons in object recognition.

ACKNOWLEDGMENTS

The work in this chapter was supported in part by the National Natural Science Foundation of China under Grants 61672336, 61303095, and 61573160, in part by "Chen Guang" project supported by Shanghai Municipal Education Commission and Shanghai Education Development Foundation under Grant 15CG43, and in part by Research Fund for the Doctoral Program of Higher Education of China under Grant 20133108120017. We thank NVIDIA Corporation for providing their GPU device for our academic research.

REFERENCES

[1] B. Alexe, T. Deselaers, V. Ferrari, What is an object?, in: Proc. CVPR, 2010, pp. 73–80.
[2] A. Sironi, V. Lepetit, P. Fua, Multiscale centerline detection by learning a scale-space distance transform, in: Proc. CVPR, 2014, pp. 2697–2704.
[3] X. Bai, X. Wang, L.J. Latecki, W. Liu, Z. Tu, Active skeleton for non-rigid object detection, in: Proc. ICCV, 2009, pp. 575–582.
[4] N.H. Trinh, B.B. Kimia, Skeleton search: category-specific object recognition and segmentation using a skeletal shape model, Int. J. Comput. Vis. 94 (2) (2011) 215–240.
[5] Z. Zhang, W. Shen, C. Yao, X. Bai, Symmetry-based text line detection in natural scenes, in: Proc. CVPR, 2015, pp. 2558–2567.
[6] P.K. Saha, G. Borgefors, G. Sanniti di Baja, A survey on skeletonization algorithms and their applications, Pattern Recognit. Lett. 76 (2016) 3–12.
[7] K. Siddiqi, A. Shokoufandeh, S.J. Dickinson, S.W. Zucker, Shock graphs and shape matching, Int. J. Comput. Vis. 35 (1) (1999) 13–32.
[8] T.B. Sebastian, P.N. Klein, B.B. Kimia, Recognition of shapes by editing their shock graphs, IEEE Trans. Pattern Anal. Mach. Intell. 26 (5) (2004) 550–571.
[9] M.F. Demirci, A. Shokoufandeh, Y. Keselman, L. Bretzner, S.J. Dickinson, Object recognition as many-to-many feature matching, Int. J. Comput. Vis. 69 (2) (2006) 203–222.
[10] Z. Yu, C.L. Bajaj, A segmentation-free approach for skeletonization of gray-scale images via anisotropic vector diffusion, in: Proc. CVPR, 2004, pp. 415–420.
[11] J.-H. Jang, K.-S. Hong, A pseudo-distance map for the segmentation-free skeletonization of gray-scale images, in: Proc. ICCV, 2001, pp. 18–25.
[12] T. Lindeberg, Edge detection and ridge detection with automatic scale selection, Int. J. Comput. Vis. 30 (2) (1998) 117–156.
[13] Q. Zhang, I. Couloigner, Accurate centerline detection and line width estimation of thick lines using the Radon transform, IEEE Trans. Image Process. 16 (2) (2007) 310–316.
[14] S. Tsogkas, I. Kokkinos, Learning-based symmetry detection in natural images, in: Proc. ECCV, 2012, pp. 41–54.
[15] A. Levinshtein, S.J. Dickinson, C. Sminchisescu, Multiscale symmetric part detection and grouping, in: Proc. ICCV, 2009, pp. 2162–2169.
[16] T.S.H. Lee, S. Fidler, S.J. Dickinson, Detecting curved symmetric parts using a deformable disc model, in: Proc. ICCV, 2013, pp. 1753–1760.
[17] N. Widynski, A. Moevus, M. Mignotte, Local symmetry detection in natural images using a particle filtering approach, IEEE Trans. Image Process. 23 (12) (2014) 5309–5322.
[18] S. Xie, Z. Tu, Holistically-nested edge detection, in: Proc. ICCV, 2015, pp. 1395–1403.
[19] J. Long, E. Shelhamer, T. Darrell, Fully convolutional networks for semantic segmentation, in: Proc. CVPR, 2015, pp. 3431–3440.
[20] X. Chen, H. Fang, T. Lin, R. Vedantam, S. Gupta, P. Dollár, C.L. Zitnick, Microsoft COCO captions: data collection and evaluation server, arXiv:1405.0312.
[21] X. Bai, L.J. Latecki, W. Liu, Skeleton pruning by contour partitioning with discrete curve evolution, IEEE Trans. Pattern Anal. Mach. Intell. 29 (3) (2007) 449–462.
[22] W. Shen, K. Zhao, Y. Jiang, Y. Wang, Z. Zhang, X. Bai, Object skeleton extraction in natural images by fusing scale-associated deep side outputs, in: Proc. CVPR, 2016.
[23] W. Shen, X. Bai, Z. Hu, Z. Zhang, Multiple instance subspace learning via partial random projection tree for local reflection symmetry in nature images, Pattern Recognit. 52 (2016) 266–278.
[24] X. Ren, Multi-scale improves boundary detection in natural images, in: Proc. ECCV, 2008, pp. 533–545.

[25] P. Dollár, C.L. Zitnick, Fast edge detection using structured forests, IEEE Trans. Pattern Anal. Mach. Intell. 37 (8) (2015) 1558–1570.
[26] W. Shen, X. Wang, Y. Wang, X. Bai, Z. Zhang, Deepcontour: a deep convolutional feature learned by positive-sharing loss for contour detection, in: Proc. CVPR, 2015, pp. 3982–3991.
[27] D.R. Martin, C. Fowlkes, D. Tal, J. Malik, A database of human segmented natural images and its application to evaluating segmentation algorithms and measuring ecological statistics, in: Proc. ICCV, 2001, pp. 416–425.
[28] E. Borenstein, S. Ullman, Class-specific, top-down segmentation, in: Proc. ECCV, 2002, pp. 109–124.
[29] P. Agrawal, R.B. Girshick, J. Malik, Analyzing the performance of multilayer neural networks for object recognition, in: Proc. ECCV, 2014, pp. 329–344.
[30] K. Simonyan, A. Zisserman, Very deep convolutional networks for large-scale image recognition, CoRR, arXiv:1409.1556.
[31] H. Blum, A transformation for extracting new descriptors of shape, in: Models for the Perception of Speech and Visual Form, MIT Press, Boston, MA, USA, 1967, pp. 363–380.
[32] C. Bishop, Pattern Recognition and Machine Learning, Springer, New York, NY, USA, 2006.
[33] P.F. Felzenszwalb, R.B. Girshick, D.A. McAllester, D. Ramanan, Object detection with discriminatively trained part-based models, IEEE Trans. Pattern Anal. Mach. Intell. 32 (9) (2010) 1627–1645.
[34] X.C. Jun Zhu, A. Yuille, DeePM: a deep part-based model for object detection and semantic part localization, arXiv:1511.07131.
[35] C.L. Zitnick, P. Dollár, Edge boxes: locating object proposals from edges, in: Proc. ECCV, 2014, pp. 391–405.
[36] V. Ferrari, T. Tuytelaars, L.J.V. Gool, Object detection by contour segment networks, in: Proc. ECCV, 2006, pp. 14–28.
[37] Y. Lin, S. Saripalli, Road detection from aerial imagery, in: Proc. ICRA, 2012, pp. 3588–3593.
[38] A. Shahab, F. Shafait, A. Dengel, ICDAR 2011 robust reading competition challenge 2: reading text in scene images, in: Proc. ICDAR, 2011.

CHAPTER

10 Characterization of trabecular bone plate–rod micro-architecture using skeletonization and digital topologic and geometric analysis

Punam K. Saha

Department of Electrical and Computer Engineering, Department of Radiology, University of Iowa, Iowa City, IA, USA

Contents

10.1 Introduction 287
10.2 Definitions and Notations 289
10.3 Skeletonization 290
10.4 Digital Topological Analysis 291
10.4.1 Surface–Surface Junction Line Extension 294
10.4.2 Detection of Junction Voxels Between Surfaces and Curves 295
10.5 Volumetric Topological Analysis 297
10.5.1 Geodesic Distance Transform 299
10.5.2 Feature Propagation and Representative 302
10.5.3 Applications 302
10.6 Conclusion 305
References 305

10.1 INTRODUCTION

Osteoporosis, linked to reduced bone mineral density (BMD) and structural degeneration of cortical and trabecular bone (TB), is associated with increased fracture risk. Osteoporotic fracture is a significant public health concern [1]. Following the bulletin of the World Health Organization [2], 40% of women and 13% of men suffer at least one osteoporotic fracture in their lifetime. The continued increase in life expectancy is predicted to increase the number of fracture incidences by three-fold [3]. Following

Skeletonization. DOI: 10.1016/B978-0-08-101291-8.00012-2

the estimate by Cooper et al. [4], the annual incidence of osteoporotic hip fractures will rise to 6.3 million by 2050. Hip fractures are typically thought to reduce life expectancy by 10–20% [5]. In individuals who experience hip fractures, 20% will die within the next year, and 20% will require permanent nursing home care. Although, effective therapies are available to treat osteoporosis [6,7], these therapies are often expensive and associated with side effects [8–10]. Accurate assessments of fracture risk, clear guidelines to initiate preventive intervention, and monitoring treatment response are of urgent needs in public health [11–13], and osteoporotic imaging plays a central role in that process.

A large number of factors, including sex, age, race, family history, body size, body mass index (BMI), hormone levels, dietary factors, steroids and other medications, physical inactivity, vitamin D deficiency, tobacco and alcohol consumption, contribute to the pathophysiology of osteoporosis. Seeman [14] reviewed the mechanisms of bone loss under various influence factors and discussed their implications to bone microstructures and biomechanics. These structural and biomechanical differences during bone loss under various influence factors may account for divergent fracture risk among individuals with similar measures of BMD [15]. It is known that BMD explains approximately 60–70% of the variability in bone strength [16]. The remainder is due to cumulative and synergistic effects of other factors, including cortical and TB micro-architecture [17].

Osteoporotic imaging is critically important in identifying fracture risk among individuals for planning of therapeutic intervention and, also, for monitoring treatment response. Dual-energy X-ray absorptiometry (DXA) is the current clinical benchmark technique for classifying BMD among postmenopausal women or older men as normal, osteopenic, or osteoporotic. Despite being the current clinical benchmark, DXA suffers from several pertinent limitations. First, DXA provides a measure of areal BMD, which is susceptible to bone size and thus overestimates fracture risks among individuals with small body frame. Also, it fails to provide metrics related to cortical bone geometry or TB micro-architecture, which are important determinants of fracture risk [18–20] and sensitive to disease progression and therapeutic intervention [10]. Recent advances in medical imaging, such as magnetic resonance imaging (MRI) [21–23], high-resolution peripheral quantitative computed tomography (HR-pQCT) [24,25], and multirow detector computed tomography (MD-CT) [26,27], allow in vivo segmentation of individual TB structures for quantitative analysis and characterization of their micro-architectural quality and assessment of bone strength and fracture risk.

Different groups have applied various methods for topologic and geometric characterization of TB micro-architecture [20,28–30]. Parfitt et al. [20] proposed a parallel interconnected trabecular plate model computing TB area, volume fractions, spacing, and number. Vesterby et al. [28] introduced the star volume measure, which is the average volume of an object region that can be seen from a point inside that region unobscured in all directions. Hildebrand et al. [29] formulated the 3-D structure model index, a pseudo-measure of global plate-to-rod ratio, based on the observation that, for plate-like structures, the rate of change of surface area with respect to thickness is less than that for rod-like structures. Feldkamp et al. [30] expressed the makeup of TB networks in terms of topological entities such as the 3-D Euler number.

There is histologic evidence confirming the relationship between the gradual conversion of trabecular plates to rods and increased fracture risk [18,31]. Kleerekoper et al. [18] observed lower mean TB plate density among individuals with osteoporotic vertebral compression fractures compared with BMD-matched controls without fractures. Recker [31] reported reduced trabecular connectivity among patients with vertebral crush fractures as compared to healthy controls with matching TB volume. Digital topological analysis is applied to characterize individual trabecular plates, rods, edges, and junctions in skeletal representations derived from in vivo imaging. This chapter overviews the principles and algorithms of digital topological analysis and presents recent advances in skeleton-based quantitative characterization of trabecular plate–rod micro-architecture at in vivo imaging. More specifically, this chapter describes the topics related to skeletonization, topological plate–rod characterization, geodesic distance transform, structure plate-width computation, and feature propagation.

10.2 DEFINITIONS AND NOTATIONS

Basic principles of the methods described in this chapter are applicable to digital objects in any three-dimensional (3-D) digital grids. However, for simplicity, the methods are formulated for 3-D cubic grids, only. A large volume of images in 3-D cubic grids is acquired in medical imaging on a daily basis. A 3-D cubic grid, or simply a *cubic grid*, is represented by Z^3, where Z is the set of integers. An element of Z^3 is often referred to as a *voxel*. Standard 26-, 18-, and 6-adjacencies are used here [32]. Two α-adjacent voxels, where $\alpha \in \{6, 18, 26\}$, are also referred to as *α-neighbors*; the set of α-neighbors of a voxel p excluding itself is denoted by $N_\alpha^*(p)$. A binary digital image or an object in a cubic grid is represented as a quadruple $\mathcal{B} = (Z^3, \alpha, \beta, B)$, where Z^3 defines the 3-D image space, and B is the set of object voxels; the parameters α and β denote the adjacency relations for object and background voxels, respectively [33]; here 26- and 6-adjacencies are used for α and β, respectively. A fuzzy digital image or a fuzzy object is denoted as $\mathcal{O} = (Z^3, \alpha, \beta, \mu_\mathcal{O})$, where $\mu_\mathcal{O} : Z^3 \rightarrow [0, 1]$ is its *membership function*. The *support* of the fuzzy object, denoted by O, is the set of all points with nonzero membership values, i.e., $O = \{p \mid p \in Z^3 \wedge \mu_\mathcal{O}(p) > 0\}$. In a fuzzy digital image, the adjacency relation α is used for voxels inside O, whereas β is used for the voxels inside the *background* $\overline{O} = Z^3 - O$.

An *α-path*, where $\alpha \in \{6, 18, 26\}$, is a nonempty sequence of voxels such that every two successive voxels are α-adjacent. Two voxels p, q are α-connected in a set of voxels S if there exists an α-path between p, q that is contained in S. An α-component in S is a maximal α-connected subset of S; S is referred to as α-connected if S includes only one α-component, i.e., every two voxels in S are α-connected. In a binary image $(Z^3, 26, 6, B)$, a cavity is a 6-component of background voxel that is surrounded by object voxels. Although it is difficult to define a tunnel, the number of tunnels in an object is related to the number of handles in the object or inside its cavities. A handle is formed when two ends of a cylinder are glued to each other, or those are glued to another connected object; for example, a doughnut or a coffee mug has a handle. A handle inside a cavity is formed when the two ends of a hollow cylinder are glued to each other; for example a hollow torus has a handle within its cavity. More precisely, the number of tunnels in an object is related to the rank of its first homology group [34]. The numbers of objects, tunnels, and cavities in an object represent its 0th, 1st, and 2nd Betti numbers, respectively.

10.3 SKELETONIZATION

Skeletonization provides a compact yet effective representation of key topologic and geometric features in an object. Essentially, skeletonization reduces the dimension of an object to generate its "medial axis" or "skeleton". Blum [35] founded the skeletonization transform using the notion of grassfire transformation [36]. Different approaches of skeletonization algorithms were discussed in [37]. A recent survey on skeletonization methods and their applications were presented in [38]. Skeletonization has been popularly used in various image processing and computer vision applications including object representation, retrieval, manipulation, matching, registration, tracking, recognition, compression, etc.

The compactness feature of skeletons facilitates efficient assessment of local structural metrics including scale, orientation, topology, geometry, etc. This specific feature of skeletons has been explored for quantitative characterization object morphology in several imaging applications. For example, the Hough transform of skeletal lines has been used to locate spicules on mammograms for malignant tumor identification [39]. Others have used skeletonization for detection and classification of different anatomical structure types, e.g., vessels, spicules, ducts, etc. [40]. Three-dimensional Voronoi skeletonization has been applied for characterization and recognition of complex anatomic shapes [41]. Saha and his colleagues [23,27,42–49] have used skeletonization for characterization of trabecular bone micro-architecture from ex vivo and in vivo imaging.

Blum's grassfire transform is the common principle of the most skeletonization algorithm available in literature. However, often, the mechanism and computational pathways adopted by various skeletonization algorithms are widely divergent. Many research groups [50–53] have adopted a continuous deformation of object boundary through some form of curve evolution process to simulate Blum's grassfire progression on an image. In such approaches, the skeleton is formed at quench points where the evolution process is interrupted by colliding curve fronts. Several researchers [41, 54–56] have explored the use of geometric features, e.g., Voronoi diagram to directly compute symmetry structures and to locate the object skeleton. A large number of researchers have embraced an iterative erosion of digital object boundary subject to topology preservation [57–61] and certain shape constraints [62–64] to simulate the

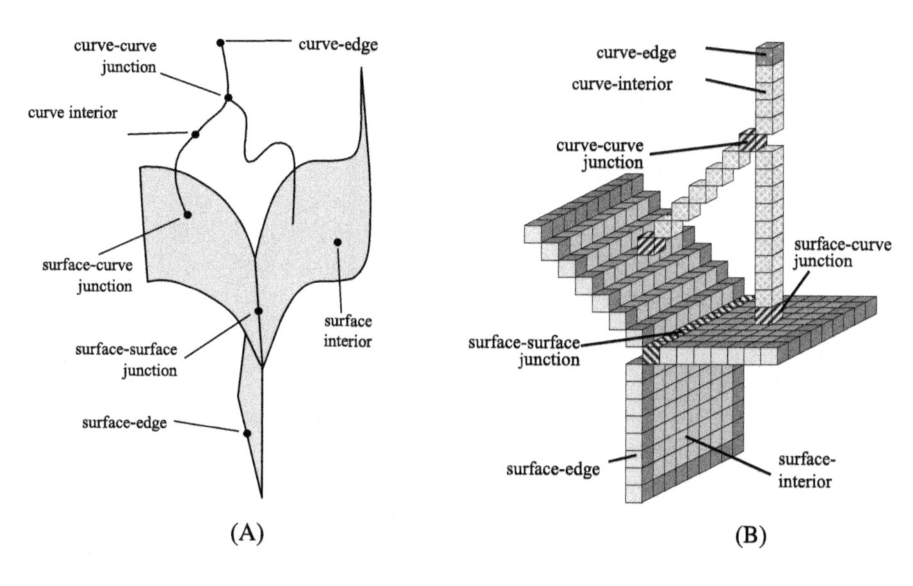

FIGURE 10.1

Topological classes of different structures in a surface skeletal representation of an object. (A) An example in the continuous space. (B) An example in a 3-D digital grid.

grassfire propagation on digital images [64–69]. Other researchers [27,70–76] have demonstrated the use of a digital distance transform [77] to directly locate digital symmetry structures in the form of centers of maximal balls (CMBs) [72] to compute the skeletal of digital objects. Descriptions of overviews of different skeletonization algorithms have been presented in Chapter 1. In this chapter, the methods related to applications of quantitative morphometric analysis of local structures are described.

10.4 DIGITAL TOPOLOGICAL ANALYSIS

Digital topological analysis or DTA [78] accurately determines the topological class (e.g., surfaces, curves, junctions) of each individual voxel on the surface skeleton of 3-D digital object. This method has been popularly applied for quantitative assessment of TB architectural quality. Before applying DTA, a fuzzy or binary segmented TB volume object is skeletonized using a suitable algorithm [38]. The principle of DTA is based on a simple observation on the skeletal representation, say $\mathcal{S}$, of a 3-D object in the continuous space. A notable observation is that the topological class of a point $p \in \mathcal{S}$ is uniquely defined by its local topological numbers, i.e., the numbers of object components, tunnels, and cavities in a sufficiently small neighborhood of p but excluding p itself. Topological classes of different structures on a surface skeleton are illustrated in Fig. 10.1. Let $\xi(p)$, $\eta(p)$, and $\delta(p)$ denote the numbers of object components, tunnels, and cavities in a sufficiently small excluded neighborhood of a skeletal point $p \in \mathcal{S}$. The topological numbers associated with different topological classes on a skeleton $\mathcal{S}$ is listed in the following [48,78,79]:

- Isolated (I) point: $\xi(\cdot) = 0$; $\eta(\cdot) = 0$; and $\delta(\cdot) = 0$.
- Surface- or curve-edge (SE or CE) point: $\xi(\cdot) = 1$; $\eta(\cdot) = 0$; and $\delta(\cdot) = 0$.
- Curve-interior (C) point: $\xi(\cdot) = 2$; $\eta(\cdot) = 0$; and $\delta(\cdot) = 0$.
- Surface-interior (S) point: $\xi(\cdot) = 1$; $\eta(\cdot) = 1$; and $\delta(\cdot) = 0$.
- Curve–curve junction (CC) point: $\xi(\cdot) > 2$; $\eta(\cdot) = 0$; and $\delta(\cdot) = 0$.
- Surface–curve junction (SC) point: $\xi(\cdot) > 1$; $\eta(\cdot) \geq 1$; and $\delta(\cdot) = 0$.
- Surface–surface junction (SS) point: $\xi(\cdot) = 1$; $\eta(\cdot) > 1$; and $\delta(\cdot) = 0$.

In this list, the local topological number $\delta(\cdot)$ does not play a discriminative role since its value is always zero in a small excluded neighborhood of a skeletal point in the continuous space. However, in a digital space, nonzero values of $\delta(\cdot)$ are common for skeletal voxels, and the parameter plays an important role in determining the topological class of a voxel on a surface skeleton. Whereas the computation of $\xi(p)$ and $\delta(p)$ is trivial in a digital object, $\eta(p)$ is computed using the following theorem [57,78,80].

Theorem 10.1. *If a voxel p has a background 6-neighbor, the number of tunnels $\eta(p)$ in its excluded neighborhood $N^*_{26}(p)$ is one less than the number of 6-components of background 18-neighbors of p that contain a 6-neighbor or zero otherwise.*

In the above theorem, nonskeletal voxels are considered as the background voxels. Unlike the continuous case, the three local topological numbers $\xi(\cdot)$, $\eta(\cdot)$, and $\delta(\cdot)$ fail to uniquely define topological classes of every voxel on a digital surface skeleton. For example, in a cubic grid, the specific combination of local topological $\xi(\cdot) = 1$, $\eta(\cdot) = 1$, and $\delta(\cdot) = 0$ may occur at a surface interior voxel as well as at an intersection of curves as shown in Fig. 10.2. More examples of ambiguities in determining topological classes of skeletal voxels from their local topological numbers may be constructed in a cubic grid and in other digital grids. To overcome this problem, a three-step approach was counseled by Saha et al. [78] to determine the unique topological class of every skeletal voxels. These three steps of the digital topological classification (DTA) are: (1) determination of the local topological type of every skeletal voxels, (2) initial classification of every skeletal voxels based on their neighborhood configuration of topological types, and (3) final topological classification after correction around junctions. Separating the comprehensive task of the DTA in three steps makes it easier to understand the function of the algorithm and more efficient to implement.

The first step of DTA determines the initial topological type based on the three local topological numbers using a look-up-table (Table 10.1). As discussed before, this step fails to uniquely determine topological classes of all skeletal voxels. Fol-

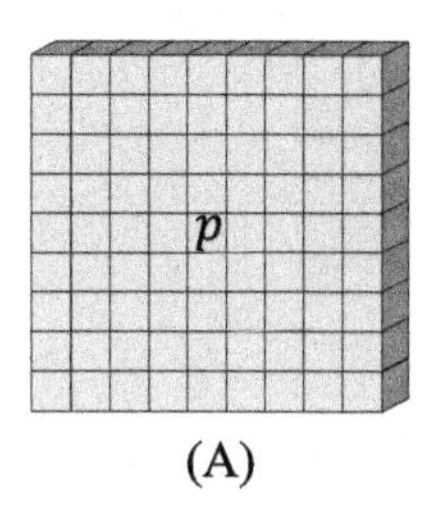

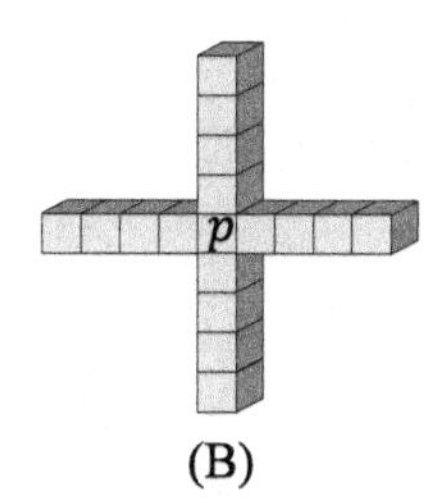

FIGURE 10.2

Examples in a digital cubic voxel grid where local topological numbers fail to uniquely identify the local topological class of different structures. Local topological numbers of p in both (A) and (B) are $\xi(\cdot) = 1$, $\eta(\cdot) = 1$, and $\delta(\cdot) = 0$. The ambiguity of local topological classes of p in (A) and (B) can be solved using skeletal voxel information over the extended neighborhood.

Table 10.1 The look-up-table for initial topological type of skeletal voxels based on their local topological numbers of components (ξ), tunnels (η), and cavities (δ). The following abbreviations are used—I: interior; C: curve; S: surface; E: edge

$\xi(\cdot)$	$\eta(\cdot)$	$\delta(\cdot)$	Initial type	Possible topological classes
0	0	0	T_1	I
1	0	0	T_2	CE or SE
2	0	0	T_3	C
> 2	0	0	T_4	CC
1	1	0	T_5	S or CC
> 1	≥ 1	0	T_6	CC, SC, or SS
1	> 1	0	T_6	CC, SC, or SS
1	0	1	T_6	CC, SC, or SS

lowing the table, each of the initial topological types T_1, T_3, and T_4 represents a unique topological class of a skeletal voxel. However, other topological types T_2, T_5, and T_6 represent ambiguous topological classes of skeletal voxels. These ambiguities are solved in the second step using a neighborhood analysis of topological types determined during the previous step. Essentially, these two steps together use the topological data of skeletal voxels over the $5 \times 5 \times 5$ neighborhood to determine the topological class of each skeletal voxel. The specific rules of determining topological classes of skeletal voxels from their initial topological types are presented in Table 10.2.

After the initial topological classification of the second step, the process is almost complete except that some corrections are needed at some surface junctions, and several surface–curve junctions are missed. Two examples where the extension

Table 10.2 The look-up-table for topological classification of skeletal voxels using neighborhood configuration of initial topological types

Topological type	Neighborhood configuration	Topological class
T_2	Exactly one skeletal voxel neighbor	CE
T_2	Multiple skeletal voxel neighbors	SE
T_5	All skeletal voxel neighbors are T_3 or T_4	CC
T_5	Not all skeletal voxel neighbors are T_3 or T_4	S
T_6	All skeletal voxel neighbors are T_3 or T_4	CC
T_6	Some (not all) skeletal voxel neighbors are T_3 or T_4	SC
T_6	No skeletal voxel neighbors are T_3 or T_4	SS

of surface–surface junction lines is needed are shown in Fig. 10.3A, B. However, in many cases, the surface–surface junction line is completely detected by Steps 1 and 2, and no postextension of junction line is needed. Therefore, a proper criterion of SS-line extension is necessary as described in Section 10.4.1. Also, using the initial two steps, surface–curve junction voxels missed, where a curve emanates from the edge of a surface as shown in Fig. 10.4. Such junction voxels are located when a component of curve voxels is adjacent to a component of surface voxels without a junction voxel gluing the two different components of topological classes.

10.4.1 SURFACE–SURFACE JUNCTION LINE EXTENSION

The step of surface–surface junction line extension is accomplished by correcting the topology of surface-edge lines around a junction. In absence of junctions, the surface-edge voxels on a skeleton forms a simple closed curve. This unique topological property of a surface edge is perturbed near junctions. Two types of such perturbations are commonly observed as shown in Fig. 10.3A, B. In Fig. 10.3A, the surface–surface junction line extends up to the edge of individual surface structures. In such cases, ideally, multiple surface-edges meet at the end of the junction line. However, the neighborhood configuration rules of Table 10.1 and Table 10.2 fail to detect the complete junction line, and the end-voxels of the junction line are misclassified as surface edge voxels. These misclassified surface edge voxels (marked with '$*$' in Fig. 10.3A) form junctions on surface edges; see Fig. 10.3A. In the case of Fig. 10.3B, a junction voxel is missed near the end of a surface-edge line. A surface–surface extension algorithm along these principles is presented below. The following notations are used. S_{SE} will denote the set of all SE voxels on a skeleton. Let p be an end voxel of an SS-line; an end voxel has at most one 26-adjacent SS voxel, which is denoted as p_{SS}. Finally, let P_S denote the set of all S voxels in $N^*(p)$.

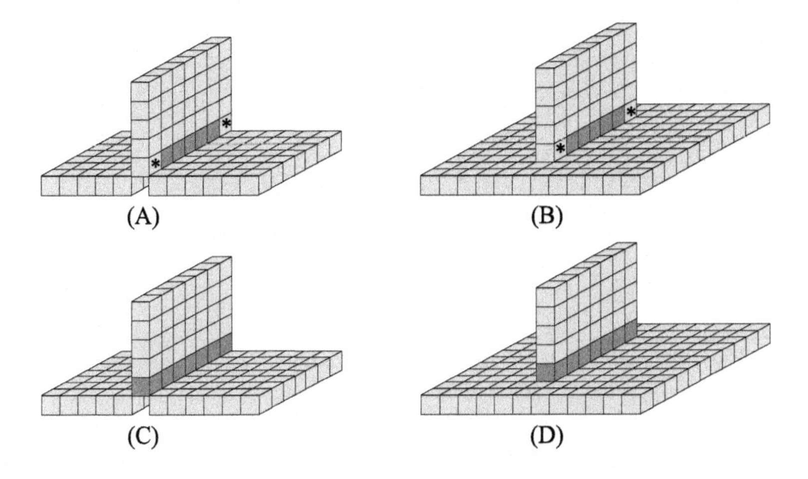

FIGURE 10.3

Corrections of topological classification around surface–surface junction lines. (A, B) Initial topological classification of junction lines computed using the neighborhood configuration rules of Table 10.1 and Table 10.2. The voxels marked with '$*$' are classified as surface-edge based on their local topological numbers. (C, D) Corrected surface–surface junction lines after applying the surface–surface junction-line extension algorithm.

begin algorithm *surface–surface junction-line extension*
for all SS-line end voxel p
  for every SE voxels $q \in N^*(p) - N^*(p_{SS})$
    if $S_{SE} \cap N^*(q)$ contains more than two components or forms tunnels
      flag q
    else if $S_{SE} \cap N^*(q)$ contains exactly one component
        AND $\exists r \in P_S \cap N^*(q) - N^*(p_{SS})$ such that $|q - r| < |q - p|$
      flag one of the r's nearest to q
  reclassify all flagged voxels (if any) as SS voxels
end algorithm *extend surface–surface junction lines*

The results of Fig. 10.3C, D are obtained by applying the above algorithm on (A) and (B), respectively. It is reemphasized that, following the above algorithm, for Fig. 10.3D, the extended SS voxel was not an SE voxel; instead, it is one of the nearest voxels in $P_S \cap N^*(q) - N^*(p_{SS})$.

10.4.2 DETECTION OF JUNCTION VOXELS BETWEEN SURFACES AND CURVES

A junction voxel must exist between mutually adjacent surface and curve structures; see Fig. 10.4 for an example. However, the neighborhood configuration rules of local topological numbers do not classify the voxel marked with '$*$' as a junction voxel;

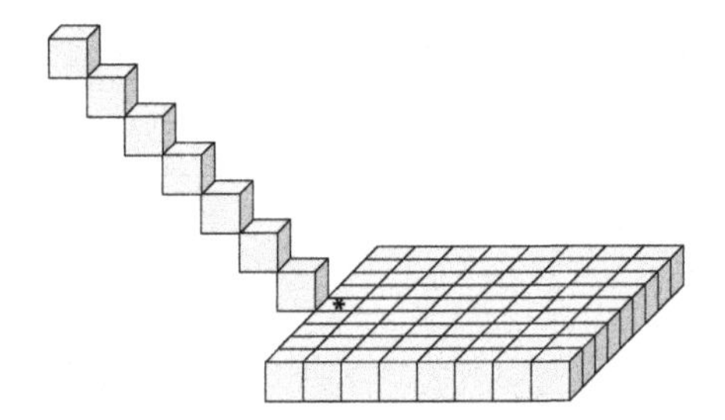

FIGURE 10.4

An example where a surface–curve junction voxel is missed. The junction voxel marked with '∗' is missed using the neighborhood configuration rules of local topological numbers. Such missing junction voxels are retrieved using a separate step for detection of junction voxels between surfaces and curves.

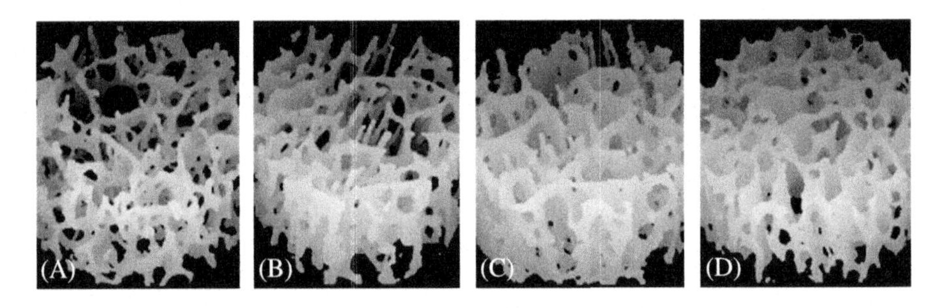

FIGURE 10.5

Plate–rod distribution from in vivo micro-MR images of distal radial trabecular bone [44]. Trabecular bone examples with highly rod-like (74 years female) (A), moderately rod-like (53 years female) (B), mostly plate-like (68 years male), and highly plate-like (60 years male) (D) trabeculae. Topological parameters parallel visual assessment of increasing plate-like trabecular bone micro-architecture with the surface-to-curve ratio increasing nearly 20-fold from (A) through (D).

instead, the voxel is marked as an S voxel. Therefore, an extra step is needed to detect such missing junctions between mutually adjacent surface and curve structures. This step works as follows. Let p be a C or CC voxels, and let P be the set of S, SC, and SS voxels in $N^*(p)$. Let $P_1, P_2, \ldots, P_m$ be the 26-components of P. For each $i = 1, \ldots, n$, if P_i does not include an SC or SS voxel, then the voxel in P_i nearest to p is reclassified as an SC voxel. The voxel marked with '∗' in Fig. 10.4 is reclassified as an SC voxel using this algorithm.

Digital topological analysis has been widely applied for characterization of trabecular bone plate/rod micro-architecture from in vivo imaging [23,43,47,81–85]; see Fig. 10.5. Digital topological classes of individual bone voxels are used to compute several topological parameters characterizing the structural integrity of a trabecular

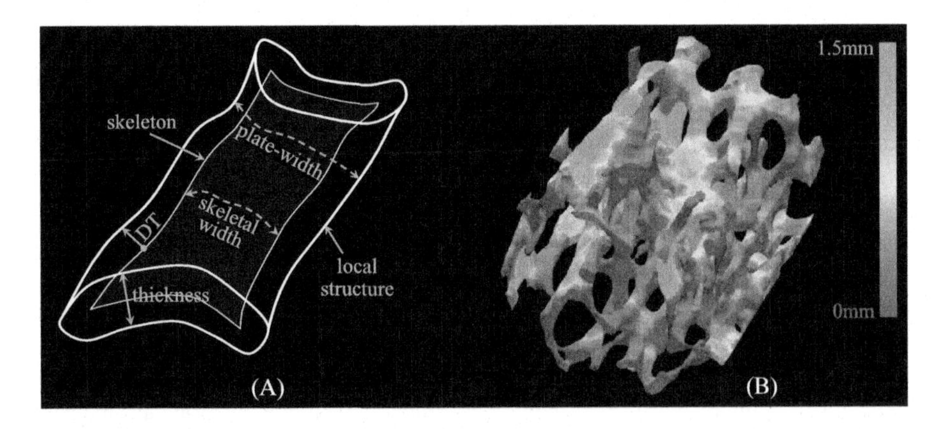

FIGURE 10.6

Illustration of the plate-width measure. (A) A schematic description of different morphologic local measures on a plate-like structure. The volumetric topological analysis or VTA algorithm computes the plate-width measure of a local structure. (B) A color-coded illustration of local plate-width measures at every TB voxels computed from a CT image of a cadaveric ankle specimen.

bone network. The specific parameter, namely, erosion index (EI) was found to be highly sensitive in most studies using DTA, and this DTA parameter is used in our experiments. EI is defined as a ratio where the numerator adds all topological parameters expected to increase during the bone erosion process (specifically, curve, curve-edge, surface-edge, profile-edge, and curve–curve junction types), whereas the denominator adds those that are expected to decrease (surface and surface–surface junction types) [48].

Malandain et al. [86] have used topological analysis to segmental discrete skeletal surfaces. Serino et al. [87] have used distance-labeled curve skeletons to decompose 3-D objects into disjoint parts. More specifically, they used distance transform and curvature analysis to locate skeletal segments. Svensson et al. [79] have used topological voxel classification to guide curve skeletonization of surface-like objects in 3-D digital images. Giblin and Kimia [88] developed a method for computing hypergraph skeletal representations of 3D objects and used such representations to derive formal classification of 3D medial axis points and their local geometry.

10.5 VOLUMETRIC TOPOLOGICAL ANALYSIS

Volumetric topological analysis or VTA [27] accurately measures local structure plate-width at every voxel in a digitized object and characterizes the local structure type on the continuum between a perfect plate and a perfect rod; see Fig. 10.6. The original VTA algorithm [27] was developed for binary objects requiring thresholding on fuzzy representations of TB images, which is a sensitive and undesired step at in vivo resolution [89]. Recently, the algorithm was modified to directly apply on fuzzy representation of an object eliminating the thresholding step [90]—a major source of error, especially, at low spatial resolution common in in vivo imaging.

The VTA method starts with the surface skeletal representation of an object and computes the geodesic manifold distance transform at curve skeletal or axial locations to determine the local structure plate-width (Fig. 10.6). Local plate-width measures are then propagated from the curve-skeleton to the surface-skeleton and then from the surface-skeleton to the entire object volume. Specifically, the VTA algorithm is accomplished in five sequential steps [27]—(1) computation of surface-skeleton of a binary or fuzzily segmented object, (2) digital topological analysis, (3) geodesic manifold distance transform, (4) local structure plate-width computation at curve-skeletal locations, (5) local plate-width feature propagation from the curve-skeleton to the surface-skeleton, and (6) plate-width feature propagation from the surface-skeleton to the entire object volume.

Basic principles of the VTA algorithm are described in Fig. 10.7. Let us assume that the surface manifold of Fig. 10.7A represents the surface skeleton of a plate-like object with varying widths along its central axis. A surface skeletal representation of an object may be obtained using a suitable skeletonization algorithm. After computing the surface skeleton, the DTA algorithm, described in Section 10.4, is applied to identify different topological structures in a surface skeleton including the surface edges shown using a thick border line in Fig. 10.7. Geodesic distance transform (GDT) is computed at every surface skeletal points. The formulation of the VTA algorithm is based on a simple observation that the GDT at a curve skeletal or axial point provides the half-plate-width of the local structure; see Fig. 10.7A. However, GDT fails to determine local plate-width at nonskeletal points, e.g., the point p in Fig. 10.7. The curve skeleton or axial line is shown as a dotted line Fig. 10.7. A feature propagation step is applied, where a nonaxial point inherits the local plate-width from its most-representative axial point. For example, the point p is assigned with the same plate-width as that of a after feature propagation as illustrated in Fig. 10.7B. Finally, another level of feature propagation is applied where the plate-width values are propagated from the surface-skeleton to the entire object volume. Saha et al. [27] recommended this two-step feature propagation method, which propagates feature in the reverse order of grassfire propagation. The feature propagation of Step 5 is accomplished using the principle established by Liu et al. [91] that is independent of scan or processing order. Lastly, it should be clarified that a VTA algorithm using GDT from surface-skeletal edges returns the value of the local skeletal width at each point; see Fig. 10.6A. To compute local width of a volumetric structure, the algorithm is modified, where the GDT value at a skeletal edge point is initialed with its distance transform or fuzzy distance transform value instead of zero. Results of intermediate steps of the VTA algorithm are presented in Fig. 10.8. More discussions on

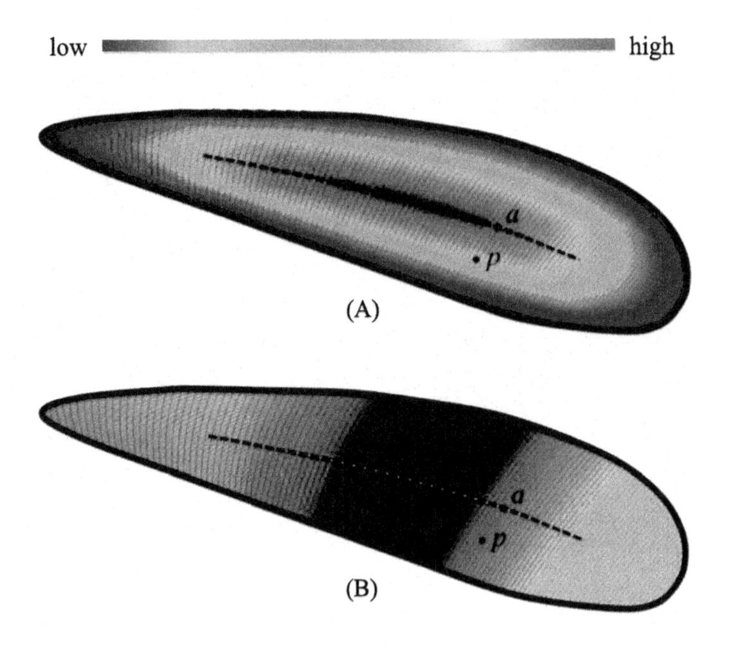

FIGURE 10.7

Basic principles of the VTA algorithm. (A) Color-coded geodesic distance transform from the surface edge, which provides the half-plate-width measure of the local structure at an axial point. However, geodesic distance transform fails to provide the knowledge of local plate-width at nonaxial points. (B) Color-coded display of local plate-width. At a nonaxial point, e.g., p, the measure of local plate-width is derived from its "most-representative" axial point, here the axial point a.

geodesic distance transform and feature propagation are presented in Sections 10.5.1 and 10.5.2.

10.5.1 GEODESIC DISTANCE TRANSFORM

The value of geodesic distance transform or GDT at a surface skeletal point is equal to its geodesic distance from the nearest surface edge point. The geodesic distance between two points on a surface skeleton is the length of the shortest path between the two points such that the entire path lies on the skeletal surface. A front-propagation approach similar to Dijkstra's algorithm [92] offers an efficient implementation of GDT computation in a digital grid. It was argued by Saha et al. [27] that GDT propagation paths should not cross a junction line. Allowing GDT fronts to propagate through a junction line leads to artefactual reduction in GDT values around the junction; see Fig. 10.9. A similar observation was noted by Arcelli et al. [71]. Crossing of a junction line by a propagation path in a digital grid is defined as follows [27]. Let S denote a surface skeleton, and let $S_{\text{junction}} \subset S$ be the set of all junction voxels

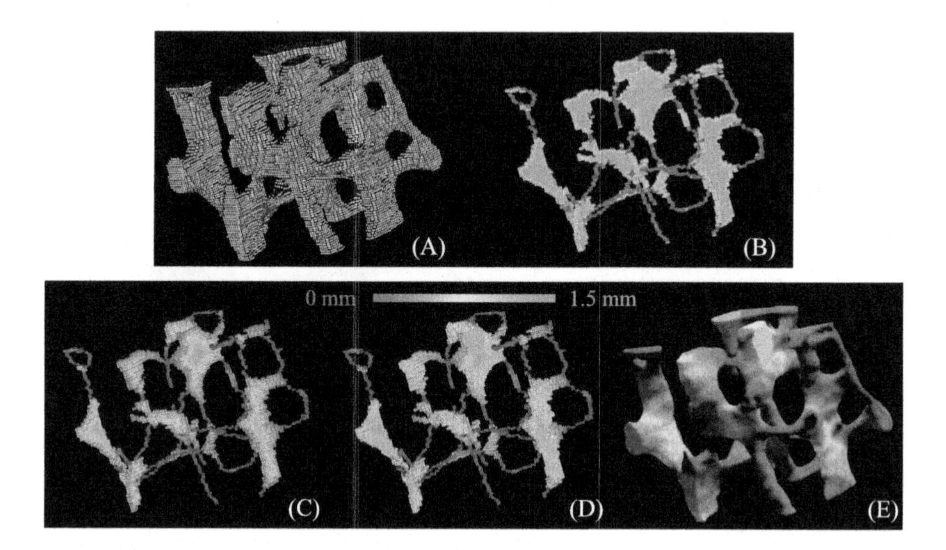

FIGURE 10.8

Results of intermediate steps in VTA. (A) A trabecular bone region from a micro-CT image of a cadaveric ankle specimen. (B) DTA-computed topological classes including plates (green), rods (red), edges (light colors), and junctions (blue) on the skeleton of (A). (C–E) Color-coded displays of geodesic distance transform (C), plate-width outcome on the surface skeleton after the first feature propagation step from curve skeleton to surface skeleton (D), and surface rendition of plate-width or VTA outcome over the entire TB volume after the second feature propagation step (E). In the panel (E), surface rendition is used instead of voxel display for better display.

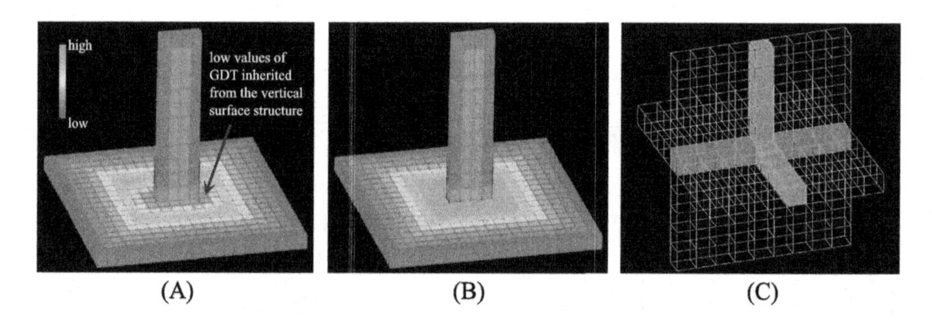

FIGURE 10.9

Crossing of junctions in GDT computation. (A) Computed GDT on a digital surface allowing propagation paths to cross junction voxels. (B) Correct GDT computation on the same digital surface of (A). (C) An example of crossing between a propagation path (red) and a junction line (blue) on a (26, 6) digital surface.

in S. A path $\pi = \langle p = p_0, p_1, \ldots, p_n = q \rangle$ between two voxels $p, q \in S$ is called a *valid propagation path in S* if—(1) $\forall i = 1, 2, \ldots, n-1$, $p_i \in S - S_{\text{junction}}$ and (2) $\nexists r \in S_{\text{junction}}$ and $0 \le i < n$ such that $|p_i - p_{i+1}| > \max |p_i - r|, |r - p_{i+1}|$, where $|p - q|$ is the Euclidean distance between two voxels p, q. A violation of the second condition of a path is considered as a "crossing" with a junction line without sharing a common voxel, because, the path $\ldots, p_i, r, p_{i+1}, \ldots$ is more natural as compared to the path $\ldots, p_i, p_{i+1}, \ldots$ in the sense that the former path accomplishes the move from p_i to p_{i+1} using shorter steps. An example of such crossing between a 26-path π (the path of red voxels) and a junction line l_{junction} (the path of blue voxels) is illustrated in Fig. 10.9C, where π and l_{junction} have no common voxel.

As mentioned earlier, GDT at a curve skeletal voxel provides the half-width of the local skeleton. However, as illustrated Fig. 10.6A, the plate-width measure captures the width of the volumetric local structure. This goal is accomplished by initializing the binary or fuzzy distance transform value at skeletal surface edge voxels as their GDT values. An algorithm for computing geodesic distance transform is presented in the following.

begin algorithm *compute_GDT*
inputs:
  S: a surface skeletal
  $S_{\text{junction}} \subset S$: the set of all junction voxels in S
  S_{edge}: the set of all surface-edge or curve (SE, CE, C, or CC) voxels
for each voxel $p \in S$
  assign $GDT(p) = \max\_value$
for each voxel $p \in S_{\text{edge}}$
  assign $GDT(p) = FDT(p)$
  push p in a queue Q
while Q is not empty
  pop a voxel p from Q
  for each voxel $q \in \mathcal{N}^*(p)$
    if $q \in S_{\text{junction}}$ AND $GDT(q) < GDT(p) + |p - q|$
      assign $GDT(q) = GDT(p) + |p - q|$
    else if $GDT(q) < GDT(p) + |p - q|$
      AND $\nexists r \in S_{\text{junction}}$ such that $|p - q| > \max |p - r|, |r - q|$
        assign $GDT(q) = GDT(p) + |p - q|$
        push q in Q
end algorithm *compute_GDT*

It may be noted that, in the above algorithm when the voxel q is a junction voxel, i.e., $q \in S_{\text{junction}}$, it can receive the value of the GDT propagation front from its neighboring voxels. However, in such situations, the voxel q is not pushed in the queue Q. Therefore, the GDT front propagation stops at q and cannot propagate further through the junction voxel. It avoids the type of crossing where a path actually includes a junction voxel. The other type of crossing, where the propagation path does not include a junction voxel but passes over a junction line, is avoided using an explicit check that $\nexists r \in S_{\text{junction}}$ such that $|p - q| > \max |p - r|, |r - q|$. It may be clarified that, although the above algorithm works for any Q, its efficiency improves significantly when the Q is maintained as an ordered queue.

10.5.2 FEATURE PROPAGATION AND REPRESENTATIVE

A common strategy for morphometric analysis of local structures in an object is to measure desired local properties, e.g., thickness, orientation, cross-sectional area, plate–rod characterization, etc., at representative sample points, e.g., skeletal voxels, and then propagate it back to the entire volume [27,91]. Properties measured at skeletal voxels benefit from symmetry, the larger context of local structure, reduced effects of digitization, and other artifacts. Also, such a strategy drastically improves computational efficiency, especially, for computationally extensive applications. This strategy is related to the salience DTs of Rosin and West [93]. It leads to the notion of "feature propagation," where regional features are computed at a set S of representative sample points, e.g., skeletal voxels, and propagated back to a larger set $X \supset S$, e.g., the entire object volume. Bonnassie et al. [94] proposed a feature propagation algorithm from skeletal voxels to the entire volume by copying feature values from a skeletal voxel p to all voxels within the maximal inscribed ball (MIB) centered at p. The nonuniqueness of MIBs containing a given voxel $q \in X$ causes the final result to be dependent of the scan order of skeletal voxels. Liu et al. [91] compared the performance among four different strategies of feature propagation, namely, (1) largest MIB, (2) smallest MIB, (3) nearest MIB, and (4) farthest circumference MIB, and demonstrated the advantages of using the farthest circumference MIB for feature propagation. The results would probably be more stable if the set of MIBs is first reduced as much as possible; see [95].

Saha et al. [49] discussed the choice of "representative points" for different local measures. For example, to compute local structure thickness, the entire surface-skeleton constitutes the set of representative points. However, to compute local structure plate-width, it is important to reduce another dimension of freedom for representative points. In other words, the local width should be initially determined at points on the curve skeleton. Also, local cross sectional area should be computed at curve skeletal points. Measures computed at curve skeletal points should be propagated to object volume using a two-step propagation strategy—from curve skeleton to surface skeleton and then from surface skeleton to curve skeleton.

10.5.3 APPLICATIONS

The measure of local structure plate-width resulted from VTA is used to compute several micro-architectural parameters of a TB network volume. The average plate-width measure PW_{VTA} of a TB region O is computed as the bone volume fraction (BVF)-weighted average of local trabecular plate-width values over O. Depending upon the imaging modality used, different methods are applied to compute BVF values at individual image voxels. In magnetic resonance (MR) imaging, Hwang et al.

[96] used an iterative deconvolution technique to diminish noise in an MR intensity histogram and then computed the BVF map analyzing the modified histogram and the original MR image. CT imaging generates Hounsfield numbers at individual voxels, which are first converted to BMD (mg/cc) values using a calibration phantom. A calibration phantom is scanned every time a bone image is acquired, and the known material density values at known locations within the phantom are used to determine the conversion function from Hounsfield numbers to BMD. Finally, the BVF at each voxel is computed using the following equation [49]:

$$BVF(p) = \begin{cases} 0 & \text{if } BMD(p) < 940 \text{ mg/cc}, \\ \frac{BMD(p)-940}{2184-940} & \text{if } BMD(p) \geq 940 \text{ mg/cc and } < 2184 \text{ mg/cc}, \\ 1 & \text{otherwise.} \end{cases}$$

In the above equation, 940 mg/cc is used as the bone marrow density, whereas 2184 mg/cc is used as the density of fully mineralized bone as determined following the results observed by Hernandez et al. [97]. The average plate-width parameter PW_{VTA} is computed as follows:

$$PW_{\text{VTA}} = \frac{\sum_{p \in O} VTA(p) BVF(p)}{\sum_{p \in O} BVF(p)},$$

where O is the target TB region, and $VTA(p)$ is the VTA-derived local plate-width measure at a given voxel p.

VTA-derived trabecular plate-width measure is used to assess plate-to-rod ratio over a TB region. At each individual bone voxel, a normalized measure of "platelikeliness" between 0 and 1 is obtained using the eccentricity between the average human TB thickness and the local plate-width value; the measure of "rodlikeliness" is computed as a complement of "platelikeliness." Platelikeliness at individual voxels is defined using the following equation:

$$\textit{platelikeliness}(p) = 1 - \frac{\text{avTB}_{\text{th}}^2}{\max(VTA^2(p), \text{avTB}_{\text{th}}^2)},$$

where avTB_{th} denotes the average human TB thickness observed using the specific imaging modality applied for bone imaging. The average TB thickness is preferred over the TB thickness at individual voxels to reduce random errors in thickness computation at individual locations and to avoid overestimation of platelikeliness at thinner trabeculae, often, associated with porous bones. Following the average human TB thickness reported in [91], 200 micron is a suitable value for the parameter avTB_{th} for TB plate–rod analysis using CT imaging. The plate-to-rod-ratio (PRR_{VTA}) in a TB region is defined as the ratio of BVF-weighted average "platelikeliness" and average "rodlikeliness" over the target region. Color-coded TB plate/rod classification at distal tibia of the left leg for one body-mass-index (BMI)-matched female volunteers are shown in Fig. 10.10. These BMI matched females were selected from four

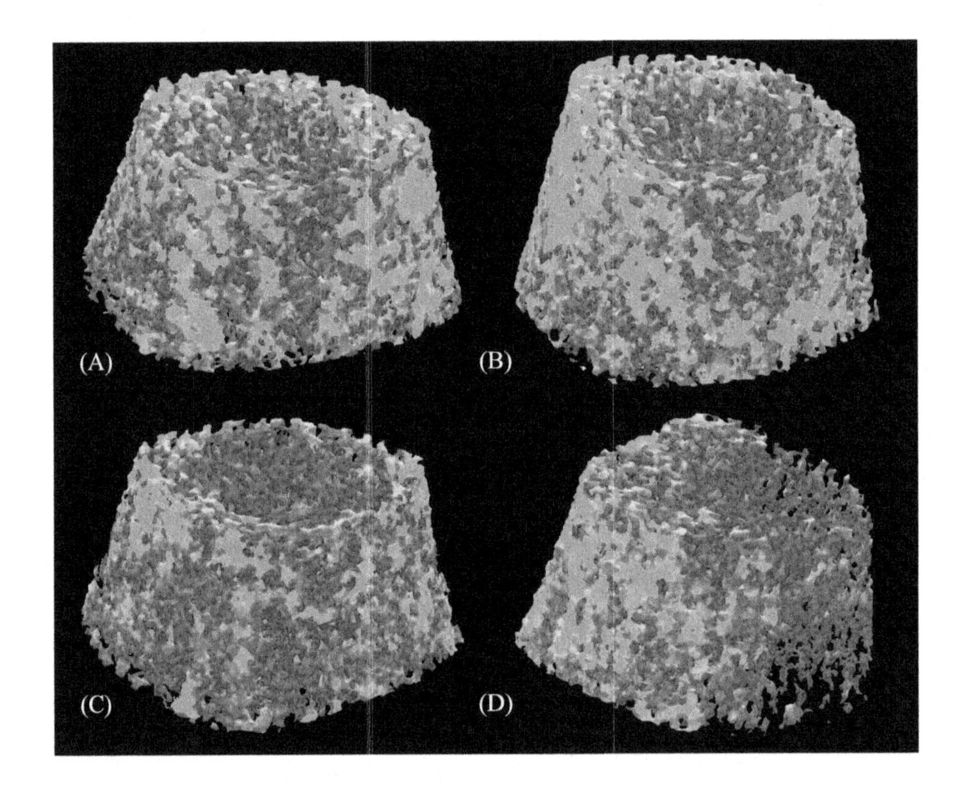

FIGURE 10.10

Color-coded display of TB plate/rod classification for a healthy female IBDS control (A) and sex- and BMI-matched athlete (B), a patient with confirmed diagnosis of cystic fibrosis (CF) (C), and another patient on continuous treatment of a selective serotonin reuptake inhibitor (SSRI) (D). The same color-coding scheme of Fig. 10.8 is used here. Both healthy control and the athlete have more TB plates colored in green as compared to the two patient participants. No apparent difference in TB is observed between the control and the athlete; between the two patients, the CF patient has some signs of nonhomogeneous bone loss.

different groups of participants—healthy young adults from the Iowa Bone Development (IBDS) group [98,99] athletes, patients with cystic fibrosis (CF), and patients on continuous treatment with a selective serotonin reuptake inhibitor (SSRI) [100]. As observed in the figure, the healthy female and the athlete have similar TB structure, whereas the two participants from the clinical groups have compromised TB structure. TB plate–rod micro-architecture among patients treated with SSRI ($N = 12$) and matched controls ($N = 97$) was compared in a pilot study [49]. It was observed in the study results that patients on SSRI have 12.5% lower average TB plate-width as compared to age-similar, sex-, height-, and weight-matched healthy controls.

10.6 CONCLUSION

Skeletonization is a common technique, which has been popularly used in methods related to quantitative assessment and characterization of various morphologic aspects of local structures an object. A large number of real-life applications utilize such methods computing morphometric features in an object or use such features to locate an object. Often, such methods use topologic and geometric analysis of skeletal structures at representative points, often, located on the skeleton, to quantitatively assess object morphometry and local structural properties in an object. In such methods, three aspects are critical—(1) determining the set of representative sample points, (2) accurate measurement of the target local morphologic feature, and (3) propagation to feature values from sample points to the entire object volume.

The topological methods discussed in this chapter for assessing local structure morphology use digital skeletal locations as representative sample points and geodesic path analysis for quantifying target morphologic features. Such methods suffer from two critical factors emerging from intrinsic limitations of digital grids—(1) error accumulation during digital path propagation and (2) structural measurement errors due to quantization of medial-surface, medial-axis, and boundary locations. Such errors are large at in vivo image resolution. An alternative approach of local morphologic assessment and analysis is geometric modeling of local structures using simple elements such as spherical harmonic representation of 3-D shape descriptors [101], spherical representation for recognition of free-form surfaces [102], and ellipsoidal representation of local structures [49,103]. Such methods suffer from model inaccuracy near junctions.

REFERENCES

[1] S. Khosla, B.L. Riggs, E.J. Atkinson, A.L. Oberg, L.J. McDaniel, M. Holets, J.M. Peterson, L.J. Melton 3rd, Effects of sex and age on bone microstructure at the ultradistal radius: a population-based noninvasive in vivo assessment, J. Bone Miner. Res. 21 (2006) 124–131.

[2] T.W.H.O. Bulletin, Aging and Osteoporosis, 1999.

[3] L.J. Melton, E.A. Chrischilles, C. Cooper, A.W. Lane, B.L. Riggs, Perspective. How many women have osteoporosis?, J. Bone Miner. Res. 7 (1992) 1005–1010.

[4] C. Cooper, G. Campion, L.J. Melton, Hip fractures in the elderly: a world-wide projection, Osteoporos. Int. 2 (1992) 285–289.

[5] L.J. Melton, Epidemiology of fractures, in: B.L. Riggs, L.J. Melton (Eds.), Osteoporosis: Etiology, Diagnosis, and Management, Raven Press, New York, 1988, pp. 133–154.

[6] A. Klibanski, L. Adams-Campbell, T.L. Bassford, S.N. Blair, S.D. Boden, K. Dickersin, D.R. Gifford, L. Glasse, S.R. Goldring, K. Hruska, Osteoporosis prevention, diagnosis, and therapy, J. Am. Med. Assoc. 285 (2001) 785–795.

[7] N.B. Watts, E.M. Lewiecki, P.D. Miller, S. Baim, National osteoporosis foundation 2008 clinician's guide to prevention and treatment of osteoporosis and the world health organization fracture risk assessment tool (FRAX): what they mean to the bone densitometrist and bone technologist, J. Clin. Densitom. 11 (2008) 473–477.

[8] S.R. Cummings, S. Eckert, K.A. Krueger, D. Grady, T.J. Powles, J.A. Cauley, L. Norton, T. Nickelsen, N.H. Bjarnason, M. Morrow, M.E. Lippman, D. Black, J.E. Glusman, A. Costa, V.C. Jordan, The effect of raloxifene on risk of breast cancer in postmenopausal women: results from the MORE randomized trial. Multiple outcomes of raloxifene evaluation, JAMA 281 (1999) 2189–2197.

[9] J.A. Cauley, F.L. Lucas, L.H. Kuller, K. Stone, W. Browner, S.R. Cummings, Elevated serum estradiol and testosterone concentrations are associated with a high risk for breast cancer. Study of osteoporotic fractures research group, Ann. Intern. Med. 130 (1999) 270–277.

[10] D.M. Black, P.D. Delmas, R. Eastell, I.R. Reid, S. Boonen, J.A. Cauley, F. Cosman, P. Lakatos, P.C. Leung, Z. Man, C. Mautalen, P. Mesenbrink, H. Hu, J. Caminis, K. Tong, T. Rosario-Jansen, J. Krasnow, T.F. Hue, D. Sellmeyer, E.F. Eriksen, S.R. Cummings, H.P.F. Trial, Once-yearly zoledronic acid for treatment of postmenopausal osteoporosis, N. Engl. J. Med. 356 (2007) 1809–1822.

[11] H.K. Genant, K. Engelke, S. Prevrhal, Advanced CT bone imaging in osteoporosis, Rheumatology (Oxford) 47 (Suppl. 4) (2008) iv9–iv16.

[12] T.M. Link, Osteoporosis imaging: state of the art and advanced imaging, Radiology 263 (2012) 3–17.

[13] H.K. Genant, C. Gordon, Y. Jiang, T.F. Lang, T.M. Link, S. Majumdar, Advanced imaging of bone macro and micro structure, Bone 25 (1999) 149–152.

[14] E. Seeman, Reduced bone formation and increased bone resorption: rational targets for the treatment of osteoporosis, Osteoporos. Int. 14 (2003) 2–8.

[15] L.C. Hofbauer, C.C. Brueck, S.K. Singh, H. Dobnig, Osteoporosis in patients with diabetes mellitus, J. Bone Miner. Res. 22 (2007) 1317–1328.

[16] P. Ammann, R. Rizzoli, Bone strength and its determinants, Osteoporos. Int. 14 (Suppl. 3) (2003), S13–S18.

[17] E. Seeman, P.D. Delmas, Bone quality – the material and structural basis of bone strength and fragility, N. Engl. J. Med. 354 (2006) 2250–2261.

[18] M. Kleerekoper, A.R. Villanueva, J. Stanciu, D.S. Rao, A.M. Parfitt, The role of three-dimensional trabecular microstructure in the pathogenesis of vertebral compression fractures, Calcif. Tissue Int. 37 (1985) 594–597.

[19] E. Legrand, D. Chappard, C. Pascaretti, M. Duquenne, S. Krebs, V. Rohmer, M.F. Basle, M. Audran, Trabecular bone microarchitecture, bone mineral density, and vertebral fractures in male osteoporosis, J. Bone Miner. Res. 15 (2000) 13–19.

[20] A.M. Parfitt, C.H.E. Mathews, A.R. Villanueva, M. Kleerekoper, B. Frame, D.S. Rao, Relationships between surface, volume, and thickness of iliac trabecular bone in aging and in osteoporosis – implications for the microanatomic and cellular mechanisms of bone loss, J. Clin. Invest. 72 (1983) 1396–1409.

[21] F.W. Wehrli, P.K. Saha, B.R. Gomberg, H.K. Song, P.J. Snyder, M. Benito, A. Wright, R. Weening, Role of magnetic resonance for assessing structure and function of trabecular bone, Top. Magn. Reson. Imaging 13 (2002) 335–355.

[22] S. Majumdar, Magnetic resonance imaging of trabecular bone structure, Top. Magn. Reson. Imaging 13 (2002) 323–334.

[23] G. Chang, S.K. Pakin, M.E. Schweitzer, P.K. Saha, R.R. Regatte, Adaptations in trabecular bone microarchitecture in Olympic athletes determined by 7T MRI, J. Magn. Reson. Imaging 27 (2008) 1089–1095.

[24] S. Boutroy, M.L. Bouxsein, F. Munoz, P.D. Delmas, In vivo assessment of trabecular bone microarchitecture by high-resolution peripheral quantitative computed tomography, J. Clin. Endocrinol. Metab. 90 (2005) 6508–6515.
[25] A.J. Burghardt, J.B. Pialat, G.J. Kazakia, S. Boutroy, K. Engelke, J.M. Patsch, A. Valentinitsch, D. Liu, E. Szabo, C.E. Bogado, M.B. Zanchetta, H.A. McKay, E. Shane, S.K. Boyd, M.L. Bouxsein, R. Chapurlat, S. Khosla, S. Majumdar, Multicenter precision of cortical and trabecular bone quality measures assessed by high-resolution peripheral quantitative computed tomography, J. Bone Miner. Res. 28 (2013) 524–536.
[26] T.M. Link, V. Vieth, C. Stehling, A. Lotter, A. Beer, D. Newitt, S. Majumdar, High-resolution MRI vs multislice spiral CT: which technique depicts the trabecular bone structure best?, Eur. Radiol. 13 (2003) 663–671.
[27] P.K. Saha, Y. Xu, H. Duan, A. Heiner, G. Liang, Volumetric topological analysis: a novel approach for trabecular bone classification on the continuum between plates and rods, IEEE Trans. Med. Imaging 29 (2010) 1821–1838.
[28] A. Vesterby, H.J. Gundersen, F. Melsen, Star volume of marrow space and trabeculae of the first lumbar vertebra: sampling efficiency and biological variation, Bone 10 (1989) 7–13.
[29] T. Hildebrand, P. Rüegsegger, Quantification of bone microarchitecture with the structure model index, Comput. Methods Biomech. Biomed. Eng. 1 (1997) 15–23.
[30] L.A. Feldkamp, S.A. Goldstein, A.M. Parfitt, G. Jesion, M. Kleerekoper, The direct examination of three-dimensional bone architecture in vitro by computed tomography, J. Bone Miner. Res. 4 (1989) 3–11.
[31] R.R. Recker, Architecture and vertebral fracture, Calcif. Tissue Int. 53 (Suppl. 1) (1993), S139–S142.
[32] P.K. Saha, R. Strand, G. Borgefors, Digital topology and geometry in medical imaging: a survey, IEEE Trans. Med. Imaging 34 (2015) 1940–1964.
[33] T.Y. Kong, A. Rosenfeld, Digital topology: introduction and survey, Comput. Vis. Graph. Image Process. 48 (1989) 357–393.
[34] P.K. Saha, A. Rosenfeld, Determining simplicity and computing topological change in strongly normal partial tilings of R^2 or R^3, Pattern Recognit. 33 (2000) 105–118.
[35] H. Blum, A transformation for extracting new descriptors of shape, in: Models for the Perception of Speech and Visual Form, vol. 19, 1967, pp. 362–380.
[36] H. Blum, R. Nagel, Shape description using weighted symmetric axis features, Pattern Recognit. 10 (1978) 167–180.
[37] K. Siddiqi, S.M. Pizer, Medial Representations: Mathematics, Algorithms and Applications, vol. 37, Springer, 2008.
[38] P.K. Saha, G. Borgefors, G. Sanniti di Baja, A survey on skeletonization algorithms and their applications, Pattern Recognit. Lett. 76 (2016) 3–12.
[39] H. Kobatake, Y. Yoshinaga, Detection of spicules on mammogram based on skeleton analysis, IEEE Trans. Med. Imaging 15 (1996) 235–245.
[40] R. Zwiggelaar, S.M. Astley, C.R. Boggis, C.J. Taylor, Linear structures in mammographic images: detection and classification, IEEE Trans. Med. Imaging 23 (2004) 1077–1086.
[41] M. Näf, G. Székely, R. Kikinis, M.E. Shenton, O. Kübler, 3D Voronoi skeletons and their usage for the characterization and recognition of 3D organ shape, Comput. Vis. Image Underst. 66 (1997) 147–161.
[42] G. Chang, S. Honig, Y. Liu, C. Chen, K.K. Chu, C.S. Rajapakse, K. Egol, D. Xia, P.K. Saha, R.R. Regatte, 7 Tesla MRI of bone microarchitecture discriminates between

women without and with fragility fractures who do not differ by bone mineral density, J. Bone Miner. Metab. 33 (2014) 285–293.
[43] S. Dudley-Javoroski, P.K. Saha, G. Liang, C. Li, Z. Gao, R.K. Shields, High dose compressive loads attenuate bone mineral loss in humans with spinal cord injury, Osteoporos. Int. 23 (2012) 2335–2346.
[44] B.R. Gomberg, P.K. Saha, H.K. Song, S.N. Hwang, F.W. Wehrli, Topological analysis of trabecular bone MR images, IEEE Trans. Med. Imaging 19 (2000) 166–174.
[45] B.R. Gomberg, F.W. Wehrli, B. Vasilic, R.H. Weening, P.K. Saha, H.K. Song, A.C. Wright, Reproducibility and error sources of micro-MRI-based trabecular bone structural parameters of the distal radius and tibia, Bone 35 (2004) 266–276.
[46] X.S. Liu, P. Sajda, P.K. Saha, F.W. Wehrli, X.E. Guo, Quantification of the roles of trabecular microarchitecture and trabecular type in determining the elastic modulus of human trabecular bone, J. Bone Miner. Res. 21 (2006) 1608–1617.
[47] X.S. Liu, P. Sajda, P.K. Saha, F.W. Wehrli, G. Bevill, T.M. Keaveny, X.E. Guo, Complete volumetric decomposition of individual trabecular plates and rods and its morphological correlations with anisotropic elastic moduli in human trabecular bone, J. Bone Miner. Res. 23 (2008) 223–235.
[48] P.K. Saha, B.R. Gomberg, F.W. Wehrli, Three-dimensional digital topological characterization of cancellous bone architecture, Int. J. Imaging Syst. Technol. 11 (2000) 81–90.
[49] P.K. Saha, Y. Liu, C. Chen, D. Jin, E.M. Letuchy, Z. Xu, R.E. Amelon, T.L. Burns, J.C. Torner, S.M. Levy, C.A. Calarge, Characterization of trabecular bone plate–rod microarchitecture using multirow detector CT and the tensor scale: algorithms, validation, and applications to pilot human studies, Med. Phys. 42 (2015) 5410–5425.
[50] B.B. Kimia, A. Tannenbaum, S.W. Zucker, Shape, shocks, and deformations I: the components of two-dimensional shape and the reaction–diffusion space, Int. J. Comput. Vis. 15 (1995) 189–224.
[51] R. Kimmel, D. Shaked, N. Kiryati, Skeletonization via distance maps and level sets, Comput. Vis. Graph. Image Process. 62 (1995) 382–391.
[52] F. Leymarie, M.D. Levine, Simulating the grassfire transform using an active contour model, IEEE Trans. Pattern Anal. Mach. Intell. 14 (1992) 56–75.
[53] K. Siddiqi, S. Bouix, A. Tannenbaum, S.W. Zucker, Hamilton–Jacobi skeletons, Int. J. Comput. Vis. 48 (2002) 215–231.
[54] J.W. Brandt, V.R. Algazi, Continuous skeleton computation by Voronoi diagram, CVGIP, Image Underst. 55 (1992) 329–338.
[55] R. Ogniewicz, M. Ilg, Voronoi skeletons: theory and applications, in: Proc. of the Computer Vision and Pattern Recognition, 1992, pp. 63–69.
[56] R.L. Ogniewicz, O. Kübler, Hierarchic Voronoi skeletons, Pattern Recognit. 28 (1995) 343–359.
[57] P.K. Saha, B. Chanda, D.D. Majumder, Principles and Algorithms for 2-D and 3-D Shrinking, Indian Statistical Institute, Calcutta, India, 1991, TR/KBCS/2/91.
[58] P.K. Saha, B.B. Chaudhuri, Detection of 3-D simple points for topology preserving transformations with application to thinning, IEEE Trans. Pattern Anal. Mach. Intell. 16 (1994) 1028–1032.
[59] P.K. Saha, B.B. Chaudhuri, B. Chanda, D.D. Majumder, Topology preservation in 3D digital space, Pattern Recognit. 27 (1994) 295–300.
[60] G. Bertrand, G. Malandain, A new characterization of three-dimensional simple points, Pattern Recognit. Lett. 15 (1994) 169–175.

[61] T.Y. Kong, Topology-preserving deletion of 1's from 2-, 3- and 4-dimensional binary images, in: Proc. of the International Workshop on Discrete Geometry for Computer Imagery, Montpellier, France, 1997, pp. 1–18.
[62] P.K. Saha, D.D. Majumder, Topology and shape preserving parallel 3D thinning – a new approach, in: Proc. of the 9th International Conference on Image Analysis and Processing, ICIAP, vol. 1310, 1997, pp. 575–581.
[63] K. Palágyi, A. Kuba, A 3D 6-subiteration thinning algorithm for extracting medial lines, Pattern Recognit. Lett. 19 (1998) 613–627.
[64] K. Palágyi, A. Kuba, A parallel 3D 12-subiteration thinning algorithm, Graph. Models Image Process. 61 (1999) 199–221.
[65] L. Lam, S.-W. Lee, C.Y. Suen, Thinning methodologies – a comprehensive survey, IEEE Trans. Pattern Anal. Mach. Intell. 14 (1992) 869–885.
[66] Y.F. Tsao, K.S. Fu, A parallel thinning algorithm for 3D pictures, Comput. Graph. Image Process. 17 (1981) 315–331.
[67] T.-C. Lee, R.L. Kashyap, C.-N. Chu, Building skeleton models via 3-D medial surface/axis thinning algorithm, CVGIP, Graph. Models Image Process. 56 (1994) 462–478.
[68] G. Németh, P. Kardos, K. Palágyi, Thinning combined with iteration-by-iteration smoothing for 3D binary images, Graph. Models 73 (2011) 335–345.
[69] P.K. Saha, B.B. Chaudhuri, D.D. Majumder, A new shape preserving parallel thinning algorithm for 3D digital images, Pattern Recognit. 30 (1997) 1939–1955.
[70] C. Arcelli, G. Sanniti di Baja, A width-independent fast thinning algorithm, IEEE Trans. Pattern Anal. Mach. Intell. 7 (1985) 463–474.
[71] C. Arcelli, G. Sanniti di Baja, L. Serino, Distance-driven skeletonization in voxel images, IEEE Trans. Pattern Anal. Mach. Intell. 33 (2011) 709–720.
[72] G. Sanniti di Baja, Well-shaped, stable, and reversible skeletons from the (3, 4)-distance transform, J. Vis. Commun. Image Represent. 5 (1994) 107–115.
[73] I. Bitter, A.E. Kaufman, M. Sato, Penalized-distance volumetric skeleton algorithm, IEEE Trans. Vis. Comput. Graph. 7 (2001) 195–206.
[74] G. Borgefors, I. Nyström, G. Sanniti di Baja, Computing skeletons in three dimensions, Pattern Recognit. 32 (1999) 1225–1236.
[75] C. Pudney, Distance-ordered homotopic thinning: a skeletonization algorithm for 3D digital images, Comput. Vis. Image Underst. 72 (1998) 404–413.
[76] D. Jin, P.K. Saha, A new fuzzy skeletonization algorithm and its applications to medical imaging, in: Proc. of the 17th Int. Conf. Imag. Anal. Proc. (ICIAP), Naples, Italy, in: Lect. Notes Comput. Sci., vol. 8156, 2013, pp. 662–671.
[77] G. Borgefors, Distance transformations in digital images, Comput. Vis. Graph. Image Process. 34 (1986) 344–371.
[78] P.K. Saha, B.B. Chaudhuri, 3D digital topology under binary transformation with applications, Comput. Vis. Image Underst. 63 (1996) 418–429.
[79] S. Svensson, I. Nyström, G. Sanniti di Baja, Curve skeletonization of surface-like objects in 3d images guided by voxel classification, Pattern Recognit. Lett. 23 (2002) 1419–1426.
[80] P.K. Saha, 2D Thinning Algorithms and 3D Shrinking, INRIA, Sophia Antipolis Cedex, France, June 1991.
[81] S.C. Lam, M.J. Wald, C.S. Rajapakse, Y. Liu, P.K. Saha, F.W. Wehrli, Performance of the MRI-based virtual bone biopsy in the distal radius: serial reproducibility and reliability of structural and mechanical parameters in women representative of osteoporosis study populations, Bone 49 (2011) 895–903.

[82] F.W. Wehrli, B.R. Gomberg, P.K. Saha, H.K. Song, S.N. Hwang, P.J. Snyder, Digital topological analysis of in vivo magnetic resonance microimages of trabecular bone reveals structural implications of osteoporosis, J. Bone Miner. Res. 16 (2001) 1520–1531.
[83] M. Stauber, R. Muller, Age-related changes in trabecular bone microstructures: global and local morphometry, Osteoporos. Int. 17 (2006) 616–626.
[84] M. Stauber, R. Muller, Volumetric spatial decomposition of trabecular bone into rods and plates – a new method for local bone morphometry, Bone 38 (2006) 475–484.
[85] M. Stauber, L. Rapillard, G.H. van Lenthe, P. Zysset, R. Muller, Importance of individual rods and plates in the assessment of bone quality and their contribution to bone stiffness, J. Bone Miner. Res. 21 (2006) 586–595.
[86] G. Malandain, G. Bertrand, N. Ayache, Topological segmentation of discrete surfaces, Int. J. Comput. Vis. 10 (1993) 183–197.
[87] L. Serino, C. Arcelli, G. Sanniti di Baja, Decomposing 3d objects in simple parts characterized by rectilinear spines, Int. J. Pattern Recognit. Artif. Intell. 28 (2014).
[88] P.J. Giblin, B.B. Kimia, A formal classification of 3D medial axis points and their local geometry, IEEE Trans. Pattern Anal. Mach. Intell. 26 (2004) 238–251.
[89] R. Krug, A.J. Burghardt, S. Majumdar, T.M. Link, High-resolution imaging techniques for the assessment of osteoporosis, Radiol. Clin. North Am. 48 (2010) 601–621.
[90] C. Chen, D. Jin, Y. Liu, F.W. Wehrli, G. Chang, P.J. Snyder, R.R. Regatte, P.K. Saha, Trabecular bone characterization on the continuum of plates and rods using in vivo MR imaging and volumetric topological analysis, Phys. Med. Biol. 61 (2016) N478.
[91] Y. Liu, D. Jin, C. Li, K.F. Janz, T.L. Burns, J.C. Torner, S.M. Levy, P.K. Saha, A robust algorithm for thickness computation at low resolution and its application to in vivo trabecular bone CT imaging, IEEE Trans. Biomed. Eng. 61 (2014) 2057–2069.
[92] E. Dijkstra, A note on two problems in connection with graphs, Numer. Math. 1 (1959) 269–271.
[93] P.L. Rosin, G.A.W. West, Salience distance transforms, Graph. Models Image Process. 57 (1995) 483–521.
[94] A. Bonnassie, F. Peyrin, D. Attali, A new method for analyzing local shape in three-dimensional images based on medial axis transformation, IEEE Trans. Syst. Man Cybern., Part B, Cybern. 33 (2003) 700–705.
[95] G. Borgefors, I. Nyström, Efficient shape representation by minimizing the set of centres of maximal discs/spheres, Pattern Recognit. Lett. 18 (1997) 465–471.
[96] S.N. Hwang, F.W. Wehrli, Estimating voxel volume fractions of trabecular bone on the basis of magnetic resonance images acquired in vivo, Int. J. Imaging Syst. Technol. 10 (1999) 186–198.
[97] C.J. Hernandez, G.S. Beaupre, T.S. Keller, D.R. Carter, The influence of bone volume fraction and ash fraction on bone strength and modulus, Bone 29 (2001) 74–78.
[98] K.F. Janz, S.M. Levy, T.L. Burns, J.C. Torner, M.C. Willing, J.J. Warren, Fatness, physical activity, and television viewing in children during the adiposity rebound period: the Iowa bone development study, Prev. Med. 35 (2002) 563–571.
[99] K.F. Janz, H.C. Medema-Johnson, E.M. Letuchy, T.L. Burns, J.M. Gilmore, J.C. Torner, M. Willing, S.M. Levy, Subjective and objective measures of physical activity in relationship to bone mineral content during late childhood: the Iowa bone development study, Br. J. Sports Med. 42 (2008) 658–663.
[100] P.K. Saha, R.E. Amelon, Y. Liu, C. Li, D. Jin, C. Chen, J.M. Fishbaugher, E.M. Letuchy, C.A. Calarge, K.F. Janz, D.B. Hornick, J. Eichenberger-Gilmore, T.L. Burns, J.C. Torner, S.M. Levy, In vivo study of trabecular and cortical bone in young adults with varying

trajectories of bone development using multi-row detector CT imaging, in: Proc. of the Annual Meeting of the American Society for Bone and Mineral Research, Baltimore, MD, 2013.

[101] M. Kazhdan, T. Funkhouser, S. Rusinkiewicz, Rotation invariant spherical harmonic representation of 3D shape descriptors, in: Proc. of the Symposium on Geometry Processing, vol. 6, 2003, pp. 156–164.

[102] M. Hebert, K. Ikeuchi, H. Delingette, A spherical representation for recognition of free-form surfaces, IEEE Trans. Pattern Anal. Mach. Intell. 17 (1995) 681–690.

[103] P.K. Saha, Tensor scale: a local morphometric parameter with applications to computer vision and image processing, Comput. Vis. Image Underst. 99 (2005) 384–413.

CHAPTER

Medial structure generation for registration of anatomical structures

11

Sergio Vera[*,†], Debora Gil[†], H.M. Kjer[‡], Jens Fagertun[‡], Rasmus R. Paulsen[‡], Miguel Á. González Ballester[§,¶]

Alma IT Systems, Barcelona, Spain[*] Computer Vision Center, Universitat Autònoma de Barcelona, Barcelona, Spain[†] Department of Applied Mathematics and Computer Science, Technical University of Denmark, Copenhagen, Denmark[‡] Department of Information and Communication Technologies, Universitat Pompeu Fabra, Barcelona, Spain[§] ICREA, Barcelona, Spain[¶]

Contents

11.1 Medial Maps for Reliable Extraction of Anatomical Medial Surfaces 314
11.2 Extracting Anatomical Medial Surfaces Using Medialness Maps 316
11.2.1 Gaussian Steerable Medial Maps 317
11.2.1.1 Nonmaxima Suppression Binarization 320
11.2.1.2 Parameter Setting 320
11.3 Validation Framework for Medial Anatomy Assessment 322
11.3.1 Synthetic Database 322
11.3.2 Medial Surface Quality Metrics 323
11.4 Validation Experiments 324
11.4.1 Medial Surface Quality 325
11.4.2 Reconstruction Power for Clinical Applications................... 327
11.5 Application to Cochlea Registration 327
11.5.1 Material and Methods 330
11.5.2 Skeletonization 330
11.5.3 Image Registration 331
11.5.3.1 Initial Rigid Alignment 331
11.5.3.2 Deformable Registration 332
11.5.3.3 Deformable Registration with Guidance from Skeleton .. 333
11.5.4 Evaluation 333
11.5.5 Results 335
11.6 Discussion 337
Acknowledgments 340
References 340

Skeletonization. DOI: 10.1016/B978-0-08-101291-8.00013-4

11.1 MEDIAL MAPS FOR RELIABLE EXTRACTION OF ANATOMICAL MEDIAL SURFACES

Many organs in the human anatomy have shapes that can be represented by genus zero surfaces (surfaces that are homeomorphic to a sphere). Nevertheless, such a generalization cannot hide the complexity of human anatomical shapes. Even with these broad common characteristics, anatomical shapes display a large amount of variability between different organs and also variability between the same organ over different patients. This large variability makes anatomical shapes challenging to process by computer techniques. Medial structures have demonstrated to be a compact shape descriptor and have shown potential to capture shape variations. Medial structures [9], such as the medial axis and the medial manifold (medial surface), completely determine the geometry of the boundary volume [22]. In the field of medical imaging, they can be used in many applications: information provided by medial surfaces has shown to improve segmentation results [43,54]; medial structures have been used to characterize pathological abnormalities [52,53] and provide detailed representations of complex organs [66]. They also provide more intuitive and easily interpretable representations of complex organs [66] and their relative positions [32]. Medial information-enhanced modeling has been used in a variety of medical imaging analysis applications, including computational neuroanatomy [69,53], 3D cardiac modeling [55], and cancer treatment planning [51,15]. In shape analysis, medial representations can provide better information than point distribution models since they can model not only the shape but also the interior variations too [67].

In any task where a medial surface is to be used to help the modelization, one must ensure that the medial surfaces have certain properties that do not hinder further computations. Ideally, the medial surface should be complete enough to capture the key anatomical shapes of an organ but simple enough so that the topology of the medial structure respond to true changes in shape and not to artifacts (spurious branches). To provide accurate meshes of anatomical geometry, the extraction of medial manifolds should satisfy three main conditions [44]:

- Homotopy: The medial manifold should maintain the same topology (number of holes and components) of the original shape.
- Medialness: The medial structure has to lie as close as possible to the center of the original object.
- Thinness: The resulting medial shape should be as thin as possible without breaking the homotopy rule. The ideal case is to have one pixel-wide structures. However, this concept is too generic, and it heavily depends on the selected connectivity.

The stability of medial manifold properties depends on the domain on which the medial manifold is computed. Existing methods compute medial structures on either the volumetric voxel domain or a tetrahedral mesh of the volume boundary.

Volumetric approaches can be classified into two big types: morphological thinning and energy-based methods. Morphological methods compute medial manifolds

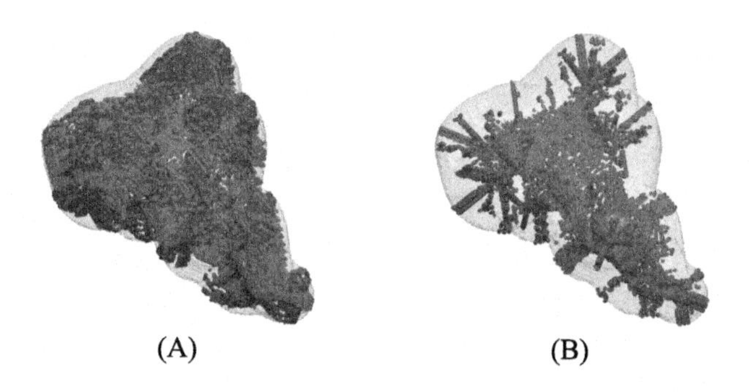

FIGURE 11.1

Medial surfaces obtained using a 6-connected neighborhood, (A), and a 26-connected neighborhood, (B).

by iterative thinning of the exterior layers of the volumetric object until more thinning breaks surface topology [11,44,49,39,26,57]. Such methods require the definition of a neighborhood set and conditions for the removal of *simple voxels*, i.e., voxels that can be removed without changing the topology of the object. Furthermore, simplicity tests alone produce (1D) medial axis. Computation of medial manifolds (medial sheets) requires additional tests to know if a voxel lies in a surface and thus cannot be deleted even if it is simple [44]. Moreover, surface tests might introduce medial axis segments in the medial surface, which is against the mathematical definition of manifold and which may require further pruning [44,3]. There are many definitions possible for simplicity and the presence or not of a medial surface voxel, also depending on the local connectivity considered when doing the tests. This means that the skeleton of a shape is not unique and that different methods will generate different medial structures (see Fig. 11.1).

Alternative methods rely on an energy map to ensure medialness on the manifold. Often, this energy image is the distance map of the object [44] or another energy derived from it, like the average outward flux [49,11], level set [47,60], or ridges of the distance map [14]. However, to obtain a manifold from the energy image, most methods rely on morphological thinning in a two-step process [11,44,49], thus inheriting the weak points of pure morphological methods.

An alternative to volumetric methods is using the Voronoi diagram tetrahedral mesh of a set of points sampled on the object boundary [16,48,2,3,20]. Voronoi methods work on a continuous domain and can naturally resolve branching medial surfaces. However, they still introduce one-dimensional spikes associated with boundary irregularities that have to be further pruned [3,20]. Also, their computational cost and quality depend on the number of vertices defining the volume boundary mesh and, thus, on the volume resolution [16]. Although some recent methods [20] are capable of efficiently dealing with surface perturbations, they are prone to introduce medial loops that distort the medial topology [37]. Finally, in the context of medical applications, the voxel discrete domain is the format in which medical data are acquired from medical imaging devices, and, thus, it is the natural domain for the implementation of image processing [27,33] and shape modeling [42,40] algorithms.

We have seen that medial surface generation methods are prone to produce noisy medial surfaces with spurial branches. Unwanted branching patterns on the medial surface makes them suboptimal for the purpose of being used in these medical imaging applications. Pruning operations will decrease these artifacts at the cost of increasing the risk of removing a relevant part of the medial structure. With all this considered, obtaining a good medial structure for usage in medical imaging means to find a compromise between the shape representation and simplicity of medial structure.

We present a method to compute medial structures that is well suited to applications of medical imaging:

1. A novel energy-based method for medial surface computation in images of arbitrary dimensions based on the combination of Gaussian and normalized operators since a medialness map followed by noniterative thinning binarization step is free of topology rules, as it is based on nonmaxima suppression (NMS) [12, Sect. 2].
2. A validation framework for fair comparison of the quality of medial surfaces: the variability in existing methods for medial surface generation makes comparisons with other methods difficult (Sections 11.3 and 11.4).
3. An application of the computation of medial axis for improved registration between human cochleas (Section 11.5).

11.2 EXTRACTING ANATOMICAL MEDIAL SURFACES USING MEDIALNESS MAPS

The computation of medial manifolds from a segmented volume may be split into two main steps: computation of a medial map from the original volume and binarization of such a map (Fig. 11.2). Medial maps should achieve a discriminant value on the shape central voxels, whereas the binarization step should ensure that the resulting medial structures fulfill the three quality conditions [44] that ensure fair representation of volumes geometry: medialness, thinness, and homotopy.

Distance transforms are the basis for obtaining medial manifolds from volumes in any dimension [44]. The distance transform, also called the distance map, is an operator that, given a binary volume of a closed domain B, computes for each volume voxel its distance to the domain boundary ∂B. By definition maximum values are achieved at the center of B. These voxels correspond to the volume's medial structure, and their values depend on the local thickness of the shape.

By the maximality property of distance maps the medial surface can be obtained using several methods such as iterative thinning [44], but it can lead to generation of spikes and other discretization artifacts due to the different neighborhood definitions available. An alternative to iterative thinning is applying a threshold th to $D(B)$. The selection of a good thresholding value of th ensuring homotopy and thinness is prob-

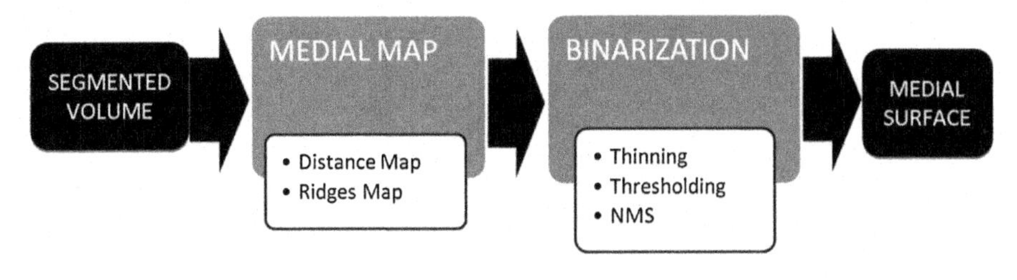

FIGURE 11.2

Schema of medial surface generation methods.

lematic. The maximum value of the distance map represents the minimum distance from the medial manifold to the object's boundary. Its value is, therefore, related to the local thickness of the object and cannot be considered as a global constant value through the whole object. On one hand, a too high threshold value is prone to generate unconnected manifold structures violating the homotopy property. On the other hand, a too small value for th means that the medial structure may not be thin. Albeit useful, the distance map is a less than an optimal medialness energy map because it is not selective enough and hinders the binarization step. Further examination of the distance map shows that its central maximal voxels are connected and constitute a ridge surface of the distance map. That is why we claim that the ridges of the distance map provide a better tool to describe the medialness of a set of shape pixels.

11.2.1 GAUSSIAN STEERABLE MEDIAL MAPS

Ridges/valleys in a digital N-dimensional image are defined as the set of points that are extrema (minima for ridges and maxima for valleys) in the direction of greatest magnitude of the second-order directional derivative [23]. In image processing, ridge detectors are based either on image intensity profiles [19] or on level sets geometry [34]. From the available operators for ridge detection we have chosen the crassness measure described in [34] because it provides (normalized) values in the range $[-N, N]$. The ridgeness operator is computed by the structure tensor of the distance map as follows.

Let D denote the distance map to the shape, and let its gradient ∇D be computed by convolution with partial derivatives of a Gaussian kernel g_σ of variance σ.

The structure tensor or second-order matrix [8] is given by averaging the projection matrices onto the distance map gradient:

$$ST_{\rho,\sigma}(D) = \begin{pmatrix} g_\rho * \partial_x D_\sigma^2 & g_\rho * \partial_x D_\sigma \partial_y D_\sigma & g_\rho * \partial_x D_\sigma \partial_z D_\sigma \\ g_\rho * \partial_x D_\sigma \partial_y D_\sigma & g_\rho * \partial_y D_\sigma^2 & g_\rho * \partial_y D_\sigma \partial_z D_\sigma \\ g_\rho * \partial_x D_\sigma \partial_z D_\sigma & g_\rho * \partial_y D_\sigma \partial_z D_\sigma & g_\rho * \partial_z D_\sigma^2 \end{pmatrix}, \tag{11.1}$$

where g_ρ is a Gaussian kernel of variance ρ, and ∂_x, ∂_y, and ∂_z are partial derivative operators. Let V be the eigenvector of the principal eigenvalue of $ST_{\rho,\sigma}(D)$ and consider its reorientation along the distance gradient $\tilde{V} = (P, Q, R)$ given as

$$\tilde{V} = \text{sign}(\langle \tilde{V} \cdot \nabla D \rangle) \cdot \tilde{V}, \tag{11.2}$$

where $\langle \cdot \rangle$ the scalar product. The ridgeness measure or NRM (normalized ridge map) [34] is given by the divergence:

$$\text{NRM} := \text{div}(\tilde{V}) = \partial_x P + \partial_y Q + \partial_z R \tag{11.3}$$

The above operator assigns positive values to ridge pixels and negative values to valley ones. The more positive the value, the stronger the ridge patterns. The main advantage over other operators (such as second-order oriented Gaussian derivatives) is that NRM $\in [-N, N]$ for N the spacial dimension of the volume. In this way, it is possible to set a threshold τ common to any volume for detecting significant ridges and, thus, points highly likely to belong to the medial surface. However, by its geometric nature, NRM has two main limitations. To be properly defined, NRM requires that the vector $\tilde{V}$ uniquely defines the tangent space to image level sets. Therefore, the operator achieves strong responses in the case of one-fold medial manifolds but significantly drops anywhere two or more medial surfaces intersect each other. Additionally, NRM responses are not continuous maps but step-wise almost binary images (Fig. 11.3). Such discrete nature of the map is prone to hinder the performance of the NMS binarization step that removes some internal voxels of the medial structure and, thus, introduces holes in the final medial surface.

Ridge maps based on image intensity are computed by convolution with a bank of steerable filters [19]. Steerable filters are given by derivatives of oriented anisotropic 3D Gaussian kernels. Let $\sigma = (\sigma_x, \sigma_y, \sigma_z)$ be the scale of the filter, and $\Theta = (\theta, \phi)$ its orientation given by the unitary vector $\eta = (\cos(\phi)\cos(\theta), \cos(\phi)\sin(\theta), \sin(\phi))$, then the oriented anisotropic 3D Gaussian kernel g_σ^Θ is given by

$$g_\sigma^\Theta = g_{(\sigma_x,\sigma_y,\sigma_z)}^{(\theta,\phi)} = \frac{1}{(2\pi)^{3/2}\sigma_x\sigma_y\sigma_z} e^{-\left(\frac{\tilde{x}^2}{2\sigma_x^2} + \frac{\tilde{y}^2}{2\sigma_y^2} + \frac{\tilde{z}^2}{2\sigma_z^2}\right)} \tag{11.4}$$

for the change of coordinates $(\tilde{x}, \tilde{y}, \tilde{z})$ given by the rotations of angles θ and ϕ that transform the z-axis into the unitary vector η:

$$\begin{pmatrix} \tilde{x} \\ \tilde{y} \\ \tilde{z} \end{pmatrix} = R_x(\theta) R_y(\phi) R_x(-\theta) \begin{pmatrix} x \\ y \\ z \end{pmatrix} \tag{11.5}$$

with the following rotation matrices:

$$R_x(\theta) = \begin{pmatrix} 1 & 0 & 0 \\ 0 & \cos(\theta) & -\sin(\theta) \\ 0 & \sin(\theta) & \cos(\theta) \end{pmatrix}, \quad R_y(\phi) = \begin{pmatrix} \cos(\phi) & 0 & \sin(\phi) \\ 0 & 1 & 0 \\ -\sin(\phi) & 0 & \cos(\phi) \end{pmatrix}. \tag{11.6}$$

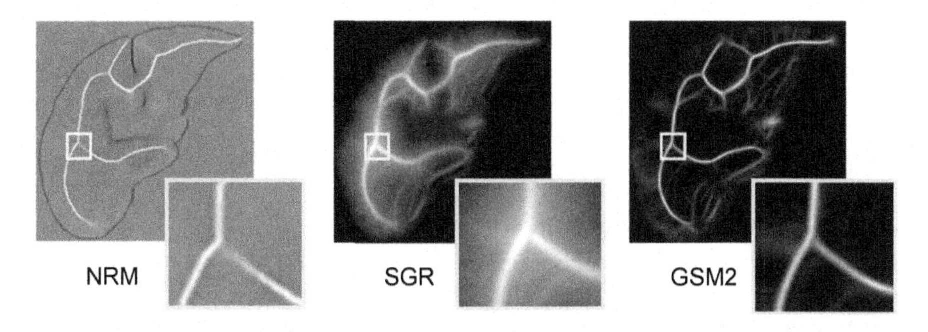

FIGURE 11.3

Performance of different ridge operators. From left to right: NRM, SGR, and GSM2.

The second partial derivative of g_σ^Θ along the $\tilde{z}$ axis constitutes the principal kernel for computing ridge maps:

$$\partial_z^2 g_\sigma^\Theta = (\tilde{z}^2/\sigma_z^4 - 1/\sigma_z^2) g_\sigma^\Theta \tag{11.7}$$

We note that by tuning the anisotropy of a Gaussian we can detect independently medial surfaces and medial axes. For detecting sheet-like ridges, the scales should be set to $\sigma_z > \sigma_x = \sigma_y$, whereas for medial axes, they should fulfill $\sigma_z < \sigma_x < \sigma_y$.

The maximum response across Gaussian kernel orientations and the scales gives the standard Gaussian ridge (SGR) medial map:

$$\mathrm{SGR} := \max_{\Theta,\sigma} \left(\partial_z^2 g_\sigma^\Theta * D \right) \tag{11.8}$$

for Θ expressing different orientations of the Gaussian kernel and σ the scales.

The main advantage of using steerable filters is that their response does not decrease at self-intersections. Their main counterpart is that their response is not normalized, so setting the threshold for binarization becomes a delicate issue [11, 36].

Given that geometric and intensity methods have complementary properties, we propose combining them into a geometric steerable medial map (GSM2):

$$\mathrm{GSM2} := \mathrm{SGR}(\mathrm{NRM}). \tag{11.9}$$

GSM2 generates medial maps with good combination of specificity in detecting medial voxels while having good characteristics for NMS binarization, which does not introduce internal holes.

The three images of Fig. 11.3 show the performance of different ridge operators at a two-dimensional branch (highlighted in the square close up). The geometric NRM (left) produces highly discriminant ridge values. However, they depend on the uniqueness of the direction surface normal, and thus its response significantly decreases at surface branches or self-intersections. Steerable Gaussian filters (center) are less sensitive to strong ridges while having increased sensitivity to small secondary noisy ridges. Finally, the combined approach GSM2 (right) inherits the strong features of each approach. It follows that it achieves a homogeneous response along ridges (induced by NRM normalization) that does not decrease at branches (thanks to the orientations provided by SGR).

11.2.1.1 Nonmaxima Suppression Binarization

Converting the medialness energy map into a binary set of voxels can be achieved in several ways. As previously stated, thresholding the intensity values of the medialness map yields a reduced set of voxels that are likely to belong to the medial manifold. However, the subset of voxels obtained using thresholding does not necessarily fulfill the property of thinness, and the homotopy heavily depends on the threshold value and the performance of the medial operator at intersections. The usage of iterative thinning schemes after thresholding can generate a thin structure [11], but at the risk of introducing spikes and different surfaces depending on the definition of simple or medial voxels and the order in which voxels are processed.

As an alternative, we propose to use nonmaxima suppression (NMS) to obtain a thin, one-voxel-wide medial surface. Nonmaxima suppression is a well-known technique for getting the local maxima of an energy map [12]. For each voxel, NMS consists in checking that the value of its neighbors in a specific direction V is lower than the actual voxel value. If this condition is not met, then the voxel is discarded. In this manner, only voxels that are local maxima along the direction V are preserved.

$$\mathrm{NMS}(x, y, z) = \begin{cases} \mathcal{M}(x, y, z) & \text{if } \mathcal{M}(x, y, z) > \max(\mathcal{M}_{V+}(x, y, z), \mathcal{M}_{V-}(x, y, z)), \\ 0 & \text{otherwise} \end{cases} \tag{11.10}$$

for $\mathcal{M}_{V+} = \mathcal{M}(x + V_x, y + V_y, z + V_z)$ and $\mathcal{M}_{V-} = \mathcal{M}(x - V_x, y - V_y, z - V_z)$.

A main requirement to apply NMS is identifying the local-maxima direction from the medial map derivatives. The search direction for local maxima is given by the eigenvector with highest eigenvalue of the structure tensor of the ridge map $ST_{\rho,\sigma}(\mathcal{M})$ given by Eq. (11.1) since it indicates the direction of highest variation of the ridge image. To overcome small glitches due to discretization of the direction, NMS is computed using trilinear interpolation.

11.2.1.2 Parameter Setting

Unlike most of existing parametric methods, the theoretical properties of GSM2 provide a natural way of setting parametric values regardless of the volume size and shape. This new method depends on the parameters involved in the definition of the map GSM2 and in the NMS binarization step.

The parameters arising in the definition of GSM2 are the derivation scale σ and the integration scale ρ of the structure tensor $ST_{\rho,\sigma}(\mathcal{M})$ used to compute NRM. The derivation scale σ is used to obtain regular gradients in the case of noisy images. The larger it is, the more regular the gradient will be at the cost of losing contrast. The integration scale ρ used to average the projection matrices corresponds to time in a solution to the heat equation with initial condition the projection matrix. Therefore, large values provide a regular extension of the level sets normal vector, which can be used for contour closing [21]. Since in our case we apply NRM to a regular distance map with well-defined completed ridges, σ and ρ can be set to their minimum values, $\sigma = 0.5$ and $\rho = 1$.

Concerning steerable filters, the parameters are the scales $\sigma = (\sigma_x, \sigma_y, \sigma_z)$ and orientations Θ, defining the steerable filter bank in Eq. (11.9). These last parameters are usually sampled on a discrete grid, so that Eq. (11.8) becomes

$$\mathrm{SGR} := \max_{i,j,k} \left(\partial_z^2 g_{\sigma_k}^{\Theta_{i,j}} * D \right) \tag{11.11}$$

for $\Theta_{i,j}$ given by $\theta_i = \{i\frac{\pi}{N}, \forall i = 1, \ldots, N\}$, $\phi_j = \{j\frac{\pi}{M}, \forall j = 1, \ldots, M\}$, and $\sigma_k = (\sigma_x^k, \sigma_y^k, \sigma_z^k) = (2^{k+1}, 2^{k+1}, 2^k)$, $k = [0, K]$. The scale depends on the thickness of the ridge, and the orientations on the complexity of the ridge geometry. The selection of the scale might be critical in the general setting of natural scenes [30]. However, in our case, SGR is applied to a normalized ridge map that defines stepwise almost binary images of ridges (see Fig. 11.3, left). Therefore, the choice of scale is not critical anymore. To get medial maps as accurate as possible, we recommend using a minimum anisotropic setting: $\sigma_z = 1$, $\sigma_x = \sigma_y = 2$. Finally, orientation sampling should be dense enough to capture any local geometry of medial surfaces. In the case of using the minimum scale, eight orientations, $N = M = 8$, are enough.

Therefore, GSM2 is given by

$$\mathrm{GSM2} = \max_{i,j} \left(\partial_z^2 g_{(2,2,1)}^{\Theta_{i,j}} * \mathrm{NRM} \right) \tag{11.12}$$

for NRM computed over $ST_{1,0.5}(\mathcal{M})$ and $\Theta_{i,j}$ computed setting $N = M = 8$.

The parameters involved in the NMS binarization step are the scales of the structure tensor $ST_{\rho,\sigma}(\mathrm{GSM2})$ and the binarizing threshold τ. Like in the case of NRM, GSM2 is a regular function the maximums of which define closed medial manifolds, so we set the structure tensor scales to their minimum values $\sigma = 0.5$ and $\rho = 1$. Concerning τ, it can be obtained using any histogram threshold calculation since GSM2 inherits the uniform discriminative response along ridges of NRM.

11.3 VALIDATION FRAMEWORK FOR MEDIAL ANATOMY ASSESSMENT

To address the representation of organs for medical use, medial representations should achieve a good reconstruction of the full anatomy and guarantee that the boundaries of the organ are reached from the medial surface. Given that small differences in algorithm criteria can generate different surfaces, we are interested in evaluating the quality of the generated manifold as a tool to recover the original shape.

Validation in the medical imaging field is a delicate issue due to the difficulties for generating ground truth data and quantitative scores valid for reliable application to clinical practice. In this section, we describe our validation framework for evaluating medial surface quality in the context of medical applications. In particular, we will generate a synthetic database with ground truth (GT) and two quality tests for assessing the quality of the medial anatomy for data with and without GT. The database can be used to benchmark algorithms using two tests.

11.3.1 SYNTHETIC DATABASE

The test set of synthetic volumes/surfaces aims to cover different key aspects of medial surface generation (see the first row in Fig. 11.4). The first batch of surfaces (labeled 'Simple') includes objects generated with a single medial surface. A second batch of surfaces is generated using two intersecting medial surfaces (labeled 'Multiple'), whereas a last batch of objects (labeled 'Homotopy') covers shapes with different numbers of holes. Each family of medial topology has 20 samples. The volumetric object obtained from a surface can be generated by using spheres of uniform radii (identified as 'UnifDist') or with spheres of varying radii (identified as 'VarDist').

Volumes are constructed by assigning a radial coordinate to each medial point. In the case of UnifDist, all medial points have the same radial value, whereas for VarDist, they are assigned a value in the range [5, 10] using a polynomial. The values of the radial coordinate must be in a range ensuring that volumes will not present self-intersections. Therefore, the maximum range and procedure this radius is assigned depend on the medial topology:

- *Simple*. In this case, there are no restrictions on the radial range.
- *Multiple*. For branching medial surfaces, special care must be taken at surface self-intersecting points. At these locations, radii have to be below the maximum value that ensures that the medial representation defines a local coordinate change [22]. This maximum value depends on the principal curvatures of the intersecting surfaces [22] and is computed for each surface. Let $\mathcal{M}$ be the medial surface, Z denote the self-intersection points, and $D(Z)$ the distance map to Z. The radial coordinate is assigned as follows:

$$R(X) = \min(R(X), \max(r_Z, D(Z))), \tag{11.13}$$

$R(X)$ being the value of the polynomial function, and r_Z the maximum value allowed at self-intersections. In this manner, we obtain a smooth distribution of the radii ensuring volume integrity.

- *Homotopy*. To be consistent with the third main property of medial surfaces [44], volumes must preserve all holes of medial surfaces. To do so, the maximum radius r_2 is set to be under the minimum of all surface holes radii.

11.3.2 MEDIAL SURFACE QUALITY METRICS

The database can be used to benchmark algorithms using two tests. The first test evaluates the quality of the medial surface generated, whereas the second one explores the capabilities of the generated surfaces to recover the original volume and describing anatomical structures. Surface quality tests start from known medial surfaces, which will be considered as ground truth. From these surfaces volumetric objects can be generated by placing spheres of different radii at each point of the surface. The newly created object is then used as input to several medial surface algorithms and the resulting medial surfaces, compared with the ground truth.

The quality of medial surfaces has been assessed by comparing them to ground truth surfaces in terms of surface distance [24]. The distance of a voxel y to a surface X is given by: $D_X(y) = \min_{x \in X} \|y - x\|$, for $\|\cdot\|$ the Euclidean norm. If we denote by X the reference surface and Y the computed one, the scores considered are:

1. *Standard Surface Distances*:

$$AvD = \frac{1}{|Y|} \sum_{y \in Y} D_X(y) \tag{11.14}$$

$$MxD = \max_{y \in Y}(D_X(y)) \tag{11.15}$$

2. *Symmetric Surface Distances*:

$$AvSD = \frac{1}{|X| + |Y|} \left(\sum_{x \in X} D_Y(x) + \sum_{y \in Y} D_X(y) \right) \tag{11.16}$$

$$MxSD = \max \left(\max_{x \in X}(D_Y(x)), \max_{y \in Y}(D_X(y)) \right) \tag{11.17}$$

Standard distances measure deviation from medialness, while differences between standard and symmetric distances indicate the presence of homotopy artifacts and presence of unnecessary medial segments.

For each family and method, we have computed quality scores statistical ranges as $\mu \pm \sigma$, for μ and σ being the average and standard deviation computed over the 20 samples of each group of shapes. The Wilcoxon signed rank test [64] has been used to detect significant differences across performances.

In medical imaging applications the aim is to generate the simplest medial surface that allows recovering the original volume without losing significant voxels. Volumes recovered from surfaces generated with the different methods are compared with ground truth volumes. Volumes are reconstructed by computing the medial representation [9] with radius given by the values of the distance map on the computed medial surfaces. Ground truth volumes are given by anatomical meshes extracted from original medical scans.

Comparisons with the original anatomical volumes are based on the average and maximum symmetric surface distances ($AvSD$ and $MxSD$ given in (11.16) and (11.17)) respectively, computed using the anatomic boundary surface and reconstructed volume boundaries, as well as the following volumetric measures:

1. *Volume Overlap Error:*

$$VOE(A,B) = 100 \times \left(1 - 2\frac{|A \cap B|}{|A| + |B|}\right) \tag{11.18}$$

2. *Relative Volume Difference:*

$$RVD(A,B) = 100 \times \frac{|A| - |B|}{|B|} \tag{11.19}$$

3. *Dice Coefficient:*

$$Dice(A,B) = \frac{2|A \cap B|}{|A| + |B|} \tag{11.20}$$

for A and B, being respectively the original and reconstructed volumes. Aside from the Dice coefficient, lower metric values indicate better reconstruction capability. Like in the case of the synthetic surfaces, for each medial surface method we have computed quality scores statistical ranges as $\mu \pm \sigma$, for μ, σ computed on the medical data set, and Wilcoxon signed rank tests.

11.4 VALIDATION EXPERIMENTS

Our validation protocol has been applied to the method described in Section 11.2.1. To provide a real scenario for the reconstruction tests we have used 14 livers from the SLIVER07 challenge [24] as a source of anatomical volumes. In order to compare to morphological methods, we have also applied it to an ordered thinning using a 6-connected neighborhood criterion for defining medial surfaces (labeled Th_6) described in [10] and a 26-connected neighborhood surface test (labeled Th_{26}) following [44]. The consistency of surface pruning is tested on a pruned version of the 26-connected neighborhood method (labeled ThP_{26}), which does not allow degenerated medial axis segments, and on the scheme (labeled Tao_6) described in [26], which alternates 6-connected curve and surface thinning with more sophisticated pruning stages.

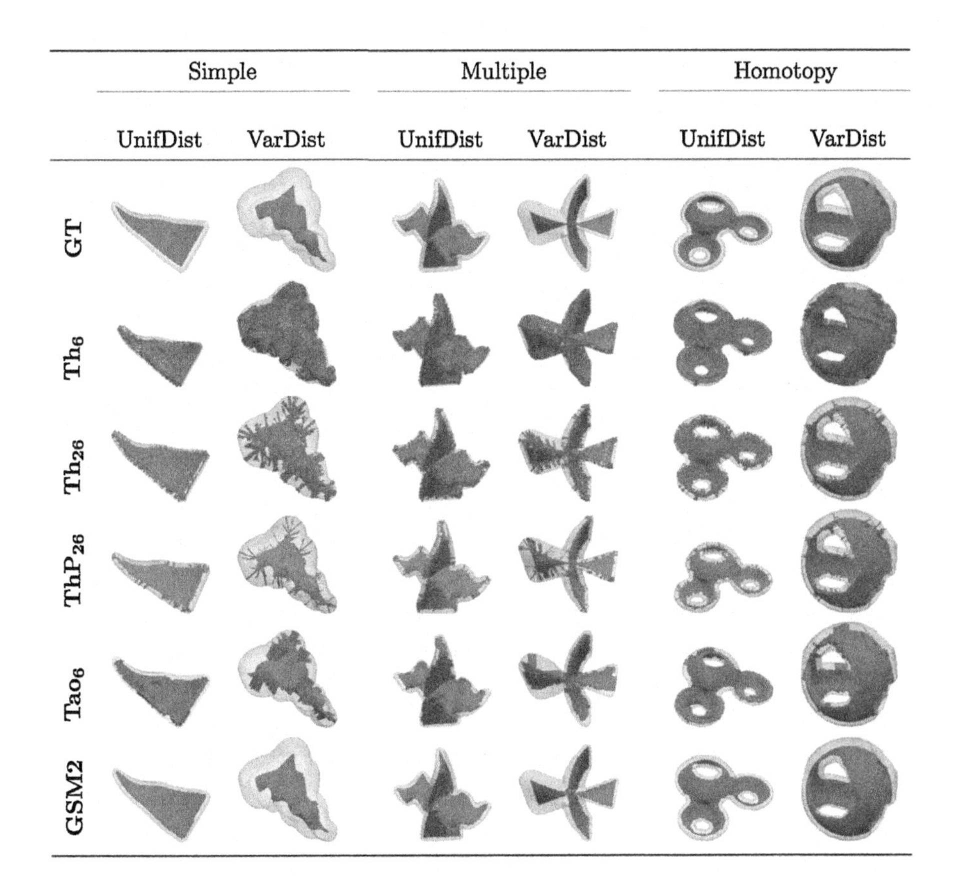

FIGURE 11.4

Medial surfaces. Examples of the compared methods for each synthetic volume family.

11.4.1 MEDIAL SURFACE QUALITY

Fig. 11.4 shows an example of the synthetic volumes in the first row and the computed medial surfaces in the remaining rows. Columns exemplify the different families of volumes generated: one (Simple in 1st and 2nd columns) and two (Multiple in 3rd and 4th columns) foil surfaces, as well as surfaces with holes (Homotopy in 5th and 6th columns). For each kind of topology, we show a volume generated with constant (1st, 3rd, and 5th columns) and variable distance (2nd, 4th, and last columns). We show medial surfaces in solid meshes and the synthetic volume in semitransparent color. The shape of surfaces produced using morphological thinning strongly depends on the connectivity rule used. In the absence of pruning, surfaces, in addition, have either extra medial axes attached or extra surface branches in the case pruning is included as part of the thinning surface tests (Tao_6). On the contrary, GSM2 medial surfaces have a well-defined shape matching the original synthetic surface.

Table 11.1 Error ranges (mean and standard deviation) for the synthetic volumes

	Simple		Multiple		Homotopy		Total
	UnifDist	VarDist	UnifDist	VarDist	UnifDist	VarDist	
GSM2							
AvD	0.28 ± 0.09	0.28 ± 0.07	0.38 ± 0.09	0.43 ± 0.18	0.37 ± 0.18	0.34 ± 0.14	0.34 ± 0.14
MxD	2.99 ± 0.50	3.50 ± 1.53	3.56 ± 0.53	4.76 ± 1.51	3.39 ± 0.48	3.70 ± 0.84	3.65 ± 1.13
AvSD	0.24 ± 0.05	0.25 ± 0.05	0.37 ± 0.32	0.37 ± 0.18	0.29 ± 0.10	0.28 ± 0.08	0.30 ± 0.17
MxSD	3.02 ± 0.46	3.66 ± 1.52	4.10 ± 2.61	4.76 ± 1.51	3.39 ± 0.48	3.70 ± 0.84	3.78 ± 1.52
Th_6							
AvD	1.52 ± 0.27	5.63 ± 2.19	1.66 ± 0.30	3.05 ± 0.75	1.56 ± 0.35	2.96 ± 1.17	2.73 ± 1.80
MxD	5.55 ± 0.26	16.21 ± 4.76	5.82 ± 0.27	10.75 ± 3.40	5.54 ± 0.20	10.17 ± 3.20	9.01 ± 4.72
AvSD	1.04 ± 0.21	4.34 ± 1.94	1.16 ± 0.24	2.24 ± 0.56	1.09 ± 0.28	2.13 ± 0.98	2.00 ± 1.48
MxSD	5.55 ± 0.26	16.21 ± 4.76	5.82 ± 0.27	10.75 ± 3.40	5.54 ± 0.20	10.17 ± 3.20	9.01 ± 4.72
Th_{26}							
AvD	0.85 ± 0.25	3.15 ± 1.34	1.00 ± 0.19	1.89 ± 0.52	0.86 ± 0.37	1.63 ± 0.84	1.56 ± 1.07
MxD	5.51 ± 0.25	16.17 ± 4.78	5.58 ± 0.19	10.64 ± 3.43	5.46 ± 0.25	10.09 ± 3.21	8.91 ± 4.75
AvSD	0.56 ± 0.14	2.02 ± 0.92	0.67 ± 0.12	1.24 ± 0.35	0.59 ± 0.22	1.05 ± 0.56	1.02 ± 0.69
MxSD	5.51 ± 0.25	16.17 ± 4.78	5.58 ± 0.19	10.64 ± 3.43	5.46 ± 0.25	10.09 ± 3.21	8.91 ± 4.75
ThP_{26}							
AvD	0.57 ± 0.20	2.24 ± 1.00	0.70 ± 0.17	1.38 ± 0.37	0.54 ± 0.24	1.11 ± 0.62	1.09 ± 0.79
MxD	5.49 ± 0.27	16.16 ± 4.78	5.58 ± 0.19	10.61 ± 3.43	5.41 ± 0.27	10.08 ± 3.23	8.89 ± 4.76
AvSD	0.41 ± 0.11	1.38 ± 0.61	0.50 ± 0.11	0.92 ± 0.24	0.41 ± 0.12	0.72 ± 0.37	0.72 ± 0.47
MxSD	5.49 ± 0.27	16.16 ± 4.78	5.58 ± 0.19	10.61 ± 3.43	5.41 ± 0.27	10.08 ± 3.23	8.89 ± 4.76
Tao_6							
AvD	0.79 ± 0.21	4.82 ± 2.05	0.86 ± 0.17	2.46 ± 1.09	0.85 ± 0.29	2.48 ± 1.20	2.04 ± 1.79
MxD	4.87 ± 0.20	17.55 ± 5.19	4.92 ± 0.17	11.10 ± 3.71	4.79 ± 0.21	11.64 ± 4.33	9.14 ± 5.68
AvSD	0.51 ± 0.14	3.92 ± 1.73	0.59 ± 0.13	2.00 ± 0.96	0.59 ± 0.27	1.99 ± 1.03	1.60 ± 1.52
MxSD	4.89 ± 0.18	17.55 ± 5.19	5.32 ± 1.42	11.10 ± 3.71	5.53 ± 3.26	11.87 ± 4.25	9.38 ± 5.73

Table 11.1 reports error ranges for the four methods and different types of synthetic volumes and the total errors in the last column. For all methods, there are no significant differences between standard and symmetric distances for a given volume. This indicates a good preservation of homotopy. Even with pruning, thinning has significant geometric artifacts (maximum distances increase) and might drop its performance for variable distance volumes due to a different ordering for pixel removal and type of surface preserved.

According to the Wilcoxon signed rank test, strategies alternating curve and surface thinning with pruning stages have worse average distances than other morphological strategies ($p < 0.0001$ for AvD and $p < 0.0001$ for $AvSD$). Given that maximum distances do not significantly differ ($p = 0.4717$, $p = 0.6932$, $p = 0.7752$ for MxD and $p = 0.9144$, $p = 0.7463$, $p = 0.6669$ for $MxSD$), it indicates an introduction of extra structures of larger size (extra surface branches in Tao_6 for the variable volumes shown in Fig. 11.4).

The performance of GSM2 is significantly better than that of other methods (Wilcoxon signed rank test with $p < 0.0001$), presents high stability across volume geometries, and produces accurate surfaces matching synthetic shapes. The small in-

Table 11.2 Mean and standard deviation of errors in volume reconstruction for each metric

	GSM2	Th_6	Th_{26}	ThP_{26}	Tao_6
Volume Error					
VOE	7.96 ± 1.70	8.84 ± 1.73	8.25 ± 1.72	7.84 ± 1.68	8.49 ± 1.77
RVD	8.49 ± 2.03	9.10 ± 2.10	8.96 ± 2.08	7.86 ± 2.23	5.91 ± 1.99
Dice	$.959 \pm .009$	$.954 \pm .009$	$.957 \pm .009$	$.963 \pm .005$	$.955 \pm .010$
Surface Dist.					
AvSD	0.80 ± 0.06	0.89 ± 0.06	0.83 ± 0.05	0.70 ± 0.11	0.83 ± 0.06
MxSD	5.61 ± 2.68	6.00 ± 2.58	5.52 ± 2.56	5.94 ± 1.45	6.42 ± 2.33

crease in errors for multiple self-crossing surfaces is explained by the presence of holes at intersections between medial manifolds. Still its overall performance clearly surpasses performance of morphological approaches.

11.4.2 RECONSTRUCTION POWER FOR CLINICAL APPLICATIONS

Table 11.2 reports the statistical ranges for all methods and measures computed for the 14 livers. There are no significant differences among methods, and best performers vary depending on the quality measure. However, our approach and the two thinnings ThP_{26} and Tao_6 have an overall better reconstruction power.

The medial surface of a healthy liver obtained with the thinning methods can be seen in Fig. 11.5, and the GSM2 medial surface in Fig. 11.6, left. In the case of thinning-based methods, medial manifolds have a more complex geometry than GSM2 and might include extra structures and self-intersections (Fig. 11.5). In medical applications, such extra structures might hinder the identification of abnormal or pathological structures. This is not the case for GSM2 surfaces as exemplified in Fig. 11.6. The oversized superior lobe on the right liver is captured by the presence of an unusual medial manifold configuration.

11.5 APPLICATION TO COCHLEA REGISTRATION

The cochlea is the inner ear structure that controls the sensation of hearing and balance, and an understanding of the anatomy and anatomical variability plays an important part in utilizing the full potential of cochlear implants [65]. Detailed anatomical models have interesting patient-specific applications since they can provide information about the type of electrode design that best suits the anatomy of the user [63] or allow improvements to the implant programming based on simulations mimicking the actual anatomical and physiological situation [13].

Image registration of the cochlea is challenging for a couple of reasons. The human cochlea is a spiral structure with outer dimensions of approximately $10 \times 8 \times 4$ mm. The size and the shape of the spiral can vary extensively. On average, the

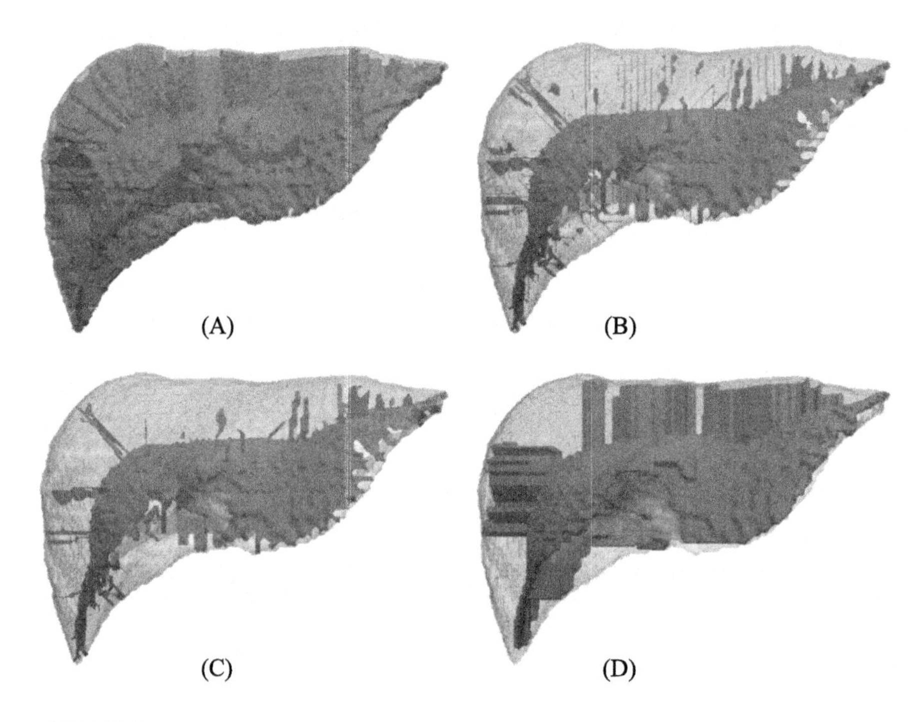

FIGURE 11.5

Medial manifolds of a healthy liver generated with morphological methods. Th_6 (A), Th_{26} (B), ThP_{26} (C), and Tao_6 (D).

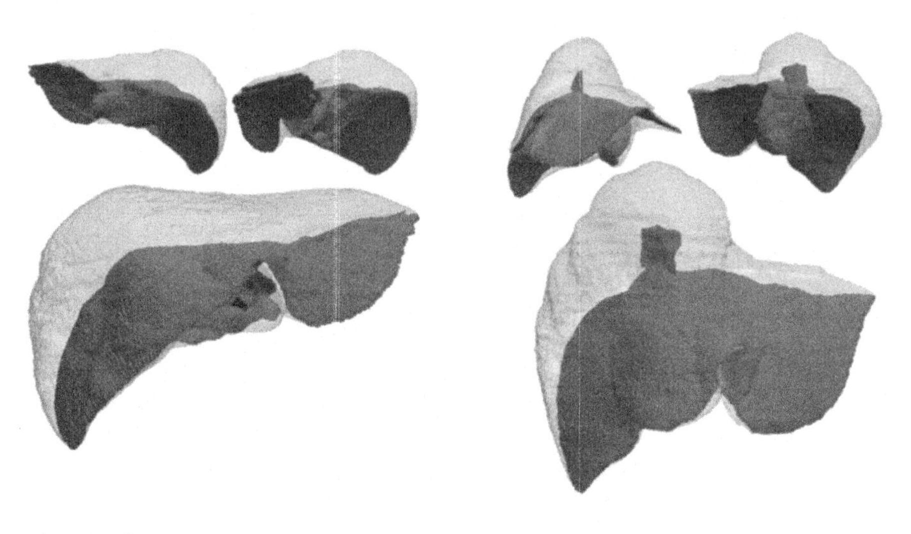

FIGURE 11.6

GSM2 medial manifolds of a healthy liver (left) and a liver with an unusual lobe (right).

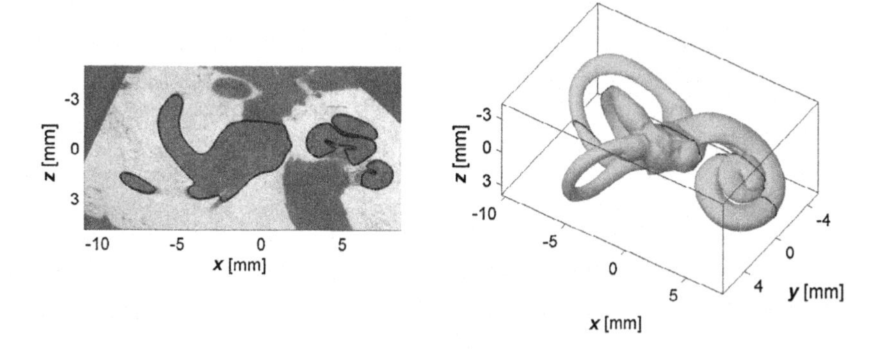

FIGURE 11.7

Left: Impression of the μCT data and segmentation. Notice the small spacing separating the cochlear turns (right side of CT image), the weak contrast toward internal cochlea borders, and the opening into the middle ear cavity (middle of the image). Right: The corresponding surface model provides an overview of the inner ear topology.

cochlea winds 2.6 turns [18] but can approach up to three full turns, corresponding to a difference in the order of 1–2 mm following the path of the spiral. The separation between the cochlear turns is typically one order of magnitude smaller. Deformations to properly align the most apical region of spiral have been difficult to model to our experience. Further, the whole spiral is a tube-like structure (see Fig. 11.7, right) with a large degree of self-similarity in the cross-sections. This lack of distinct features makes it difficult to identify corresponding anatomical positions across samples.

The desired registration model should not just expand or compress the apical part of the spiral to align two samples, but rather model a change along the entire spiral. Essentially, the model should be able to handle very local deformations while still adhering to the global structure of the samples. This type of behavior is usually not native to nonrigid registration models without some kind of prior or regularization included.

Modifications to a registration model to include such prior knowledge have been studied previously. A way of introducing anatomical shape priors is the use of a statistical shape model [7,25]. However, building statistical shape models is in itself a labor intensive task rivaling if not surpassing the task of the registration, as the prerequisite for building the model is data that is already registered to have correspondences.

A multitude of physical constraints have also been proposed as regularizations. For example, local tissue rigidity can be enforced in specified areas [50], or conditions of incompressibility or volume preservation can be applied [45]. However, finding the suitable physical constraint for a registration task is not straightforward, as this is case- and application-dependent.

In [6] an articulated skeleton model was preregistered to intra-mouse data studies in order to recover large pose-differences between data acquisitions. The presented application is narrow in its scope, but the registration methodology of using landmark correspondences as regularization is more generally applicable, and thus we adopt this approach for this work.

In this section, we explore the potential of using the skeleton of the cochlea as anatomical prior in free-form registrations using a B-spline transformation model. The skeleton provides a global description of shape in a simplified and structured form. Matching based on skeleton similarity could provide a global anatomical guidance or regularization to a locally defined free-form image registration procedure with a high resistance to noise compared to using only the image intensity similarity.

The use of skeleton similarity in image registrations should be applicable to many different problems, and there are many published methods and approaches for finding and matching the skeletons for differing types of data and geometries [56,59]. Skeleton correspondence has been seen in image registration tasks before, relating to, for instance, 2D/3D multimodal registration [31] and matching of vessels in time-series angiography data [61]. More related to our approach is the work of [58], where multiple different shape features were calculated from surface objects and transformed into vector-valued 2D feature images, which were aligned with a classic image registration formulation. Skeleton features were used for global alignment in the coarser levels of the registration. Our strategy is similar although the prior will be included into the registration model differently.

11.5.1 MATERIAL AND METHODS

A collection of 17 dried temporal bones from the University of Bern were prepared and scanned with a Scanco Medical μCT100 system. The data was reconstructed and processed to obtain image volumes of 24-micron isotropic voxel sizes containing the inner ear (Fig. 11.7, left).

Image segmentation: The border of the inner ear was segmented in all datasets semiautomatically using ITK-SNAP [68].

The semiautomatic tool in the segmentation software was critical for achieving smooth and rounded segmentations in data with that kind of resolution and for reducing the amount of manual work. A surface model was generated for each dataset using marching cubes [35] followed by a surface reconstruction [41] to obtain a well-formed triangular mesh (Fig. 11.7, right).

11.5.2 SKELETONIZATION

To avoid working with a genus 3 surface, we exclude the vestibular system and focus only on a skeleton of the spiral shaped cochlea. We propose to use a set of corresponding pseudo-landmarks $\mathcal{Z}^{\mathrm{LM}}$ of the cochleae obtained from a parametric "curved skeleton" that will improve registration results.

We compute the medial axis of the skeleton by using the GSM2 method on the binary segmentations of each cochlea. The computed centerline of the cochlea runs close to the *spiral lamina ossea*, an internal feature of the cochlea that is tied to the perception of different frequencies of sound. We manually define the cochlear apex landmark (A_i) at the extreme of the coclea in each dataset.

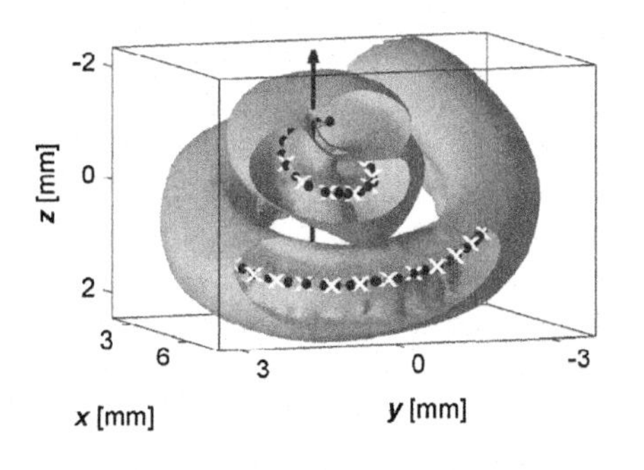

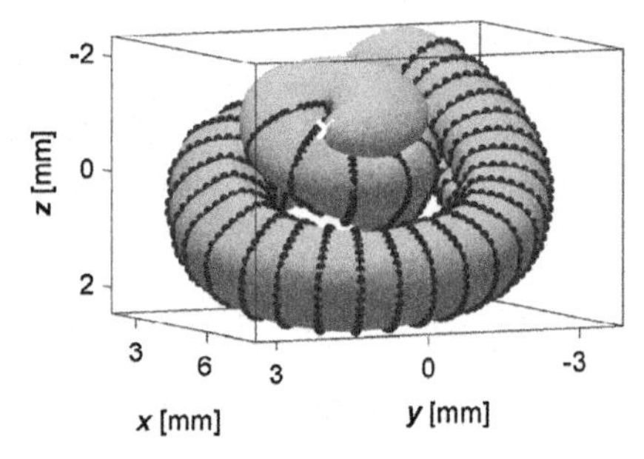

FIGURE 11.8

The cochlear skeletonization. Left: White 'x' annotations are sampled skeleton information. Right: cochlear apex (A_i) in white. Points on the surface represent parametric pseudo-landmarks.

We generate a naive parametric model of the cochlea. First, we create a parametric description of the cochlea skeleton by sampling 37 corresponding positions on the skeleton with equal arc-length ($\mathcal{Z}_i^{\mathrm{S}}$). Secondly, we extract planar surface cross-section at each of the points p in $\mathcal{Z}_i^{\mathrm{S}}$. The cross-section plane is determined by the tangent of the skeleton at p. Each cross-section of the surface mesh is then parameterized using 40 points. These cross-sectional points, together with the apex landmark (A_i), provide a set $\mathcal{Z}_i^{\mathrm{LM}}$ of 1481 corresponding surface pseudo-landmarks (Fig. 11.8, right) to be included in a registration model. Finding the cochlea cross-section in the apical region of the cochlear can potentially lead to some ambiguity since they could intersect with themselves. To avoid this, the skeleton cross-sections in the apical turn were not included.

11.5.3 IMAGE REGISTRATION

The registration procedure follows a common work-flow. One dataset was chosen as the reference, to which the remaining moving datasets were registered in two steps: rigid initialization followed by the deformable registration, both detailed in the following subsections.

11.5.3.1 *Initial Rigid Alignment*

There are many approaches for finding rigid transformations. The chosen procedure is independent from the skeleton information and is the same no matter the chosen deformable registration model. In that way, later comparisons of registration results are not affected by the initialization. The whole initialization procedure relies solely upon the extracted surface meshes, but the calculated rigid transformations were also applied to the gray-scale volumes and their segmentations.

Translation: Let $p_{(i,j)}$ be the jth vertex position of dataset i. A translation was applied so that the center of mass is placed in position (0, 0, 0), i.e., the mean vertex position $\bar{p}_i$ was subtracted from all vertices. This places all datasets in a coordinate system where the inner ear center of mass of each dataset is in the origin.

Rotation: Let Σ_i be the 3×3 covariance matrix of the mesh vertex positions of dataset i (after the translation). The eigenvectors W_i of Σ_i provides a rotation matrix that transforms the data to the principal component directions. This essentially corresponds to fitting an ellipsoid to the point cloud and aligning the axes.

Check directions: This alignment procedure is robust due to the asymmetry of the inner ear shape (Fig. 11.7, right). However, the sign of a principal direction in the ith dataset could potentially be opposite compared to that of the reference. To handle this, we make a simple check. The bounding box of the reference and of the moving point cloud is divided into a coarse grid. We use the sum of squared grid vertex-density difference between the two as a check metric. If the axis-flip results in a lower metric, then the flip is made to the moving dataset. Although there is no guarantee for this to work in all cases, it has worked well for our data. In principle, any kind of rigid alignment could be used instead of the one suggested here.

11.5.3.2 ***Deformable Registration***

The nonrigid image registration follows the formulation and framework of `elastix` [29].

The registration is done between the segmentations rather than the gray-scale volumes for two reasons. First, the μCT data contain smaller artifacts and certain weakly contrasted edges that were handled during the segmentation. Second, the registration should not be influenced by the anatomical differences in the surrounding bone structure.

The registration of the moving dataset I_M toward the reference I_F is formulated as a (parametric) transformation T_μ where the vector μ containing the p-parameters of the transformation model are found as an optimization of a cost function $\mathcal{C}$:

$$\hat{\mu} = \arg\min_{\mu} \mathcal{C}(T_\mu, I_F, I_M) \tag{11.21}$$

The transformation model used in this paper is the cubic B-spline in a multiresolution setting. We apply image smoothing with a Gaussian kernel to both the fixed and moving image. For each level of resolution, the spacing between grid points and the width of the smoothing kernel follow a decreasing scheme, starting with a coarse registration that is gradually refined. The following scheme was chosen by experimentation:

Control point grid spacing (isotropic, voxels)

$$[144, 72, 48, 48, 36, 24, 18, 12, 6].$$

Width of Gaussian kernel (isotropic, voxels)

$$[10, 10, 1, 1, 1, 1, 1, 1, 1].$$

The width of the kernel was deliberately kept narrow in most levels to avoid that small and sharp features would be blurred out (for instance, the separation of the cochlear turns). Two of the levels have the same values to overcome limitations of maximum deformation per step.

The cost function used in this "basic" registration setup:

$$\mathcal{C}_1 = \alpha \cdot \mathcal{S}_{\text{Sim}}(\mu, I_F, I_M) + (1-\alpha) \cdot \mathcal{P}_{\text{BE}}(\mu), \tag{11.22}$$

where α is a weight parameter in the interval [0, 1]. The similarity term $\mathcal{S}_{\text{Sim}}$ is chosen as the sum of squared differences (SSD). The term $\mathcal{P}_{\text{BE}}$ is the energy-bending regularization used to penalize strong changes and foldings in the transformation [46]. The weighting of the similarity term was chosen 0.9 by experimentation. Increasing α would provide more freedom for deformation of the shapes, but also increase the risk of having nonplausible anatomical results.

The optimization is solved using adaptive stochastic gradient descent [28]. The maximum number iterations was set to 2500. To reduce the computational burden of the optimization, only a subset of voxels is sampled for the evaluation. For each iteration, 2^{14} random coordinate points were sampled. These settings were fixed for all resolutions.

11.5.3.3 *Deformable Registration with Guidance from Skeleton*

The free-form registration setup remains largely the same when a skeleton is included to make comparisons fair. The cost function is modified to include a landmark similarity term [6]:

$$\mathcal{C}_2 = \alpha \cdot \mathcal{S}_{\text{Sim}}(\mu, I_F, I_M) + \beta \cdot \mathcal{S}_{\text{CP}}(\mu, \mathcal{Z}_F, \mathcal{Z}_M) + (1-\alpha-\beta) \cdot \mathcal{P}_{\text{BE}}(\mu), \tag{11.23}$$

where α and β are weightings in the interval [0, 1] and fulfilling $\alpha + \beta \leq 1$. The landmark similarity term $\mathcal{S}_{\text{CP}}(\mu, \mathcal{Z}_F, \mathcal{Z}_M)$ uses the Euclidean distance between the set of corresponding landmarks $\mathcal{Z}_F$ and $\mathcal{Z}_M$. In this way, intensity-based image registration is guided with features extracted from the anatomical skeleton (i.e., using $\mathcal{Z}_i^{\text{LM}}$ from Section 11.5.2). By experimentation the weightings were set to $\alpha = 0.8$ and $\beta = 0.11$. The landmark similarity is kept small in order not to force the alignment, and the ratio between image similarity and bending energy regularization is kept similar to the previous setup $\mathcal{C}_1$ (Eq. (11.22)). Settings for the transformation model and optimizer were unchanged from the previous registration model.

11.5.4 EVALUATION

We are interested in comparing the 16 registration results of model 1 (Eq. (11.22)) and model 2 (Eq. (11.23)) using a number of different image- and mesh-based metrics.

Image-based evaluation: Let $I_i(\mu)$ be the moving segmentation volume after application of the resulting transformation. We compare the Dice score [17] to the segmentation of the reference dataset I_{Ref}:

$$\text{DSC} = \frac{2 \cdot \left|I_{\text{Ref}} \bigcap I_i(\mu)\right|}{|I_{\text{Ref}}| + |I_i(\mu)|} \tag{11.24}$$

Mesh-based evaluation: We define the surface-based scores as follows. Let $\mathbb{S}_{\text{Ref}}(\mu)$ be the reference surface mesh after application of the resulting transformation. There is no direct point correspondence between the reference and the ground truth surfaces $\mathbb{S}_i$, and they each contain a varying number of vertices. Metrics are therefore based on the closest points, i.e., the minimum Euclidean distance from a point p to any of the points q in the other surface $\mathbb{S}$:

$$d(p, \mathbb{S}) = \min_{\forall q \in \mathbb{S}} \left(\|p - q\|_2\right). \tag{11.25}$$

The mean surface error $d_{\bar{s}}$ of each sample is defined as the average of all the closest point distances:

$$d_{\bar{s}} = \frac{1}{N_{\text{Ref}} + N_i} \left(\sum_{\forall p \in \mathbb{S}_{\text{Ref}}(\mu)} d(p, \mathbb{S}_i) + \sum_{\forall p \in \mathbb{S}_i} d(p, \mathbb{S}_{\text{Ref}}(\mu)) \right), \tag{11.26}$$

where N_{Ref} and N_i are the total numbers of points in the reference and the moving surface, respectively.

The Hausdorff distance d_H is the maximum of all the closest point distances:

$$d_H = \max \left\{ \max_{\forall p \in \mathbb{S}_{\text{Ref}}(\mu)} d(p, \mathbb{S}_i), \max_{\forall p \in \mathbb{S}_i} d(p, \mathbb{S}_{\text{Ref}}(\mu)) \right\}. \tag{11.27}$$

The above-mentioned metrics are very generic and will hardly be able to reflect and evaluate the change in the registration model that we intend to explore. We therefore include two additional scores, apex error and torque.

First, we calculate the Euclidean distance between apexes of the target data and of the reference:

$$d_A = \left\|A'_{\text{Ref}}(\mu) - A_i\right\|_2. \tag{11.28}$$

The apex is one of the few locations on the cochlea that can be placed relatively precisely. Even though an arc-length distance might be more correct to use, the Euclidean apex error should be indicative of the registration model behavior in the apical region, even though this point is also included in the registration model.

Secondly, we look at the differences in the vector deformation fields obtained by the registration models. The cochlear samples have a different number of turns, and we wish to evaluate the registration models ability to capture this rotational behavior of the anatomy. Our postulation and assumption are that this ability of the registration model should correlate with the "torque" τ on the central axis of the cochlear exerted by the deformation field.

Table 11.3 Statistics of registration evaluation metrics, reported as the mean ± std. Model 1 is the nonrigid image registration model, and Model 2 the nonrigid image registration model incorporating a skeleton prior. Surf. Error, Hausdorff, and Apex Error metrics are expressed in mm, and Avg. Torque in mm^2

	Dice Score	Surf. Error	Hausdorff	Apex Error	Avg. Torque
M1	0.96 ± 0.01	0.040 ± 0.01	0.69 ± 0.24	1.01 ± 0.59	−0.04 ± 0.09
M2	0.95 ± 0.01	0.045 ± 0.01	0.73 ± 0.35	0.69 ± 0.52	−0.53 ± 0.28

Let the force vector $\vec{F}_p$ on the vertex p in the reference mesh be defined simply as the vector between the vertex position before and after application of the registration transformation:

$$\vec{F}_p = p(\mu) - p.$$

Further, we can calculate the perpendicular arm from the central axis to the mesh vertex $\hat{v}_p$. This vector is normalized to unit length, so that the vertices farther from the axis will not contribute with a greater torque.

The scalar projection of the force vector F_p onto the unit arm that is perpendicular to both the central axis and $\hat{v}_p$ is then the acting force contributing to the torque:

$$F_p = \vec{F}_p \cdot \left(\vec{n} \times \hat{v}_p\right).$$

Using this local vertex torque force leads to our definition of the torque of the registration:

$$\tau = \frac{1}{N_{\text{Ref}}} \sum_{\forall p \in \mathbb{S}_{\text{Ref}}} F_p = \frac{1}{N_{\text{Ref}}} \sum_{\forall p \in \mathbb{S}_{\text{Ref}}} (p(\mu) - p) \cdot \left(\vec{n} \times \hat{v}_p\right). \tag{11.29}$$

11.5.5 RESULTS

The registrations were done on a desktop with a quad-core 3.6 GHz processor, 64 GB RAM, running elastix v4.7. The average time per registration was approximately 0.8 hours, and we observed no notable difference in run times or convergence speed between the two registration models.

The statistics of the different metric scores are presented in Table 11.3. Fig. 11.9 elaborates on the sample-wise apex error and torque metric, and Figs. 11.10 and 11.11 show the qualitative difference between the registration models.

The general metrics (DSC, $d_{\bar{s}}$, d_H) show a small decrease in performance accuracy for model 2.

From Fig. 11.9 we observe that the apex errors of model 1 grow more or less proportionally to the discrepancy in cochlear turns. The torque is close to zero on average. These observations reflect that model 1 only adapts very locally and behaves indifferently with regards to the turning of the target shape. That is, the resulting cochlear shapes after registration have little variation in the turns.

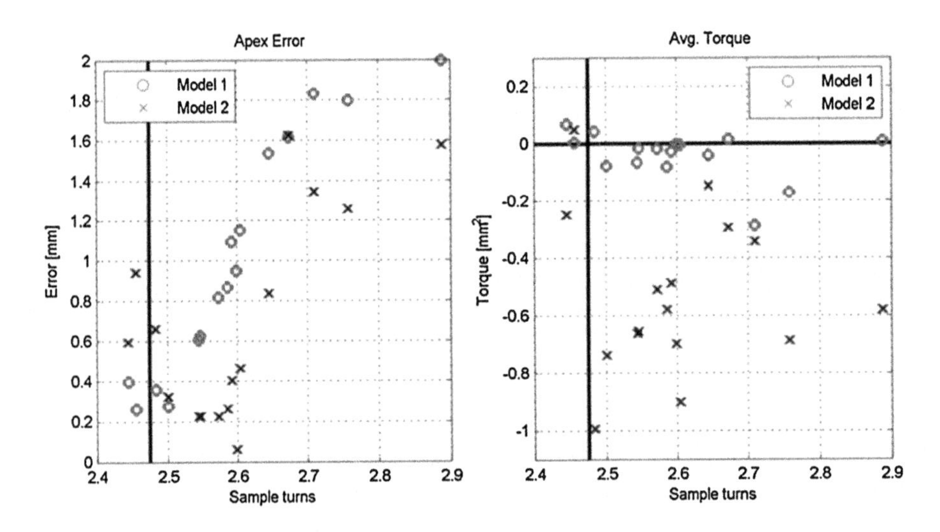

FIGURE 11.9

Sample-wise apex error (Left) and average torque (Right) plotted against the number of cochlear turns of the target samples. Vertical black line indicates the number of turns in the reference sample.

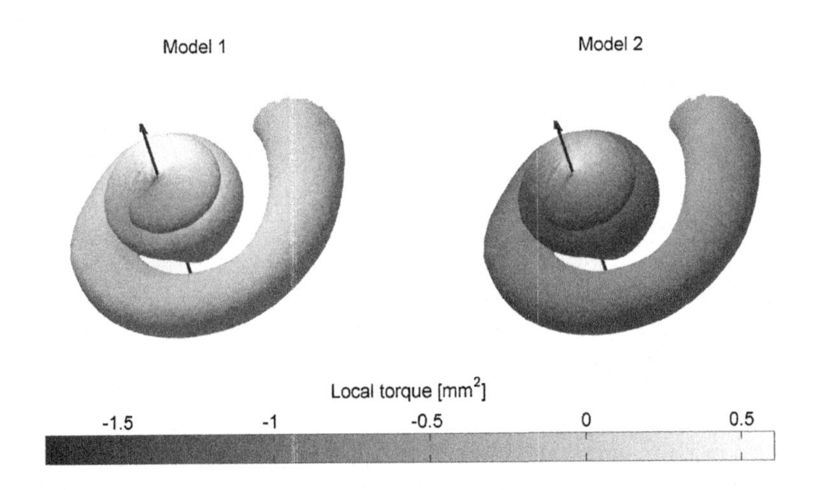

FIGURE 11.10

Qualitative difference in the local torque acting on the cochlea central axis (black vector). The target sample has 2.60 turns, compared to the 2.46 of the reference (the shown surface). Positive direction of the central axis is defined from the cochlea base toward the apex.

FIGURE 11.11

The visual difference between registration models. The reference surface is deformed using either model 1 (right, light-gray) or model 2 (left, dark-gray) to align with the target sample (middle, gray). The surfaces have been moved apart to avoid overlap between shapes.

The apex errors are seen to be generally lower for model 2. Note that the apex landmark used to calculate this error was a part of the optimization procedure. That the error is reduced is therefore no surprise, and it is a biased metric for considering the model accuracy and precision. However, it provides a summarizing pseudo-measure of how much more turning registration model 2 on average is able to capture, which is further illustrated in Fig. 11.11. For very large differences in cochlear turns, it would seem that both registration models have trouble with aligning the apexes.

The torque of model 2 is in most of the cases negative. This indicates vector fields pointing more tangentially in the direction of the spiral toward to the apical region. This would be the expectation as most of the target samples have more turns than the reference. The torque is not a measure of accuracy nor precision. The torque merely provides a simple quantification of the overall rotation of the cochlear shape. It further gives a good way of illustrating the differences between the registration models as demonstrated in Fig. 11.10.

11.6 DISCUSSION

Medial manifolds are powerful descriptors of shapes. The method presented in this chapter allows the computation of medial manifolds without relying in morphological methods on neighborhood or surface tests. Additionally, it can be seamlessly implemented regardless of the dimension of the embedding space.

The performance of our method has been compared to current morphological thinning methods in terms of the quality of medial manifolds and their capability to recover the original volume. For the first experiment, a battery of synthetic shapes covering different medial topologies and volume thickness has been generated. For the second one, we have used a public database of CT volumes of livers, including pathological cases with unusual deformations.

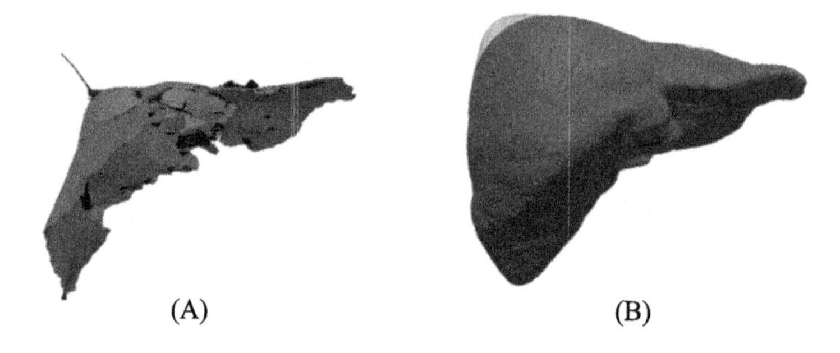

FIGURE 11.12

Impact of pruning in reconstructed volumes: medial manifolds (A) and reconstructed volumes (B).

The proposed method has several advantages over thinning strategies. It performs equally across medial topologies and volume thickness. The resulting medial surfaces are of greater simplicity than the generated by thinning methods. Although having this minimalistic property, the resulting medial manifolds are suitable for locating unusual pathological shapes and properly restoring original volumes. We conclude that our methodology reaches the best compromise between simplicity in geometry and capability for restoring the original volumetric shape.

Any simplification of a medial surface results in a drop in reconstruction quality as illustrated in the images of Fig. 11.12. The images show a medial surface of a liver with a pruned version removing the top branch on the top. Fig. 11.12B shows the volumes reconstructed using the pruned surface (dark color) and the complete one (light color). In this case, the pruned surface cannot reconstruct the external part of the superior lobe of the liver. This drop in accuracy is hard to relate to the simplification process because the branching topology of thinning-based medial manifolds is not always related to the anatomy curvature (concavity–convexity pattern). A main advantage of GSM2 medial surfaces is that their branches are linked to the shape concavities due to the geometrical and normalized nature of the operator. In this context, GSM2 manifolds can be simplified (pruned) ensuring that the loss of reconstruction power will be minimum [62].

Regarding computational efficiency, our method is up to 5 times faster than thinning strategies. Unlike parallelization of topological strategies that require special treatment of topological constrains [5,39], our code is straightforward to parallelize, even on GPU.

We have shown also how medial information can be used to improve registration of complex anatomy, in this case, the cochlea. The Dice score, surface error, and Hausdorff distance serve as very general metrics for evaluating the local adaptability of the registration models. Further, they indicate the general accuracy and precision that we are achieving with the data. The performance with model 2 was decreased on these scores. It would seem that we are trading some local adaption for guiding the model with the landmarks. The determination of the skeleton inherently carries some uncertainties. By introducing the landmarks into the registration model extra noise is added to the procedure. It may happen that a poor skeleton estimate is drawing the spiral in the wrong direction. By providing a more robustly determined skeleton that additionally could fully reach the most apical turn, we expect that the performance of model of 2 could be increased.

Aspects of the skeletonization and its influence can be studied furthermore. For instance, the number of landmarks used to represent the skeleton. By experimentation we found an amount of cross-sections that works, but the number of landmarks per cross-section could potentially be reduced. However, the primary concern is the current lack of information in the most apical cochlear turn. For this to be included, it would be interesting to look into other skeletal representations. That would in turn potentially require a different way of measuring the similarity of skeletons and possibly an extension to the registration framework to accommodate this. It holds an interesting research potential since both the field of skeletonization and image registration are well-researched areas, but so far joining the two have received little focus. A reason might be the challenge in automatically obtaining consistent skeletons from volumetric data. In this work the skeletons were based on the surface models (i.e., the data segmentation), which in many cases are difficult and/or time-consuming to obtain. Ideally, the skeletons should be extracted from volumetric gray-scale data similarly to the works [1,4].

Using the B-spline grid as the transformation model in the registration has limitations. Choosing a fluid- or optical flow-based model [38] could potentially be more suited for this kind of spiral anatomy. Alternatively, the performance of the B-spline approach can perhaps be improved with some data preprocessing. If the cochlea were unfolded, possibly based on the skeleton cross-sections, then it would be in a space more suited for a B-spline grid transformation. Along the same line of thinking, the deformation control points could be placed in a noncubic grid structure favoring the spiral nature of the data. However, these suggestions may be difficult to realize and involve adapting the registration method to one very specific task or anatomy. In this and potentially other cases, finding a skeleton and including it into the a registration model may be an easier or more feasible approach. The results reflect that it is possible to modify and regularize the registration by using skeleton similarity as a prior, even though there is room for improvements in our methodology.

The registration parameters used in this work were manually determined. A set of parameters that works well on all data samples while running within a reasonable time frame can be difficult to find. Regarding the choice of metric weights, an interval of $\alpha = 0.7$–0.9 would seem to be the most appropriate for model 1. Higher α increases the flexibility of the model, which is needed for capturing the cochlear turning. However, increasing beyond 0.9 made some cases fail. In particular, the behavior of the deformations in the semicircular canals performed poorly. The same holds for model 2. For having a fair comparison between the registration models, the same relative weight of the image similarity and bending energy metric was kept. Having $\beta < 0.15$ was found to be reasonable. Forcing more weight on the landmarks could result in too strong deformations in some cases, and going much lower counters the idea of having the landmarks. Variable metric weights throughout the resolutions were also tested for model 2, that is, a scheme where a strong weighting was placed on the landmarks in the initial resolutions and then gradually reduced. It worked well in some cases only, so to keep the registration models comparable, the fixed weightings scheme was used. Regarding the optimization only, the default optimizer and automatically determined settings were used. A number of samples in the range of 2^{14}–2^{17} and a maximum number of iterations between 1000–2500 seemed to produce stable results. Tweaking of registration parameters could result in minor changes of the performance scores, but the same tendencies of the registration models would be observed.

The local torque forces (Fig. 11.10) provides the most qualitative view of the differences between the registration models. There is no ground truth torque, but it illustrates that the normal registration model is very local in its adaption, whereas model 2 provides more turning in the region where the skeleton is defined. Ideally, we could have shown a more convincingly stronger negative correlation (Fig. 11.9) between the differences in the cochlear turns and the average size of the torque. However, we have a low number of samples, and the registration also has to deal with general differences in the size and orientation of the samples apart from the turning. In future work, the torque could perhaps even be used as a regularization in the registration model, where it could favor a constant torque in the B-spline grid points near the spiral.

ACKNOWLEDGMENTS

This research has been supported by the European Commission (GA304857, HEAR-EU), the Spanish Ministry of Economy and Competitiveness (TIN2013-47913-C3-1-R, DEFENSE, and DPI2015-65286-R) and the Generalitat de Catalunya (2014-SGR-1470 and Serra Húnter Program).

REFERENCES

[1] Sasakthi S. Abeysinghe, Matthew Baker, Wah Chiu, Tao Ju, Segmentation-free skeletonization of grayscale volumes for shape understanding, in: IEEE International Conference on Shape Modeling and Applications 2008, Proceedings, 2008, pp. 63–71.

[2] Nina Amenta, Marshall Bern, Surface reconstruction by Voronoi filtering, Discrete Comput. Geom. 22 (1998) 481–504.

[3] Nina Amenta, Sunghee Choi, Ravi Krishna Kolluri, The power crust, unions of balls, and the medial axis transform, Comput. Geom. 19 (2–3) (2001) 127–153.

[4] Emilio Antúnez, Leonidas Guibas, Robust extraction of 1D skeletons from grayscale 3D images, in: Proceedings – International Conference on Pattern Recognition, 2008. ISSN 10514651.

[5] B. Gilles, A parallel thinning algorithm for medial surfaces, Pattern Recognit. Lett. 16 (9) (1995) 979–986.

[6] Martin Baiker, Marius Staring, Clemens W.G.M. Löwik, Johan H.C. Reiber, Boudewijn P.F. Lelieveldt, Automated registration of whole-body follow-up microCT data of mice, Med. Image Comput. Comput. Assist. Interv. 14 (Pt 2) (2011) 516–523, http://dx.doi.org/10.1007/978-3-642-23629-7_63, ISSN 03029743, 16113349.

[7] Floris F. Berendsen, Uulke A. van der Heide, Thomas R. Langerak, Alexis N.T.J. Kotte, Josien P.W. Pluim, Free-form image registration regularized by a statistical shape model: application to organ segmentation in cervical MR, Comput. Vis. Image Underst. 117 (9) (2013) 1119–1127, http://dx.doi.org/10.1016/j.cviu.2012.12.006, ISSN 10773142, 1090235x, 1090235X.

[8] J. Bigun, G. Granlund, Optimal orientation detection of linear symmetry, in: Proc. of the IEEE 1st International Conference on Computer Vision, 1987, pp. 433–438.

[9] H. Blum, A Transformation for Extracting Descriptors of Shape, MIT Press, 1967.

[10] S. Bouix, K. Siddiqi, Divergence-based medial surfaces, in: ECCV, 2000, pp. 603–618.

[11] Sylvain Bouix, Kaleem Siddiqi, Allen Tannenbaum, Flux driven automatic centerline extraction, Med. Image Anal. 9 (3) (2005) 209–221.

[12] J. Canny, A computational approach to edge detection, IEEE Trans. Pattern Anal. Mach. Intell. 8 (1986) 679–698.

[13] Mario Ceresa, Nerea Mangado Lopez, Hector Dejea Velardo, Noemi Carranza Herrezuelo, Pavel Mistrik, Hans Martin Kjer, Sergio Vera, Rasmus Reinhold Paulsen, Miguel Ángel González Ballester, Patient-specific simulation of implant placement and function for cochlear implantation surgery planning, in: Proc. MICCAI, 2014, pp. 49–56.

[14] S. Chang, Extracting skeletons from distance maps, Int. J. Comput. Sci. Netw. Secur. 7 (7) (2007).

[15] Jessica R. Crouch, Stephen M. Pizer, Senior Member, Edward L. Chaney, Yu chi Hu, Gig S. Mageras, Marco Zaider, Automated finite-element analysis for deformable registration of prostate images, IEEE Trans. Med. Imaging 26 (10) (Oct. 2007) 1379–1390.

[16] Tamal K. Dey, Wulue Zhao, Approximate medial axis as a Voronoi subcomplex, in: Proc. of the Seventh ACM Symposium on Solid Modeling and Applications, ISBN 1-58113-506-8, 2002, pp. 356–366.

[17] Lee R. Dice, Measures of the amount of ecologic association between species, Ecology 26 (3) (1945) 297–302, ISSN 00129658. URL http://www.jstor.org/stable/1932409.

[18] Elsa Erixon, Herman Högstorp, Karin Wadin, Helge Rask-Andersen, Variational anatomy of the human cochlea: implications for cochlear implantation, Otology & Neurotology 30 (1) (2009) 14–22, http://dx.doi.org/10.1097/MAO.0b013e31818a08e8.

[19] W.T. Freeman, E.H. Adelson, The design and use of steerable filters, IEEE Trans. Pattern Anal. Mach. Intell. 13 (9) (1991) 891–906.

[20] J. Giesen, B. Miklos, M. Pauly, C. Wormser, The scale axis transform, in: SCG, 2009, pp. 106–115.

[21] Debora Gil, Petia Radeva, Extending anisotropic operators to recover smooth shapes, Comput. Vis. Image Underst. (ISSN 1077-3142) 99 (1) (July 2005) 110–125.

[22] A. Gray, Tubes, Birkhäuser, 2004.

[23] R. Haralick, Ridges and valleys on digital images, CVGIP, Graph. Models Image Process. 22 (10) (1983) 28–38.

[24] T. Heimann, B. van Ginneken, et al., Comparison and evaluation of methods for liver segmentation from CT datasets, IEEE Trans. Med. Imaging 28 (8) (2009) 1251–1265.

[25] Tobias Heimann, Hans-Peter Meinzer, Statistical shape models for 3D medical image segmentation: a review, Med. Image Anal. 13 (4) (2009) 543–563, http://dx.doi.org/10.1016/j.media.2009.05.004. ISSN 13618415, 13618423.
[26] Tao Ju, Matthew L. Baker, Wah Chiu, Computing a family of skeletons of volumetric models for shape description, in: Geometric Modeling and Processing, 2007, pp. 235–247.
[27] F. Khalifa, A. El-Baz, G. Gimel'farb, R. Ouseph, Shape-appearance guided level-set deformable model for image segmentation, in: ICPR, 2010.
[28] Stefan Klein, Josien P.W. Pluim, Marius Staring, Max A. Viergever, Adaptive stochastic gradient descent optimisation for image registration, Int. J. Comput. Vis. 81 (3) (2009) 227–239, http://dx.doi.org/10.1007/s11263-008-0168-y. ISSN 09205691, 15731405.
[29] Stefan Klein, Marius Staring, Keelin Murphy, Max A. Viergever, P.W. Josien, Pluim. elastix: a toolbox for intensity-based medical image registration, IEEE Trans. Med. Imaging 29 (1) (January 2010) 196–205.
[30] Tony Lindeberg, Feature detection with automatic scale selection, Int. J. Comput. Vis. 30 (2) (1998) 79–116.
[31] A. Liu, E. Bullitt, S.M. Pizer, 3D/2D registration via skeletal near projective invariance in tubular objects, Med. Image Comput. Comput. Assist. Interv. – MICCAI'98 1496 (1998) 952–963. ISSN 03029743.
[32] Xiaofeng Liu, Marius G. Linguraru, Jianhua Yao, Ronald M. Summers, Organ pose distribution model and an MAP framework for automated abdominal multi-organ localization, in: Medical Imaging and Augmented Reality, in: Lect. Notes Comput. Sci., vol. 6326, 2010, pp. 393–402.
[33] L.L. Dinguraru, J.A. Pura, et al., Multi-organ segmentation from multi-phase abdominal CT via 4D graphs using enhancement, shape and location optimization, in: MICCAI, vol. 13, 2010, pp. 89–96.
[34] A.M. Lopez, F. Lumbreras, J. Serrat, J.J. Villanueva, Evaluation of methods for ridge and valley detection, IEEE Trans. Pattern Anal. Mach. Intell. 21 (4) (1999) 327–335.
[35] William E. Lorensen, Harvey E. Cline, Marching cubes: a high resolution 3D surface construction algorithm, SIGGRAPH Comput. Graph. (ISSN 0097-8930) 21 (4) (August 1987) 163–169, http://dx.doi.org/10.1145/37402.37422, URL http://doi.acm.org/10.1145/37402.37422.
[36] G. Malandain, S. Fernández-Vidal, Euclidean skeletons, Image Vis. Comput. 16 (5) (1998) 317–327.
[37] N. Faraj, J.-M. Thiery, T. Boubekeur, Progressive medial axis filtration, in: SIGGRAPH, 2013.
[38] Francisco P.M. Oliveira, Joao Manuel, R.S. Tavares, Medical image registration: a review, Comput. Methods Biomech. Biomed. Eng. 17 (2) (2014) 73–93, http://dx.doi.org/10.1080/10255842.2012.670855. ISSN 10255842, 14768259.
[39] Kálmán Palágyi, Attila Kuba, A parallel 3d 12-subiteration thinning algorithm, Graph. Models Image Process. 61 (4) (1999) 199–221.
[40] H. Park, P.H. Bland, C.R. Meyer, Construction of an abdominal probabilistic atlas and its application in segmentation, IEEE Trans. Med. Imaging 22 (4) (2003) 483–492.
[41] R.R. Paulsen, J.A. Baerentzen, R. Larsen, Markov random field surface reconstruction, IEEE Trans. Vis. Comput. Graph. (ISSN 1077-2626) 16 (4) (July 2010) 636–646, http://dx.doi.org/10.1109/TVCG.2009.208.
[42] T. Peters, K. Cleary (Eds.), Image-Guided Interventions: Technology and Applications, Springer, 2008.

[43] Stephen M. Pizer, P. Thomas Fletcher, Joshi Sarang, A.G. Gash, J. Stough, A. Thall, Gregg Tracton, Edward L. Chaney, A method and software for segmentation of anatomic object ensembles by deformable M-Reps, Med. Phys. 32 (5) (2005) 1335–1345.
[44] C. Pudney, Distance-ordered homotopic thinning: a skeletonization algorithm for 3D digital images, Comput. Vis. Image Underst. 72 (2) (1998) 404–413.
[45] T. Rohlfing, C.R. Maurer, D.A. Bluemke, M.A. Jacobs, Volume-preserving nonrigid registration of MR breast images using free-form deformation with an incompressibility constraint, IEEE Trans. Med. Imaging 22 (6) (2003) 730–741, http://dx.doi.org/10.1109/TMI.2003.814791. ISSN 02780062, 1558254x.
[46] Daniel Rueckert, L.I. Sonoda, C. Hayes, D.L.G. Hill, M.O. Leach, D.J. Hawkes, Nonrigid registration using free-form deformations: application to breast MR images, IEEE Trans. Med. Imaging 18 (8) (1999) 712–721, http://dx.doi.org/10.1109/42.796284. ISSN 02780062, 1558254x.
[47] H.M. Sabry, A.A. Farag, Robust skeletonization using the fast marching method, in: ICIP, vol. 1, 2005, pp. 437–440.
[48] Damian J. Sheehy, Cecil G. Armstrong, Desmond J. Robinson, Shape description by medial surface construction, IEEE Trans. Vis. Comput. Graph. 2 (1996) 62–72, http://doi.ieeecomputersociety.org/10.1109/2945.489387.
[49] Kaleem Siddiqi, Sylvain Bouix, Allen Tannenbaum, Steven W. Zucker, Hamilton–Jacobi skeletons, Int. J. Comput. Vis. 48 (3) (2002) 215–231.
[50] Marius Staring, Stefan Klein, Josien P.W. Pluim, A rigidity penalty term for nonrigid registration, Med. Phys. 34 (11) (2007) 4098–4108, http://dx.doi.org/10.1118/1.2776236. ISSN 00942405.
[51] Joshua V. Stough, Robert E. Broadhurst, Stephen M. Pizer, Edward L. Chaney, Regional appearance in deformable model segmentation, in: Proc. of the 20th Int. Conf. on Information Processing in Medical Imaging, vol. 4584, 2007, pp. 532–543.
[52] M. Styner, G. Gerig, et al., Statistical shape analysis of neuroanatomical structures based on medial models, Med. Image Anal. 7 (3) (Sep. 2003) 207–220.
[53] M. Styner, J.A. Lieberman, et al., Boundary and medial shape analysis of the hippocampus in schizophrenia, Med. Image Anal. 8 (3) (2004) 197–203.
[54] H. Sun, A.F. Frangi, H. Wang, et al., Automatic cardiac MTI segmentation using a biventricular deformable medial model, in: MICCAI, vol. 6361, Springer, 2010, pp. 468–475.
[55] Hui Sun, Brian B. Avants, Alejandro F. Frangi, Federico Sukno, James C. Gee, Paul A. Yushkevich, Cardiac medial modeling and time-course heart wall thickness analysis, in: MICCAI, in: Lect. Notes Comput. Sci., vol. 5242, 2008, pp. 766–773.
[56] H. Sundar, D. Silver, N. Gagvani, S. Dickinson, Skeleton based shape matching and retrieval, in: SMI 2003: Shape Modeling International 2003, Proceedings, 2003, pp. 130–139.
[57] Stina Svensson, Ingela Nyström, Gabriella Sanniti di Baja, Curve skeletonization of surface-like objects in 3D images guided by voxel classification, Pattern Recognit. Lett. 23 (12) (2002) 1419–1426.
[58] Lisa Tang, Ghassan Hamarneh, SMRFI: shape matching via registration of vector-valued feature images, in: IEEE Conference on Computer Vision and Pattern Recognition (CVPR), 2008, pp. 1–8. ISSN 10636919.
[59] Johan W.H. Tangelder, Remco C. Veltkamp, A survey of content based 3D shape retrieval methods, in: Proceedings – Shape Modeling International SMI 2004, 2004, pp. 145–156.
[60] A. Telea, J. van Wijk, An augmented fast marching method for computing skeletons and centerlines, in: Symposium on Data Visualisation, 2002, pp. 251–259.

[61] B.C.S. Tom, S.N. Efstratiadis, A.K. Katsaggelos, Motion estimation of skeletonized angiographic images using elastic registration, IEEE Trans. Med. Imaging (ISSN 0278-0062) 13 (3) (Sep. 1994) 450–460.
[62] Sergio Vera, Miguel A. González, et al., Optimal medial surface generation for anatomical volume representations, in: Lect. Notes Comput. Sci., vol. 7601, 2012, pp. 265–273.
[63] Sergio Vera, Frederic Perez, Clara Balust, Ramon Trueba, Jordi Rubió, Raul Calvo, Xavier Mazaira, Anandhan Danasingh, Livia Barazzetti, Mauricio Reyes, Mario Ceresa, Jens Fagertun, Hans Martin Kjer, Rasmus Reinhold Paulsen, Miguel Ángel González Ballester, Patient specific simulation for planning of cochlear implantation surgery, in: Lecture Notes in Computer Science, Springer, ISBN 978-3-319-13908-1, 2014, pp. 101–108.
[64] F. Wilcoxon, Individual comparisons by ranking methods, Biom. Bull. 1 (6) (1945) 80–83.
[65] Blake S. Wilson, Michael F. Dorman, Cochlear implants: a remarkable past and a brilliant future, Hear. Res. 242 (1–2) (2008) 3–21, http://dx.doi.org/10.1016/j.heares.2008.06.005. ISSN 03785955, 18785891.
[66] J. Yao, R.M. Summers, Statistical location model for abdominal organ localization, Med. Image Comput. Comput. Assist. Interv. 12 (Pt 2) (2009) 9–17.
[67] P.A. Yushkevich, H. Zhang, J.C. Gee, Continuous medial representation for anatomical structures, IEEE Trans. Med. Imaging 25 (12) (2006) 1547–1564.
[68] Paul A. Yushkevich, Joseph Piven, Heather C. Hazlett, Rachel G. Smith, Sean Ho, James C. Gee, Guido Gerig, User-guided 3D active contour segmentation of anatomical structures: significantly improved efficiency and reliability, NeuroImage 31 (3) (2006) 1116–1128.
[69] Paul A. Yushkevich, Hui Zhang, Tony J. Simon, James C. Gee, Structure-specific statistical mapping of white matter tracts, NeuroImage 41 (2) (2008) 448–461.

CHAPTER

12 Skeleton-based fast, fully automated generation of vessel tree structure for clinical evaluation of blood vessel systems

Kristína Lidayová*, Hans Frimmel†, Chunliang Wang‡, Ewert Bengtsson*, Örjan Smedby‡

*Centre for Image Analysis, Division of Visual Information and Interaction, Uppsala University, Uppsala, Sweden**

Division of Scientific Computing, Uppsala University, Uppsala, Sweden†

School of Technology and Health, KTH Royal Institute of Technology, Stockholm, Sweden‡

Contents

12.1 Introduction ... 346

12.2 Medical Context ... 347

12.2.1 Clinical Evaluation of the Peripheral Arterial System ... 347

12.2.2 Need for Image Postprocessing ... 349

12.3 Background ... 350

12.3.1 Previous Methods ... 350

12.3.2 From Colon to Blood Vessels ... 352

12.4 Fast Skeleton-Based Generation of the Centerline Tree ... 354

12.4.1 Parameter Selection ... 355

Fixed Parameters ... 355

Changeable Parameters ... 357

12.4.2 The Artery Nodes Detection ... 359

12.4.3 The Artery Nodes Connection ... 364

12.4.4 Anatomy-Based Analysis ... 365

12.4.5 Vascular Segmentation Based on the Vascular Centerline Tree ... 366

12.5 Validation of the Vascular Centerline Tree ... 367

12.5.1 Computed Angiographic Images of the Lower Limbs ... 368

12.5.2 Evaluation of Automatically Selected Changeable Parameters ... 368

12.5.3 Reference and United Skeleton ... 369

Process of Preparing the Reference Skeleton ... 370

Skeletonization. DOI: 10.1016/B978-0-08-101291-8.00014-6

Process of Preparing the United Skeleton 370
12.5.4 Skeleton Evaluation 371
Overlap and Detection Metrics 371
Distance Error Metric 374
12.5.5 Computation Time 374
12.6 Adapting the Algorithm to Other Applications 375
12.7 Discussion and Conclusion 376
12.8 Future Development of the Method 379
Acknowledgments 379
References 379

12.1 INTRODUCTION

Segmentation of tubular structures, such as blood vessels, is a specific segmentation challenge. Thanks to their characteristic tubular shape, the segmentation can be facilitated and accelerated by first identifying the centerlines, where finding the centerlines of vessels can be seen as a special case of an image skeleton extraction algorithm. A common way of generating a vessel skeleton is to first obtain a volumetric segmentation of the vasculature and later perform a thinning process where *simple points* are removed until the 1D representation of the 3D object is reached. Skeletons created this way should fulfill the criteria of being a subset of the original object, one-voxel wide, being topologically equivalent to the object and allowing reconstruction of the object [9,26].

However, in this chapter, we work with the reversed idea of constructing the skeleton straight from the medical images and subsequently use this skeleton to obtain the volumetric segmentation of the vasculature. In this chapter, we use the term skeleton to refer to a rather concise representation of the vasculature that consists of a set of connected polyline segments. Our skeleton fulfills the criteria of having the same topology as the vascular tree, being thin and roughly centrally located. More precisely, the polyline vertices are centrally located within the blood vessel; however, the connection between them is a straight line and does not necessarily lie in the center of the vessel. The resulting centerline is a subset of the original object, and it is possible to reconstruct the vascular tree from it. However, the difference is in the reconstruction method, which needs to be more advanced than a simple distance-based expansion. The reconstruction method we are using to reproduce an accurate segmentation of the vascular tree combines level sets with an implicit 3D model of the vessels [54]. The method takes an approximated centerline tree as an input and generates a 3D vessel model. As a byproduct, it returns also a new centerline tree, which is now properly centered within the vessel.

Our skeleton extraction method constructs a vascular skeleton from medical data acquired by computed tomography (CT) scanner. The method is based on the idea of reducing the dimensionality of the problem before performing the actual analysis. Although the input data is a 3D volume, the problem is simplified into a set of 1D processes. Such an approach reduces the computational time by an order of magnitude compared to the commonly used seed-point generation and propagation centerline extraction methods. In addition, our method is fully automatic and allows computing multiple centerlines in parallel. This idea was originally presented for colon centerline extraction [15,16] and is currently part of a computer-aided diagnosis system for computed tomography colonoscopy [24].

12.2 MEDICAL CONTEXT

The clinically most important disease of the vascular system is atherosclerosis, which is the pathological process underlying most cases of myocardial infarction, peripheral vascular disease [36], and cerebrovascular disease (stroke). It is currently controversial whether atherosclerosis should be seen as a lipid storage disorder or an inflammatory disease [29], but the disease is characterized by a thickening of the arterial wall resulting in a reduction of the lumen (internal cavity) of the arteries with impairment of the blood flow [46]. In late stages of atherosclerosis, it is also commonly associated with calcifications in the arterial wall.

The term peripheral artery disease is used to designate atherosclerosis of the abdominal aorta, iliac, and lower-extremity arteries, leading to stenosis (narrowing) or occlusion of the vessels [36]. It has been found to be present in 12% of the adult population, in both sexes [23], and is thus a major public health problem. In older age groups, it is more common, up to around 40% in selected populations [34]. The first clinical symptom is a pain in the legs upon exercise (claudication), caused by insufficient circulation to the muscles (ischaemia), which can later progress to rest pain (critical ischaemia), and in advanced cases, there may occur ulcerations or gangrene, which sometimes necessitate amputation.

12.2.1 CLINICAL EVALUATION OF THE PERIPHERAL ARTERIAL SYSTEM

A clinically working physician can usually establish the diagnosis of peripheral artery disease by history and clinical examination, including blood pressure measurements in both arms and ankles, which are often summarized as a ratio (ankle brachial index) [23]. However, whenever drug and life-style interventions are not adequate and revascularization therapy (either surgical or through a minimally invasive catheter procedure [19]) is considered, a detailed description of the state of each major artery in the arterial tree of the lower arteries is needed. Such a description is offered by several medical imaging methods.

The oldest way of imaging the arteries is by invasive *catheter angiography*, which involves the introduction of a catheter into the arterial system and the injection of an iodinated radiopaque contrast medium. Nowadays the image acquisition is always digital, and subtraction of precontrast images is performed, hence the term *digital subtraction angiography* (DSA). It offers superior spatial resolution but gives a two-dimensional projection of the three-dimensional anatomy (with the exception of rotational angiography, which is mostly used for intracranial vessels [50]), and its clinical use is limited by potential medical complications of the catheter and contrast medium [22]. Costs also tend to be greater than for alternative methods. In most modern hospitals, invasive angiography is almost exclusively performed in immediate association with a catheter-based interventional procedure. An advantage is a possibility to supplement the imaging with pressure measurements carried out through the catheter, which are often considered to be the most reliable indication of the haemodynamic importance of a stenosis [2].

Ultrasonography is capable of identifying not only stenoses but also arterial wall thickening [37]. With the use of Doppler flow measurements, the resulting flow disturbance may also be quantified [18]. An important advantage is the absence of ionizing radiation and potentially toxic contrast media. In clinical practice, ultrasound examinations are commonly used for selecting candidates for revascularization, but they are less useful for planning the intervention since the presentation modes do not readily correspond to what the surgeon or interventionist will see during the procedure. For clinical follow-up after intervention, ultrasonography plays an important role. A completely different application of ultrasound is the use of intravascular techniques for vessel wall imaging via a catheter in conjunction with interventional procedures [57].

In *computed tomography angiography* (CTA), X-rays and iodine-based contrast media are employed just like in catheter angiography, but the injection is made intravenously, which reduces patient discomfort and complication risks, and the image acquisition is performed by CT, resulting in a 3D (voxel) dataset [41]. With modern multislice CT scanners and spiral acquisition, the voxels are approaching isotropic (cubic) shape. By carefully timing the acquisition to the first passage of the contrast medium through the arteries, a selective visualization of the arterial tree is obtained. In addition to the arterial phase, image acquisition is sometimes performed in the venous phase. The examination is fast and rather easily standardized and may, in addition to the vessels, provide information about the surrounding tissues. The main factors limiting the use of CTA are the radiation dose, which is, of course, multiplied when several phases are acquired, and potential adverse effects of the contrast medium, in particular in patients with impaired renal function. In elderly patients with normally functioning kidneys, however, these factors are a minor concern.

Magnetic resonance angiography (MRA) can be performed with several techniques. The currently most popular technique resembles CTA in that it uses intravenous contrast injection and image acquisition synchronized to the arterial phase and possibly one or more venous phases [40,39]. An advantage is the fact that the patient is not exposed to any ionizing radiation, and the number of phases acquired may be increased without any additional risk. In MRA, the contrast medium is gadolinium-based, and complications are less frequent than with iodine-based contrast-agents. However, in patients with severe impairment of the renal function, a rare, serious complication called nephrogenic systemic fibrosis (NSF) may occur [49]. To avoid injecting gadolinium-based contrast agents in these patients, a number of noncontrast-based MRA methods are available [31].

The tomographic angiography techniques CTA and MRA both give a detailed 3D map of the arterial tree with resolution adequate for most clinical needs. The choice between the methods is often based on local availability and competence. In CTA, the composition of the vessel wall can be studied in the same dataset as the lumen [1] with MRA, a similar assessment can be made by adding dedicated sequences to the examination [59]. Calcifications are seen only in CTA, but both severe calcifications and metallic implants can sometimes cause disturbing artefacts. In MRA, metallic implants often generate severe artefacts that make a diagnostic assessment of stenoses impossible. With certain implants, including cardiac pacemakers, safety concerns preclude any MRI examinations.

12.2.2 NEED FOR IMAGE POSTPROCESSING

The primary purpose of a CTA or MRA examination is often to give the interventionist an intuitive image of the vascular anatomy and the presence of stenoses, occlusions, and collateral vessels that may have grown to provide an alternative way for the blood flow. As long as the focus is on *visualization*, a simple rendering algorithm is usually preferred. The two most commonly used are the *maximum intensity projection* (MIP) [32] and the *direct volume rendering* (DVR) [13]. In both cases, rays are simulated through the volume or part of the volume; in MIP the maximum voxel value along the ray is selected for presentation, whereas DVR uses algorithms including various optical concepts such as emittance, reflectance, and absorption [12].

In CTA, the presence of skeletal structures with equal or higher intensity than the contrast-filled vessels may obscure the image. One solution is to cast the ray through a limited volume restricted to a few mm above and below a curved surface representing the curved centerline of each artery (curved planar reformatting, CPR). Obtaining these centerlines by manual interaction is feasible, but often much too time-consuming to fit into a pressed clinical workflow.

In clinical cases where the decision whether to perform an intervention is difficult, and, even more, in research concerning the efficiency of various therapeutic measures, there is an urgent need for *quantitative measurements* of arterial disease or, in other words, an *imaging biomarker* [38] for atherosclerosis. Such measurements are almost always based on a segmentation of the lumen or the vascular wall, and the numeric results usually represent the diameter or the cross-sectional area of the lumen in the stenosis [3]. A more sophisticated approach is to use the segmentation result as an input in calculations predicting blood flow and pressure in the vessels using computational fluid dynamics (CFD) [60]. In this case, the result of an intervention may also be predicted.

For segmentation of the arterial tree in angiographic voxel data, a considerable number of algorithms have been proposed [7,25,27,48]. Varying degrees of user interaction are required, and in many of the algorithms, centerlines serve as a starting point for segmenting the lumen and identifying stenoses [42]. There is thus a great need for fast methods for extracting at least approximate centerlines of the arterial tree. For widespread use in clinical routine, time efficiency is extremely important. Regardless of how much interaction is required, most working radiologists are not likely to accept computation times above 1 or 2 min. This goal is currently not met with most algorithms proposed so far. A possible exception could be if the segmentation process is initiated completely automatically once the image data arrive in the PACS (picture archiving and communication system), so that only limited editing work requires the user's attention [56].

12.3 BACKGROUND

In this section, we first give a more general background and introduce previous methods and approaches that have been done in the field of vascular segmentation. In the second part, we focus more specifically on our current method.

12.3.1 PREVIOUS METHODS

Numerous image analysis methods have been proposed for vascular segmentation. Since no general purpose segmentation method is suitable for all applications, this problem still attracts the interest of many researchers. In the literature, two main approaches can be recognized [48].

The first approach first focuses on extracting the vascular skeleton straight from medical images, usually computed from two-dimensional vessel cross sections. Subsequently, this vascular skeleton is used for finding the vessel boundaries. Methods following this approach are called *skeleton-based* or *indirect* methods.

The second approach extracts the vessel boundaries from medical images without the need of prior skeleton extraction. Optionally, the obtained volumetric representation of the vessels might be used for skeletonization either by using iterative thinning, mesh contraction (shrinking), or methods based on fast marching. The extracted skeleton can be used for detecting and evaluating artery diseases, such as stenoses detection [35] or visualization of aneurysms and vessel wall calcifications [47]. The methods based on this approach are often called *direct* or *nonskeleton-based* methods.

In this section, we focus on skeleton-based methods because our main contribution uses this approach. Within these methods, we give more attention to algorithms that are able to extract the whole vascular centerline tree as the presented method was tested on a dataset of the lower limbs (scans from the abdominal aorta to the feet). We also mention methods that provide information about their execution time as the aim of our method is to be fast enough for clinical use. It is hard to compare different machines, datasets, or degrees of optimization used in different methods. Nevertheless, we present an overview to give a rough idea of the execution times. For a detailed survey on different vessel segmentation algorithms, we recommend the reader the works [25,27].

Most centerline extraction algorithms are semiautomatic, based on one or more user-defined seed-points and their propagation. Particularly, popular are minimal-path approaches identifying the minimum-cost path as the vessel centerline. The cost function can be efficiently calculated by using various operators such as medialness filters [4,8,20], vesselness filters [14], or optimally oriented flux [6]. The main disadvantage of an algorithm based on propagation from a seed is that the propagation usually stops in a presence of vessel occlusion, calcification, or other vascular pathologies, leaving behind many undetected vascular parts. The solution is to add new seeds to the omitted vascular parts; however, obtaining the final vascular skeleton from such data of a nonhealthy patient is tedious and hard work.

A centerline extraction algorithm intended for use in catheter simulation was presented in [11]. This algorithm is semiautomatic and requires two user-defined points (start- and endpoint). A coarse-to-fine approach is then used to find a centerline between the points. First, a coarse centerline is created by Dijkstra's shortest path algorithm [10]. Thereafter, the centerline is refined by an active contour model using polyhedra placed along the model. The algorithm was tested on a reasonably big real CT dataset with $512 \times 512 \times 386$ voxels. However, the length of the centerline extracted from this volume is less than 100 voxels. Depending on the number of polyhedra surface points, the centerline extraction took 2.3–31.4 seconds for 12–272 surface points respectively. The method is very suitable for catheter simulation. However, the requirement of start- and endpoints does not allow extraction of complete centerline trees, which is limited only to single centerline extraction.

A fully automated method of arterial centerline extraction for the planning of transcatheter aortic valve replacement (TAVR) was recently proposed in [17]. The method uses the intensity and shape information of target arteries to construct a cost function representing the speed of the wave propagation. Afterwards, the Dijkstra algorithm [10] is applied to extract the arterial centerlines. The algorithm was validated on CTA datasets that are routinely acquired prior to the TAVR procedures. The size of one case in the dataset was approximately $512 \times 512 \times 1000$ voxels, with a computational time to process each CTA dataset around 53 ± 11 seconds. From these data, the method extracts arterial centerlines in the entire femoral access route, which includes the aorta segment and both left and right femoral segments. Unfortunately, even though the authors mention that their method extracts arterial centerlines from whole-body CTA images, they validated it only on the aorta and femoral arteries.

The work presented by Li et al. [28] extends the shortest path computation using one additional dimension, a vessel radius. The vessel centerline representation is then shifted from a 3D to a 4D curve. As a result, the centerline paths are better centered. In addition, the vessel surface can be directly extracted without any postprocessing, as the information of the radius for each centerline point is part of the 4D curve. However, the computation time is a drawback. For a 3D CT image of the coronary artery of size $110 \times 90 \times 80$ voxels, the aorta centerline extraction "takes less than 2 minutes." The method has also the limitation that it extracts the centerline only between two user-defined endpoints.

In contrast to techniques that can only detect a single centerline, some algorithms can propagate outwards from only one seed point and detect multibranch centerlines simultaneously. Gülsün and collaborators [20] proposed such an algorithm. The algorithm uses a graph-based optimization together with multiscale medialness filters and extracts the centerlines by computing the minimum-cost path. The method has been tested mainly on coronary artery CTA data, where the full coronary artery trees were extracted in 21 seconds on average. Furthermore, the algorithm works well also for different types of vessels obtained from different imaging modalities. It was very interesting to see a visual result of centerlines extracted from peripheral vessels obtained from CTA. This is the same region and the same modality we are using in our experiments. Visually, the centerlines obtained by Gülsün's method and by our method seem to be of similar length and cover similar regions. However, it would be really interesting to know if the volume was obtained by scanning a healthy or sick patient, what was the approximate volume size, and how long did the method take to process the volume.

An algorithm proposed in [55] performs segmentation and skeletonization of coronary arteries based on a fuzzy connectedness tree. The algorithm also works well outside the coronary region. With limited user interaction, it is possible to extract the full body vascular skeleton; however, it still takes approximately 20 min. This algorithm is used to evaluate our proposed method and will be in detail explained in a reference skeleton preparation in Section 12.5.

This brief review shows that there are many methods and approaches dealing with vascular skeleton extraction. However, these methods are not fast enough to permit interactive clinical use. The aim of our work is to fill this gap and present a method that is able to extract a whole-body centerline tree in reasonable time and hence accelerate the vascular segmentation.

12.3.2 FROM COLON TO BLOOD VESSELS

Back in 2001, automatic segmentation of a single human colon in CT took around 10 min on, at the time, a modern computer workstation. With colorectal cancer classified as the second leading cause of cancer-related deaths among men and women in the United States [44], early detection was critical to reduce the risk of developing colon cancer [51]. Population screening was crucial to this end. For this reason, *computed tomographic colonography* (CTC), a technique for detecting polyps by use of a computed tomography (CT) scan of the colon, was introduced as a possible complement to optical colonoscopy. As the total evaluation time should desirably be kept under five minutes [45], automated segmentation and computer-aided analysis were considered. This, in turn, demanded faster and more accurate automated segmentation methods.

In CTC volumetric images of the human colon, a computed centerline is an anatomically attempting guide, making it possible to fly through the colon using a virtual camera, measuring global size, and local properties along the colonic path. Furthermore, such a centerline could be useful in order to segment the colonic pathway itself. Also, in conjunction with anatomical measures, a centerline could be used to distinguish the colonic pathway itself from other objects in the CTC image. Such observations might seem as a deadlock, as a centerline commonly is computed based on a segmentation of the object it represents.

However, as the colonic structure, in broad terms, is a tubular structure with repetitive folds perpendicular to the colonic pathway, local distance maxima available at the center of the colonic pathway are expected. Such local maxima can easily be verified by first thresholding the air of a CTC image using moderate threshold settings and then computing the distance transform and finally detecting the maxima within the resulting image. By connecting the local maxima using a statistic clustering approach, one-dimensional paths are created for each air-filled region in the CTC image, after which anatomical evaluation can automatically discern which path, centerline, represents the colonic pathway. Knowing the path of the colon, segmentation is straightforward as local operation can be performed during the process.

The method can be seen as first generating a 1D line representation of each air-filled region within the image, then performing an anatomic evaluation of the 1D structures to determine the colonic pathway amongst the structures, and finally to transform the result to a voxel representation. As the main part of the computations will be done in a 1D representation with just a fraction of data points as compared to the CTC image, there simply are not so many computations to be done reducing the computational burden. This is when we introduced the *centerline-based segmentation* (CBS) concept as presented in [16]. Instead of computing the distance transform (DT), local maxima were in practice computed using a 1D scanline algorithm reducing the computation time. The final segmentation was done using a custom-made region growing algorithm where the air was first coarsely thresholded, after which the actual colonic wall was detected using a so-called onion-shaped kernel, both giving precision to the border and avoiding excessive leakage in cases where the scan resolution introduced artificial holes in the colonic wall.

At the time, approximately 20% of false-positive findings in computer-aided detection (CAD) schemes originated from the small bowel and stomach [33,58]. CBS had an average computation time for the segmentation of 14.8 seconds as compared to around 10 min for the previous state of the art, the *knowledge-guided segmentation* (KGS) algorithm [33]. The sensitivity of our new approach was, on average, 96%, and the specificity was estimated to be 99%. A total of 21% of the voxels segmented using KGS were disregarded by CBS. However, 96% of these voxels represented extracolonic structures, and only 4% represented the colon, giving virtually only the segmented colon as a result. This algorithm is currently part of a computer-aided diagnosis system for computed tomography colonoscopy [24].

As presented in Section 12.2, the segmentation of the vascular tree is an important topic of today. By inspecting the vascular tree anatomy, it is straightforward to conclude that it should be possible to segment it using a similar approach as that used in CBS [16] for the colon. There are differences, however. Folds are not expected within the tubular structures, bifurcations occur, and some paths can be narrow by nature.

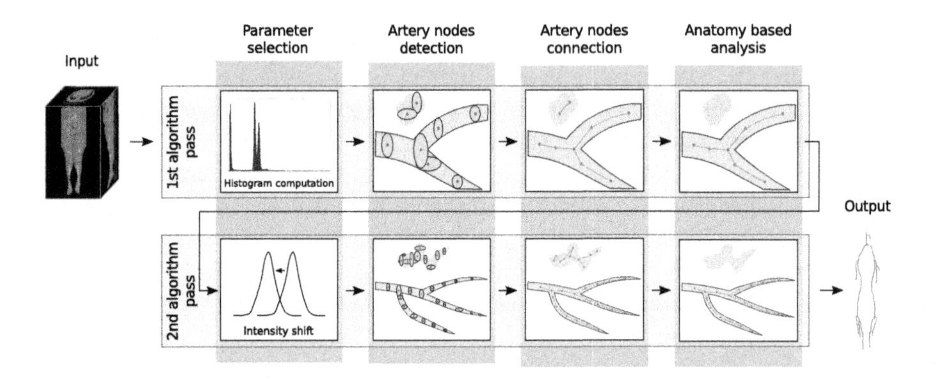

FIGURE 12.1

Flow chart of the proposed method. A 3D CTA volume is processed in two passes where each pass consists of four main steps: parameter selection, node detection, node connection, and anatomy-based analysis. The first pass detects centerlines of large arteries, whereas the second pass detects centerlines of small arteries. The result is a graph where detected nodes serve as vertices and the connections between them are edges.

Furthermore, there are different kinds of challenges on top of the standard cases, e.g., calcifications within the vascular tree as compared to stool within the colon.

Thus, two major parts need to be adjusted for the new application: First, the detection of local maxima must be rewritten to focus on vascular structures; in fact, more than a single filter will be needed in order to pick up both thin structures of a voxel or less in diameter and the main arteries. Secondly, the anatomic definition must be adapted to the objects in search for. However, as the vascular tree is a tree structure and not a single centerline, the clustering algorithm must allow such trees to be created; a centerline is simply a particular case of a centerline tree where all unnecessary branches have been removed from the tree. The details of this method will be described more in-depth in Section 12.4.

12.4 FAST SKELETON-BASED GENERATION OF THE CENTERLINE TREE

In this section, we present our method for generation of the centerline tree. The method takes as input a 3D unprocessed CTA scan and produces a graph as output. Nodes in this graph are centrally located artery voxels found by a set of criteria and the connections between them represent graph edges. The algorithm consists of four main parts that are repeated in two passes. An outline of the proposed algorithm is presented in Fig. 12.1.

Each algorithm pass consists of four main steps: parameter selection, nodes detection, nodes connection, and anatomy-based analysis. The parameter selection step automatically selects appropriate parameters for the analysis based on vascular morphology and intensity histogram of the data. The artery nodes detection step applies different filters to detect voxels that are centrally located within the arteries. The locations of these voxels are then connected, using the distance and intensity information, as nodes into a graph structure. After this artery nodes connection step, some graphs that do not belong to the arteries still remain. The anatomy-based analysis identifies and removes these false graphs returning only one or more graphs corresponding to the vascular tree.

All four steps are repeated in each one of two algorithm passes. During the first pass, the algorithm focuses on extracting the centerline of large arteries. During the second pass, additional centerlines of finer structures are added to the graph. Performing two passes is important for better distinguishing between true and false artery nodes. The true artery nodes are voxels that are truly part of the arteries, and false artery nodes are voxels that are detected by the filter but lie outside the arteries. False artery nodes often arise from data noise or are detected on the boundary of compact bone tissue and inside of spongy bone tissue. Both noise and bone tissue have similar intensities and sizes as the nodes in smaller arteries. By initially focusing on arteries having a larger radius, we can eliminate many of these false nodes and create thereby a basis of reliable nodes. In the second pass, we relax our filter criteria and focus on detecting arteries having only small and tiny radius. This will result in a large amount of detected nodes. By keeping only nodes and graphs that are connected to the graphs generated in the first pass we can successfully remove the false nodes and graphs, reaching the final centerline tree. In the following subsections, we will look in more detail into each algorithm step.

12.4.1 PARAMETER SELECTION

The first step of the proposed method sets the parameters to appropriate values. All parameters can be divided into two groups. Parameters that are based on vascular morphology are part of the *fixed parameter group*. Their values need to be configured only once and can be used without any change for other datasets that depict the same vascular structures, e.g., vasculature of lower limbs, coronary artery, and others. Such parameters depend on the size and shape of the vasculature they depict. The second group are *changeable parameters*, consisting of intensity ranges for relevant tissues such as fat $[\theta^f_{low}, \theta^f_{high}]$, muscles $[\theta^m_{low}, \theta^m_{high}]$, and blood $[\theta^b_{low}, \theta^b_{high}]$. These intensities depend on various factors and therefore have to be adapted for each patient separately. We propose a way to set them automatically.

Fixed Parameters

To select the size of the arteries, we aim to detect at each algorithm pass we use a minimum and a maximum radius parameter $[min\_radius, max\_radius]$. In the first algorithm pass, where the focus is on detecting larger arteries, we set the minimum radius to 2 mm and maximum radius to 25 mm, as arterial diameters above 50 mm are very rare and only found in aneurysms. In the second pass, we aim to detect

Table 12.1 Fixed parameters used in the algorithm listed with their corresponding values

Parameter name	Value used in 1st round	Value used in 2nd round
min_radius	2 mm	limited by the voxel size
max_radius	25 mm	2.5 mm
cutting_thresh	5 times the radius of the artery at a bifurcation point	
short_graph_thresh	1/5 of the leg length	does not exist
gradient_thresh	0.6	0.4

finer arteries with a radius smaller than 2.5 mm. The lower radius limit, in this case, is limited by the size of the voxel in the database. However, these values can be modified depending on the requested artery size.

The artery walls rarely have the shape of a completely regular tube, and they commonly contain small bulges and irregularities. Some nodes are detected inside these bulges close to artery surfaces. When they are connected to other nodes, short false artery branches are created. These spurious branches are lying inside the artery but do not correspond to real artery bifurcations. We are removing them by cutting off all branches shorter than *cutting_thresh* parameter. In this work, we use *cutting_thresh* equal to 5 times the radius of the artery at a bifurcation point. Note that it is inevitable to cut off some true but very short arterial branches.

The majority of the remaining false artery nodes were detected in anatomical structures that resemble the arteries in terms of intensity and cross-sectional shape. For the CT scan of the lower limbs, these structures are mainly spongy trabecular bone tissue in the knee region and surface of the compact cortical bone tissue in the pelvic region. These regions are in general shorter than a fifth of the leg length. By removing all graphs being shorter than a fifth of the leg length we keep only true artery graphs. The threshold parameter for removing short graphs is named *short_graph_thresh*. It is applied during the anatomy-based analysis at the end of the first round and will be explained in detail in Section 12.4.4.

A final parameter, *gradient_thresh*, sets a boundary on the smoothness of the perimeters of the elliptical cross-sections we are interested in detecting. Larger arteries usually have more uneven elliptical cross-sections compared to smaller vessels, which have smooth and regular cross-sections. We have a measure to calculate the smoothness that ranges from 0 to 1, where 0 represents an ideal ellipse, and 1 represents a very uneven cross-section perimeter. Therefore, in the first pass, the *gradient_thresh* parameter is set to 0.6, and in the second pass, we lower the value to 0.4. The parameter will be detailed in Section 12.4.2, in the algorithm step, where it is first used.

All fixed parameters (Table 12.1) were set based on the training datasets containing 19 cases. First, a range of reasonable values was roughly estimated for each parameter, and then we tested them and visually compared the result with the centerline extraction done by a radiologist. Thereafter the specific values were chosen. For

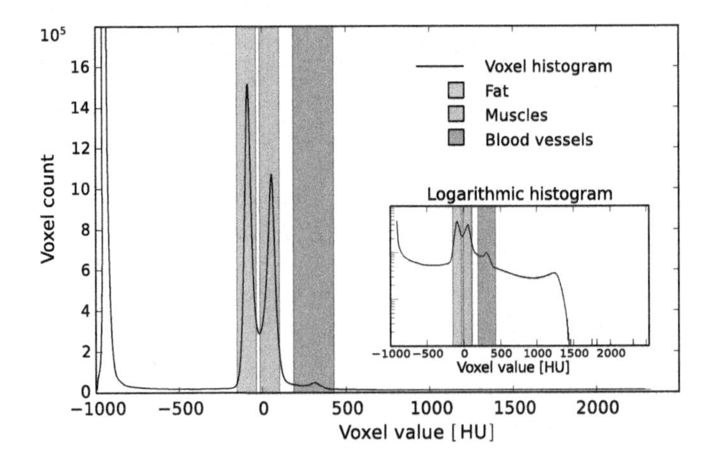

FIGURE 12.2

The histogram of a single case is shown with three relevant ranges detected automatically (fat, muscle, and blood vessels).

the evaluation, these parameters were used without further modification in another 25 cases from the test dataset.

Changeable Parameters

In CT, the voxel intensity is expressed in Hounsfield units (HU) with water corresponding to 0 HU and air to −1000 HU [21]. Whereas the Hounsfield scale is relatively standardized between different machines for most soft tissues like fat or muscles, the iodine concentration, and thus the HU number, of contrast-mixed blood in the arteries may vary considerably between scans due to variations in hemodynamics and timing between injection and image acquisition. To estimate the Hounsfield values corresponding to the blood vessels, a smoothed histogram of CTA volume intensities is produced (Fig. 12.2). This histogram can, in general, be described as the succession of the following peaks:

- an obvious peak reflecting air inside the volume and background artifacts is always present in the neighborhood of −1000 HU;
- the peaks of interest slightly below and slightly above the value 0 (corresponding to water), belonging to fat and muscles, respectively;
- the blood vessel intensities are defined by a small peak located above the muscle peak.

There is no visible peak representing bone tissue. As bone tissue consists of spongy trabecular bone and compact cortical bone with typical tabular values 300–500 HU and 600–3000 HU, respectively, the intensities are too spread out to be visible on a classical histogram. The peak is clearly visible, though, if we plot a logarithmic histogram. See the subimage of Fig. 12.2.

Parameters relevant to the method are the intensities corresponding to fat, muscles, and the blood peak. To find these intensity ranges, a sum of Gaussian curves is fitted to the image histogram using the criterion of nonlinear least squares fitting. First of all, a Gaussian curve is fitted to the position of the highest peak. This Gaussian is then subtracted from the original smoothed histogram, thereby creating a new histogram curve. The same process is repeated, but instead of the original histogram, we now use the new histogram curve. Two detected Gaussian curves represent the fat and muscle peaks. As the blood peak is usually very small, we have to specify the intensity range where the peak location is expected and detect the exact position as a maximum of the updated histogram within this range. A Gaussian curve fitted to this position then represents the blood peak. The range $\pm 3\sigma$ around the Gaussian's mean determines the intensity values for each peak. In cases where the intensity ranges overlap, the border between the peaks is determined by the position of the original histogram's minimum between the two peaks.

If some of the peaks are not visible or are too flat, resulting in the Gaussian curve failing to fit the data, a fallback range of values will be used: $[-300, 0]$, $[0, 100]$, and $[100, 700]$ ranges for fat, muscles, and blood intensities, respectively. The fallback ranges are particularly useful when the method is applied to a cropped CT volume, where all tissues are not necessarily present in sufficient quantities. The fallback HU intervals for fat and muscle are, compared to the typical tabular values cited in the literature, somewhat more inclusive. However, no other tissue than fat is likely to occur between -300 and 0 HU, and no substantial amounts of any other tissue than muscle is likely to be found between 0 and 100 HU. The fallback value for the upper blood threshold is set to 700 HU, which is deliberately set higher than the tabular value for blood (usually up to 500 HU). The reason for this is the high proportion of arteries containing calcifications. The calcifications are expressed by high HU numbers and thus inevitably reduce the number of voxels with standard blood intensity, causing the blood peak to be low in the histogram. These abnormal situations are automatically detected by our method, and the user is informed in such cases.

The objective of our algorithm is to detect centerline of arteries. Since the arteries are surrounded mainly by muscles, how to set the intensity threshold between the muscles and blood is important and should be set with care. Fig. 12.3 demonstrates that the intensities calculated by our automatic method correspond to proper tissues in a clinical case. Voxels located around the bone boundaries and voxels corresponding to spongy (trabecular) bone have the same intensity as blood, and therefore no conventional thresholding method would be sufficient for artery segmentation.

The values for changeable parameters as defined in the preceding paragraphs are used in the first pass. For the second pass, the objective is to extract the centerlines of smaller arteries, and therefore the intensity ranges need to be properly adjusted. Arteries with smaller diameters may cross only through part of the voxel, whereas the rest of the voxel space is filled with muscle tissue. Therefore the intensity of this voxel does not attain as high intensity value as the voxel that lies fully within an artery. Due to this effect, known as a partial volume effect, small arteries tend to have lower intensities, and the intensity ranges for muscles and blood have to be

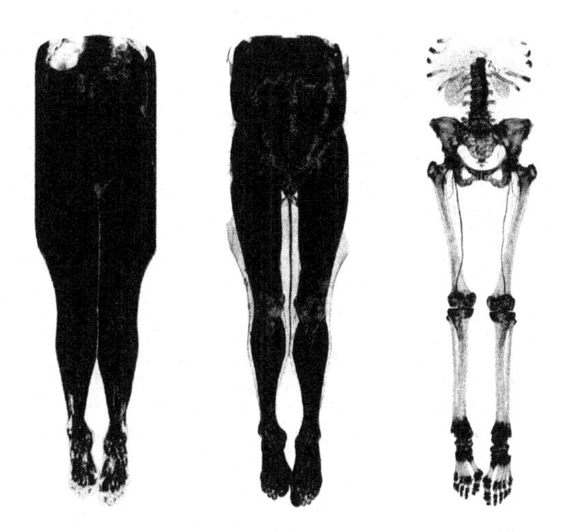

FIGURE 12.3

For a single case, Hounsfield values for fat, muscles, and arteries ([−146, −44], [−7, 109], and [187, 425], respectively) are displayed, based on the automatic classifier.

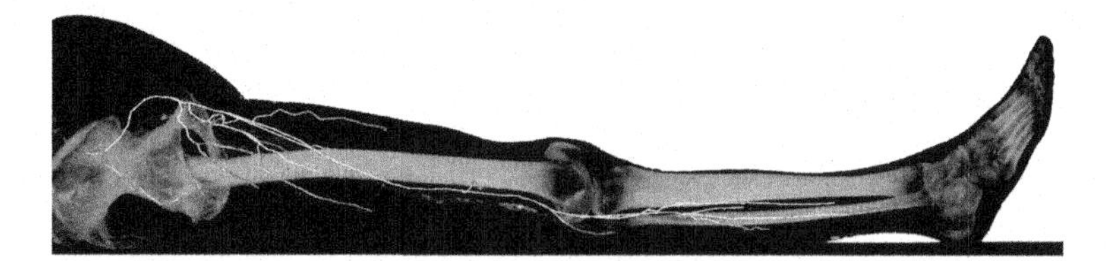

FIGURE 12.4

Example of decreasing intensity related to artery width decrease. A 3D vascular skeleton with higher intensities shown by yellow and lower by red color is overlaid on a 2D maximum intensity projection of CTA of the lower limbs.

shifted towards lower intensities too. The nonuniform nature of the HU intensities is demonstrated in Fig. 12.4.

12.4.2 THE ARTERY NODES DETECTION

Arteries intersect two-dimensional slices of the input volume in elliptical-like cross-sections. In the centers of these cross-sections, plateaus of blood intensities, are artery nodes. The detection of these artery nodes is performed by applying several filters to the original data. First, a simple intensity-based filter is used to detect a set of potential artery nodes. The set of potential artery nodes then needs to be cleaned up from the false artery nodes that occur because of the partial volume effect, noisy original data, or bones that resemble arteries. The majority of these false artery nodes is removed by specialized filters. For each of the two algorithm passes, a different set of filters is used as each pass focuses on detecting different parts of the vasculature.

In detail, the artery nodes detection procedure works as follows. The method scans all axial, sagittal, and coronal slices of the volume. Depending on the diameter of the smallest artery that needs to be detected in the particular algorithm pass, the algorithm may repeatedly skip some of the voxels within each slice. By performing this subsampling using a satisfactory sampling rate we decrease the execution time and (in theory) do not miss any artery node. The satisfactory sampling rate f_s is

$$f_s = \frac{min\_diameter}{2} = min\_radius. \tag{12.1}$$

For better understanding of the subsampling process, consider a situation where we want to detect arteries with diameter 5 mm and larger. The satisfactory sampling rate, in this case, would be 2.5 mm, which is for our datasets equivalent to approximately 3 voxels. The algorithm scans all axial, sagittal, and coronal planes; however, within each plane only every 3rd voxel inside every 3rd row needs to be checked. The sampling rate of 3 voxels is enough to not miss any artery with the diameter of size 5+ mm.

Thus, only a subset of voxels in each volume slice is analyzed. If any of these voxels matches two basic criteria during this scanning process, it is selected as a potential artery node. The basic criteria are:

1. The voxel has an intensity from the previously estimated blood vessel range $[\theta^b_{low}, \theta^b_{high}]$.
2. On the volume slice, the voxel is located in the center of an elliptical-like area of similar intensities surrounded by voxel with lower intensities.

A candidate voxel that matches the first blood vessel intensity criterion is verified for the second criterion by casting four rays in four main directions (up, down, left, and right) from its position. Along each beam, we are looking for at least three consecutive voxels of intensity smaller than blood, θ^b_{low}. Three voxels are needed to exclude the possible presence of noise. The first one of the three voxels then represents the edge of the blood intensity plateau in that particular direction. The candidate voxel is verified if it is well centered between two edges, first in one direction (horizontal) and then in the other direction (vertical). In case that the candidate voxel is not centrally located, the method omits this voxel and considers another one that lies, instead, in the middle between two horizontal and two vertical edges. The distances between the centrally located voxel and the edge voxels define the major a and minor b axes of an ellipse. This ellipse is called bounding ellipse since it limits, or bounds, the area of similar intensities around the central voxel. Another two auxiliary ellipses, inner and outer ellipses, are needed for further analysis and therefore are introduced here. Both ellipses are concentric, i.e., share the same center, with the bounding ellipse and lie in the same slice. The inner ellipse has axes of sizes $a - 1.5$ and $b - 1.5$, whereas the outer one has axes of sizes $a + 1.5$ and $b + 1.5$, all in mm. Fig. 12.5 shows

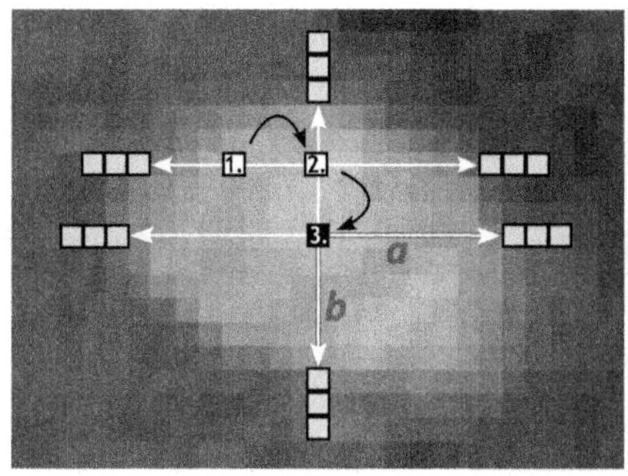

(A) Process of selecting centrally located voxel

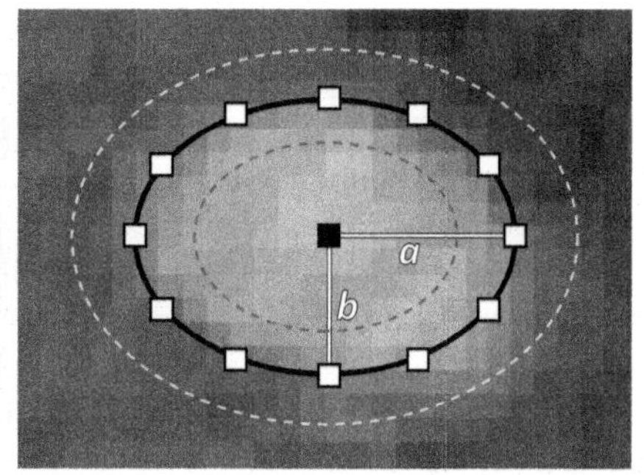

(B) Three relevant ellipses used in advanced filters

FIGURE 12.5

(A) The method found a voxel (white square no. 1) that has an intensity within blood intensity range. This voxel is not well-centered between two horizontal edges, so the voxel lying truly in the middle (white square no. 2) became a new potential central voxel. The same verification is repeated iteratively for vertical and horizontal directions until a truly centrally located voxel (black square no. 3) is found. (B) Centrally located candidate voxel (black square) with black ellipse corresponding to the elliptical area of similar intensities. a and b represent the major and minor ellipse axis, and n white squares correspond to n sampling voxels lying on the ellipse border. The inner ellipse is shown in red (dark), and the outer ellipse in yellow (bright) color, both dashed. The images show intensities in the HU range $[-154, 436]$.

all relevant ellipses, the bounding ellipse in black and the inner and outer ellipses in red and yellow, respectively.

The basic set of filters is set to be inclusive, and therefore it also detects many false artery nodes, which must be removed later. For that purpose, a different set of filters is used. In addition to the central candidate voxel, these filters use also voxels from the local neighborhood. These neighborhood voxels are all lying on the perimeter of the ellipses as introduced earlier and are chosen by sampling the perimeter at, approximately, constant sampling rate. The sampling rate has the effect of selecting fewer voxels for smaller ellipses and more voxels for bigger ellipses. The neighborhood voxels sampled along the perimeter of the bounding ellipse are illustrated in Fig. 12.5 by white squares. In order for candidate voxel to pass through the filter set, the candidate voxel together with the neighboring voxels has to satisfy a characteristic appearance of a vessel and its surrounding.

The first characteristic is that the inside space of an artery, also called the artery lumen, is homogeneously filled with blood. Therefore the average intensity of the voxels sampled along the perimeter of the inner ellipse (red ellipse in Fig. 12.5) should be from the predefined blood vessel intensity range $[\theta^b_{low}, \theta^b_{high}]$.

The artery neighborhood, on the other hand, is heterogeneous and can contain three types of tissue: fat, muscles, or bones. In discrete data, however, the partial volume effect will give the surface of the bone similar intensities as the artery. So, in the very close neighborhood of the artery, only intensities from the range between fat and blood $[\theta^f_{low}, \theta^b_{high}]$ are acceptable. If an artery would be so close that it would be touching the bone, a consequence is that the voxels on the outer ellipse of the true artery node would have blood intensities on the side of the ellipse that is touching the bone. However, if the candidate voxel would not be lying inside the artery, but rather be a false artery node detected at the bone boundary, some of the voxels on the outer ellipse would have considerably higher intensity that corresponds to the bone tissue. For this reason, the second filter checks if both the maximum and minimum intensity of the voxels sampled along the perimeter of the outer ellipse (yellow ellipse in Fig. 12.5) are within the predefined fat-blood intensity range $[\theta^f_{low}, \theta^b_{high}]$. This was the description of the second characteristic.

The third characteristic is related to the shape of the vascular cross-section. The aorta and its main branches are very elastic, which helps them to balance sudden blood pressure changes during the heart beat. They stretch during systole and rebound during diastole to keep the blood pressure constant. Large arteries also contain a large subendothelial layer, which grows with age or disease conditions and creates arteriosclerosis (on CT images visible as high intensities within the vessels). Small arteries, on the contrary, maintain their shape, and arteriosclerosis is less common in them. Two-dimensional cross-section of small arteries, therefore, is usually circular, and high intensities are not present inside their arterial lumen. Cross-section of a large artery tends to be more elliptical, occasionally containing arteriosclerosis. For false artery nodes, the cross-sectional shape of the structure where the nodes were detected can have diverse shape, far from being an ellipse. Therefore, inspecting the shape of the bounding ellipse is an indication of a true or a false artery node. We perform this by computing the gradient at each position of the voxels on the perimeter of the bounding ellipse (black ellipse in Fig. 12.5), normalize it to the maximum gradient, average the gradient vectors, and finally compute the magnitude ranging from 0 to 1. A null vector corresponds to an ideally shaped ellipse with equal magnitudes of oppositely oriented gradient vectors; a unit vector, conversely, suggests that all gradient vectors are orientated in the same direction. By using a $gradient\_thresh$ value the method thresholds artery nodes based on the gradient magnitude and orientation into true and false artery nodes. As a reminder, the $gradient\_thresh$ value was set to 0.6 for larger arteries and 0.4 for smaller arteries. The idea is to be more strict for small vessels and more benevolent for irregularities for large arteries.

The last characteristics is to distinguish between large and small arteries. We want to detect only large arteries in the first pass and only smaller arteries in the second pass, so this filter is controlling the node radius and allow one to pass only the nodes we are wishing to detect in the current pass.

To summarize, the set of filters, designed to detect large arteries in the first algorithm, pass:

1. $I^I_{avg} \in [\theta^b_{low}, \theta^b_{high}]$, where $I^I_{avg} = \frac{1}{N}\sum_{n=1}^{N} I^I_n$ is the average intensity of N voxels from the perimeter of the inner ellipse.

2. $I^O_{min} \wedge I^O_{max} \in [\theta^f_{low}, \theta^b_{high}]$, where $I^O_{min} = \min\{I^O_n\}^N_{n=1}$ is the minimum intensity, and $I^O_{max} = \max\{I^O_n\}^N_{n=1}$ is the maximum intensity of N voxels from the perimeter of the outer ellipse.
3. $\Gamma^B \leq gradient\_thresh$, where $\Gamma^B = |\frac{1}{N}\sum^N_{n=1} \nabla^B_n|$ is the magnitude of averaged local gradient vectors normalized to the maximum gradient on the voxels from the perimeter of the bounding ellipse.
4. $a \leq max\_radius$ and $b \leq max\_radius$, where a and b are major and minor axes of the bounding ellipse, respectively.
5. $a \geq min\_radius$ and $b \geq min\_radius$.

For the exact parameter values, see Table 12.1.

The set of filters designed to find tiny arteries is similar to the set presented for large arteries. The difference is that the first filter is skipped and only filters nos. 2–5 are performed in this set. The reason for that is that the inner ellipse in small arteries is extremely small (usually, 1–2 mm or smaller). Similarly, if the major a or minor b axes are smaller than 1 mm, then filter no. 3 is skipped. Also the values of fixed parameters have different values designed for detecting small arteries in the second algorithm pass (for the exact parameter values, see Table 12.1). The set of filters in the second algorithm pass can by summarized as:

1. Filter no. 1 is skipped because the inner ellipse is too small in this algorithm pass.
2. $I^O_{min} \wedge I^O_{max} \in [\theta^f_{low}, \theta^b_{high}]$, where $I^O_{min} = \min\{I^O_n\}^N_{n=1}$ is the minimum intensity and $I^O_{max} = \max\{I^O_n\}^N_{n=1}$ is the maximum intensity of N voxels from the perimeter of the outer ellipse.
3. $\Gamma^B \leq gradient\_thresh$, where $\Gamma^B = |\frac{1}{N}\sum^N_{n=1} \nabla^B_n|$ is the magnitude of averaged local gradient vectors normalized to the maximum gradient on the voxels from the perimeter of the bounding ellipse. If the major a or minor b axis is smaller than 1-mm filter no. 3 is skipped.
4. $a \leq max\_radius$ and $b \leq max\_radius$, where a and b are major and minor axes of the bounding ellipse, respectively.
5. $a \geq min\_radius$ and $b \geq min\_radius$.

Finally, all the successfully passing nodes are sorted by their z coordinate values, in inferior direction from head to toes, and saved as artery nodes together with data on their respective positions and radii. The radius r for each node is calculated by using an area-preserving transformation from an ellipse to a circle, defined by

$$r = \sqrt{ab}, \tag{12.2}$$

where a and b are the previously mentioned major and minor axes of the bounding ellipse, respectively. The nodes that are within the radius distance to another already saved node are discarded. To decrease the execution time, the verification of proximity to another database node is checked even before the filtering is performed.

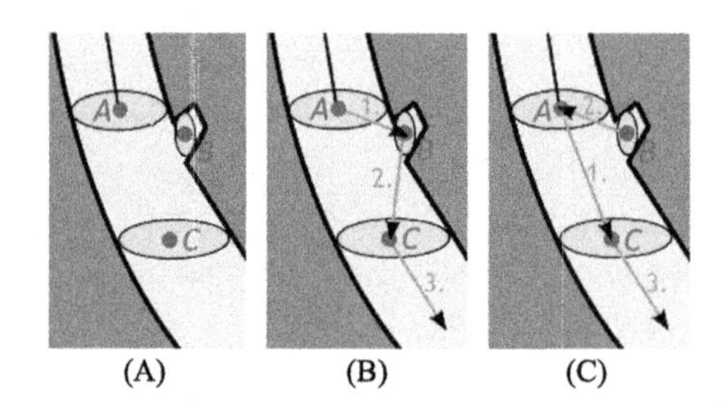

FIGURE 12.6

Example of three artery nodes; (A) before the connection, (B) connection of the nodes without using the 2nd condition, (C) connection of the nodes with using the 2nd condition.

12.4.3 THE ARTERY NODES CONNECTION

To obtain a continuous centerline tree for the entire vasculature, the previously detected nodes need to be connected. The method links artery nodes when they satisfy specific conditions. This is done gradually, until no more links can be created.

At the beginning, each node is a separate graph. Nodes are iteratively connected with the closest nodes until all possible connections are established. Connectivity follows 4 conditions that have to be satisfied to establish a link from node A in graph G_A to node B in graph G_B.

1. $G_A \neq G_B$.
2. Node B has a radius greater than or equal to that of node A.
3. Every voxel on the link from A to B has an intensity from the predefined blood vessel range $[\theta^b_{low}, \theta^b_{high}]$.
4. There is no other node C for which the Euclidean distance between nodes A and C is less than between nodes A and B that fulfills conditions 1–3.

Connectivity criterion no. 2 serves as a way of sorting the subset of the closest nodes. It will ensure to make connections to nodes considered more important as they have a bigger radius which eliminates the side paths close to the artery perimeter. Let us consider an example shown in Fig. 12.6A, where three artery nodes need to be connected. As the nodes are saved in memory according to their increasing z-coordinates, they will be processed in the order A, B, C. Fig. 12.6B shows how these nodes would be connected if condition 2 were not satisfied:

1. A would connect to B ($A \rightarrow B$); B is the closest node to A, and link between them lies inside the artery.
2. $B \rightarrow C$; B is already connected to A, so it would connect to the second closest node C.
3. C would connect to some other following node.

Fig. 12.6C shows how the artery nodes would be connected using condition 2:

1. $A \to C$; although B is the closest node, it has a smaller radius than A so A would connect to the second closest node C.
2. $B \to A$; A is the closest node to B, A has bigger radius and link between them lies inside the artery.
3. C would connect to some other following node.

After all connections between the nodes are created, very short branches, like the connection between node A and node B in Fig. 12.6C will be removed. For each tree node (defined as a node having three or more connections), the implicit length of all its branches is calculated. For each branch, we make a decision to keep or remove the branch with the goal to only keep longer branches that are clinically relevant. Branches whose lengths do not exceed the radius of the artery at a bifurcation node by a factor of five are not considered relevant, but a result of noise, and are therefore removed.

The noise or body tissue, such a spongy trabecular bone or articular cartilage, has similar intensity and structure compared to the tiny arteries of interest. They will also be detected and connected. Such erroneous nodes will be removed in the following anatomy-based analysis step.

12.4.4 ANATOMY-BASED ANALYSIS

The anatomy-based analysis identifies and removes complete graphs or branches of graphs that do not belong to the arteries. Such graphs occur mainly in an articular cartilage of the knee or in the pelvic region due to the remaining false positive nodes. These graphs typically consist of many zigzagged branches that are concentrated in space. On the other hand, graphs being part of the artery structure have straight or subtly corrugated branches. When an artery passes right alongside a bone, the true and false artery graphs might be connected (Fig. 12.7). By measuring, at each line node, the angle formed between its two edges, it is possible to distinguish between true and false artery branches. If an average angle per branch is smaller than $\frac{3\pi}{4}$ [43], then this branch has to be removed (Fig. 12.7A). The angle $\frac{3\pi}{4}$ corresponds to a deviation greater than 45° from the plane angle.

In cases where an artery contains calcification, high-intensity voxels are present inside the artery. Connectivity criterion no. 3 will fail in such areas, and thus a connection between two graphs will not be created although the segments are part of the same artery tree. In the anatomy-based analysis, we identify this situation and add such connections to the graphs. For each end node A_0, we search for the presence of another end node B_0 from a different graph in its neighborhood. End node A_0 is connected to exactly one node A_1, just as end node B_0 is connected only to B_1. If $\measuredangle A_1 A_0 B_0 \geq \frac{3\pi}{4}$ and $\measuredangle A_0 B_0 B_1 \geq \frac{3\pi}{4}$ and every voxel on the link from A_0 to B_0 has an intensity within the range $[\theta^b_{low}, max\_intensity]$, and the connection between the two graphs is created (Fig. 12.7B).

After the zigzagged graphs were removed, some shorter nonartery graphs might still remain (Fig. 12.7C). By connecting the graphs separated by the calcification,

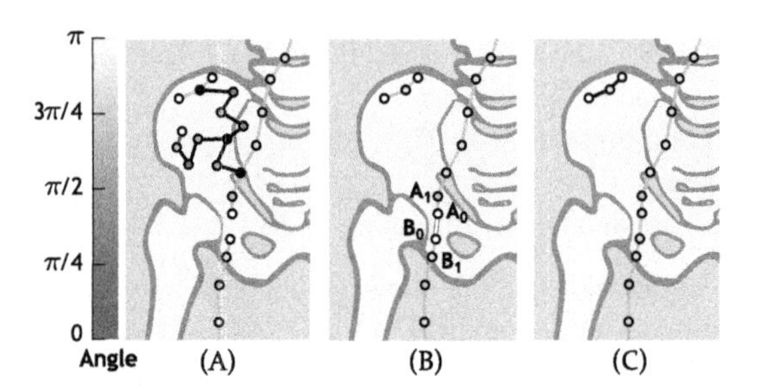

FIGURE 12.7

An artery passes right alongside the hip bone. False positive nodes from the bone are connected with artery nodes. End nodes are shown with white color, tree nodes with black, and line nodes use a linear gradient from yellow to red to express an angle between their two edges. (A) Branches having an average angle per line node smaller than $\frac{3\pi}{4}$ are shown with black lines and will be removed, (B) two end nodes A_0 and B_0 with calcification between them are shown with a white line and will be connected, (C) short graphs not being connected with the rough graph structure are shown with black lines and will be removed.

some real artery graphs were prolonged. Therefore, a good criterion for the final differentiation of artery graphs from nonartery ones is the length of the graph (by length we mean the distance between the min and max z-coordinate of the graph). All graphs shorter than $short\_graph\_thresh$ are removed. This value should be smaller than the expected length of the large arteries that we want to preserve and greater than the length of the trabecular bone regions where the nonartery graphs often occur. For our data, a fifth of the leg length worked reasonably well. This criterion is used only at the rough level of the algorithm. At the fine level of the algorithm, the graphs not having a connection with the previously built rough graph structure are removed.

12.4.5 VASCULAR SEGMENTATION BASED ON THE VASCULAR CENTERLINE TREE

The final vessel segmentation is done using an implicit model-guided level set method. In conventional model-based segmentation methods, statistical shape models are used to limit the propagation of active contours. However, creating statistical shape models for vessels is difficult due to large anatomical variation between different subjects. In [54], we proposed a skeleton-based tubular shape model to guide the level set segmentation. In this framework, a vessel model is defined as a cylinder with a varying radius $R(x)$, with x lying in the centerline L of the vessel. $R(x)$ is defined as a piecewise linear function reconstructed via robust linear regression. A vessel tree can then be represented by a union of several cylinders. In contrast to the conventional mesh-based vasculature modeling, our cylinder tree model is embedded in a 3D image as a signed distance map d_m where $d_m(x)$ is zero if x is on the surface of the model, negative if it is inside the model and positive if it is outside. The

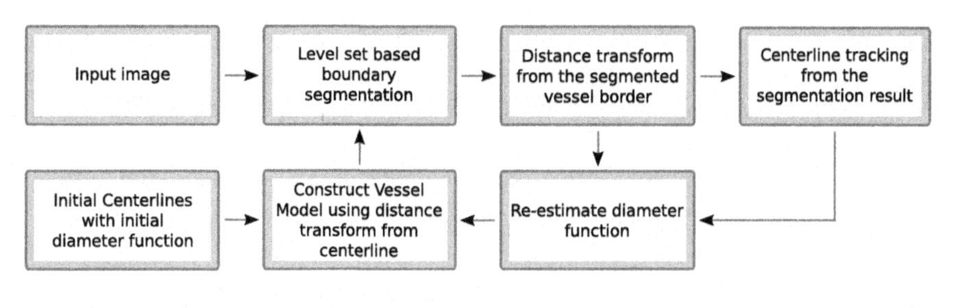

FIGURE 12.8

Flow chart of the implicit model-guided level set method.

distance map is generated by a distance transform from the centerlines while setting the points on the centerline to be $-R(x)$ as initial values.

Fig. 12.8 summarizes the implicit model-guided vessel segmentation framework. A vessel segmentation process starts by creating an implicit 3D vessel model using the preliminary centerlines generated from the previous steps. The initial vessel radius is assumed to be 1 voxel uniformly for the entire vessel tree. This model is incorporated in the level set propagation to regulate the growth of the vessel contour, so the zero level set cannot go more than a number of voxels away from the current vessel model. Meanwhile, the level set is also driven by a threshold-based image term and curvature-based internal term. After evolving the level set, new centerlines are extracted from the shape given by the zero level set through a fast marching scheme, and the diameter of vessels is reestimated in order to generate a new vessel model. The new radius function $R(x)$ is estimated for every vessel branch by fitting a smoothed curve to the distance function measured from the center points to the current zero level set $d_l(x)$ with $x \in L$. Using the new centerlines and radius functions, a new 3D tree model is created as aforementioned and the level set propagation is then resumed. The steps of level set propagation and vessel model reestimation are iteratively repeated until convergence. A fast level set algorithm [53] is used to speed up the segmentation process.

12.5 VALIDATION OF THE VASCULAR CENTERLINE TREE

The skeleton resulting from our method for fast vascular centerline tree extraction was validated against two other skeletons. One is called a reference skeleton, and the other one a united skeleton; both will be explained in detail in Section 12.5.3. For validation, we used three different metrics: an overlap rate M_o, a detection rate M_d, and an average distance error $\bar{D}_{err}$, clarified in Section 12.5.4.

The method was tested and evaluated on CTA scans of the lower limbs. A scan of the lower limbs depicts the area from the abdominal aorta to the feet. The size of such a volume is approximately $512 \times 512 \times 1700$ voxels, which offers a challenge to process it reasonably fast. Only a few algorithms proposed in the literature were tested on such big volumes. In addition, these scans were taken from the clinical routine. This means that patients were recommended to take CTA of the lower limbs because of suspicion of the lower limb peripheral arterial disease (PAD), such as arterial stenosis or occlusion. In the following subsection, we discuss the technical details of the dataset.

12.5.1 COMPUTED ANGIOGRAPHIC IMAGES OF THE LOWER LIMBS

The CTA data of the lower limbs were acquired from the clinical routine, where we identified 47 consecutive cases referred during a 17-week period for CTA with a clinical question of arterial stenosis or occlusion. By taking consecutive cases from the clinical routine we ensure that the characteristics of the sample agree with those of the relevant population [5]. Out of the 47 cases, three were excluded due to incomplete image data or severely deranged anatomy (such as amputations). The remaining 44 cases were divided into two sets: a training and a test dataset. The training data consists of 19 individual cases that were employed during the algorithm development for tuning the parameter settings. Once the parameters were set, the method was evaluated, without further modifications, using the test data consisting of 25 cases. CTA from the abdominal aorta to the feet was performed on a Siemens Somatom Flash scanner, using settings of 80 kV and 228 mAs. The protocol included automated injection of 75–100 ml of iopromide (Ultravist) 370 mg J/ml at a rate of 3.7–4.0 ml/s, followed by 70 ml of isotonic saline injected at the same rate, and images were acquired in the arterial phase. The arterial phase image acquisition means that the contrast medium has been injected but not yet reached the veins so that only the arteries were visible. Therefore, all our results depict arteries. However, the same approach should, in principle, be applicable in the venous phase, when both arteries and veins are filled with contrast, or in subtracted datasets highlighting only the veins. The average size of the volume in the training dataset was $512 \times 512 \times 1798$ voxels with an average voxel size of $0.72 \times 0.72 \times 0.7$ mm. For the test dataset, the average size of the volume was $512 \times 512 \times 1711$ voxels with an average voxel size of $0.78 \times 0.78 \times 0.7$ mm. For both datasets, the bit depth was 16 bits.

12.5.2 EVALUATION OF AUTOMATICALLY SELECTED CHANGEABLE PARAMETERS

In Section 12.4.1, we described an automatic way to properly select the intensity values of three tissues of interest: fat, muscle, and blood for each CTA scan. In this section, we evaluate the automatic selection. In most cases (17 out of 25 tested CTA datasets), the histogram of volume intensities showed clear peaks for each relevant tissue, and Gaussian curves fitted them nicely. However, in eight cases, large calcifications were present in the arteries. This resulted in fewer voxels having blood intensities and, on the contrary, more voxels having bone intensities. The blood intensity peak became too flat and the Gaussian curve fitted to this flat peak became

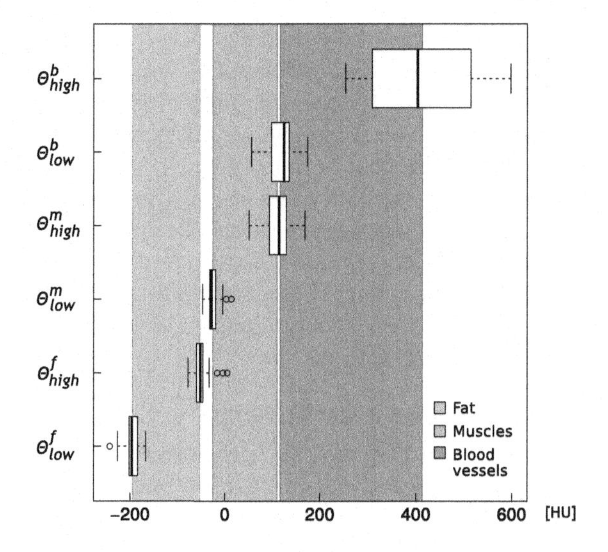

FIGURE 12.9

The median intensity values with ranges are shown for fat $[\theta^f_{low}, \theta^f_{high}]$, muscles $[\theta^m_{low}, \theta^m_{high}]$, and blood $[\theta^b_{low}, \theta^b_{high}]$ from left to right using box plots. The average intensity ranges are depicted with color bands.

extremely wide. As stated in Section 12.4.1, these situations are expected to happen, and θ^b_{low} is then automatically replaced by the position of the original histogram's minimum between the muscle and the blood peak, as the muscle and blood ranges are overlapping. Likewise, for θ^b_{high}, the predefined fallback value of 700 HU is used. The average intensity ranges (excluding the eight special cases) detected for the three tissues of interest (fat, muscles, and blood vessels, respectively) were $[-195, -51]$, $[-25, 111]$ and $[117, 414]$, which correspond to typical tabular values. The median values with respective box plots are shown in Fig. 12.9.

12.5.3 REFERENCE AND UNITED SKELETON

Both reference and united skeletons are needed for proper evaluation of our results. The reference skeleton is created by an experienced radiologist as described in the next section. Performing manual arterial tree skeleton extraction for all 25 CTA scans of the lower limbs would be out of our resource range, so the radiologist was using a semiautomatic software. Even with the help of the software, it was very tedious and almost impossible to detect the complete arterial skeleton of a human body. As it was not ensured that the radiologist did not omit some smaller arteries, the skeleton created by the radiologist cannot be named ground truth skeleton. We rather refer to it as the reference skeleton.

The united skeleton is a union of the resulting skeleton and the reference skeleton followed by necessary corrections performed by the radiologist. This process is explained in detail in the next section. The need of building a united skeleton comes from the fact that some centerline segments detected by our resulting skeleton were not detected by the radiologist in the reference skeleton. These segments would, therefore, be regarded as nonarterial wrongly detected branches. However, by performing a visual inspection the radiologist realized that most of these branches are correct artery branches which were earlier omitted by the radiologist. Conversely, there were some centerline segments that were missed by our algorithm. To provide proper and not misleading evaluation, our resulting skeleton and the reference skeleton were compared to the united skeleton.

Process of Preparing the Reference Skeleton

The reference skeleton was created using an open-source CTA analysis tool that is implemented as a plugin to the PACS workstation software OsiriX [52]. The centerline tracking is an interactive process that can be divided into two steps, vessel segmentation and centerline tracking. In the first step, a fuzzy connectedness method is used to separate the contrast-enhanced vasculature from other high-intensity structures, such as bones. This requires the radiologist to mark two sets of seed points in the images. The software then computes the connectivity between any given points to the seed points using relatively sophisticated fuzzy affinity rules. By comparing the connectivity to different seed points, the software will assign the membership of the most strongly connected seed point to these points. In healthy cases, where there are no severe artifacts and occluded vessel segments, the separation can be done with very few seed points. However, in cases where some vessel segments become occluded, the radiologist had to manually mark those regions and some connected segments on both ends to make sure that the segmented vessels are complete. The next step was skeletonization using the segmented vessel region, which is also based on the fuzzy connected theory, but combined with the vesselness filter [52]. From any distal points, a centerline was created by tracing from those end points back to the root seed points in the connectivity map while following the ridge of the vesselness map. The root seeds were put in the uppermost part of the abdominal aorta of each patient. As the vessel segmentation step may contain errors, the traced centerlines needed to be manually inspected. Wrongly marked centerline branches were removed or cropped during the inspection. The average processing time for creating the reference centerlines was around 20 min. The reference skeleton was subsequently loaded as an initial centerline tree into a level-set segmentation algorithm described in Section 12.4.5. A 3D arterial tree model, called reference segmentation, was generated.

Process of Preparing the United Skeleton

The united skeleton was prepared by computing a union of the skeleton detected by our method and the reference skeleton. The reference skeleton was free from nonartery branches; however, the resulting skeleton could contain some of these spurious branches. Therefore, all nonartery branches were removed by the radiologist using a semiautomatic utility written specifically for this purpose. It the end, only a valid arterial skeleton remained. Using the same segmentation algorithm as described in Section 12.4.5, a united segmentation was created from the initial united skeleton.

12.5.4 SKELETON EVALUATION

Three evaluation metrics, namely an overlap rate M_o, a detection rate M_d, and average distance error $\bar{D}_{err}$, were used to evaluate the detected skeleton.

The *first metric* M_o determines the overlap of the detected skeleton with the reference vascular segmentation. This measure is important in order to know if centerlines detected by our algorithm are real artery centerlines.

The *second metric* M_d determines the proportion of the reference centerline tree that was successfully detected by our detected centerline tree. This rate informs us about how much of the overall real artery centerlines were detected by our algorithm.

The *third metric* $\bar{D}_{err}$ measures the average distance error between the detected skeleton and the reference skeleton. This measure tells us how far on average they are from each other.

Overlap and Detection Metrics

In this subsection, we explain how M_o and M_d were computed. The detected vascular skeleton consists of a set of connected polyline segments. The polyline vertices are the detected artery nodes, and the straight lines between them were established during the artery nodes connection. For evaluation purposes, the detected centerline tree has to be voxelized. A set of voxels that was created by the voxelization of detected skeleton is denoted N_d. Similarly, a set of voxels N_r was created by voxelization of the reference skeleton. We define another subset called C_d of all those voxels from the voxelized detected skeleton that lie inside the voxelized reference segmentation $C_d = N_d \cap S_r$, where S_r denotes set of voxels from the voxelized reference segmentation.

The detected skeleton is, in general, less centered and thus can contain fewer voxels than the reference skeleton within the same area, and this can affect the results. Therefore, we need to introduce yet another two sets of voxels $N_{d\_centered}$ and $C_{d\_centered}$. $N_{d\_centered}$ is made by voxelizing the detected skeleton that was properly centered, and $C_{d\_centered}$ is defined as voxels lying inside the voxelized reference segmentation that are also a subset of $N_{d\_centered}$, $C_{d\_centered} = N_{d\_centered} \cap S_r$. All sets and subsets introduced in this paragraph are visually explained on a drawing of an artery in Fig. 12.10. The left subfigure shows N_d and $N_{d\_centered}$. $N_{d\_centered}$ is created from detected segmentation S_d by thinning. The middle artery branch has a sinuous shape. This shape depicts an exaggerated situation and explains why $N_{d\_centered}$ is needed for calculation of the M_d metric. The middle subfigure shows N_r together with reference segmentation S_r, and the right subfigure depicts the subsets C_d and $C_{d\_centered}$.

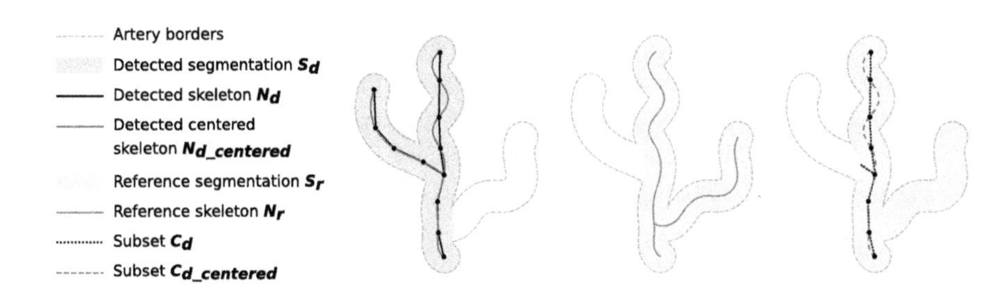

FIGURE 12.10

Drawing of an artery; (left) a situation where two artery branches were extracted and one artery branch was missed by the presented algorithm is shown; (middle) a similar situation where another two artery branches were extracted and one artery branch was missed by the radiologist is shown; (right) subsets $C_d = N_d \cap S_r$ and $C_{d\_centered} = N_{d\_centered} \cap S_r$ are shown.

The overlap rate M_o is similar to the quantity assessment in [30] and is then calculated as

$$M_o = \frac{|C_d|}{|N_d|}. \tag{12.3}$$

The detection rate M_d is determined as

$$M_d = \frac{|C_{d\_centered}|}{|N_r|}. \tag{12.4}$$

We measured the overlap and the detection rates for all 25 skeletons extracted from the test dataset for the proposed method as compared to their respective reference skeletons. The average overlap rate $\bar{M}_o$ was 62%, and the average artery detection $\bar{M}_d$ was 64%. The fairly low average overlap rate $\bar{M}_o$ suggests that many centerline segments found by our system were not in the reference segmentation. Visual inspection of the detected skeleton, however, showed that a high proportion of these centerline segments were actually correct artery centerlines but were omitted by the radiologist. In addition, some centerline segments were in fact missed by our algorithm.

Therefore, additional comparisons were performed. The rates M_o and M_d were measured for the detected skeleton as compared to the united skeleton and the reference skeleton as compared to the united skeleton. For the proposed method, the average overlap rate $\bar{M}_o$ was 97%, and the average artery detection rate $\bar{M}_d$ was 71%. For the reference skeleton, $\bar{M}_o$ was 100%, and $\bar{M}_d$ was 72%. The mean values are reported in Table 12.2, and median values with respective box plots are shown in Fig. 12.11. Fig. 12.12 shows an example of centerline tree result after the rough and fine level.

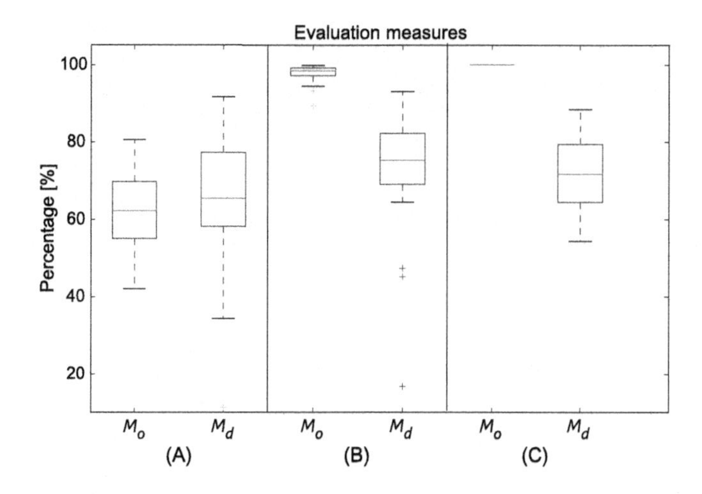

FIGURE 12.11

The overlap rate M_o and detection rate M_d, (A) proposed method compared to the reference skeleton, (B) proposed method compared to the united skeleton, (C) reference skeleton compared to the united skeleton.

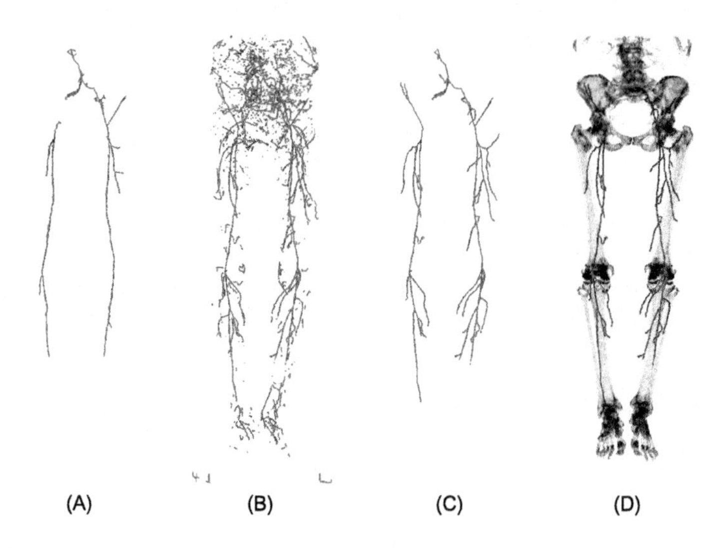

FIGURE 12.12

Example of skeleton result, (A) after performing the first pass of the algorithm, (B) after performing the second pass of the algorithm and before anatomy-based analysis, (C) final skeleton when all segments not connected with the rough graph have been removed, (D) final skeleton shown together with a CTA data with visible intensities for blood range. The example demonstrates how the skeleton is interrupted because of vascular occlusion but continues again distal to the occlusion.

Table 12.2 The overlap and detection rate for three different comparisons

Comparison	M_o	M_d
Proposed method to the reference skeleton	62.4%	64.0%
Proposed method to the united skeleton	97.4%	71.3%
Reference skeleton to the united skeleton	100%	72.2%

Distance Error Metric

In this subsection, the average distance error metric is explained in more detail. This measure shows the average distance between our detected skeleton and the reference skeleton. The distance from each voxel of the detected skeleton to the closest voxel of the reference skeleton was calculated. To avoid a bias caused by false artery skeleton segments and those true artery skeleton segments in the detected skeleton that were not included in the reference skeleton, we calculated this measure only on those parts of the detected skeleton that overlapped the reference segmentation. The resulting average distance error of 1.35 mm was measured. This result corresponds with the fact that only the artery nodes are guaranteed to be located at the center of the vessel. The voxelized edges between the nodes are, based on the connectivity criterion no. 3, inside the artery but not necessarily centered in the middle. However, as our proposed method is intended to be part of a complete arterial tree segmentation, it is sufficient if the centerline is roughly centrally located. In the complete arterial tree segmentation, the resulting skeleton is used as a seed for segmentation algorithm [54] explained in detail in Section 12.4.5. This segmentation algorithm generates a 3D artery model and a new correctly recentered centerline tree. Regardless of the initial distance deviation of the detected skeleton from the true artery centerline, the end segmentation result would be the same. In situations where curved planar reformatting needs to be used, it is possible to use splines between the artery nodes.

12.5.5 COMPUTATION TIME

The main goal of the presented algorithm is to provide whole-body vascular skeleton in reasonable time to permit interactive clinical use. Using an Intel Core i7-2760QM CPU@2.40 GHz and a C++ single-threaded implementation, the average running time of the proposed method was 88 seconds per CTA volume of mean matrix size 512 × 512 × 1711 voxels. The first pass of the algorithm took an average time of 35 seconds, whereas the second pass took an average time of 53 seconds. An advantage of the proposed method is that the detection of vascular nodes and their connection can be easily parallelized. Using an OpenMP implementation on 4 CPU cores with two threads per core, the running time on the same test dataset dropped to an average of 29 seconds and thus gives a total parallel speedup of 3.03.

12.6 ADAPTING THE ALGORITHM TO OTHER APPLICATIONS

The same algorithm can, in principle, be applied to segmentation of other anatomical tree structures. However, for this to work, it will be necessary to make some modifications, mainly in the parameter setting step. In this section, we will give a brief idea how to adapt the algorithm to three main situations that one can think of: to detect centerlines of healthy vessels of different sizes (large, medium, small), to detect centerlines of diseased vessels (severe calcification, occlusion, etc.), and to detect centerlines of other nonvascular tree structures, e.g., the airway tree.

In the presented algorithm, we have shown that by using different values for *min_radius* and *max_radius* we can extract the vascular skeleton of various artery sizes. In the algorithm, large arteries are extracted in the first pass as explained earlier and small arteries in the second algorithm pass. Using at least two algorithm passes is important for the removal of false artery branches that takes place at a later stage in the anatomy-based analysis step.

In theory, the idea of two passes can be extended to three or more algorithm passes where each one could be specialized to detect, for example, a different vascular disease. The first and second passes would remain the same. A third pass could then be added in order to detect arteries with severe calcifications. An interior of such artery is formed by blood and calcium that built up and hardened over time. To detect nodes inside the arteries with calcifications, sometimes at a length of several centimeters, can be done by making the basic criteria for node detection more inclusive and detect all voxels within the intensity range $[\theta^b_{low}, max\_intensity]$. Apart from detecting calcifications, as a side effect, nodes also will be found inside the bones; however, those nodes will not be connected to the trustworthy set of centerlines from the first phase and will, therefore, be removed in the anatomy-based analysis step. Another additional pass could focus, for example, on detecting the very tiny arteries that provide an oxygen supply to the distal parts of the body in the presence of an occlusion.

The most general case is to use the algorithm for the detection of anatomical tree structure other than vascular structures. In such case, the algorithm might need to be modified at different steps, not only in the parameter selection step. To verify this concept, we modified the algorithm to extract the skeleton of an airway tree. For this purpose, a CT piglet animal model was used. The airway tree is filled with air as compared to arteries that are filled with blood. Also, the close neighborhood of the airways consists of different tissue than the close neighborhood of arteries, and thus there are different tissues that are relevant in this case. The airway tree is smaller as compared to the vascular tree, and its branches are shorter. This observation was taken into account when setting the parameters *short_graph_thresh* and *cutting_thresh*. In the nodes detection step, we could skip the filter no. 3, which uses the parameter *gradient_thresh* since no other structures than airways have similar intensity inside the lungs and there was no need to analyse the cross-sectional shape. The nodes connection and anatomy-based analysis steps worked well without any modification for the airway tree extraction example. Fig. 12.13 shows that by

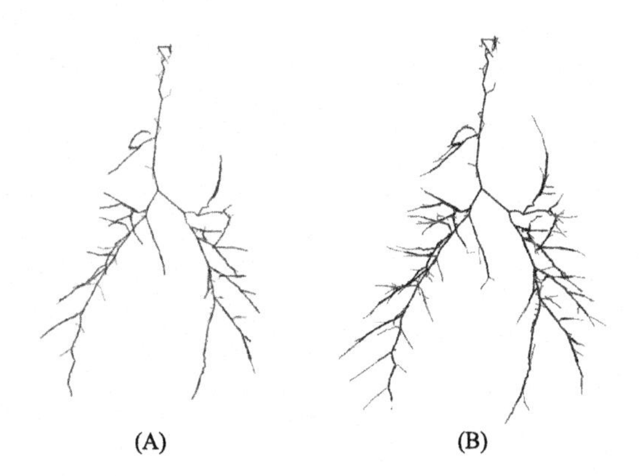

FIGURE 12.13

Extracted airway centerline tree, (A) after the first pass, shown in blue color, (B) after the second, final, pass, shown in red color.

modification of the presented algorithm it is possible to extract centerline tree of the airway tree.

12.7 DISCUSSION AND CONCLUSION

In this chapter, we presented a fully automated algorithm for fast extraction of the vessel skeleton. This algorithm has several advantages.

First, the method is fully automatic. All *changeable parameters* needed for the algorithm are extracted from the data by fitting Gaussian curves to the image histogram. We have observed that if the proportion of the arteries containing calcifications is high, then the Gaussian curves fail to fit the data. These situations are expected to happen, and the algorithm automatically switches to fall back parameters. An analysis of the histogram can also reveal abnormalities, such as hip prosthesis or metallic wires in the body. A hip prosthesis can be identified by a small peak around intensity 3000 HU. Metallic wires or other small metallic objects are much smaller in size than hip prosthesis. Therefore their presence is revealed if an obvious peak of similar intensities is visible in the logarithmic histogram instead. All *fixed parameters* are based on conventional size and shape of vascular branches we want to detect and are slightly different for different parts of vasculature. In this chapter, we have provided parameter values that work well for CTA images of the lower limbs. However, by modifying these parameters we can focus on the detection of different vascular morphology, different vascular pathologies, and various anatomical tree structures. It is even possible to detect a 2D surface of internal organs by using the same algorithmic principle. This is briefly explained in Section 12.8.

Second, our algorithm is fast enough to permit interactive clinical use. The algorithm forms the basis for a future clinical tool, enabling the user to identify and quantify stenoses and occlusions in the named arteries of the lower limbs. The current computation time of 29 seconds for finding a vascular skeleton in a CTA volume of a mean matrix size of $512 \times 512 \times 1711$ voxels using a multithread implementation is perfectly acceptable for achieving this goal. However, the implementation is not optimized, and, for example, traversing the volume is done multiple times. Thus, a streamline implementation is expected to reduce the computation time. Unfortunately, there are not many methods dealing with vascular skeleton extraction that are tested on a CTA dataset of the lower limbs, a full body CTA dataset, or any other comparatively large and challenging CTA volume. In the work [20] that was tested on the lower limbs, the authors do not mention the execution time for constructing the centerline. The only time information in the paper refers to the full coronary artery trees and claims that these can be extracted in 21 seconds in average on a 3.2 GHz PC. If we relate this time to our proposed method, then reducing the volume size by half should lower the execution time by half and reach approximately 15 seconds. Without using a multithread implementation, the same execution time of 20 seconds can be reached by processing a volume of size $120 \times 120 \times 410$ voxels. In a paper [11], an execution time of 2.3–31.4 seconds is declared for processing a real CTA volume of size $512 \times 512 \times 386$ voxels. However, the length of the extracted centerline is less than 100 voxels ≈ 7 cm.

Third, the proposed algorithm is selective in detecting skeletons that are truly vascular skeletons. The average overlap rate $\bar{M}_o$ of our resulting skeleton compared to the united skeleton was 97%. The average detection rate $\bar{M}_d$, measuring what proportion of the true artery skeleton was detected by our proposed algorithm, was 71%. This is only 1% less than the average detection rate $\bar{M}_d$ for the reference skeleton compared to the united skeleton. To understand these figures, we performed a close visual inspection of the detected skeleton, comparing it to the reference skeleton. Both methods were successful in detecting all main artery branches. The major difference was observed in cases where vascular occlusion was present. Whereas our resulting centerline was interrupted and continued again once the occlusion disappeared, the reference centerline was still continuous. The connection was traced through tiny arteries that were hardly visible in the CT scan, and therefore our method failed to detect them. However, these small arteries, which are the only way for the blood to reach the more distal parts of the leg, are important, and we plan to look into this in our future research. On the other hand, our method detected some additional artery branches that were omitted by the radiologist. In current clinical practice, the number of arteries described by the radiologist is limited by the time available for the manual review work rather than by the interest of the referring vascular surgeon. Therefore, a method describing more arteries than is currently clinically feasible is of great interest. Fig. 12.14 shows two examples from a clinical routine where occlusion and arterial stent are clearly visible in maximum intensity projection images. It should be borne in mind that our patient material represents the more diseased end of the spectrum of arterial disease, where vascular surgeons are considering surgery or

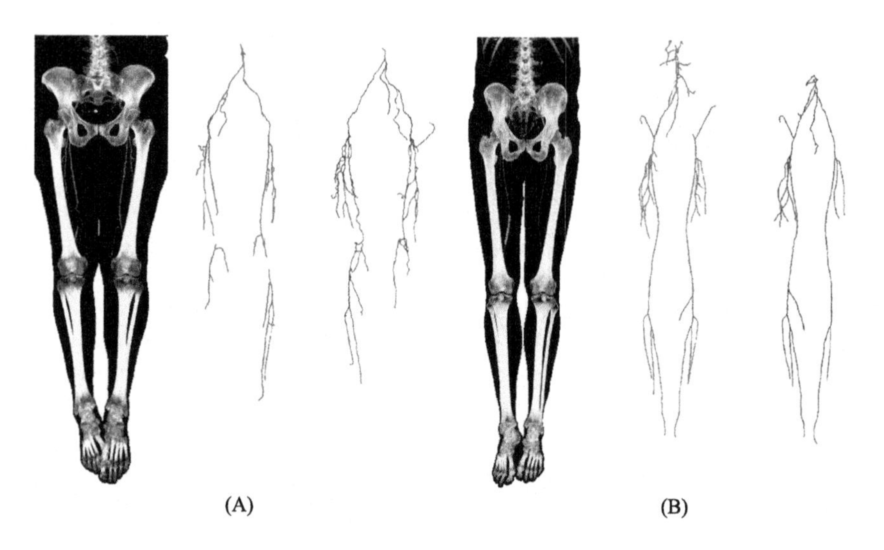

FIGURE 12.14

Maximum intensity projection, resulting skeleton in red color and reference skeleton in blue color are shown for two cases from the clinical routine, (A) contains an occluded segment in the femoral artery, (B) contains an arterial stent.

other interventions. It is, however, relevant to evaluate novel proposed methods even in these rather difficult circumstances. We do not know whether an evaluation in a less diseased patient cohort would give more advantageous results.

The major limitations of our method are the following: The position of the centerline is not accurately centered in the middle of the artery, only the artery nodes are ensured to be located at the center. For the curved planar reformatting, it is advisable to calculate splines between the nodes. However, for our purpose, the speed gained by the linear interpolation is more important than accuracy, as the resulting arterial skeleton is used as seed for a segmentation algorithm [54], which generates the same 3D artery model for any roughly centrally located centerline tree. Another consideration is that, for some fixed parameters ($cutting\_thresh$, $short\_graph\_thresh$), it is hard to determine one specific threshold value giving the best results for each dataset. It is a trade-off between keeping the false positives and removing the true negatives. We used stricter values and rather removed some artery branches and graphs than kept the nonartery ones. Possibly, the $short\_graph\_thresh$ would need to be adjusted for CTA scans of other body parts than the lower limbs. The rationale is that the lower limbs contain longer segments of large arteries compared, e.g., to CTA of the cranial arteries in the head. Similarly, suitable min_radius and max_radius should be chosen based on the expected size of the vessels in the targeted clinical application.

12.8 FUTURE DEVELOPMENT OF THE METHOD

Our plan for future work includes increasing the speed of the proposed algorithm. Further, we plan to focus on cases where vascular occlusion is present. Also, we would like to expand this algorithm from CTA to other imaging modalities, mainly to MRA.

Furthermore, the principle of using statistical clustering for generating these representative structures of objects is actually not limited to 1D structures, and computation time will stay in the same order of magnitude even in the 2D case. However, again there is a caveat: generating a 2D structure will not give anatomic guides of the same kind as those of a centerline or a center tree structure. If, for example, part of a surface is missing, then what will the actual transform between the surface and a 3D object be? With this in mind, we currently have a working surface generating algorithm able to segment and distinguish lungs in a full-body volumetric image. Computation time is around 5 seconds on an average laptop (single core used). The same algorithm is also capable of segmenting the colon instead of the lungs, or both simultaneously, if set up to do so.

ACKNOWLEDGMENTS

This work was funded by the Swedish Research Council (grant no. VR-NT 2014-6153).

REFERENCES

[1] S. Achenbach, F. Moselewski, D. Ropers, M. Ferencik, U. Hoffmann, B. MacNeill, K. Pohle, U. Baum, K. Anders, I.K. Jang, W.G. Daniel, T.J. Brady, Detection of calcified and noncalcified coronary atherosclerotic plaque by contrast-enhanced, submillimeter multidetector spiral computed tomography: a segment-based comparison with intravascular ultrasound, Circulation 109 (1) (2004) 14–17.

[2] H.V. Anderson, G.S. Roubin, P.P. Leimgruber, W.R. Cox, J.S. Douglas Jr., S.B. King III, A.R. Gruentzig, Measurement of transstenotic pressure gradient during percutaneous transluminal coronary angioplasty, Circulation 73 (6) (1986) 1223–1230.

[3] Malin Andersson, Karl Jägervall, Per Eriksson, Anders Persson, Göran Granerus, Chunliang Wang, Örjan Smedby, How to measure renal artery stenosis – a retrospective comparison of morphological measurement approaches in relation to hemodynamic significance, BMC Med. Imaging 15 (1) (2015) 1–11.

[4] Stephen R. Aylward, Elizabeth Bullitt, Initialization, noise, singularities, and scale in height ridge traversal for tubular object centerline extraction, IEEE Trans. Med. Imaging 21 (2) (2002) 61–75.

[5] Mette Bjørn, Connie Brendstrup, Sven Karlsen, Jan E. Carlsen, Consecutive screening and enrollment in clinical trials: the way to representative patient samples?, J. Card. Fail. 4 (3) (1998) 225–230.

[6] Sylvain Bouix, Kaleem Siddiqi, Allen Tannenbaum, Flux driven fly throughs, in: Proceedings 2003 IEEE Computer Society Conference on Computer Vision and Pattern Recognition, vol. 1, IEEE, June 2003, pp. 449–454.

[7] Katja Bühler, Petr Felkel, Alexandra La Cruz, Geometric Methods for Vessel Visualization and Quantification – A Survey, Springer, Berlin, Heidelberg, 2004, pp. 399–419.

[8] Hasan Ertan Çetingül, Mehmet Akif Gülsün, Hüseyin Tek, A unified minimal path tracking and topology characterization approach for vascular analysis, in: Medical Imaging and Augmented Reality, Springer, 2010, pp. 11–20.

[9] Nicu D. Cornea, Deborah Silver, Patrick Min, Curve-skeleton properties, applications, and algorithms, IEEE Trans. Vis. Comput. Graph. 13 (3) (2007) 530–548.

[10] Edsger W. Dijkstra, A note on two problems in connexion with graphs, Numer. Math. 1 (1) (1959) 269–271.

[11] Jan Egger, Zvonimir Mostarkic, Stefan Großkopf, Bernd Freisleben, A Fast Vessel Centerline Extraction Algorithm for Catheter Simulation, IEEE Computer Society, 2007, pp. 177–182.

[12] T. Todd Elvins, A survey of algorithms for volume visualization, ACM SIGGRAPH Comput. Graph. 26 (3) (1992) 194–201.

[13] Elliot K. Fishman, Derek R. Ney, David G. Heath, Frank M. Corl, Karen M. Horton, Pamela T. Johnson, Volume rendering versus maximum intensity projection in CT angiography: what works best, when, and why, Radiographics 26 (3) (2006) 905–922.

[14] Alejandro F. Frangi, Wiro J. Niessen, Koen L. Vincken, Max A. Viergever, Multiscale vessel enhancement filtering, in: Medical Image Computing and Computer-Assisted Intervention – MICCAI98, Springer, 1998, pp. 130–137.

[15] Hans Frimmel, Janne Näppi, Hiroyuki Yoshida, Fast and robust computation of colon centerline in CT colonography, Med. Phys. 31 (11) (2004) 3046–3056.

[16] Hans Frimmel, Janne Näppi, Hiroyuki Yoshida, Centerline-based colon segmentation for CT colonography, Med. Phys. 32 (8) (2005) 2665–2672.

[17] Xinpei Gao, Shengxian Tu, Michiel A. de Graaf, Liang Xu, Pieter Kitslaar, Arthur J.H.A. Scholte, Bo Xu, Johan H.C. Reiber, Automatic extraction of arterial centerline from whole-body computed tomography angiographic datasets, in: Computing in Cardiology Conference, 2014, IEEE, Sept. 2014, pp. 697–700.

[18] Robert W. Gill, Pulsed Doppler with B-mode imaging for quantitative blood flow measurement, Ultrasound Med. Biol. 5 (3) (1979) 223–235.

[19] A. Gruntzig, D.A. Kumpe, Technique of percutaneous transluminal angioplasty with the Gruentzig balloon catheter, Am. J. Roentgenol. 132 (4) (1979) 547–552.

[20] M. Akif Gülsün, Hüseyin Tek, Robust vessel tree modeling, in: Medical Image Computing and Computer-Assisted Intervention – MICCAI 2008, Springer, 2008, pp. 602–611.

[21] J.C.P. Heggie, N.A. Liddell, K.P. Maher, Applied Imaging Technology, 4th ed., St. Vincent's Hospital Melbourne, 2001.

[22] S.J. Hessel, D.F. Adams, H.L. Abrams, Complications of angiography, Radiology 138 (2) (1981) 273–281.

[23] W.R. Hiatt, Medical treatment of peripheral arterial disease and claudication, N. Engl. J. Med. 344 (21) (2001) 1608–1621.

[24] C. Daniel Johnson, Abraham H. Dachman, CT colonography: the next colon screening examination?, Radiology 216 (2) (2000) 331–341.

[25] Cemil Kirbas, Francis K.H. Quek, A review of vessel extraction techniques and algorithms, ACM Comput. Surv. 36 (2) (2004) 81–121.

[26] T. Yung Kong, Azriel Rosenfeld, Digital topology: introduction and survey, Comput. Vis. Graph. Image Process. 48 (3) (1989) 357–393.

[27] David Lesage, Elsa D. Angelini, Isabelle Bloch, Gareth Funka-Lea, A review of 3D vessel lumen segmentation techniques: models, features and extraction schemes, Med. Image Anal. 13 (6) (2009) 819–845.

[28] Hua Li, Anthony Yezzi, Vessels as 4-D curves: global minimal 4-D paths to extract 3-D tubular surfaces and centerlines, IEEE Trans. Med. Imaging 26 (9) (2007) 1213–1223.

[29] P. Libby, P.M. Ridker, A. Maseri, Inflammation and atherosclerosis, Circulation 105 (9) (2002) 1135–1143.

[30] C.T. Metz, M. Schaap, A.C. Weustink, N.R. Mollet, T. van Walsum, W.J. Niessen, Coronary centerline extraction from CT coronary angiography images using a minimum cost path approach, Med. Phys. 36 (12) (2009) 5568–5579.

[31] M. Miyazaki, M. Akahane, Non-contrast enhanced MR angiography: established techniques, J. Magn. Reson. Imaging 35 (1) (2012) 1–19.

[32] S. Napel, M.P. Marks, G.D. Rubin, M.D. Dake, C.H. McDonnell, S.M. Song, D.R. Enzmann, R.B. Jeffrey Jr., CT angiography with spiral CT and maximum intensity projection, Radiology 185 (2) (1992) 607–610.

[33] Janne Näppi, Abraham H. Dachman, Peter MacEneaney, Hiroyuki Yoshida, Automated knowledge-guided segmentation of colonic walls for computerized detection of polyps in CT colonography, J. Comput. Assist. Tomogr. 26 (4) (2002) 493–504.

[34] A.B. Newman, K. Sutton-Tyrrell, L.H. Kuller, Lower-extremity arterial disease in older hypertensive adults, Arterioscler. Thromb. Vasc. Biol. 13 (4) (1993) 555–562.

[35] Ingela Nyström, Örjan Smedby, Skeletonization of volumetric vascular images—distance information utilized for visualization, J. Comb. Optim. 5 (1) (2001) 27–41.

[36] J.W. Olin, B.A. Sealove, Peripheral artery disease: current insight into the disease and its diagnosis and management, Mayo Clin. Proc. 85 (7) (2010) 678–692.

[37] P. Pignoli, E. Tremoli, A. Poli, P. Oreste, R. Paoletti, Intimal plus medial thickness of the arterial wall: a direct measurement with ultrasound imaging, Circulation 74 (6) (1986) 1399–1406.

[38] Jeffrey William Prescott, Quantitative imaging biomarkers: the application of advanced image processing and analysis to clinical and preclinical decision making, J. Digit. Imaging 26 (1) (2013) 97–108.

[39] M.R. Prince, T.M. Grist, J.F. Debatin, 3D Contrast MR Angiography, Springer-Verlag, Berlin, 1999.

[40] M.R. Prince, E.K. Yucel, J.A. Kaufman, D.C. Harrison, S.C. Geller, Dynamic gadolinium-enhanced three-dimensional abdominal MR arteriography, J. Magn. Reson. Imaging 3 (6) (1993) 877–881.

[41] G.D. Rubin, P.J. Walker, M.D. Dake, S. Napel, R.B. Jeffrey, C.H. McDonnell, R.S. Mitchell, D.C. Miller, Three-dimensional spiral computed tomographic angiography: an alternative imaging modality for the abdominal aorta and its branches, J. Vasc. Surg. 18 (4) (1993) 656–664.

[42] M. Schaap, C.T. Metz, T. van Walsum, A.G. van der Giessen, A.C. Weustink, N.R. Mollet, C. Bauer, H. Bogunovic, C. Castro, X. Deng, E. Dikici, T. O'Donnell, M. Frenay, O. Friman, M. Hernandez Hoyos, P.H. Kitslaar, K. Krissian, C. Kuhnel, M.A. Luengo-Oroz, M. Orkisz, Ö. Smedby, M. Styner, A. Szymczak, H. Tek, C. Wang, S.K. Warfield, S. Zambal, Y. Zhang, G.P. Krestin, W.J. Niessen, Standardized evaluation methodology and reference database for evaluating coronary artery centerline extraction algorithms, Med. Image Anal. 13 (5) (2009) 701–714.

[43] Ö. Smedby, N. Högman, S. Nilsson, U. Erikson, A.G. Olsson, G. Walldius, Two-dimensional tortuosity of the superficial femoral artery in early atherosclerosis, J. Vasc. Res. 30 (4) (1993) 181–191.
[44] American Cancer Society, Cancer Facts & Figures, The Society, 2004.
[45] Amnon Sonnenberg, Fabiola Delcò, John M. Inadomi, Cost-effectiveness of colonoscopy in screening for colorectal cancer, Ann. Intern. Med. 133 (8) (2000) 573–584.
[46] H.C. Stary, Natural history and histological classification of atherosclerotic lesions: an update, Arterioscler. Thromb. Vasc. Biol. 20 (5) (2000) 1177–1178.
[47] Matúš Straka, Michal Červeňanský, Alexandra La Cruz, Arnold Köchl, Eduard Gröller, Dominik Fleischmann, et al., The VesselGlyph: Focus & Context Visualization in CT-Angiography, IEEE, 2004.
[48] Jasjit S. Suri, Kecheng Liu, Laura Reden, Swamy Laxminarayan, A review on MR vascular image processing: skeleton versus nonskeleton approaches: part II, IEEE Trans. Inf. Technol. Biomed. 6 (4) (December 2002) 338–350.
[49] H.S. Thomsen, P. Marckmann, V.B. Logager, Nephrogenic systemic fibrosis (NSF): a late adverse reaction to some of the gadolinium based contrast agents, Cancer Imaging 7 (2007) 130–137.
[50] W.J. van Rooij, M.E. Sprengers, A.N. de Gast, J.P. Peluso, M. Sluzewski, 3D rotational angiography: the new gold standard in the detection of additional intracranial aneurysms, Am. J. Neuroradiol. 29 (5) (2008) 976–979.
[51] D.J. Vining, Virtual colonoscopy, Gastrointest. Endosc. Clin. N. Am. 7 (2) (1997) 285–291.
[52] Chunliang Wang, Hans Frimmel, Anders Persson, Örjan Smedby, An interactive software module for visualizing coronary arteries in CT angiography, Int. J. Comput. Assisted Radiol. Surg. 3 (1–2) (2008) 11–18.
[53] Chunliang Wang, Hans Frimmel, Örjan Smedby, Fast level-set based image segmentation using coherent propagation, Med. Phys. 41 (7) (2014) 073501.
[54] Chunliang Wang, Rodrigo Moreno, Örjan Smedby, Vessel segmentation using implicit model-guided level sets, in: Proceedings of MICCAI Workshop 3D Cardiovascular Imaging: A MICCAI Segmentation Challenge – MICCAI, 2012.
[55] Chunliang Wang, Örjan Smedby, Coronary artery segmentation and skeletonization based on competing fuzzy connectedness tree, in: Medical Image Computing and Computer-Assisted Intervention – MICCAI, vol. 4791, Springer, 2007, pp. 311–318.
[56] Chunliang Wang, Örjan Smedby, Integrating automatic and interactive methods for coronary artery segmentation: let the PACS workstation think ahead, Int. J. Comput. Assisted Radiol. Surg. 5 (3) (2010) 275–285.
[57] P.G. Yock, D.T. Linker, Intravascular ultrasound. Looking below the surface of vascular disease, Circulation 81 (5) (1990) 1715–1718.
[58] Hiroyuki Yoshida, Janne Näppi, Peter MacEneaney, David T. Rubin, Abraham H. Dachman, Computer-aided diagnosis scheme for detection of polyps at CT colonography 1, Radiographics 22 (4) (2002) 963–979.
[59] C. Yuan, K.W. Beach, L.H. Smith Jr., T.S. Hatsukami, Measurement of atherosclerotic carotid plaque size in vivo using high resolution magnetic resonance imaging, Circulation 98 (24) (1998) 2666–2671.
[60] Christopher K. Zarins, Charles A. Taylor, James K. Min, Computed fractional flow reserve (FFTCT) derived from coronary CT angiography, J. Cardiovasc. Trans. Res. 6 (5) (2013) 708–714.

Index

Symbols

0-cell, 229, 245, 249
  single, 231, 233
0-cliques, 185, 188, 198, 199
0-faces, 183, 223, 224
0-surface, 190, 195, 199
1-cells, 229
1-cliques, 185, 188, 196, 198, 199
1-faces, 183, 223, 224
1-isthmus, 182, 189, 190, 195, 196, 200
1-neighbors, 183, 198
1-surface, 190, 191, 195, 196, 199
2-cells, 232, 233
2-cliques, 185, 188, 196, 198, 199
  critical, 186, 187
2-faces, 183, 186, 224
  disjoint, 224
2-isthmus, 182, 189, 190, 195, 196, 199, 200, 207
2-neighbors, 183, 184
3-clique, 185, 188, 198
  critical, 185–187
α-cuts, 72, 74

A

Adjacency relation, 51, 220, 289
AFMM, 60, 61
Airway trees, 160, 169, 172, 173, 375, 376
Algorithm pass, 354, 355, 363
  second, 363, 375
Algorithms, 5, 6, 8, 25, 153, 154, 167, 301, 354
  asymmetric, 182
  benchmark, 322, 323
  centerline extraction, 29, 175, 351
  curve skeletonization, 13
  Dijkstra's, 100, 162, 299
  efficient, 20, 44, 153, 197
  fully predicate-kernel-based iterative, 14
  iterative boundary peeling, 14
  Ma and Sonka's, 25
  most skeletonization, 152, 290
  multiscale skeletonization, 44, 56
  parallel skeletonization, 24–26, 30
  parallel thinning, 182, 186, 212–217, 253
  skeletonization, 5, 6, 8, 9, 14, 15, 25, 29, 105, 124–126, 147, 148, 175, 290
  surface skeletonization, 30, 155
  symmetric, 182
Analysis
  anatomy-based, 354–356, 365, 373, 375
Anatomic shapes, 28, 114, 115, 290
Animation, 26, 27, 152, 153, 169
Apex errors, 334–337
Apexes, 334, 336, 337
Applications, 29, 31, 43, 45, 72, 73, 92, 151, 169, 283, 316, 348
  object recognition, 262
  potential, 214, 263, 273
  real-life, 153, 154, 305
Approaches, 7, 9, 10, 72–74, 77, 110, 155, 156, 200–202, 213, 253, 327, 349
  dilation-based, 83
  direct, 81, 82
  distance-based, 72, 73, 82, 84
  graph-based, 10
  main, 71, 72, 84, 350
  similar, 74, 77, 100, 353
Approximation, 10, 12, 73, 93, 100, 140, 142–145
Arterial centerlines, 174, 351
Arterial phase, 348, 368
Arterial tree, 347–350
Arterial tree segmentation
  complete, 374
Arterial wall, 347
Arteries, 347–349, 355–366, 368, 371, 372, 374–378
  large, 354, 355, 362, 363, 366, 375, 378
  smaller, 355, 358, 362, 369
Arteriosclerosis, 362
Artery branches, 372, 377, 378
Artery centerlines
  real, 371
Artery nodes, 359, 360, 363, 364, 366, 374, 378
  false, 355, 356, 359, 361, 362
  potential, 359, 360
  true, 355, 362

Skeletonization. DOI: 10.1016/B978-0-08-101291-8.00023-7

Atherosclerosis, 347, 349
Augmented fast-marching method, *see* AFMM
Average intensity, 361, 362, 369
Average voxel size, 368
Axes
  central, 298, 334–336
Axial points, 74, 298, 299
  most-representative, 298, 299

B

Background, 19, 71, 91, 125, 126, 273, 277, 289, 350
Background 6-neighbor, 20, 21, 292
Backprojecting, 94, 95, 106
Backprojects, 94, 95
Bertrand's definition, 213, 214, 217, 253, 254
Birth date, 200, 201
Blood, 355, 358, 360–362, 368, 369, 375, 377
Blood peak, 358, 369
Blood vessels, 76, 346, 352, 357, 369
BMD, 287, 288, 303
BMI, 288, 303
Body mass index, *see* BMI
Body-centered cubic grids, 214
Bone, 72, 358, 359, 361, 362, 365, 366, 370, 375
Bone mineral density, *see* BMD
Bone strength, 28, 288
Bone volume fraction, *see* BVF
Boundary
  entire, 59, 153, 223, 227, 232, 233
  object's, 24, 124, 148, 317
  smoothed, 107
  vessel, 350
Boundary points, 5, 10, 58–60, 73, 84, 105, 108, 154
  initial, 59
Box
  bounding, 134, 137, 277, 279, 332
Branch points, 22, 124, 127, 129, 130, 132, 133, 146, 148
  clusters of, 125, 129, 130, 143, 148
  delimiting, 22, 23, 132
  fusion of close, 125, 127
Branch significance, 124, 166, 167
  skeletal, 153, 166, 167
Bronchoscopy, 28, 170–172
BVF, 302, 303

C

CAD, 15, 25, 95, 132, 182, 187–189, 194, 207, 324, 340, 353
Calcifications, 347, 349, 351, 354, 358, 365, 366, 375, 376
Candidate voxel, 20, 360–362
Cartesian arrays
  2-dimensional, 213–216, 221, 234, 245
  3-dimensional, 213–216, 221, 234, 239, 245
  4-dimensional, 213–217, 228, 234, 245
  n-dimensional, 216, 221, 225, 228, 235
Cartesian arrays complexes, 217, 220, 228, 234, 239, 253, 254
Catheter, 347, 348
Cavities, 19, 47, 160, 169, 190, 289–293
CBMBs, 9, 16, 72, 74–76, 79, 291
CBS, 353
Cells, 216, 219–221, 225, 227, 231, 237, 238, 241, 242, 245, 247, 248, 253
  free pair of, 231, 232, 246, 249
  maximal, 217
Centerline segments, 157, 370, 372
Centerline tracing, 156, 158, 170, 174
Centerline tree, 157, 354
  detected, 371
Centerline-based segmentation, *see* CBS
Centerlines, 28, 157, 162–164, 277, 279, 346, 351–355, 358, 370, 371
  computed, 174, 330, 352
Centers of maximal balls, *see* CMBs
Central voxels, 21, 162, 316, 360
Centroids, 129–131, 143, 147, 148
CFD, 349
Circumscribing, 55, 56
Classifiers, 267, 268
Clinical applications, 327, 378
Clinical routine, 350, 368, 377, 378
Cliques, 182, 185–189, 193–195, 197–200, 237, 242
  element of, 239, 240, 244
  essential, 198–200
  maximal subset of, 244
  multiple-voxel, 201

Cliques *(cont.)*
proper subset of, 243
Close branch points, 124, 125, 127, 129
Closest nodes, 364, 365
Cluster, 124, 129, 130
CMBs, 16, 17, 22, 72–75, 161, 163–166, 291
strong, 163, 164
CMDs, 126, 133, 134
CMIBs, 152, 154, 155, 159
Cochlea, 327, 330, 331, 334, 336, 338, 339
Cochlear, 329, 331, 333–337, 339, 340
Collection, 120, 183, 214, 225, 226, 241, 243, 330
Collision impact, 163–166
Collision-impact values, 19, 163, 164
Colonic pathway, 352, 353
Colonoscopy, 28, 156, 170, 171, 352
Complexes, 183, 185, 213, 219, 221, 222, 233, 235, 239, 246, 247, 249
Complexity, 10, 56, 60, 66, 206, 207
Components, 4, 20, 45, 52, 54, 79, 83, 292–295
weak, 220, 221, 243–245
Computation time, 350, 351, 353, 374, 377, 379
Computational approaches, 5, 30, 152
Computational complexity, 167, 168
Computational fluid dynamics, *see* CFD
Computed tomographic colonography, *see* CTC
Computed tomography angiography, *see* CTA
Computed tomography colonoscopy, 347, 353
Computed tomography, *see* CT
Computer-aided detection, *see* CAD
Computing curve skeletons, 154, 156, 158, 159
Computing multiscale, 44, 67
Concatenations, 127, 131–135
Connected components, 52, 78, 82, 83, 96, 129, 165, 184, 230, 231
Connection
artery nodes, 355, 364, 371
Connectivity, 65, 102, 364, 370
Constraint set, 26, 187, 189, 191, 194, 201
Contrast medium, 348, 368
Conventional methods, 169, 171
Convex polytope, 218, 219, 221–223, 226, 227, 231–234, 243, 246
Convolutional layers, 261, 265, 266
Coordinates
spatial, 138–144, 149
Cost function, 67, 158, 332, 333, 351
Crisp, 71–76, 78, 82, 84
Crisp sets, 72–75, 82
Crisp skeleton, 73, 74
Critical cliques, 182, 186, 187, 194, 195, 197, 199
Critical kernels, 26, 181, 186, 213, 214, 216, 217, 239, 242, 245, 253
definition of, 217, 253, 254
Critical kernels of images, 213, 214
Crossing, 21, 299–301
CT, 172, 346, 348, 352, 357
CTA, 348, 349, 352, 359, 368, 378, 379
CTA datasets, 351, 377
CTA volume, 354, 374, 377
CTC, 352
CTC image, 353
Curvature, 111, 124, 125, 141, 142, 148
Curve evolution process, 5, 12, 290
Curve skeleton hierarchical structure, 94
Curve skeleton points, 44, 154
Curve skeletonization
applications of, 26, 153, 170
guide, 155, 297
minimum-cost path-based, 158, 160, 165, 167
Curve skeletonization methods, 154, 175
Curve skeletons, 44, 47–50, 53, 67, 90, 92–94, 98, 106, 152–158, 191, 192, 195
computed, 170
distance-labeled, 297
initial, 159
meaningful, 169
pure, 4
symmetric, 195
thin, 192
Curve structures, 295, 296
Curve-skeleton points, 98
Curves, 92, 94, 127, 152, 279, 280, 291, 293
continuous, 8, 11, 12
geodesic, 104
precision/recall, 271, 273, 276, 281

Curves (*cont.*)
simple, 94, 147, 148, 189, 192
simple closed, 190, 199, 200, 294
Cut-sets, 100, 102, 104
Cut-space, 96, 100, 105, 116
Cut-space partitioning, 102, 116

D

d-cliques, 185, 188, 194, 202, 204, 206
d-neighbors, 183, 185
Dataset, 66, 262, 263, 331
SK506, 262–264, 271–273
SYMMAX300, 263, 264, 273, 274
test, 357, 368, 372, 374
WH-SYMMAX, 263, 264, 273–275
Detection, 101, 133, 196–198, 263, 290, 354, 359, 374, 375
artery nodes, 355, 360
noisy branch, 157, 158, 169
object proposal, 263, 277, 283
road, 260, 262, 277, 281, 283
Detection rates, 367, 371–374, 377
Digital images, 8, 12, 76, 155, 161, 289, 291, 297
fuzzy, 289
Digital objects, 6, 9, 14, 29, 124, 152, 153, 166, 292
binary, 12, 17, 161
fuzzy, 14, 15, 17, 161, 163, 166
segmented, 161
Digital subtraction angiography, *see* DSA
Digital topological analysis, *see* DTA
Dilations, 9, 76, 79–81
Direct volume rendering, *see* DVR
Discs, 124, 126, 129, 133, 134
Distance
boundary-to-curve-skeleton, 94
closest point, 334
geodesic, 104, 155, 299
linearized, 144, 145
symmetric surface, 323, 324
Distance functions, 72, 73, 367
Distance information, 126, 147
Distance map, 315–317, 322, 324, 367
Distance transform, *see* DT
Distance values, 15, 27, 52, 55, 125–128, 133, 139–146, 148, 149
distribution of, 128, 142
DSA, 347
DT, 8, 12, 14, 15, 126, 127, 153, 155, 161, 162, 316, 353
DT value, 17, 162, 163
DTA, 289, 291, 292, 296, 298
Dual-energy X-ray absorptiometry, *see* DXA
DVR, 349
DXA, 288

E

Edge detection, 263
Edge-neighbors, 125–127, 144
Edges, 27, 51, 114, 218, 223, 229, 230, 289, 293, 300, 354, 360
Ellipses
bounding, 360–363
inner, 360–363
outer, 360–363
Endvertices, 229, 230
Erosion, 30, 47, 75–77, 79, 152, 155
Errors, 30, 298, 305, 326, 327, 335, 337, 370
average distance, 367, 371, 374
Essential cell, 216, 217, 237, 251, 253
critical, 216, 217, 253
Euclidean distance, 142
Euclidean skeletons, 45
Euler characteristic, 220, 221, 227, 229, 231, 233
Evaluation of skeleton extractors, 272, 275
Execution time, 350, 360, 363, 377
Experimental results, 169, 187, 283
Experiments, 191, 207, 262, 267, 271, 277, 279, 283, 297, 352
Expression, 79–81
Extension, 52, 67, 73, 74, 77–80, 100, 136, 182, 293, 339
surface–surface junction line, 294
Extraction
curve-skeleton, 94
Extremes, 133, 141, 145, 146

F

F-measure, 271, 272, 274, 275, 277
Facets, 220–222, 225, 231, 242, 245–252
border, 220, 234, 235
Fandisk, 108, 109
fCMB, 161
FCN, 261, 270, 283

FDT, 9, 17, 161, 301
Feature points, 45, 46, 49–51, 93, 97, 100, 105, 106, 108, 118
Feature propagation, 289, 298, 299, 302
Feature transform, 45, 106, 108
  extended, 108, 109
Features
  single-point, 53, 55, 59
Filtered skeletons, 198, 202–205
Filters, 261, 318, 355, 359–363, 375
  particle, 263, 271, 272, 274
Fire-fronts, 7, 12, 17, 18, 23, 153, 163
Foreground complex, 216, 222, 233, 235, 237–239, 242, 245, 247
FSDS, 269–272, 274, 275, 279
Fully convolutional network, *see* FCN
Fuzzy center of maximal ball, *see* fCMB
Fuzzy dilations, 76, 79, 80
Fuzzy distance transform, *see* FDT
Fuzzy distance transform value, 298, 301
Fuzzy numbers, 80, 81, 84
Fuzzy objects, 9, 17–19, 73, 82, 83, 155–158, 166, 289
Fuzzy sets, 9, 17, 71–80, 82, 84
Fuzzy skeleton, 72, 73, 76, 84
Fuzzy skeletonization, 9, 14, 16, 152
Fuzzy SKIZ, 9, 72, 80, 82–84

G

Gaussian curves, 358, 368, 376
Gaussian kernel, 262, 318, 319, 332
GDT, 154, 155, 289, 298–301
GDT computation, 155, 299, 300
GDT values, 298, 299, 301
Generation
  text line proposal, 277, 280, 283
Geodesic distance transform, *see* GDT
Geometric features, 4, 151, 152, 169, 170, 173, 175, 290
Geometric steerable medial maps, *see* GSM2
Grassfire propagation, 5, 7, 11, 13, 17, 152–154, 163, 291, 298
Groundtruth, *see* GT
Groundtruth skeleton map, 262, 265, 267, 270, 271
Groundtruths, 263, 264, 271, 273, 279, 280
GSM2, 319–321, 325–327, 338
GT, 322, 323

H

Hamilton–Jacobi, *see* HJ
HED, 263, 269–272, 274, 275
Heptoroid, 117, 118
Hereditarily simple sets, 214, 215, 235, 236, 242, 243, 250, 253
Histogram, 95, 101, 102, 357, 358, 368, 376
Hit-or-miss transformation, *see* HMT
HJ, 12, 60, 62, 63
HMT, 77, 78
Homotopy, 6, 73, 123, 314, 316, 322, 323, 325, 326
Horse, 108, 109, 117, 118, 263
Human airway trees, 28, 168, 172, 173

I

𝕀-attachment set, 222, 227, 234
𝕀-core, 238, 239
𝕀-critical, 239, 240, 243, 244
𝕀-critical cell, 239–243, 252, 253
𝕀-essential, 237, 238, 240, 245, 251, 252
𝕀-essential cell, 237–239, 243, 244, 251, 252
𝕀-essential proper faces, 238
𝕀-essential subcomplex, 251–253
𝕀-induced cliques, 237, 238, 242
𝕀-regular, 239, 241, 251, 252
𝕀-regular cell, 226, 239, 241, 243, 251
IFT, 44, 48, 50, 51, 53, 62–67
IFT algorithm, 55, 64, 66
IMA, 60, 62, 63
Image foresting transform, *see* IFT
Image processing, 5, 26, 71, 151, 152, 157, 316, 317
Image registration, 169, 327, 330, 331, 339
Importance metric, 44, 47, 57, 62–64, 66, 67, 99
Influence
  zones of, 8, 9, 52, 71, 78–83, 94, 130, 143, 147, 148
Integer medial axis, *see* IMA
Intensities, 157, 349, 351, 355–362, 364, 365, 375, 376
  lower, 358–360
  similar, 355, 360–362, 365, 375
Internal branches, 127, 132, 134, 135, 147
Intersection, 21, 80, 216, 218, 222, 230, 234, 237, 240, 243

Isthmus persistence, 183, 200
Isthmuses, 187, 189–191, 195–197, 200, 201, 203, 207
Iterative thinning, 77, 315, 316, 350
Iterative thinning process, *see* ITP
ITP, 60, 62, 63

J

Junction line, 294, 295, 299–302
Junction voxels, 155, 294–296, 299–302
Junctions, 8, 9, 22, 27, 289, 291, 292, 294, 300

K

k-isthmuses, 191, 195, 197, 199, 202, 204
$\mathcal{K}$-neighborhoods, 185, 198, 199
Key topological properties, 212, 214, 216
KGS, 353
Knowledge-guided segmentation, *see* KGS

L

Legs, 98, 103, 104, 137, 269, 347, 377
Length
  leg, 356, 366
  shortest-path, 54–56, 59
Livers, 324, 327, 328, 337, 338
  healthy, 327, 328
Local topological numbers, 291–293, 295, 296

M

Magnetic resonance angiography, *see* MRA
Magnetic resonance imaging, MRI, 288
MAT, 45, 47, 92, 147
Maximal $\mathbb{I}$-critical cells, 242–245
Maximal inscribed balls, *see* MIBs
Maximum intensity projection, *see* MIP
MBS, 60, 62–67
Measure
  global, 44, 47, 92
  local, 44, 46, 47, 297, 302
  skeleton importance, 156
Medial axes, 4, 26, 43, 60, 67, 73, 74, 194, 195, 290, 314–316, 319
Medial axis transform, *see* MAT
Medial geodesic function, *see* MGF
Medial manifolds, 314, 315, 317, 320, 327, 328, 337, 338
Medial maps, 316, 319, 321
Medial structures, 314–318
Medial surface generation methods, 316, 317
Medial surfaces, 47, 60, 314–316, 318, 319, 321–325, 327, 338
Medialness, 314–317, 323
Medical imaging, 28, 90, 157, 288, 289, 314, 316
MGF, 47, 48, 50, 56, 59, 64, 67, 93, 98, 99, 156
MGF metric, 60, 64, 67, 98–100
MIBs, 4, 6, 7, 156, 302
Minima rule, 93
Minimal nonsimple sets, 25, 26, 181, 182, 212–215, 220, 235, 236, 242–245, 253
  possible, 215
Minimal paths, 72, 73, 157
Minimum-cost path method, 157, 169–171
Minimum-cost paths, 156–158, 161–164, 351, 352
  centered, 157, 158, 163, 164
MIP, 349, 359, 378
MRA, 348, 349, 379
MRI, 288
Multiscale, 23, 48, 53, 60, 61, 63, 67, 94, 262
Multiscale curve skeletonization, 24
Multiscale importance, 49
Multiscale methods, 44, 64, 65
Multiscale regularization, 13, 48, 49, 60
Multiscale skeleton computation, 56, 58
Multiscale skeletonization, 23, 27, 44, 56, 58
Multiscale skeletonization methods, 49, 58, 62, 64, 67
Multiscale skeletons, 4, 23, 24, 44, 47–50, 57, 60, 62, 63, 65, 67, 93
Muscles, 347, 355, 357–359, 361, 368, 369

N

Natural images, 260, 262, 275, 277, 282, 283
Nephrogenic systemic fibrosis, *see* NSF
NETS, 222–234, 237, 239–244, 246–253
NETS axioms, 217, 225, 226, 228, 232
NETS definitions, 213, 217, 229, 245, 251
NETS polyhedra, 213, 217, 222–224, 227–229, 233–235, 253

NETS polyhedra (*cont.*)
properties of, 217, 223, 228, 229, 231
valid definition of, 217, 253
NETS-collapsing, 245, 251
NETS-collapsing sequence, 247–249, 252
Network architecture, 264, 266
NMS, 316, 320
Node connection, 354
Node detection, 354, 375
Nodes connection, 354, 355, 371, 375
Nodes detection, 354, 355, 360, 375
Nonempty convex polytopes, 218, 219, 224–226, 240, 243
Nonempty subcollection, 226, 240, 242, 243
Nonempty topologically simple, *see* NETS
Nonmaxima suppression, *see* NMS
Normalized ridge map, *see* NRM
NRM, 318, 319, 321
NSF, 348

O

Object pixels, 124, 127, 131, 134, 135
Object recognition, 140, 147, 260, 283
Object skeleton extraction, 260, 262, 264
Object skeletonization, 9, 260
Object skeletons, 263, 271, 273, 290
Object voxels, 20, 21, 25, 161, 289, 290
Occlusion, 347, 349, 368, 373, 375, 377
Optimization, 332, 333, 340
Osteoporosis, 287, 288

P

P-simple points, 26, 181, 182, 195, 215
Parallel skeletonization scheme, 25, 26
Parameter selection, 354, 355
Part segmentation
symmetric, 275, 276, 283
Part-based segmentation, 90–96, 104–106, 108–110, 112, 114
Partition, 90, 94, 100, 102, 104, 105, 147
Patch-based segmentations, 90, 92, 95, 106–110, 114, 117
Patches, 95, 96, 108, 110–112, 114–117, 119
Peaks, 101, 102, 357, 358, 376
Per-pixel classification problem, 262, 263, 265
Performance evaluation, 29, 271
Peripheral branches, 22, 23, 124, 125, 131, 132, 134, 135, 137, 147
Persistence, 196, 201–204, 207
Pixels, 4, 8, 9, 12, 124–128, 144, 145, 205
delimiting, 127, 129, 132, 141
nonskeleton, 265, 267
normal, 128
simple, 127
skeletal, 28, 125, 127–130, 141, 142, 147–149
skeleton, 141, 143, 261, 263, 265–267, 269–271, 276
Platelikeliness, 303
Point-cloud skeletons, 65, 95, 96, 98, 103, 104, 118
Point-clouds, 8, 10, 90, 92, 98, 104, 119, 332
Polygonal approximation, 10, 124, 125, 138, 140–149
Polyhedra, 217, 221–228, 231–234, 238, 253
Polytopal complexes, 25, 215, 217, 219, 231, 240, 245, 253, 254
Positives
false, 271, 272, 378
Precision, 271, 272, 275, 337, 338, 353
Problems
unknown-scale, 260, 261, 263
Propagation path, 299–301
Proper faces, 216, 218, 219, 223, 224, 238, 239
Pruning, 4, 15, 19, 22, 124, 125, 127, 131, 135–140, 148, 325, 326
skeleton, 46, 72, 140, 200

Q

Quench points, 5, 7, 15, 153, 154, 290
Quench voxels, 15, 17, 18
strong, 163, 166
Queue
priority, 54, 55, 58, 59

R

Reference segmentation, 370, 372, 374
voxelized, 371
Reference skeleton, 367, 369–374, 377, 378
Regularization, 13, 23, 63, 67, 96, 329, 330, 340
Representative points, 302, 305

Ridges, 29, 72–74, 262, 315, 317, 319–321, 370

S

SBB, 95, 108, 115, 116
SC, 293, 296
Scapula, 66, 115, 117, 118
SDF (shape-diameter function), 94
SE, 292–294, 301
SE voxels, 294, 295
Segment borders, 93, 95, 106, 111, 115, 116
  smooth, 106
Segment boundaries, 94–96
Segmentation, 90, 91, 93–95, 112, 260, 329, 332, 334, 346, 352, 353
  patch-type, 109
  unified, 107, 110, 112, 114, 117
  vascular, 350, 352, 366
  volumetric, 346
Segmentation algorithm, 161, 371, 374, 378
Segmentation groundtruth, 278
Segmentation methods, 90, 93, 117, 119
  part-based, 109, 116, 118
  patch-based, 91, 93, 95, 107, 108, 119, 120
  skeleton-based, 90, 91, 95
Segmentation problems, 91
Segmentation results
  symmetric part, 278
Segmentation techniques
  curve-skeleton-based part, 105
Segmentation types, 107, 110, 117
Segments, 27, 95, 102, 105, 144, 147, 149
  valid, 110
Selection function, 187–189
Selective serotonin reuptake inhibitor, *see* SSRI
Sequence
  simple, 235, 236, 247, 250
Sets
  attachment, 21, 233
SGR, 319–321
Shape-diameter function, *see* SDF
Shortest-and-straightest geodesics, *see* SSGs
Shortest-path length computation, 55, 59, 60
Side output layers
  scale-associated, 262, 265, 266
Significance, 18, 22, 23, 72, 124, 133, 157, 166–168
Significance factor, 15, 23, 24
Simple voxels, 183–186, 315
Simplification, 50, 53, 57, 58, 125, 338
Single-point feature transform, 51–54, 56, 58
Singularity points, 153, 155, 156
Skeletal branches, 10, 12, 17, 159, 160, 162–167
  current, 164, 165
  new, 10, 158, 159, 164, 165, 168
Skeletal points, 4, 7, 94, 97, 98, 102, 104, 292, 298, 299, 302
Skeletal representations, 24, 90, 289, 291, 297, 298, 339
Skeletal segments, 10, 23, 297
Skeletal voxels, 167, 187–189, 292–294, 301, 302
Skeleton boundary backprojection, *see* SBB
Skeleton branches, 22, 124, 125, 127, 140, 141, 148
  jaggedness of, 124, 125
  noisy peripheral, 124
  peripheral, 137
  spurious, 22, 200, 201
Skeleton extraction, 260–266
  vascular, 352, 377
Skeleton extraction methods, 261, 262, 269, 271–273, 275, 277
Skeleton extraction results, 271, 273, 274, 276
Skeleton junction points, 45
Skeleton map
  scale-associated groundtruth, 261, 266, 267
Skeleton points, 47, 73, 92, 98–100, 102–104, 123
  spurious, 98, 200
Skeleton regularization, 44, 45
Skeleton segments, 142, 276
  local, 260, 263
Skeleton similarity, 330, 339
Skeleton stack function, *see* SSF
Skeleton structure, 125, 138, 140
  hierarchical, 132
Skeletonization, 3, 6, 14, 26–29, 44, 45, 55, 91, 94, 124, 151, 290, 305, 339
  Blum's, 165, 174

Skeletonization (*cont.*)
curve, 13, 28, 152, 153, 157, 160, 161, 169, 171, 173–175
final, 15
foundation of, 5, 152
parallel, 24, 25
Skeletonization methods, 4, 6, 62, 175, 181
asymmetric curve, 207
known multiscale, 67
path-based curve, 169
point-cloud, 119
Skeletons, 3–7, 10, 12–15, 17, 19, 22, 23, 27–29, 43–46, 57, 71–79, 90, 92, 94, 123, 124, 127, 128, 130, 131, 135–140, 146, 147, 153, 182, 200, 207, 290, 330, 346, 367, 369
simplified, 47, 60, 63, 98, 102
using surface, 95, 114, 117
vessel, 346, 376
Small arteries, 354, 358, 362, 363, 375, 377
Spels, 4, 5, 8, 19, 29, 49–59
Spiral, 327, 329, 330, 337, 339, 340
Spoke vectors, 45, 92
SSF, 203–205
SSGs, 100
SSRI, 304
Standard Gaussian ridge, *see* SGR
Stenoses, 28, 347–349, 377
arterial, 368
Step-cost function, 161, 162, 164
Straight-line segments, 129, 140–142, 144, 147, 149
Structure tensor, 317, 320
Subcomplexes, 26, 216, 219, 222, 239, 246, 247, 249, 252
Surface–surface junction lines, 294, 295

T

TB region, 302, 303
Termination criterion, 153, 158, 165–167
Text line proposals, 277, 280, 282, 283
Text lines, 277, 279, 282
Thinning, 77, 78, 184–186, 188, 326
asymmetric, 187
symmetric, 200, 203
Thinning algorithms, 182, 187, 212, 213, 215, 216
designing parallel, 213, 216, 253
parallel, 215
topological soundness of parallel, 215, 253
Thinning process, 63, 94, 191, 200, 203, 346
Thinning scheme, 182, 187, 194, 207, 320
Topological classes, 27, 291–294
Topology preservation, 5, 6, 21, 24–26, 154, 155, 181, 182
Torque, 334, 335, 337, 340
Tree, 49, 57, 157, 168, 229, 230, 233, 354
Tunnels, 4, 19–21, 160, 169, 290–293

U

Ultimate asymmetric skeletons, 189, 190
Ultrasonography, 348
UnifDist, 322, 326
Unit thickness, 123–125
United skeleton, 367, 369, 370, 372–374, 377

V

Validation, 322, 367
Valleys, 12, 29, 101, 102, 317, 318
VarDist, 322, 326
Vascular centerline tree, 350, 366, 367
Vascular skeleton, 346, 350, 359, 375, 377
Vasculature, 346, 355, 360, 376
VCS, 96–98, 101, 102, 104, 112, 114, 116, 118, 119
VCS method, 100, 102, 107, 108, 118
Vector field, 12, 13, 49, 53
Vessel model, 346, 366, 367
Volumetric topological analysis algorithm, *see* VTA algorithm
Voronoi diagram, 5, 8, 10, 124, 290
Voronoi skeletonization, 10, 11, 23, 28
Voronoi skeletons, 10, 23
Voxel cut-space segmentation, *see* VCS
Voxel grids, 14, 30, 152
Voxel representations, 5, 91, 92, 353
Voxel set, 182, 194, 195
Voxels, 4, 8, 15, 17, 152, 161, 183–196, 198, 199, 289, 296, 355, 358, 360, 361
background, 21, 289, 292
branch-end, 164, 165
distinct, 183, 185
individual, 27, 194, 291, 303
internal, 15, 318

Voxels (*cont.*)
neighbor, 161
neighboring, 17, 207, 301, 361
single, 182, 184, 185, 188, 190, 195, 198–200
surface-edge, 294
VTA algorithm, 297–299

W

Water-sheds, *see* WS
Weight layers, 262, 263, 265, 268
WS, 79

Y

Y-intersection curves, 43, 45, 92

Z

Zig-zag straightening, 125, 127–129, 135–140
Zig-zags, 125, 127, 128, 131, 148
Zone of influence, 129–131, 143, 146–148

Printed in the United States
By Bookmasters